**Current Plant Science:**
**Novel Concepts and Innovative Strategies: Chittaranjan Kole**

1. Next Generation Crop Improvement for Agricultural Sustainability and Food Security
   *Belaghihalli N Gnanesh* and *Chittaranjan Kole (Editors), 2026*
2. Next Generation Crop Production for Agricultural Sustainability and Food Security
   *Belaghihalli N Gnanesh, Megha N Parajulee* and *Chittaranjan Kole (Editors), 2026*
3. Next Generation Crop Protection for Agricultural Sustainability and Food Security
   *Belaghihalli N Gnanesh, Megha N Parajulee, Ankush M Raut* and *Chittaranjan Kole (Editors), 2026*

Current Plant Science: Novel Concepts and Innovative Strategies: Chittaranjan Kole

# Next Generation Crop Improvement for Agricultural Sustainability and Food Security

*Editors:*

**Belaghihalli N Gnanesh**
Indian Council of Agricultural Research
JSS Krishi Vigyan Kendra
Suttur, Mysore, Karnataka, India

**Chittaranjan Kole**
Prof. Chittaranjan Kole Foundation for Science and Society
Kolkata, India

CRC Press
Taylor & Francis Group
Boca Raton London New York

CRC Press is an imprint of the
Taylor & Francis Group, an **informa** business
A SCIENCE PUBLISHERS BOOK

Cover page designed by Prof. Chittaranjan Kole

First edition published 2026
by CRC Press
2385 NW Executive Center Drive, Suite 320, Boca Raton FL 33431

and by CRC Press
4 Park Square, Milton Park, Abingdon, Oxon, OX14 4RN



*CRC Press is an imprint of Taylor & Francis Group, LLC*



*Library of Congress Cataloging-in-Publication Data (applied for)*

ISBN: 978-1-032-94743-3 (hbk)
ISBN: 978-1-032-95285-7 (pbk)
ISBN: 978-1-003-58410-0 (ebk)

DOI: 10.1201/9781003584100

Typeset in Times New Roman
by Prime Publishing Services

# Message

***'Next Generation Crop Improvement for Agricultural Sustainability and Food Security'*** is the outcome of hard work and incisive knowledge acquired over the years by Dr B.N. Gnanesh and Prof. Kole. We are pleased that this valuable work will serve as a perennial source of guiding principles for Indian farmers.

The book addresses the urgent need for innovative solutions in crop improvement, driven by climate variability, population growth, and nutritional insecurity. While traditional breeding methods remain important, the book emphasises the integration of advanced technologies like genomic selection, genome editing, and artificial intelligence to enhance crop resilience and yield quickly and precisely.

It explores various transformative technologies and their potential to develop climate-resilient crops and improve nutritional content. Additionally, the book provides practical strategies for tackling hidden hunger through bio-fortification and nutrient enhancement. Overall, it presents valuable resources and scalable solutions for ensuring global food security and sustainability in agriculture.

We wish the authors all the very best in their future endeavours.

His Holiness Sri Shivarathri Deshikendra Mahaswamiji

**Chancellor of the JSS Academy of Higher Education and Research (JSS AHER), Mysuru and JSS University, Noida**

# Preface

The global challenges of climate change, population growth, rising malnutrition, environmental degradation, and declining arable land have intensified the need for innovative, sustainable, and resilient agricultural systems. "Next-Generation Crop Improvement for Agricultural Sustainability and Food Security" brings together contemporary advancements and interdisciplinary insights aimed at transforming crop improvement to meet these emerging demands.

This comprehensive volume features 13 meticulously curated chapters spanning various topics crucial for modern agriculture. **Chapter 1** focuses on **recent advancements in stress management in cereals**, highlighting molecular and agronomic strategies to enhance tolerance against abiotic and biotic stressors. Following this, the role of **cutting-edge genome editing technologies**, such as CRISPR/Cas systems, is explored for their potential in precise and efficient crop enhancement in **Chapter 2**.

Recognizing the need for speed and precision, **Chapter 3** delves into **accelerated breeding approaches**, offering insights into methods that shorten breeding cycles, particularly for sustainable rice farming. The application of **genomic selection** is examined in **Chapter 4** as a cornerstone for developing climate-smart, nutrient-rich crops, vital for ensuring food and nutritional security.

**Chapter 5** is dedicated to **genetic innovations in wheat,** and **Chapter 6** on **biofortification** emerges as a critical pathway to combat hidden hunger, with novel methods to elevate iron, zinc, and vitamin content in staple crops, providing concrete examples of how modern breeding can address global malnutrition and promote sustainability. **Millets**, long underappreciated and celebrated for their resilience and nutritional value, positioning them as vital crops for future food systems, are highlighted in **Chapter 7**.

Significant emphasis is also placed on **underutilized vegetable crops for food and nutritional security** in **Chapter 8. Grafting innovations in vegetable crops** are dicussed extensively in Chapter 9. Cutting-edge technologies such as **plasma treatments** and **bioengineering of starch and amylose–lipid complexes** are discussed for their novel applications in both food and non-food sectors (**Chapters 10 and 11)**.

**Technological integration** is equally pivotal: **Chapters 12 and 13** shed light on the transformative power of **high-throughput phenotyping** and machine learning, to optimize trait selection and resource allocation. The integration of **bioinformatics and machine learning** showcases how digital agriculture can drive precision, efficiency, and sustainability in crop improvement.

This book is intended for researchers, academicians, policy-makers, and students in the fields of plant science, agronomy, biotechnology, and agricultural engineering. By bridging classical breeding with modern innovations, we hope to inspire integrated approaches that can reshape agricultural landscapes and build a climate-resilient food system for generations to come. The editors acknowledge the help from all the staff of Taylor & Francis Group, LLC and CRC Press at all stages.

Mysuru, Karnataka, India — **Belaghihalli N Gnanesh**
Kolkata, West Bengal, India — **Chittaranjan Kole**

# Contents


*Message* *v*
*Preface* *vii*
*List of Contributors* *xi*
*Abbrevations* *xvii*

1. Recent Advancements and Insights on Various Stress Management Techniques in Cereals 1
2. Advancements in Genome Editing Technologies for Enhancing Crop Improvement 31
3. Accelerated Breeding Approaches for Enabling Sustainable Rice Farming 46
4. Navigating the Development of Next-Generation Crop Varieties Using Genomic Selection for Food and Nutritional Security 78
5. Genetic Innovations in Wheat for Sustainable Agriculture and Global Food Security 101
6. Improvement Strategies Addressing Malnutrition: Augmenting Bioavailability and Biofortification in Food Crops 120
7. Millets: An Emerging and Promising Climate-Resilient Nutri-Cereals of the Globe 141
8. Underutilized Vegetable Crops for Food and Nutritional Security 154
9. Grafting Innovations in Vegetable Crops to Enhance Nutritional Quality for Food Security 180
10. Advancements in Plasma Technology for Enhancing Crop Resilience to Environment Stress 198
11. Bioengineering Starch and Amylose-lipid Complex: From Genetic to Sustainable Industrial Application 218
12. High-Throughput Phenotyping Tools for Sustainable Agriculture 232
13. Utilizing Bioinformatics Tools and Machine Learning Techniques to Advance Sustainable Crop Improvement 259

Index 281

About the Editors 283

# List of Contributors


**Anilkumar C**
ICAR-Central Rice Research Institute, Cuttack, India. &
Department of Agronomy and Plant Genetics, University of Minnesota, MN, USA.

**Arvindkumar**
College of Horticulture, Bidar, UHS, Bagalkot, Karnataka, India.

**Ashvinkumar Katral**
ICAR-Indian Agricultural Research Institute, New Delhi, India.

**Basana Gowda G**
ICAR-Central Rice Research Institute, Cuttack, India. &
ICAR-National Research Center for Pomegranate, Solapur, India.

**Basangouda Gonal**
CSB-Central Sericultural Research & Training Institute (CSR&TI), Pampore 192 121, Jammu & Kashmir (UT), India.

**Basavaraj PS**
ICAR-National Institute of Abiotic Stress Management, Baramati, India.

**Belaghihalli N Gnanesh**
Genetics & Plant Breeding, ICAR JSS – Krishi Vigyan Kendra, Suttur 571129, Mysore, Karnataka, India.

**Chaitra Kulkarni**
Department of Biotechnology and Crop Improvement, KRCCH, University of Horticultural Sciences, Arabhavi 591218, Bagalkot, India.

**Chhattar Pal**
CSB-Regional Sericultural Research Station (RSRS), Sahaspur, Dehradun, Uttarakhand, India.

**Danishta Aziz**
Division of Entomology, Faculty of Agriculture, Sher-e-Kashmir University of Agricultural Sciences & Technology – Kashmir (SKUAST-K), Wadura, Sopore 193 201, Jammu & Kashmir (UT), India.

**Deepak Das**
Agricultural Biotechnology Division, National Agri-Food and Biomanufacturing Institute, Mohali 140306, Punjab, India.

**Dinesh Arya**
BRIC-National Agri-Food & Biomanufacturing Institute (NABI), Mohali 140306, Punjab, India. &
Regional Centre for Biotechnology (RCB), NCR Biotech Science Cluster, 3rd Milestone, Faridabad–Gurgaon Expressway, Faridabad 121001, Haryana, India.

**Dwijesh Chandra Mishra**
ICAR – Indian Agricultural Statistics Research Institute, PUSA, New Delhi 110012, India.

**Gami RA**
ICAR-AICRP on Sorghum, Centre for Millets Research, S.D. Agricultural University, Deesa 385 535, Gujarat, India.

**Ganapathy KN**
ICAR-IIMR, Hyderabad 500030, India.

**Gulab Khan Rohela**
CSB-Central Sericultural Research & Training Institute (CSR&TI), Pampore 192 121, Jammu & Kashmir (UT), India.

**Gurjeet Singh**
Texas A&M Agrilife Research Center at Beaumont, Texas 77713, USA.

**Harish Pal Bhati**
Department of Botany, Shanti Devi Government College, Jewar, Gautam Buddha Nagar 203209, U.P., India.

**Himanshu Sharma**
Amity School of Biological Sciences, Amity University Punjab, Sector 82A, IT City, Mohali 140306, Punjab, India.

**Jae-Yean Kim**
Division of Applied Life Science (BK21 Four Program), Plant Molecular Biology and Biotechnology Research Center, Gyeongsang National University, Jinju 52828, South Korea. & Nulla Bio Inc., 501 Jinju-daero, Jinju 52828, South Korea. &
Division of Life Science, Gyeongsang National University, 501 Jinju-daero, Jinju 52828, South Korea.

**Jeevan B**
ICAR-Central Rice Research Institute, Cuttack, India.

**Jitendra Kumar**
BRIC-National Agri-Food & Biomanufacturing Institute (NABI), Sector 81, Mohali 140306, Punjab, India.

**Joy Roy**
Agricultural Biotechnology Division, National Agri-Food and Biomanufacturing Institute, Mohali 140306, Punjab, India.

**Joy Roy**
BRIC-National Agri-Food & Biomanufacturing Institute (NABI), Mohali 140306, Punjab, India. &
Regional Centre for Biotechnology (RCB), NCR Biotech Science Cluster, 3rd Milestone, Faridabad–Gurgaon Expressway, Faridabad 121001, Haryana, India.

**Kugashiya KG**
Centre for Millets Research, S.D. Agricultural University, Deesa 385 535, Gujarat, India.

**Likhithashree TR**
Department of Genetics and Plant Breeding, University of Agricultural Sciences, GKVK, Bengaluru, Karnataka, India.

**Lokeshkumar BM**
ICAR-Central Soil Salinity Research Institute, Karnal, India.

**Mahantesha BN Naika**
Centre for Biotechnology Research, Department of Biotechnology and Crop Improvement, College of Horticulture, University of Horticultural Sciences, Bagalkot 587104, Karnataka, India.

**Mahesh Pujar**
Pigeonpea Breeding, Accelerated Crop Improvement, International Crops Research Institute for Semi-Arid Tropics, Patancheru, Hyderabad, Telangana, India.

**Mamatha YS**
ICAR – Indian Agricultural Statistics Research Institute, PUSA, New Delhi 110012, India.

**Manisha Godara**
Agricultural Biotechnology Division, National Agri-Food and Biomanufacturing Institute, Mohali 140306, Punjab, India.

**Mohammed Saba Rahim**
Molecular Genetic and Genomics Laboratory, Biotechnology Division, CSIR-Institute of Himalayan Bioresource Technology, Palampur 176061, Himachal Pradesh, India.

**Mohana Kumara P**
Department of Biotechnology and Crop Improvement, KRCCH, University of Horticultural Sciences, Arabhavi 591218, Bagalkot, India.

**Naresh Bomma**
Pigeonpea Breeding, Accelerated Crop Improvement, International Crops Research Institute for Semi-Arid Tropics, Patancheru, Hyderabad, Telangana, India.

**Nethra N**
AICRP on Seed (Crops), National Seed Project, University of Agricultural Sciences, GKVK, Bangalore 560065, Karnataka, India.

**Patil JV**
ICAR-IIMR, Hyderabad 500030, India.

**Pavan PR**
Department of Fruit Science, KRCCH, University of Horticultural Sciences, Arabhavi 591218, Bagalkot, India.

**Pawan Saini**
CSB-Central Sericultural Research & Training Institute (CSR&TI), Pampore 192 121, Jammu & Kashmir (UT), India.

**Plabani Roy**
CSB-Central Sericultural Research & Training Institute (CSR&TI), Pampore 192 121, Jammu & Kashmir (UT), India.

**Pooja Saini**
Department of Genetics, Plant Breeding and Biotechnology, Dr. Khem Singh Gill College of Agriculture, Eternal University, Baru Sahib 173 101, Himachal Pradesh, India.

**Pradeep Kumar Sharma**
Ch. Charan Singh University, Meerut, 250004, UP, India.

**Pradeep Singh**
Agricultural Biotechnology Division, National Agri-Food and Biomanufacturing Institute, Mohali 140306, Punjab, India.

**Prakash I Gangashetty**
Pigeonpea Breeding, Accelerated Crop Improvement, International Crops Research Institute for Semi-Arid Tropics, Patancheru, Hyderabad, Telangana, India.

**Prashant Kumar**
Department of Plant Sciences and Plant Pathology, Montana State University, Bozeman, Montana, USA.

**Rachna Shashikant Jadhav**
Faculty of Applied Sciences and Biotechnology, Shoolini University, Solan 173229, Himachal Pradesh, India.

**Raghavendra Gunnaiah**
University of Horticultural Sciences, Bagalkote 587104, Karnataka, India.

**Raghu S**
ICAR-Central Rice Research Institute, Cuttack, India.

**Rahul Kumar**
Ch. Charan Singh University, Meerut, 250004, UP, India.

**Rahul Mahadev Shelake**
Division of Applied Life Science (BK21 Four Program), Plant Molecular Biology and Biotechnology Research Center, Gyeongsang National University, Jinju 52828, South Korea.

**Ram Kumar Sharma**
Molecular Genetic and Genomics Laboratory, Biotechnology Division, CSIR-Institute of Himalayan Bioresource Technology, Palampur 176061, Himachal Pradesh, India.

**Rameswar Prasad Sah**
ICAR-Central Rice Research Institute, Cuttack, India.

**Ratnakar M Shet**
University of Horticultural Sciences, Bagalkote 587104, Karnataka, India.

**Ruksana**
Eternal University, Baru Sahib, Sirmaur 173101, Himachal Pradesh, India.

**S Syam**
CSB-Central Sericultural Research & Training Institute (CSR&TI), Pampore 192 121, Jammu & Kashmir (UT), India.

**Sangeeta Dayal**
KVSCOS, Subharti University, Meerut, 250005, India.

**Sankalpa KB**
Food Process Engineering, DSLD CHEFT, Devihosur, Haveri, India.

**Sharon Roja Kavuri**
Department of Genetics and Plant Breeding, Acharya N.G. Ranga Agricultural University, Lam, Guntur, Andhra Pradesh, India.

**Sheshshayee MS**
University of Agricultural Sciences, Bengaluru 560065, Karnataka, India.

**Shivam Bhardwaj**
CSB-Central Sericultural Research & Training Institute (CSR&TI), Pampore 192 121, Jammu & Kashmir (UT), India.

**Shoeb Ahmed**
Department of Genetics and Plant Breeding, Ch. Charan Singh University, Meerut 250004, U.P., India.

**Shridhar Ragi**
ICAR-Indian Agricultural Research Institute, New Delhi, India.

**Shruthi HB**
Pigeonpea Breeding, Accelerated Crop Improvement, International Crops Research Institute for Semi-Arid Tropics, Patancheru, Hyderabad, Telangana, India.

**Sowmya KS**
AICRP on Seed (Crops), National Seed Project, University of Agricultural Sciences, GKVK, Bangalore 560065, Karnataka, India.

**Suhas Gorakh Karkute**
Division of Vegetable Improvement, ICAR-Indian Institute of Vegetable Research, Varanasi, India.

**Suma Krishnappa**
Department of Genetics and Plant Breeding, University of Agricultural Sciences, GKVK, Bengaluru, Karnataka, India. &
Pigeonpea Breeding, Accelerated Crop Improvement, International Crops Research Institute for Semi-Arid Tropics, Patancheru, Hyderabad, Telangana, India.

**Sunitha NC**
ICAR-Central Rice Research Institute, Cuttack, India. &
University of Agricultural Sciences, Bangalore, India.

**Swarup Nagrale**
Agricultural Biotechnology Division, National Agri-Food and Biomanufacturing Institute, Mohali 140306, Punjab, India.

**Virupakshi Hiremata**
University of Agricultural Sciences, Bengaluru 560065, Karnataka, India.

**Vishvjeet S Patel**
ICAR-Central Soil Salinity Research Institute, Karnal, India

**Zakiya Saif**
College of Temperate Sericulture (CoTS), Sher-e-Kashmir University of Agricultural Sciences & Technology – Kashmir (SKUAST-K), Mirgund 193 121, Jammu & Kashmir (UT), India.

# Abbrevations

| | |
|---|---|
| 2D | 2-Dimensional |
| 2D BLUP | Two-dimensional best linear unbiased predictor |
| 3D | 3-Dimensional |
| 4D | 4-Dimensional |
| ABA | Abscisic Acid |
| ABI | Applied Biosystems |
| AFLP | Amplified fragment length polymorphism |
| AFP | Anti-freeze protein |
| AI | Artificial intelligence |
| AM | Arbuscular mycorrhizal |
| AM | Amylose |
| AML | Amylose-lipid complex |
| AP | Amylopectin |
| ARP/RAP | Rice automatic phenotyping platform |
| AS-PCR | Allele-specific polymerase chain reaction |
| BER | Blossom-end rot |
| BILs | Backcross inbred lines |
| BLUP | Best linear unbiased prediction |
| BOAA | β-N-oxalyl amino alanine |
| BP | Breeding population |
| BR | Brassinosteroids |
| BRR | Bayesian ridge regression |
| BV | Biological value |
| Ca | Calcium |
| CBF | C-repeat binding factor |
| CL | Coleoptile length |
| CMS | Cell membrane stability |
| CNNs | Convolutional neural networks |
| CRISPR-Cas9 | Clustered regularly interspaced short palindromic repeats/CRISPR-associated protein 9 |
| CSSL | Chromosomal segment substitution line |
| CT | Computed tomography |
| Cu | Copper |
| DArT | Diversity array technology |
| DArTseq | Diversity arrays technology sequencing |
| DB | Dry biomass |
| DH | Doubled haploids |

| | |
|---|---|
| DIC | Détente Instantannée Contrôlée in French |
| DIRT | Digital imaging of root traits |
| DNA | Deoxyribonucleic acid. |
| DON | Deoxynivalenol |
| DSS | Decision support system |
| DW | Dry weight |
| EAA | Essential amino acid |
| EIT | Electrical impedance tomography |
| ERF | Ethylene responsive factors |
| EST-SSR | Expressed sequence tag-simple sequence repeat |
| ETC | Electron transport chain |
| FAO | Food and agriculture organization |
| FDM | Functional domain markers |
| Fe | Iron |
| FHB | Fusarium head blight |
| FW | Fresh weight |
| G×E interaction | Genotype by environment interaction |
| GA | Gibberellic acid |
| GA | Genome analyser |
| GABA | Gamma-aminobutyric acid |
| GBLUP | Genomic best linear unbiased prediction |
| GBS | Genotyping by sequencing |
| GC | Guard cells |
| GDP | Gross domestic product |
| GEBV | Genome estimated breeding value |
| GEO | Gene expression omnibus |
| GFD | Grain filling duration |
| GH | Growth hormone |
| GI | Glycemic index |
| GIS | Geographic information system |
| GMO | Genetically modified organisms |
| GP | Genomic prediction |
| GPC | Grain protein content |
| GPR | Ground penetrating radar |
| GPS | Grain per spike |
| gRNA | guide RNA |
| GS | Genomic selection |
| GW | Grain width |
| GWAS | Genome-wide association studies |
| GWD1 | Glucan water dikinase 1 |
| GWPS | Grain weight per spike |
| GY | Grain yield |
| HCN | Hydrogen cyanide |
| HDR | Homology-directed repair |
| HI | Harvest index |
| HMWG | High molecular weight-glutenins |

| | |
|---|---|
| HP | Hybrid polysaccharide |
| HSP | Heat shock proteins |
| HSW | Hundred seed weight |
| HTP | High-throughput phenotyping |
| HTST | High-throughput sequencing technologies |
| HTTPs | High-throughput phenotyping platforms |
| IAA | Indole-3-acetic acid |
| IoT | Internet of things |
| IPCC | Intergovernmental panel on climate change |
| IR | Immediate release |
| ISBP | Insertion site-based polymorphism |
| ISSR | Inter-simple sequence repeat |
| JA | Jasmonic acid |
| K | Potassium |
| KASP | Kompetitive allele-specific pcr |
| KEGG | Kyoto encyclopedia of genes and genomes |
| KNPS | Kernel number per spike |
| KRiShI | Knowledgebase for rice sheath blight information |
| LASSO | Least absolute shrinkage and selection operator |
| LCC | Leaf chlorophyll content |
| LD | Linkage disequilibrium |
| LiDAR | Light detection and ranging technology |
| LLSR | Linear least square regression model |
| lncRNA | long non-coding RNAs |
| lpa | low phytic acid |
| LR | Leaf rolling |
| LRR | Leucine-rich repeat |
| MABB | Marker-assisted backcross breeding |
| MAF | Minor allele frequency |
| MAGIC | Multi-parent advanced generation inter crosses |
| MAS | Marker assisted selection |
| MCMC | Markov chain monte carlo |
| MeJA | Methyl jasmonates |
| Mg | Magnesium |
| ML | Machine learning |
| Mn | Manganese |
| MNVG | Median normalized vector growth |
| MT | Metric ton |
| MTA | Marker trait association |
| MTGS | Multi-trait genomic selection |
| N | Nitrogen |
| NBS | Nucleotide-binding site |
| NDVI | Normalized difference vegetation index |
| NG | Non-grafted |
| NGPP | Number of grains per panicle |
| NGPS | Number of grains per spike |
| NGS | Next generation sequencing |

| | |
|---|---|
| NGV | Next generation varieties |
| NHEJ | Nonhomologous end-joining |
| NIL | Near-isogenic lines |
| NIR | Near infrared |
| NIRS | Near infrared spectroscopy |
| NN | Neural networks |
| NO | Nitric oxide |
| NPU | Net protein utilization |
| NSC | Non-structural carbohydrate |
| NUE | Nitrogen use efficiency |
| P | Phosphorous |
| PAR | Photosynthetically active radiation |
| PBS | Polybutylene succinate |
| PCL | Poly-ε-caprolactone |
| PCR | Polymerase chain reaction |
| PD | Product development |
| PDB | Protein data bank |
| PGPR | plant growth-promoting rhizobacteria |
| PGR | Plant genetic resources |
| PHB-HV | Polyhydroxybutyrate-hydroxyvalerat |
| PHS | Pre-harvest sprouting |
| PLA | Polylactic Acid |
| POEAS | Plant ontology enrichment analysis server |
| PS I | Photosystem I |
| PS II | Photosystem II |
| PUE | Phosphorus use efficiency |
| PUFAS | Proteins of Unknown Functions Annotation Server |
| PVA | Polyvinyl alcohol |
| PVE | Phenotypic variation explained |
| QPM | Quality protein maize |
| QTL | Quantitative trait loci |
| RADAR | Radiation detection and ranging |
| RAPD | Random amplification of polymorphic DNA |
| RdDM | RNA-directed DNA methylation |
| RDW | Root dry weight |
| RFLP | Restriction fragment length polymorphisms |
| RFR | Random forest regression |
| RGB | Red-green-blue |
| RIL | Recombinant inbred lines |
| RKHS | Reproducing kernel Hilbert space |
| RNA | Ribonucleic Acid |
| ROS | Reactive oxygen species |
| rrBLUP | Ridge regression best linear unbiased prediction |
| RS | Resistant starch |
| RSA | Root system architecture |
| RUBISCO | Ribulose-1,5-bisphosphate carboxylase/oxygenase |

| | |
|---|---|
| RUE | Radiation use efficiency |
| SA | Salicylic acid |
| SAM | S-Adenosyl methionine |
| SBEs | Starch branching enzymes |
| SCAR | Sequence characterized amplified region. |
| SDW | Shoot dry weight |
| sEIT | Spatially resolved Electrical Impedance Tomography |
| SG | Self-grafted |
| SH | Seedling height |
| SHI | Spike harvest index |
| siRNAs | Small interfering RNAs |
| SNP | Single nucleotide polymorphism |
| SNPP | Spike number per plant |
| SOLiD | Sequencing using oligonucleotide ligation and detection |
| SPAD | Soil plant analysis development |
| SRA | Sequence read archive |
| SRAP | Sequence-related amplified polymorphism |
| SS | Starch synthase |
| SSRs | Simple sequence repeats |
| STS | Sequence-tagged site |
| SVM regression | Support vector machine regression |
| SW | Spike weight |
| TA | Titratable acidity |
| TALEN | Transcription activator-like effector nucleases |
| TE | Transposable elements |
| TF | Transcription factors |
| TFBS | Transcription factor binding sites |
| TGW | Thousand-grain weight |
| TILLING | Targeting induced local lesions in genomes |
| ToF | Time of flight |
| TP | Training population |
| TPS | Thermoplastic starch |
| UAV | Unmanned aerial vehicles |
| VIs | Vegetation indices |
| VRT | Variable rate technology |
| WC | Water content |
| WGS | Whole genome sequence |
| WUE | Water use efficiency |
| XRF | X-ray fluorescence |
| ZFN | Zinc finger nucleases |
| Zn | Zinc |

# CHAPTER 1

# Recent Advancements and Insights on Various Stress Management Techniques in Cereals

Deepak Das[1], Swarup Nagrale[1], Prashant Kumar[2], Rachna Shashikant Jadhav[3], Manisha Godara[1], Himanshu Sharma[3,4], Pradeep Singh[1*] and Joy Roy[1*]

## 1. Introduction

Agriculture is a cornerstone of human civilization, yet it faces unparalleled challenges in the form of environmental stresses that threaten crop productivity as well as food security. Multiple abiotic stressors act as limiting factors that impact plants' growth and development while having an adverse effect on the plant overall. Abiotic stress factors are often interrelated with each other (Zhang et al., 2023). These stresses are aggravated by climate change, global warming, and various other environmental issues which lead to extreme weather conditions, thereby making it increasingly difficult for plants to maintain optimum growth conditions for themselves (Mora et al., 2015). Abiotic stress not only adversely impacts plants at the cellular level but also at the physiological level, leading to complex alterations in their metabolic pathways, disruption of water and ion homeostasis, and increased risk of oxidative damage to cellular components.

Abiotic stress is a significant threat to agriculture and environmental health globally. These stress factors disrupt plant growth and reduce crop yields, posing challenges to food security and ecosystem stability. Abiotic stress of drought and salinity reduces crop yield and affects land irrigation, respectively. Abiotic factors also contribute to land degradation, biodiversity loss, and reduced ecosystem services such as pollination and nutrient cycling. Soil salinization, desertification, and nutrient depletion weaken ecosystems, increasing the vulnerability to climate change (Muluneh, 2021). Addressing abiotic stress involves developing stress-tolerant crops, applying sustainable agronomic practices such as precision farming, and restoring ecosystems. These strategies can enhance resilience in both agriculture and natural ecosystems, ensuring sustainable productivity and conservation. Managing abiotic stress is essential for food security, environmental protection, and long-term ecosystem stability (Mittler, 2006).

[1] Agricultural Biotechnology Division, National Agri-Food and Biomanufacturing Institute, Mohali, Punjab, India.
[2] Department of Plant Sciences and Plant Pathology, Montana State University, Bozeman, Montana, USA.
[3] Faculty of Applied Sciences and Biotechnology, Shoolini University, Solan, Himachal Pradesh, India.
[4] Amity School of Biological Sciences, Amity University Punjab, Sector 82A, IT City, Mohali, Punjab, India.
* Corresponding author: ps111186@gmail.com; joykroy@nabi.res.in

DOI: 10.1201/9781003584100-1

## 1.1 Drought Stress

Water deficit, low humidity, and reduced soil moisture can mimic drought-like conditions and affect plant growth and physiology. Recurring droughts lead to major agriculture losses that threaten food security. Even mild or transient periods of water limitation lead to changes in plant development. Such conditions are known to cause reduced water availability, leading to stomatal closure, reduced photosynthesis, rolling of the leaves, osmotic dysregulation, and impaired nutrient uptake (Seleiman et al., 2021).

Water deficit conditions can occur without a complete drought, merely in water scarcity conditions, and immediately affect plant water status by lowering the soil's moisture, reducing the plant's uptake for water, and eventually lowering leaf water potential. These conditions also minimize water losses caused due to transpiration which leads to partial stomatal closure, regardless of the degree of stomatal closure response varying from one crop species to another. Although stomatal closure conserves water in the plant, it limits the carbon dioxide uptake as well, leading to reduced photosynthesis and thereby decreased plant growth (Tardieu & Simonneau, 1998). In water scarcity, partial stomatal closure restricts the influx of carbon dioxide in the leaves, while increased mesophyll resistance further impairs carbon dioxide diffusion to chloroplasts (Flexas et al., 2004). As per research (Chaves et al., 2008), reduced chlorophyll content in drought-like conditions damages photosystem II and reduces overall photosynthetic ability of the plant.

It has been observed that in the absence of an optimum amount of water, plants often adjust themselves osmotically by accumulating macromolecules (e.g., sugars, proline, etc.) to maintain cell turgor and protect the plant cell (Bartels & Sunkar, 2005). The accumulation of osmolytes helps stabilize proteins and membranes under drought stress and maintains the metabolic integrity of plants. Osmotic adjustments not only help plants to survive transient and water deficit conditions but also impact plants' development negatively due to reduced metabolic potential.

Among the plant hormones that regulate growth, abscisic acid (ABA) is crucial in regulating response of plants to water constraints. ABA level increases even in mild stress due to the absence of water, leading to stomatal closure, which alters the gene expression aimed at conserving water and enhancing drought tolerance (Finkelstein, 2013). As plants try to exploit deeper soil layers, root-to-shoot growth ratio is also affected in times of water scarcity (Bhattacharya, 2021). This shift allows plants to maximize water uptake during water unavailability, even though it may negatively affect the energy available for the aboveground growth of plants. Root system alterations vary from one plant to another and depend on water availability.

## 1.2 Salinity Stress

Salinity stress arises due to elevated levels of soluble salts, primarily sodium chloride (NaCl), in the soil or water, leading to disruptions in plant's physiological and biochemical processes. It triggers osmotic stress and ion toxicity, ultimately disturbing cellular ion balance and reducing plant's growth and productivity (Hasanuzzaman & Fujita, 2022). Salinity stress affects the plant's osmosis, ionic stress, and oxidative damage. This impacts the plant overall at the cellular and the tissue levels. Increased salt concentrations in the soil lower the water potential, making water uptake tough for plants, thereby leading to osmotic stress, reduced cell expansion, stomatal closure, and eventually decreased photosynthesis and plant growth (Munns & Tester, 2008). Ions such as sodium and chloride can enter the plant cells individually and disrupt ion homeostasis. Excessive sodium intake leads to nutrient imbalance, especially with potassium that is essential for various enzymatic processes (Zhu, 2001). In such oxidative-stress-like conditions, plants are known to activate their antioxidant defense systems to mitigate oxidative stress (Hasegawa et al., 2000).

In reduced salinity-stress-like conditions, plants activate various transporters such as SOS1 (salt overly sensitive 1) for maintenance of low cytosolic sodium levels. Sodium is sequestered into vacuoles via tonoplast-localized sodium–hydrogen antiporter channels to prevent ion toxicity

(Apse et al., 1999). During salinity, abscisic acid, ethylene, and jasmonic acid (JA) are crucial for modulating plant responses. Plant adaptation to salinity stress involves stomatal closure through ABA signalling to preserve water and maintain internal water balance (Hedrich & Shabala, 2018). Additionally, multiple transcription factors (TF) pathways such as Dehydration-responsive element binding (DREB), SOS, etc. are triggered to mitigate the effects of salinity stress (Roy et al., 2014).

## 1.3 Extreme High and Low Temperatures

Extreme temperatures are a crucial abiotic factor influencing overall plant growth and development. As temperature varies, plants experience both heat and cold stress that impacts their physiological, biochemical, and molecular mechanisms. Extreme temperatures can destabilize cellular structures, enzymes, and membranes, thereby impairing plant metabolism, growth, and development. Increased temperature affects kernel size and pollen viability, and impairs reproduction (Hatfield & Prueger, 2015).

Higher temperature is referred to as heat stress, which occurs when temperatures exceed the plant's optimum temperature requirements. Prolonged exposure to elevated temperatures can lead to reduced yield and, in severe cases, plant death. In plants that face prolonged periods of higher temperature, photosynthesis is most affected. Ribulose-1,5-bisphosphate carboxylase/oxygenase (RuBisCO) enzyme plays a key role in carbon fixation, and due to the enzyme being affected by higher temperatures, carbon fixation and eventually the plant's photosynthetic efficiency decrease (Wahid et al., 2007). Elevated temperatures not only denature the chloroplast membranes but also denature the proteins, causing their aggregation. Plants have evolved a mechanism of producing heat shock proteins (HSPs) to handle higher temperatures. HSPs act as molecular chaperones that act to refold the denatured peptide and protect against heat-induced damage (Kotak et al., 2007). Alterations in membrane lipid fluidity disrupt membrane stability and cellular ion homeostasis, thereby making the plant more susceptible to oxidative stress (Hasanuzzaman & Fujita, 2022).

On the contrary, lower temperatures cause as much damage to plant growth as much as higher temperatures. Low-temperature stress encompasses both chilling (0–15°C) and freezing (below 0°C) conditions. Cold stress causes solidification of the membrane lipids, reducing fluidity in them, disrupting their membrane integrity and cellular compartmentalization, and making them vulnerable to damage (Uemura et al., 2005). Enzymatic activities reduce during lower temperatures, causing metabolism, photosynthesis, and respiration to decrease, and the accumulation of carbohydrates to increase (Fürtauer et al., 2019). Colder temperatures also induce reactive oxygen species (ROS) production, increasing oxidative harm to proteins, lipids, and nucleic material. As seen in heat stress, cold stress affects antioxidant systems in plants (Zhao et al., 2016).

Plants respond to temperature extremes by producing HSPs under high-temperature stress and antifreeze proteins (AFPs) under low-temperature stress. Plants possess a natural defense mechanism that activates the C-repeat binding factor (CBF) pathway, enhancing the expression of cold tolerance genes to aid in survival and adaptation to low temperatures (Mittler et al., 2012). Additionally, CBF transcription factors participate in signalling by binding to C-repeat/Dehydration-Responsive Element (CRT/DRE) regulatory elements in the promoter regions of cold-responsive genes, thereby inducing the expression of protective proteins that enhance plant's cold tolerance (Chinnusamy et al., 2007). At the epigenetic level, DNA methylation and histone modification change in order to regulate the plant system under extreme temperatures.

## 1.4 Heavy Metal Toxicity

The presence of toxic heavy metals severely hampers plant growth and development. Nonessential metals such as cadmium (Cd), lead (Pb), chromium (Cr), mercury (Hg), and arsenic (As) can cause severe damage to plants even at low levels. Similarly, essential metals such as copper (Cu), zinc (Zn), and nickel (Ni) become harmful in excess. Heavy metals from the plant surroundings compete with essential nutrients for absorption through root systems, leading to nutrient deficiency. Cadmium and lead are known to compete with calcium and magnesium, disrupting nutrient uptake and ultimately hindering plant growth (Clemens, 2006). Heavy metal toxicity damages chloroplast structure by inhibiting photosynthesis, disrupting the electron transport chain (ETC) and the Calvin cycle, and impairing stomatal function, ultimately reducing carbon fixation (Burzyński & Kłobus, 2004). Heavy metals disrupt key metabolic enzymes by attaching to sulfhydryl groups or replacing essential metal ions at enzyme active sites, disrupting the plant's cellular functions such as respiration and antioxidant defense (Sachdev et al., 2021).

### 1.5 Nutrient Deficiency

Apart from external abiotic factors, nutrient deficiency too is a critical element that negatively affects plant's overall growth. Essential macronutrients (such as nitrogen, phosphorus, calcium, magnesium, and sulfur) and micronutrients (such as iron, copper, zinc, manganese, boron, and molybdenum) are required for a plant's optimal functioning and growth. A plant may face deficiency of these nutrients due to poor soil conditions, improper agricultural activity, or external factors. When a plant faces deficiency of any of these nutrients, the growth is affected at physiological and cellular levels (de Bang et al., 2021).

Each nutrient deficiency triggers its own set of unique symptoms and responses, highlighting its critical role in maintaining plant health (van Maarschalkerweerd & Husted, 2015). Nitrogen is a vital macronutrient, as it supplies amino acids and nucleic acids, and maintains chlorophyll in the plants. When nitrogen deficiency occurs, vegetative growth is hindered, resulting in chlorosis due to decreased chlorophyll levels. Additionally, a shift in carbon and nitrogen metabolism leads to carbohydrate accumulation and reduced protein synthesis (Kant et al., 2010). Phosphorus deficiency affects root development and causes anthocyanin accumulation, which turns dark green leaves purple, affects flower and seed formation, and delays the maturity of plants (Vance et al., 2003). The deficiency of another macronutrient potassium causes leaf necrosis and curling, as potassium is important for osmoregulation, enzyme activation, and photosynthesis. Potassium-deficient plants are also known to exhibit necrotic spots on older leaves, typically leaf curling, and interveinal chlorosis (Pettigrew, 2008).

Micronutrients although required in lesser concentrations, as compared to macronutrients, their deficiency negatively impacts plant growth. Iron is an essential micronutrient for the synthesis of chlorophyll and several enzymatic reactions. Iron deficiency results in interveinal chlorosis, causing the leave veins to turn yellow, due to reduced chlorophyll concentrations (Briat et al., 2015). Zinc is important for protein synthesis, enzyme activation, auxin production, and DNA transcription regulation. Zinc deficiency causes stunted plant growth, leads to the production of smaller leaves, and even weakens plants' defense mechanism against pathogens by impairing the synthesis of protective enzymes, thus making plants susceptible to several infections (Broadley et al., 2007).

## 2. Physiological Responses of Cereal Crops to Abiotic Stress

Plants constantly encounter abiotic stresses which affect their growth and survival. To adapt, they activate several protective physiological responses. For instance, during drought and salinity, stomatal closure by plants is triggered to reduce water loss and produce osmolytes to retain cellular water. Antioxidant systems are also activated to combat the damaging effects of ROS generated during stress (Wang et al., 2019). Plant hormones (ABA, JA, and ethylene) regulate these responses by adjusting gene expression. Nutrient deficiencies, such as low nitrogen or phosphorus, lead to changes in root architecture and nutrient remobilization. Plants shift their metabolism toward stress defense, focusing on survival over growth. These responses are crucial for plant resilience, though prolonged or intense stress can still lead to damage.

## 2.1 *Stomatal Regulation*

Stomata are found mostly on the abaxial of leaves, bordered by guard cells (GCs) that regulate gas exchange between plants and surrounding atmosphere. These pores let in carbon dioxide ($CO_2$) for photosynthesis and also release water content during transpiration. When plants are under stress, the guard cells respond by adjusting anion flow, either expelling anions or storing them in vacuoles leading to a drop in turgor pressure, which causes the guard cells to shrink and close the stomata (Auler et al., 2022). Essentially, stomatal opening and closing are regulated by turgor-driven changes in guard cells (Rui et al., 2024).

Interestingly, stomata serve as gateways for pathogens into the plant (Melotto et al., 2017). Their function is influenced by a wide range of abiotic stress factors, as well as internal cues such as hormones and microbial activity (Wachendorf & Küppers, 2017). Because of this, plants have evolved advanced mechanisms to regulate stomatal movement. For example, even when conditions are favorable for photosynthesis, plants might close their stomata to minimize water loss or prevent pathogen entry (Melotto et al., 2017).

The regulation of stomatal movement involves a variety of molecular players (ion channels, secondary messengers, etc.), which work together to mediate the plant's response to its environment (Assmann & Jegla, 2016). Stress conditions, like drought and salinity, disrupt plant function, prompting plants to close their stomata quickly to conserve water. This rapid closing is driven by the activation of outward potassium (K+) and anion channels in guard cells. Beyond this, stomatal regulation involves metabolite reorganization, post-translational modifications, and gene expression (Cominelli et al., 2005). These mechanisms ensure plants can adjust stomatal openings dynamically against stress conditions. When plants detect drought, intracellular signals activate the ABA pathway, boosting drought tolerance (Yang et al., 2020). ABA receptors are crucial for initiating this process, and various receptors have been identified (Lee et al., 2013). Overall, finely tuning stomatal closure is essential for plants to reduce water loss and survive in stressful environments.

## 2.2 *Osmotic Adjustment*

The role of osmotic adjustment in helping bacterial and yeast cells adapt to harsh and osmotic conditions has been understood for over a century. Some early reports have highlighted the significance of solutes in maintaining plant cell turgor and aiding their adaptation to challenging environments (Hsiao et al., 1997). Osmotic adjustment and compatible solute accumulation are well-known for helping plants adapt to dehydration, mainly by maintaining turgor and safeguarding specific cellular functions.

Osmoregulation sustains photosynthesis, with a considerable portion of the assimilates being redirected to root growth. Deeper root systems or osmotic adjustments allow plants to access water from dry soils while protecting cellular membranes and enzymes from stress (Lemcoff et al., 1998). Under salt- and mannitol-induced osmotic stress, alterations in xylem pressure were observed despite a stable transpiration rate (Zhu et al., 2010).

In wheat, osmotic adjustment is a key component of drought resistance, involving higher soluble sugar and potassium (K+) content, as well as greater proline levels (Kocsy et al., 2004). Drought-resistant cultivars show lower $CO_2$ exchange rates but exhibit greater osmotic adjustment as well as water-use efficiency. When seedlings are exposed to salt stress, levels of sodium, proline, and soluble sugar rise, while relative growth rates, potassium concentration, and leaf osmotic potential fall (Almansouri et al., 1999). It is demonstrated that osmotic adjustment in barley genotypes was positively correlated with yield and stomatal conductance under drought stress (González et al., 2008).

## 2.3 Photosynthetic Regulation

Plant productivity is shaped by various physiological and molecular processes, with stable photosynthetic performance being crucial for healthy growth and development. The chloroplast, where both light and dark reactions of photosynthesis occur, is particularly sensitive to abiotic stress. These stresses can reduce stomatal conductance and trigger oxidative stress, which in turn decreases the activity of the enzyme RuBisCO, hindering the photosynthetic process (Kohli et al., 2017).

Any disturbances to key components such as photosystem I (PS I), photosystem II (PS II), and other pigments and electron transport systems can hinder the entire photosynthesis (Ashraf & Hafeez, 2004). The photosynthetic apparatus is highly stress-sensitive and among the first cellular systems affected. High temperatures enhance cyclic electron flow around PS I and modify thylakoid proton conductance, thereby influencing PS I activity (Bukhov & Carpentier, 2000). Moderate heat stress can have a major impact on the cytochrome complex and PS I, which is more susceptible to damage than PS II when temperatures rise (Ashraf & Hafeez, 2004). Additionally, PS II and the stroma can become oxidized, causing a redox imbalance in the photosynthetic electron transport chain (Zhang & Sharkey, 2009).

High temperatures disrupt stomatal conductance, intracellular $CO_2$ levels, and leaf water status. When stomata close under these conditions, $CO_2$ concentrations shift, further inhibiting photosynthesis. Additionally, temperature changes directly influence vapor pressure deficit, affecting plant hydraulic conductance and water delivery to leaves (Yang et al., 2012). During drought, plants initially respond by closing their stomata to conserve water, which also limits $CO_2$ uptake and disrupts nonstructural carbon transport, leading to carbon starvation that affects photosynthesis and other processes (Steppe et al., 2015). This response reduces adenosine triphosphate (ATP)/nicotinamide adenine dinucleotide phosphate hydrogen (NADPH) and carbohydrate availability for other functions, stunting growth and impairing respiration (Pinheiro & Chaves, 2010). Drought disrupts nicotinamide adenine dinucleotide phosphate (NADP+) regeneration, causing electron leakage and over-reduction in the electron transport chain (Nishiyama & Murata, 2014). The oxygen-evolving center of PS II suffers, reducing its activity and chlorophyll fluorescence (Skotnica et al., 2000). Severe drought not only restricts photosynthesis but also degrades essential proteins, lowering RuBisCO levels and hindering $CO_2$ assimilation. Both stomatal conductance and metabolic disruptions significantly contribute to the decline in photosynthesis in cereals.

## 2.4 Hormonal Regulation

To withstand abiotic stresses, plants adopt adaptive strategies that integrate morphological, physiological, and biochemical responses, ensuring their growth and productivity. A key role in these adaptive strategies is played by phytohormones, tiny signalling molecules that, even in extremely small amounts, control the majority of physiological and metabolic processes in plants.

### 2.4.1 Abscisic Acid

ABA signalling starts when ABA binds to its receptors, PYRABACTIN RESISTANCE1 (PYR1)/PYR1-LIKE (PYL)/REGULATORY COMPONENTS OF ABA RECEPTORS (RCAR), forming a complex with protein phosphatases type 2C (PP2C). This interaction activates SNF1-related protein kinase 2 (SnRK2) proteins, triggering a cascade of phosphorylation events that modulate plant stress responses (Ma et al., 2009). SnRK2 proteins specifically activate transcription factors such as ABA responsive element binding proteins (AREB)/auxin-responsive transcription factors (ARFs), inducing the expression of stress-responsive genes (Furihata et al., 2006).

Recent studies in wheat have identified several SnRK2 homologs, such as TaSnRK2.3, TaSnRK2.4, and TaSnRK2.8, which play roles in ABA-mediated stress signalling. For instance, TaSnRK2.8 overexpression triggers the gene expression that regulates ABA biosynthesis and signalling, as well as cold-response genes such as CBF1, CBF2, and CBF3 (Zhang et al., 2010).

TFs play a crucial role in the ABA pathway by controlling the expression of stress-related genes. In wheat, basic leucine zipper (bZIP) protein such as WLIP19, which are induced by ABA, drought, and cold, enhance the expression of cold- and drought-responsive genes, leading to increased stress tolerance (Kobayashi et al., 2008). Likewise, the APETALA2/ethylene responsive factor (AP2/ERF) family of transcription factors, including TaAIDFa and WDREB2, respond to both osmotic and cold stresses, emphasizing the intricate crosstalk between various stress-signalling pathways (Egawa et al., 2006).

The myeloblastosis (MYB)/myelocytomatosis (MYC)family of TFs also plays a crucial role in ABA signalling. For example, overexpression of TaMYB32 and TaMYB33 in Arabidopsis improves salt tolerance, potentially by increasing proline accumulation, which helps plants maintain osmotic balance (Qin et al., 2012). Another key aspect of ABA signalling in wheat is the involvement of WRKY-type transcription factors, which are induced by multiple stresses. For example, TaWRKY2 and TaWRKY19 enhance drought and salt stress tolerance by activating downstream stress-responsive genes (Niu et al., 2012).

Overall, ABA signalling is an extremely complex process involving several proteins and genes working together to help wheat plants survive harsh environmental conditions. Future research will likely focus on identifying key ABA signalling components that can be targeted for breeding or genetic engineering to improve crop resilience (Zhang et al., 2010). This could facilitate the development of wheat cultivars with improved resilience to the growing challenges of climate change and resource constraints.

### *2.4.2 Auxin*

Auxin, a plant hormone derived from tryptophan, is most commonly found as indole-3-acetic acid (IAA). Under normal conditions, auxin signalling functions through a system where auxin binds to receptor peptides such as translocated intimin receptor (TIR) and auxin signaling F-box proteins (AFBs), triggering the degradation of repressor proteins (Aux/IAA), which then allows ARFs to activate growth-related genes. This hormone's pathway is fine-tuned, with miRNA also contributing to the regulation of some auxin receptors and genes during growth and in response to stress (Ma et al., 2015).

However, during times of environmental stress, plants tend to downregulate auxin activity as part of a strategy to conserve resources for survival. For instance, wheat exposed to cold stress shows reduced levels of auxin in leaves due to either decreased production or impaired transport via PIN proteins (Kosová et al., 2013). Interestingly, in frost-sensitive wheat varieties, higher auxin levels post-acclimatization were linked with poorer fitness, further indicating that reduced auxin is beneficial under these conditions (Kosová et al., 2013). Similarly, wheat plants subjected to drought or salinity exhibit reduced grain filling, largely attributed to lower auxin levels during reproductive growth (Abid et al., 2017).

Moreover, studies have demonstrated that drought reduces the auxin receptors' expression and transporters in wheat roots, regardless of the plant's stress tolerance level (Krugman et al., 2011). Heat stress impairs auxin biosynthesis in reproductive organs such as anthers, which is related to increased male sterility via pollen abortion. Remarkably, applying auxin to anthers under heat stress conditions can rescue male fertility (Sakata et al., 2010). While auxin is crucial for plant development, manipulating its levels broadly across the plant can have unintended consequences. However, targeting specific tissues, such as anthers or seeds, with stress-inducible promoters to regulate auxin production could help maintain reproductive success even under harsh conditions.

### *2.4.3 Cytokinin*

Cytokinins, another group of hormones essential for plant growth, work in contrast to auxins by primarily promoting cell division and proliferation. They play a key role in maintaining undifferentiated, rapidly dividing cells in the shoot apical meristem, essential for early embryogenesis and endosperm development, which is critical for proper grain formation (Abid et al., 2017). Cytokinins also protect cells from damage causing ROS and help preserve chlorophyll during stress, such as during shading or leaf senescence. They accomplish this by inhibiting chlorophyll-degrading enzymes and enhancing antioxidant activity, thereby preserving chlorophyll levels and sustaining photosynthesis under stress (Zavaleta-Mancera et al., 2007).

Cytokinin signalling begins when the hormone is perceived by receptor proteins, known as histidine kinases, present on the plasma membrane (PM) or the endoplasmic reticulum. This triggers a cascade involving the phosphorylation of various proteins, ultimately activating response regulators in the nucleus (Osakabe et al., 2013). Cytokinin signalling is known to oppose ABA signalling, which is significant in plant's response toward abiotic stress. The balance between these hormones is crucial for regulating stress responses, with enzymes such as cytokinin oxidase/dehydrogenases (CKXs) playing a role in cytokinin breakdown, a process often regulated by ABA (Kosová et al., 2013).

During stress conditions, cytokinin production is typically reduced to prioritize ABA signalling and conserve energy for survival. For example, cold-tolerant wheat cultivars reduce cytokinin levels earlier than cold-sensitive ones, leading to better stress adaptation (Kosová et al., 2013). Suppressing cytokinin biosynthesis and signalling in plants can improve drought tolerance by limiting water loss and delaying senescence.

Despite the need to suppress cytokinin in most tissues under stress, developing grains could benefit from sustained cytokinin production. Cytokinin aids in endosperm proliferation, helping to maintain grain filling during stress conditions, a process that can be further supported by applying cytokinin externally (Abid et al., 2017). As with auxins, large-scale manipulation of cytokinin signalling could have unintended side effects, but targeted interventions—such as tissue-specific cytokinin biosynthesis or selective suppression of signalling pathways—could help boost stress tolerance and grain yield simultaneously. Most research on cytokinin interactions in wheat relies on comparative studies between stress-resistant and susceptible varieties, or extrapolations from model organisms such as rice, Arabidopsis, or tobacco. Future studies should focus on identifying and manipulating cytokinin-responsive genes in wheat to enhance its durability to environmental stress (Kosová et al., 2013).

### *2.4.4 Gibberellic Acid*

Gibberellic acids (GAs) are a diverse group of plant hormones that play a necessary role in governing growth and development, particularly by influencing cell elongation and division. These compounds are tetracyclic diterpenoid carboxylic acids whose biosynthesis involves various GA dioxygenases. One important dioxygenase, GA2ox, helps maintain the balance of GA levels by converting active GA into its inactive form (Colebrook et al., 2014). When plants perceive GAs, a signalling cascade is triggered that leads to the targeted deterioration of DELLA proteins, which are transcriptional repressors associated with GA responses. This pathway is vital for the growth of wheat seedlings in Mediterranean regions, where the upper soil layers can be extremely dry and deficient in water (Amram et al., 2015).

Wheat mutants that are insensitive to GAs, known as *Rht* mutants, display a dwarf phenotype, which hampers their ability to emerge successfully from deeper sowing—an essential adaptation in these arid soil conditions (Amram et al., 2015). Ecological stresses can limit plant growth by altering the levels of endogenous GAs; for instance, mutants with lower GA content often show increased salt tolerance. Research indicates that GAs are suppressed during drought as well as salt stress, either due to decreased biosynthesis or increased degradation (Llanes et al., 2016). Studies with GA-inhibiting compounds suggest that improving drought tolerance in crops can also enhance biomass as well as yield (Plaza-Wüthrich et al., 2016).

Notably, various staple crops have dwarf cultivars that display stress-tolerant traits because of their lower GA levels. When these varieties are treated with exogenous GA, their stress-tolerant characteristics diminish (Vettakkorumakankav et al., 1999). This suggests that engineering crop varieties could focus on manipulating GA2ox genes to optimize GA levels, rather than eliminating GAs entirely, which could have negative consequences for crop health. For instance, GA2ox genes' ectopic expression in rice crops led to a moderate reduction in GA levels, resulting in plants with stunted growth that were more resistant to lodging, utilized water more efficiently, and exhibited greater tolerance to abiotic stresses, along with an increased number of productive tillers (Lo et al., 2017).

These findings highlight the intricacies and trade-offs associated with manipulating single hormonal pathways for crop improvement. Some research has examined specific GA sensitivity genes in dwarf wheat cultivars and their impacts on yield and crop emergence (Haque et al., 2011).

### 2.4.5 Brassinosteroids

Brassinosteroids (BRs) are crucial plant steroid hormones that contribute to growth, development, and stress adaptation by modulating cell elongation, division, and vascular differentiation. Studies have shown that BR foliar application enhances wheat growth, photosynthesis, and $CO_2$ fixation, significantly increasing leaf and shoot biomass. BRs also improve stress tolerance by interacting with other hormones. The application of active BR compound 2,4-epibrassinolide (EB) reduced the stress-induced extent of ABA and indole-3-acetic acid (IAA), without changing cytokinin levels (Avalbaev et al., 2010). Under drought, EB-treated seeds showed enhanced protection, and EB also elevated cytokinin levels by inhibiting cytokinin oxidase activity (Yuldashev et al., 2012).

BRs mitigate damage from heavy metal and salinity stress by boosting antioxidative enzyme activity along with proline levels (Hayat et al., 2014). They also reduce heavy metal accumulation in wheat grown in microbes-affected soils. While much research has focused on BR application, molecular-level studies, especially involving BR receptors like BRI1, remain limited in cereals. Transgenic Arabidopsis expressing TaBRI1 showed improved thermotolerance and growth, and wheat homologs of BIN2, TaSK1, and TaSK2 regulate BR signalling, affecting plant growth (Bittner et al., 2015).

### 2.4.6 Salicylic Acid

Salicylic acid (SA) is a key plant hormone that regulates growth, development, and stress responses, especially under abiotic stress conditions. Exogenous application of SA strengthens the antioxidant defense system by minimizing oxidative damage induced by drought, heat, and salinity.

Studies have shown that lower SA concentrations (0.5–1.0 mM) are effective in promoting antioxidant production, protecting photosystem II, and improving photosynthesis under stress (Khan et al., 2013). However, high SA levels (e.g., 2.5 mM) may cause oxidative damage, leading to membrane degradation and electrolyte leakage (Chen et al., 2016). SA also reduces ethylene production during heat stress, helping maintain photosynthesis by regulating proline levels and osmotic balance (Khan et al., 2013). While SA's role in stress tolerance is well-documented, its signalling pathways remain unclear, especially regarding its interactions with other hormones and oxidative processes.

### 2.4.7 Jasmonic Acid

Traditionally, JA has been examined for its role in defending plants against biotic stress. However, recent research is revealing its significance in abiotic stress tolerance, particularly through methyl jasmonate (MeJA) or JA applications. Applying 2 mM JA to seedlings significantly increased salt tolerance by boosting antioxidant production, and reducing harmful ROS levels (Qiu et al., 2014). JA enhanced key antioxidants such as superoxide dismutase and catalase, protecting cellular health from oxidative damage.

JA also alleviates drought stress by modulating the ascorbate–glutathione pathway (AsA–GSH pathway). In seedlings of wheat, JA improved the enzyme activity involved in Asa–GSH pathway, increasing drought tolerance. Interestingly, nitric oxide (NO) seems crucial for JA signalling, as drought tolerance was reduced when a NO inhibitor was applied (Shan et al., 2015). JA's role in abiotic stress tolerance is becoming clearer, especially its ability to activate antioxidant systems and coordinate with other hormones such as ABA. Future research will likely explore how JA signalling can address multiple stresses simultaneously (Ahmad et al., 2016).

### *2.4.8 Ethylene*

Ethylene, produced from S-adenosylmethionine (SAM), is a gaseous hormone responsible for plant stress responses and late-stage growth processes such as senescence and fruit ripening (Kosová et al., 2013). Ethylene signalling occurs via membrane-bound histidine kinases, which activate ethylene responsive factors (ERFs). Ethylene has an antagonistic relationship with ABA, helping plants balance responses to mild stress while maintaining photosynthesis and normal metabolic functions (Wilkinson et al., 2012). In extreme conditions, ERFs such as TaERF1 and TaERF3, induced by both ABA and ethylene, may enhance stress tolerance (Rong et al., 2014).

During drought, upregulation of 1-aminocyclopropane-1-carboxylic acid (ACC) synthase (ACS) leads to increased ACC, inhibiting starch synthesis and reducing grain filling. Heat stress can also raise ACC levels, leading to kernel abortion and premature senescence, though the ethylene inhibitor 1-methylcyclopropene (1-MCP) has shown success in reducing these symptoms (Hays et al., 2007). Ethylene's role in salt stress is less clear, though certain ERFs improve salt, drought, and cold tolerance by enhancing photosynthesis and reducing harmful $H_2O_2$ (Rong et al., 2014). Inhibiting ethylene responses and promoting polyamine production may enhance stress resilience, especially during drought and salt conditions (Gharbi et al., 2017). Future research is required to better control ethylene signalling for improved plant stress management.

## *2.5 Antioxidant Defense*

These stresses can disrupt homeostasis and ion distribution within plant cells, which ultimately leads to osmotic stress and an increase in ROS levels (Herbette et al., 2011). The buildup of ROS can severely damage cell organelles, causing membrane peroxidation, compromising cell membranes, degrading biological macromolecules, and ultimately resulting in cell death. A plant's ability to combat the detrimental effects of ROS is crucial as a tolerance mechanism to various stresses.

The antioxidant defense system in plants comprises various enzymes that either suppress the Haber–Weiss reaction or drive the Foyer–Halliwell–Asada pathway, reducing $H_2O_2$ by utilizing NADPH's reducing potential. Enzymatic antioxidants such as superoxide dismutase (SOD), catalase (CAT), ascorbate peroxidase (APX), monodehydroascorbate reductase (MDHAR), dehydroascorbate reductase (DHAR), and glutathione reductase (GR) help alleviate oxidative stress, while nonenzymatic antioxidants include water-soluble glutathione (GSH) and ascorbic acid, along with carotenoids and tocopherols (Gill et al., 2015).

## *2.6 Leaf Senescence*

Plants adopt a range of strategies to cope with environmental stresses, and these strategies differ depending on the plant's developmental stage and species (Mickelbart et al., 2015). One such method, known as the "escape" strategy, involves plants accelerating their aging process, that is, senescence and shedding leaves called abscission to reduce their canopy size when faced with stress (Kooyers, 2015). While this helps the plant focus on ensuring the survival of the next generation by directing resources to seed production, it comes at a cost for annual crop species. Accelerated senescence can lead to significant agricultural yield losses, translating into substantial economic setbacks for farmers.

During senescence, plants effectively recycle nutrients from older tissues to their reproductive organs, a process governed by a complex network of physiological and molecular mechanisms (Munné-Bosch, 2008). For example, in rice, OsNAC2 regulates the expression of ABA-dependent genes, triggering leaf senescence by activating chlorophyll breakdown genes, which lowers stress tolerance (Shen et al., 2017). Overexpressing OsWRKY23 in Arabidopsis speeds up leaf senescence in darkness (Jing et al., 2009), while OsWRKY47 boosts seed yield and drought tolerance but accelerates drought-induced senescence when knocked out (Raineri et al., 2015).

### *2.7 Secondary Metabolite Production*

One of the key ways plants cope with the ever-changing environments is through secondary metabolism, which allows them to interact with other species and adjust to various environmental stresses. The biosynthesis of these protective substances is very well programmed, occurring in specific plant tissues at the right time, depending on environmental factors. Among these secondary metabolites, flavonoids are powerful antioxidants that are involved in plant protection from stress. For instance, plants like rice (*Oryza sativa*) respond to heat and salt stress by producing more flavonoids such as quercetin and kaempferol, which boost their resilience (Jan et al., 2021). Similarly, short-wavelength radiation triggers the production of flavanol glycosides, further enhancing the plant's defense mechanisms (Ferreyra et al., 2021).

It has been shown that wheat leaves exposed to cold temperatures experience a sharp increase in compounds such as putrescine, while other metabolites like spermidine and spermine fluctuate (Gondor et al., 2016). Salinity stress also prompts changes in secondary metabolites. In maize, for example, total phenolic and flavonoid levels rise, helping the plant to better cope with salinity stress (AbdElgawad et al., 2016). In rice, salinity stress not only improves grain quality by increasing antioxidant activities but also boosts anthocyanin and total phenolic content (Chunthaburee et al., 2015). However, under drought conditions, maize sees a reduction in phenolic compounds and a decline in biomass (Vardharajula et al., 2011).

## 3. Molecular Mechanisms Governing Stress Perception and Signal Transduction

Signal transduction in cereal plants plays a role in their ability to respond to abiotic stresses. These environmental challenges can significantly affect growth and yield, making it essential for crop plants to recognize and adapt to adverse conditions. Plants utilize four primary signal transduction pathways that can interact with one another: signalling of reactive species, calcium-dependent pathways, plant hormonal pathway, and protein phosphorylation/dephosphorylation through kinase cascades (Tiwari & Lata, 2018).

### *3.1 Signal Transduction in Response to Drought*

Osmotic stress signal transduction involves molecules such as calcium ($Ca^{2+}$), NO, and reactive sulpfur species, which activate mitogen-activated protein kinase (MAPK) cascades, calcium-dependent protein kinases (CDPK), and calcineurin B-like interacting protein kinases (CIPKs) (Lin et al., 2021). The MAPK cascade is essential in response against drought, due to its importance in ABA signalling and regulation of ROS. In rice, OsMKK4, OsMKK1, OsMPK5, OsMPK7, and OsMPK4 show higher expression in drought conditions (Xue et al., 2009), while in wheat, TaMKKK16, TaMKK1, and TaMPK8 show changes in expression during drought (Wang et al., 2016). The OsMKK10.2–OsMPK3 complex has also been known to elevate drought tolerance through ABA signalling in rice (Ma et al., 2017), although the interplay between MAPK cascades and ABA remains unclear (Lin et al., 2021). Additionally, MAPKs in rice, such as MAPK5, MAPK7, and MAPK8, are induced by drought, while MAPK4 is repressed. MAPK kinases also regulate transcription factors such as OsWRKY30, further enhancing drought tolerance (Hadiarto & Tran, 2010).

CDPKs and CIPKs are key families involved in stress responses. For instance, in rice, OsCIPK12 increases proline and soluble sugars, improving drought resistance (Saijo et al., 2000). The HSPs play significant roles in stress-tolerating mechanisms. In rice, 74 HSPs are categorized into four types: sHSP, HSP70, HSP90, and HSP100 (Xiang et al., 2013), though only a subset is activated under drought stress. In wheat and barley, dehydrins (Dhn), part of the late embryogenesis abundant (LEA) protein family, are also expressed during drought. Barley Dhn genes, like Dhn1, Dhn3, Dhn5, Dhn7, and Dhn9, positively affect drought tolerance by maintaining chlorophyll content, osmotic adjustment, and biomass while reducing oxidative damage (Kosová et al., 2014).

LEA proteins, identified for their hydrophilic nature, stabilize membranes and proteins under stress. Barley's HVA1 gene, encoding a group 3 LEA protein, has been transferred to rice, enhancing its tolerance to drought (Ali et al., 2021). Overexpressing HVA1 wheat gene, as well as OsLEA6 and OsLEA3-1 in rice, has been reported for improvement of drought tolerance (Hand et al., 2011). Phosphatidylinositol metabolism is another essential pathway for drought survival, generating phospholipid signalling molecules involved in root growth, pollen development, and stress responses.

NO serves as a key signalling molecule in stress adaptation, acting through cyclic guanosine monophosphate (cGMP) and guanylate cyclase activation. NO also regulates ROS levels by activating antioxidant enzymes. Exogenous NO application enhances drought resistance in barley by increasing proline levels and activating antioxidant enzymes (Gan et al., 2015). Moreover, NO interacts with various growth hormones as well as hydrogen sulfide ($H_2S$) in drought responses. These complex signalling pathways enable plants to manage oxidative damage, improve osmotic regulation, and boost their resilience under drought conditions.

## *3.2 Signal Transduction in Salinity Stress*

Plants respond to salinity by detecting disruptions in ion and osmotic balance, though no specific sensor for high cytosolic Na+ levels has been identified. Consequently, the exact mechanism of Na+ detection remains unclear, leaving the perception of salt stress signals largely unexplained. However, the well-characterized SOS pathways are vital for salt stress signalling in plants, including cereals crops. Additionally, the MAPK cascade plays a crucial role in transmitting stress signals to various transcription factors, which, in turn, activate salt-responsive genes (Choudhary et al., 2021).

The SOS pathway for salt stress tolerance is crucial, with key proteins working to remove excess Na+ ions from plant cells. The plasma membrane protein SOS1 is activated through the SOS2–SOS3 kinase complex. SOS3, a $Ca^{2+}$ sensor, detects salt-induced $Ca^{2+}$ signals, which are enhanced by ABA. In wheat, salt-tolerant genotypes exhibited markedly increased expression of SOS1, SOS2, and SOS3, leading to better Na+ efflux and improved K+/Na+ balance, which helps protect plants from salt damage (Sathee et al., 2015). Overexpression of wheat TaSOS1 and TaSOS1-974 in tobacco enhanced salt tolerance by reducing membrane damage and electrolyte leakage (Jiang et al., 2021). Comparable findings were observed in Arabidopsis with overexpression of a durum wheat SOS1 variant, confirming that the C-terminus of SOS1 functions as an autoinhibitory domain released by SOS2 phosphorylation (Feki et al., 2013). In rice, the SOS genes (OsSOS1, OsSOS2, and OsSOS3) were upregulated under salt stress, particularly in roots, but rice still accumulated more harmful Na+ in shoots compared to barley, which maintained normal metabolism despite lower SOS gene expression (Fu et al., 2018). This highlights the species-specific responses to salinity stress.

In response to salt stress, plant utilizes two key mechanisms: the classification of Na+ into vacuoles and Na+ transport regulation via high-affinity K+ transporters (HKT). The tonoplast Na+/H+ antiporter (NHX) helps store excess Na+ in vacuoles, supported by H+-ATPase and H+-pyrophosphatase activities. In wheat, NHX1 expression increases under saline conditions, and overexpression of TaNHX2 in eggplant and sunflower improves salt tolerance by reducing ROS and malondialdehyde (MDA) while enhancing antioxidant enzyme activity (Tiwari et al., 2020). Barley shows higher expression of HvNHX5 in roots, while rice shows higher NHX expression in shoots, possibly contributing to higher Na+ concentration in rice shoots (Fu et al., 2018). Additionally, overexpression of OsNHX1 and HvNHX2 enhances salt tolerance in rice and Arabidopsis, respectively, indicating that vacuolar Na+ compartmentalization is important for improving salt tolerance (Chen et al., 2007).

HKT proteins, primarily class 1 Na+ transporters, help in the removal of Na+ from the xylem vessel sap, confining it to roots and protecting shoots. In rice, salinity-tolerant varieties show higher expression of OsHKT1 in leaf sheaths and OsHKT1 in roots, reducing Na+ in leaf blades, unlike the sensitive varieties (Kobayashi et al., 2017). In maize, higher expression of ZmHKT2 in the salt-sensitive genotypes leads to increased Na+ uptake in roots and reduced K+ in leaves, worsening salt tolerance compared to the resistant genotype (Rizk et al., 2020). During salt stress, chloride ions ($Cl^-$) can interfere with phosphorus (P) content in plants, potentially disrupting the MAPK pathway. The wheat MAPK member TaMPK4 functions as a positive regulator against salinity. Overexpressing TaMPK4 in tobacco resulted in larger plants with increased dry mass, number, and area of the leaf, whereas knockout tobacco plants were smaller with lower biomass. These TaMPK4-overexpressing plants also had prominent K+ and osmolyte levels and lower Na+ concentrations compared to wild-type plants. In Arabidopsis, overexpressing ZmSIMK1, a maize MAPK, improved salt tolerance by enhancing seed germination rate along with seedling growth in NaCl conditions (Gu et al., 2010). Conversely, overexpressing the TaTMKP1 improved salt tolerance by promoting germination and reducing oxidative damage through increased antioxidant enzyme activities (Zaidi et al., 2016).

Alternatively, rice OsMAPK33 functions negatively to regulate salt tolerance. Its expression decreases after salt stress induction, and overexpression leads to increased sensitivity to salinity due to disrupted ion homeostasis (Lee et al., 2011). Additionally, hydrogen sulphide ($H_2S$) is reported for salt tolerance improvement in several cereals by maintaining ion balance and reducing oxidative stress, which reflects the lower ROS and MDA levels. $H_2S$ application also boosts antioxidant enzyme activity and can enhance photosynthetic efficiency and metabolic processes in rice (Wei et al., 2021).

## 3.3 Signal Transduction in Response to Heat Stress

To develop heating stress (HS) adaptive strategies, plants first need to sense HS, primarily through their PM (Hayes et al., 2021). Macromolecular sensors, including G-protein-coupled receptors (GPCRs), receptor-like kinases, and NADPH-oxidases, are localized at the PM and play crucial roles in stress perception (Bokszczanin et al., 2013). GPCRs detect various secondary signals from stressors such as HS by binding to ligands, which activates heterotrimeric G-proteins (Tuteja, 2009). This GPCR-G protein association then stimulates small Ras-related guanosine-5'-triphosphate (GTP)-binding proteins, leading to a $Ca^{2+}$ inositol triphosphate-mediated signalling pathway (Priya et al., 2019).

HS-induced PM fluidization activates $Ca^{2+}$ channels, increasing cytosolic $Ca^{2+}$ levels through specific channels such as cyclic nucleotide-gated channels (CNGCs) (Saidi et al., 2009). This influx of $Ca^{2+}$, along with elevated cytosolic cyclic adenosine monophosphate (cAMP) levels, triggers multiple HS signalling cascades, including CDPKs, which regulate transcriptional factors (Saidi et al., 2009). ROS also play a vital role in HS signalling, with respiratory burst oxidase homologs (RBOHs) generating $H_2O_2$ under stress (Yao et al., 2016). The activity of RBOHs is regulated by intracellular $Ca^{2+}$ levels and can trigger signalling that leads to HS gene expression (Bokszczanin et al., 2013).

Additionally, NO accumulates transiently in response to HS, facilitating crosstalk between NO and ROS that regulates HS signalling and protein modifications (Ohama et al., 2017). This suggests a feedback loop involving NO, ROS, and $Ca^{2+}$ signalling, part of a complex gene regulatory network (GRN) responding to HS (Ohama et al., 2017). PM phospholipids, such as phosphatidylinositol 4,5-bisphosphate (PIP2) and phosphatidic acid (PA), also play a central role in HS sensing and signal perception (Hayes et al., 2021).

Key components in HS signalling include protein kinases (PKs) that mediate phosphorylation and phosphatases (PPs), helping in dephosphorylation. Calmodulins (CaMs) and CDPKs are crucial for detecting $Ca^{2+}$ signals and activating heat shock transcription factors (HSFs) for the regulation of HS-responsive genes (Zhang et al., 2009). For example, the rice gene OsCaM1-1 enhances HS tolerance in transgenic Arabidopsis by upregulating HSFs and HSPs (Wu et al., 2012). The interaction between calcium/calmodulin binding protein-kinases (CBKs) and HSFs can modulate gene expression, demonstrating the importance of these pathways in plant HS adaptation.

Recent research has highlighted the proximal relationship among temperature plus light sensing in plants, revealing shared receptors, molecular regulators, and signalling pathways (Perrella et al., 2022). This interplay is crucial for understanding HS signalling. Notably, phytochrome interacting factor 4 (PIF4) serves as a key hub for temperature signalling, promoting the expression of auxin and brassinosteroid biosynthesis genes, suggesting a role in thermomorphogenesis (Perrella et al., 2022). Mild HS also increases PIF4 expression by disrupting the evening complex (EC) binding to target genes (Huang & Nusinow, 2016).

Phytochrome B (PhyB) is a hypersensitive sensor that integrates light and temperature signals, transitioning between active (Pfr) and inactive (Pr) forms depending on light conditions. The temperature-sensitive reversion of PhyB from Pfr to Pr under mild HS enhances HS perception by alleviating the inhibition of PIF4, initiating thermomorphogenesis (Perrella et al., 2022). Early flowering 3 (ELF3), part of the EC regulating the circadian clock, senses temperature through a prion-like domain that facilitates phase separation at mild temperatures. This mechanism supports thermosensing and restricts the expression of growth-promoting factors such as PIF4 (Huang & Nusinow, 2016). In addition to PIF4, PIF7 also promotes thermomorphogenesis by activating auxin biosynthetic genes in response to elevated temperatures (Fiorucci et al., 2020). PIF7 protein levels increase rapidly during mild HS, driven by enhanced translation due to a specific RNA hairpin formation in its 5' untranslated region (Chung et al., 2020). Further investigation is required to inspect how these receptors interact with light and HS signals in cereal plants.

## 4. Advances in Plant Breeding Techniques for Abiotic Stress Tolerance in Cereals

The increase in population and consumption expectations in emerging economies will lead to inevitable food shortages if agricultural productivity does not return to previous rates or at least keep pace with population growth (Khush, 1999). To address such issues, breeding and designing climate-resistant crops have become essential strategies to lessen the dangerous effect of changing climate in agriculture. For this, the ultimate goal is to generate crop varieties that can combat abiotic stresses while preserving yield potential and nutritional quality at their best. These efforts aim to secure food production under adverse climatic conditions.

### *4.1 Traditional Breeding Approaches for Stress Tolerance*

Traditional breeding methods have been the cornerstone for creating crop varieties that can withstand climate challenges. Traditional breeding techniques involve selection and crossing of parents with traits of interest to develop offspring with desirable characteristics. The stress timing, duration, and severity are three essential aspects determining successful stress-resistance breeding. Selection, hybridization, and mutation are the most commonly used breeding methods that have been utilized by breeders to successfully incorporate stress tolerance in various crop species (Dwivedi et al., 2016). Selection allows the discovery and propagation of plants with biotic and abiotic stresses, in particular ecosystems. Hybridization involves crossing genetically distinct plants to integrate beneficial traits from diverse parental lines. This process results in hybrid offspring with enhanced stress tolerance and performance. Through traditional breeding methods, researchers have successfully developed crop varieties such as rice, wheat, maize, and barley that can tolerate abiotic stress. However, one of the challenges with this approach is accurately identifying the traits that are strongly linked to stress tolerance (Raina et al., 2022).

### *4.1.1 Mutation Breeding*

Breeding crops via the conventional approach is labor-intensive and time-constraining, often failing to produce the desired variation. Therefore, researchers always tried to find various ways to get desired variations within short durations in different crops. For such problems, mutation breeding serves as an efficient approach for producing mutants and more genetic variability. Mutation breeding, a technique that uses radiation or chemicals to cause genetic changes, has contributed to developing climate-resilient crop varieties (Hafeez et al., 2023). Developing mutant varieties with improved abiotic stress tolerance does not cause any alterations in genetic content (Goyal et al., 2021). Mutations can lead to the introduction of novel traits such as stress tolerance, which can be selected and employed in breeding programs (Zafar et al., 2019). Zhefu 802 and Basmati 370 are the mutant rice varieties developed by gamma radiation and these rice varieties showed enhanced tolerance to cold stress (Ahloowalia et al., 2004). Therefore, mutation-assisted plant breeding techniques can be employed to design crop varieties that can tackle the hurdles posed by the changing global climate and the uncertainty surrounding plant production.

### *4.1.2 Molecular Breeding*

Molecular breeding techniques have transformed traditional breeding with precision and efficiency for trait selection by using several molecular markers that facilitate breeders to select and introgress several stress-tolerance related genes more efficiently at the molecular level. Molecular markers (eg. simple sequence repeats, SSRs and single nucleotide polymorphism, SNPs) are the specific regions in the genome that can be detected easily and are linked to particular traits of interest. Marker-assisted selection (MAS) helps breeders to identify plants having desired traits more effectively (Lebedev et al., 2020). Genetic resources such as landraces and wild crop relatives are key components of molecular breeding. These resources contain valuable stress-tolerance genes that can be introduced into cultivated crop varieties. Breeders can transfer targeted genes from genetic sources into elite cultivars by combining conventional breeding techniques with molecular markers. MAS has been applied to enhance the genetic diversity that can be utilized by breeders for developing crops that are resilient to changing climate conditions (Hafeez et al., 2023). Khan et al. (2013) successfully improved the submergence tolerance of BT7 cultivars by using the marker assisted backcrossing (MABC) method that was controlled by major Submergence 1 (SUB1) gene.

## *4.2 Genomic Approaches for Identifying Stress Tolerance Traits*

Advancements in genomic technologies have created new opportunities for creating climate-resilient crops. The genomic approach enables the researchers to analyze the structure, function, and evolution and interpret them. Based on the methodologies and outcomes, genomics is classified into three groups, that is, functional, structural, and comparative genomics. Identification of genomic areas are linked to stress tolerance using genome-wide association studies (GWAS) and quantitative trait locus (QTL) mapping (Mohammadi et al., 2020; Varshney et al., 2011). To find genetic differences associated with stress tolerance, a huge set of genetic markers are used in diverse populations.

Genomics-based selection has transformed plant breeding by predicting breeding value. Genetic selection helps in choosing plants based upon their genetics, rather than exclusively on phenotypic characteristics. This strategy accelerates the process of breeding by reducing the time needed for phenotypic selection and evaluation. Several QTLs associated with drought tolerance, including the qDTY12.1 locus, have been identified by GWAS (Sanchez et al., 2022). Similarly, the genomic region in maize has been identified through QTL mapping that provides a target for genetic improvement.

## 4.3 Genetic Engineering Approaches for Enhancing Stress Resilience in Crops

Traditional breeding is a common approach to help plants cope with abiotic stress, but it has its limitations. It's a time-consuming process and restricts cross-breeding to plants within the species that are similar and are closely related, limiting cross-breeding options. Genetic engineering stands as a powerful solution to overcome the limitations of traditional breeding, marking a major advancement in creating crops resistant to various stresses (Aziz & Masmoudi, 2024). Various transformation techniques are used, including direct methods such as biolistic, protoplast transformation, and electroporation, where the gene target is directly introduced into the plant. In the indirect method of transformation, the targeted gene is incorporated with a binary vector and transferred into the plant's DNA through *Agrobacterium tumefaciens* or *Agrobacterium rhizogenes* as carriers.

### 4.3.1 Direct Gene Transfer into Plant Cells

Direct gene transfer is a method that introduces a foreign genetic material into plant cells without using any kind of biological vector. This technique enables the direct introduction of genes in the targeted plant avoiding the natural infection process used by vectors. Particle bombardment, protoplast transformation, and electroporation are some of the techniques for direct gene transfer (Godara et al., 2023).

#### 4.3.1.1 Particle Bombardment

Particle bombardment, or biolistic transformation, is a widely used technique for genetically modifying plants and other organisms. Developed by John Sanford and colleagues in 1987, this method involves delivering substances such as DNA into cells using high-speed microprojectiles, with "biolistic" stemming from "biological" and "ballistic" (Sanford et al., 1987). In this process, genetic material is mixed with spermidine and calcium chloride, then coated onto gold or tungsten particles. To transform plants, stress-resistant genes are first isolated and cloned into a plasmid that includes marker genes for screening and promoter. The transformed plasmid is then delivered into targeted cells via particle bombardment. The bombarded cells are regenerated into transgenic plants, which are later evaluated for gene expression in laboratories, green-net houses, as well as field settings. Soybean was one of the first plants to be transformed using this technique. More recently, the AmSOD gene was introduced into the callus of high-yielding rice variety Pusa Basmati 1 to enhance salt tolerance (Sarangi et al., 2019).

#### 4.3.1.2 Protoplast Transformation

Protoplast transformation utilizes polyethylene glycol (PEG)-treated naked DNA within the plant protoplast for integration and transformation in the presence of divalent cations. PEG-mediated transformation disrupts the dynamic of the cell membrane by increasing its permeability, thereby introducing the desired gene into the targeted cell. This method usually uses the leaf mesophyll for protoplast purification. The leaf sample is treated with protease enzyme and cellulose which degrades the wall of the cell (Sahab et al., 2018). The digested protoplasts are then purified using a mannitol solution. Purified protoplast is then mixed with the DNA solution having a concentration of 60 μM in the presence of PEG/$Ca^{2+}$. Allowing incubation of the mixture enables DNA to penetrate into the protoplast and further into the nucleus for integration into the genome. The transformed protoplasts then spread over the medium suitable for their growth and DNA-containing protoplasts are selected using suitable markers. While being straightforward and relatively low-cost, the method faces challenges when regenerating plants from protoplasts, requiring careful calibration of chemical levels and DNA based on the target plant (Das et al., 2022).

#### 4.3.1.3 Electroporation

Electroporation introduces foreign DNA into cells by applying strong electrical impulses, which form pores in the cell membrane. These impulses cause the hydrophobic lipid bilayer to become polarized, with a plus (+ve) charge on the membrane outside and a minus (−ve) charge on the membrane inside. This charge difference opens temporary pores, allowing foreign DNA to enter the target cells or tissues. For electroporation, an electroporator is used, comprising a pulse power supply, electroporation cuvettes, and electrodes. The two crucial factors that differentiate the pulse being delivered are the time constant and field strength (kV $cm^{-1}$). To transfect different cell types with optimal efficiency, these variables can be changed. The electroporation method produces stable transgenics and is precise, generative, quick, very effective, and simple to utilize. However, one of its main disadvantages is that high electric voltage might cause the cells to die (Godara et al., 2023).

### *4.3.2 Indirect Gene Transfer into Plant Cells*

This is another technique for gene transfer that utilizes biological vectors such as *A. tumefaciens* rather than using any physical and chemical methods for the transfer of foreign DNA material. In such a technique, genetic material is combined with vector molecules that assist in genomic integration (Ravikumar et al., 2014).

#### 4.3.2.1 Agrobacterium-Mediated Transformation

*A. tumefaciens* is a soil-borne bacterium responsible for causing crown gall disease in plants. It transfers T-DNA into the targeted plant cell mediated by the type-IV secretion system (T4SS). This natural gene transfer system can be embraced for transformation by replacing the mutation-causing genes from the Ti-plasmid, that is, T-DNA, containing area along with the gene of interest (Figure 1). This method enabled the insertion of the AtDREB1A transcription factor into rice, resulting in transgenic plants highly resistant to drought stress during both vegetative and reproductive stages (Ravikumar et al., 2014). To enhance plant resistance to abiotic stress, researchers have also successfully used gene pyramiding, which combines multiple genes via Agrobacterium-mediated transformation (Ahmad et al., 2010).

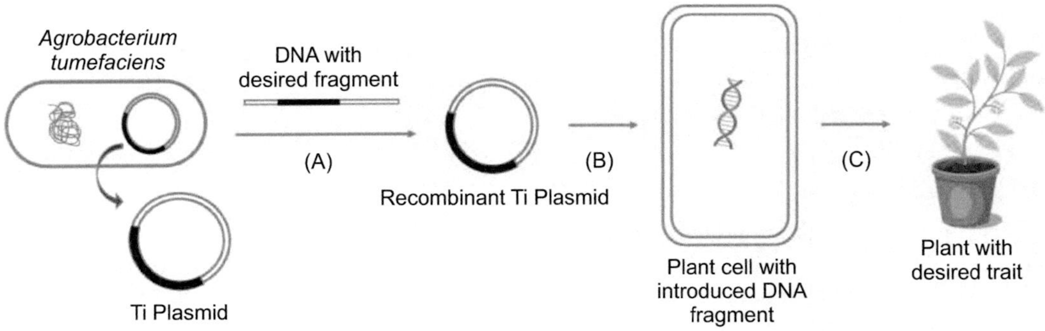


**Fig. 1** Illustrative representation of Agrobacterium-mediated transformation toward abiotic stress. (A) Introduction of desired gene fragment into plasmid utilizing REase and DNA ligase. (B) Introduction of DNA fragment into plant cell. (C) Plant regeneration.

#### 4.3.2.2 RNAi Technology

RNA interference (RNAi) silences specific genes using double-stranded RNA, triggering cellular pathways that prevent the expression of genes through small noncoding RNAs during the post-transcriptional or translational level. This approach has proven effective in engineering cereal crops for improved tolerance to several abiotic stressors. For instance, in rice, overexpressing miR156 downregulates SPL9, increasing drought along salinity resistance (Cui et al., 2011). Similarly, the knockdown of miR166 reduces transpiration, thus contributing to drought tolerance (Zhang et al., 2018). Likewise in wheat crops, under abiotic stress lower TaNAM expression helps to improve nutrient allocation and grain yield (Guttieri et al., 2013).

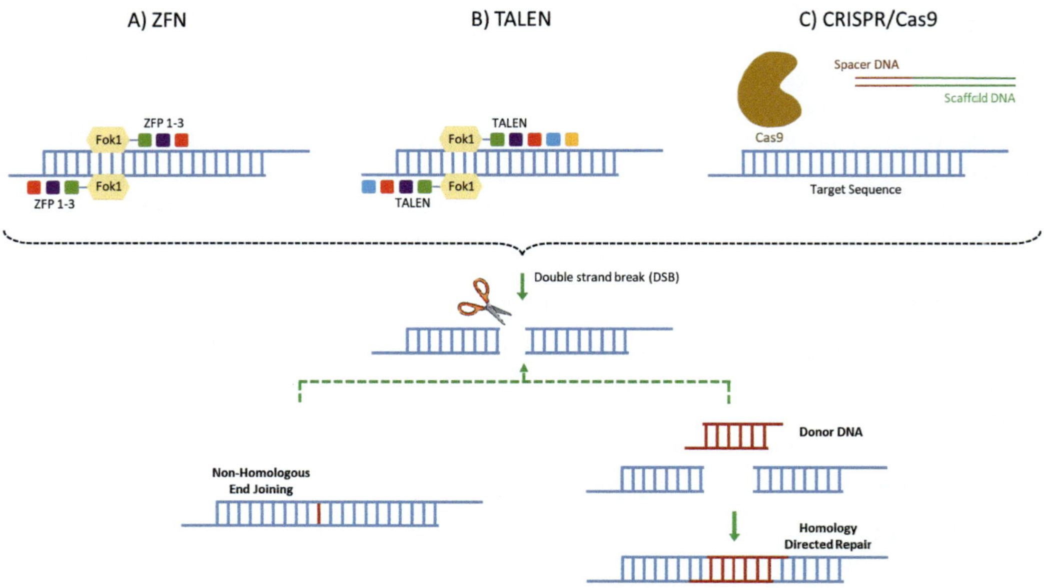


**Fig. 2** Illustration for genome editing techniques causing double-stranded break to DNA and its repair via NHEJ or HDR pathway. (A) Zinc finger nucleases (ZFN). (B) Transcription activator-like effector nucleases (TALENs). (C) Clustered regularly interspaced palindromic repeats/CRISPR-associated protein 9 (CRISPR/Cas9).

### *4.3.3 Genome Editing*

Genome editing technologies hold immense potential for plant breeding by enabling the insertion, deletion, and replacement of DNA at specific target regions within the genome of many crops to achieve a range of aims. Zinc finger nucleases (ZFNs), transcription activator-like effector nucleases (TALENs), and RNA-guided clustered regularly interspaced palindromic repeats (CRISPR)/Cas are the most used genome editing tools (Figure 2). Ran et al. (2018) have achieved gene modification in allohexaploid wheat to edit *Acetohydroxyacid synthase* gene using ZFN. Similarly, TALENs have been utilized to alter the *Mildew resistance locus O* (TaMLO1) gene (Wang et al., 2014).

#### 4.3.3.1 Zinc Finger Nucleases (ZFNs)

ZFNs consist of zinc finger proteins (ZFPs) attached to the FokI restriction enzyme domain, which creates precise cuts in targeted DNA regions. These cuts are then restored by either homology-directed repair (HDR) or nonhomologous end joining (NHEJ), resulting in targeted mutations such as insertions, deletions, or transgene insertions. However, ZFNs have seen limited use in improving cereal crops (Ran et al., 2018).

#### 4.3.3.2 Transcription Activator-Like Effector Nucleases (TALENs)

TALENs function similarly to ZFNs, with a domain for DNA-binding connected to FokI enzyme, creating specific double-strand breaks in DNA (Wang et al., 2014). Adjusting the binding domain allows targeting a variety of DNA sequences for customized gene editing. Despite its potential, TALEN technology hasn't been extensively applied to study or improve cereal crops for abiotic stress tolerance.

#### 4.3.3.3 CRISPR/Cas9

CRISPR/Cas is recognized as a highly effective and widely accepted genome editing tool, especially valued for its precision in targeting specific genes. This RNA-guided nuclease system has been transformative in crop improvement, allowing researchers to modify target genomes with high accuracy. While early studies focused on model plants such as tobacco and Arabidopsis, CRISPR/Cas is now being extensively utilized for cereals, including maize, rice, sorghum, and wheat (Table 1), contributing to enhanced responses to various abiotic stresses (Shen et al., 2017; Xu et al., 2015). OsRR22 is a gene that has been targeted to confer salt stress tolerance in rice using CRISPR/Cas9 (Zhang et al., 2019). Similarly, ARGOS8 in maize was targeted to achieve drought stress tolerance through CRISPR/Cas9 (Shi et al., 2017). In recent findings, Alam et al. (2022) identified OsbHLH024 as a negative regulator of salt stress in rice, with its mutant, osbhlh024, displaying a salt-tolerant phenotype. Additionally, Ogata et al. (2020) discovered that OsERA1 enhances rice tolerance to drought by influencing ABA signalling despite being a negative regulator for root growth. In crops such as wheat, Brandt et al. (2020) developed a gene knockout protocol for three different genes, that is, GW2-B (a negative regulator of grain width), PinB-D (the gene responsible for grain hardness), and ASN2-A (an asparagine synthetase gene from acrylamide formation pathway). Similarly in sorghum, two genes were investigated by CRISPR/Cas9 via bombardment (Liu et al., 2019).

**Table 1** List of gene and protein families in abiotic stress management in recent studies.

| *Crop* | *Type of study* | *Targeted gene* | *Gene transfer method* | *References* |
|---|---|---|---|---|
| Wheat (*Triticum aestivum*) | Genome editing | TaGASR7A1, TaGASR7 B1, and TaGASR7D1 | Biolistic | (Hamada et al., 2018) |
| | Gene editing | TaABCC6, TaNFXLI, and TansLTP | Protoplast transformation | (Cui et al., 2019) |
| | knockout | GW2-B, PinB-D, and ASN2-A | Protoplast tansformation using ribonucleoprotein (RNP). | (Brandt et al., 2020) |
| | Functional genomics | TaPDS | Agrobacterium-mediated | (Howells et al., 2018) |
| | Gene editing | TaLox2 and TaUbiL1 | Electroporation | (Bhowmik et al., 2018) |
| Sorghum (*Sorghum bicolor*) | Gene editing | CAD and PDS | Particle bombardment | (Liu et al., 2019) |
| | Knockout | SbFT and SbGA20x5 | Agrobacterium-mediated | (Char & Yang, 2019) |
| | Knockout | Sb-CENH3 | Agrobacterium-mediated | (Che et al., 2018) |
| Rice (*Oryza sativa*) | Knockout | OsRR22 | Agrobacterium-mediated | (Zhang et al., 2019) |
| | Gene editing | NbPDSI | Agrobacterium-mediated | (Raitskin et al., 2019) |
| | Gene editing | OSPIN5b, gs3, and OSMYB30 | Agrobacterium-mediated | (Zeng et al., 2020) |
| | Gene editing | OsHSA1 | Agrobacterium-mediated | (Qiu et al., 2014) |
| | Knockout | OsNAC006 | Agrobacterium-mediated | (Wang et al., 2016) |
| | Gene editing | OsSAPK2 | Electroporation | (Lou et al., 2017) |
| | Knockdown | TMS5 | Agrobacterium-mediated | (Zhou et al., 2016) |
| Maize (*Zea mays*) | Gene editing | ARGOS8 | Biolistic | (Shi et al., 2017) |

## 5. Innovative Technologies for Monitoring and Predicting Abiotic Stress

With the changes in the global temperature and climate, the impact of plant stresses is increasing day by day. So, it becomes important to predict and monitor the plant stresses. New advances in plant phenomics have opened many doors in the area of stress biology and its evaluation. In the case of the impact of stress on plants, accurate plant phenomics is the key component. Advancements in the area of deep learning have developed rapidly in the area of plant stress phenomics. Traditional approaches to monitoring the disease are very time-consuming, laborious, and not suitable for large-scale farming. With that chemical analysis using chromatography and spectroscopy is also laborious, high sample preparation, and money-consuming (Hall et al., 2022). Key advancements in the area of artificial intelligence and image sensing have shown potential for the prediction of stress on plants and their tissues. Other than that, unmanned aerial vehicle (UAV)-based thermal imaging also has been widely used in abiotic stress biology. High-imaging capturing using thermal imaging could be a novel promising approach to estimating the effect of abiotic stress and crop management. Automated artificial imaging with the drone system has boosted the phenotyping of plant health and agriculture (Maraveas, 2024).

Computer-based approaches and thermal imaging combined could be a useful tool for helping in the abiotic stress management and prediction of the severity of the stress in the plants. New satellites with hyperspectral sensors and multispectral sensors are also very helpful and can detect small changes in the plant health and vegetation indices. There are some reports where remote sensing and high-throughput imaging were used in cereal plant stress identification. In a study, Elsayed et al. (2017) estimated the wheat water stress and grain yield using remote sensing under different irrigation systems, whereas in a different study, Siddiqui et al. (2014) revealed the effect of salt stress using high-throughput infrared imaging.

There are some limitations for artificial intelligence (AI) in stress biology phenomics. The AI algorithms are sometimes biased by the quality of training data, on the basis of which algorithms perform. Biased quality data and limited data supply for algorithms sometimes hinder the better prediction of the stress-related results (Sundravadivelu et al., 2023). Other than that, biological systems are very dynamic and sometimes varietal differences in the crops are also very complex and dynamic, so stress responses can be varied. Capturing all the stress responses and their interactions with other physiological processes can be very challenging for using in AI and image sensing process (Al-Tamimi et al., 2022). The performance of the imaging system can be biased by environmental factors such as light, humidity, and rain (Gill et al., 2022; McIsaac et al., 2021). These environmental factors create lots of noise in the database generation and also reduce the chances of accurate stress and its impact. Some studies suggest that using music to condition plants and relieve stress could be a promising, eco-friendly way to prepare crops for future challenges and reduce yield losses (Bhandawat et al., 2020).

New technologies and human intelligence with satellite imaging are reshaping agricultural practices in the area of stress biology and enhancing resilience against various climate-changing factors. These tools are providing powerful tools to farmers and researchers to mitigate and understand the effects of stress. These tools have the potential to revolutionize agriculture and ensure the increasing demand for food security globally.

## 6. Conclusion and Future Prospects

Mankind is facing an extraordinary period of climate change over the last decade, showing an accelerated rise in global temperature. Because of such detrimental effects, 2023 was observed as the hottest year in the past 174 years (Morris et al., 2024). In the present scenario, increasing population, limited natural resources, and nonavailability of cultivable land can also be the factors for affecting crop productivity in upcoming years. These facts bring to notice the effect of abiotic stress on food availability for upcoming generations. Cereals are the major component of the human diet worldwide and about half of the consumed calories are taken directly from the cereal grains. Several abiotic stresses such as heat, salt, drought, flood, etc., are causing serious loss in crop yield and quality, which are getting worse because of climate change, as it is the most major contributor to yield reduction in cereal crops. To cope with such threatening climatic conditions, the quick need for the hour is the development and design of cereal crops that can withstand such weather conditions with improved quality and yield.

Plant breeding techniques have notably boosted crop yield, quality, and abiotic stress resistance in the last century. However, developing cultivars or varieties using plant breeding is much more time-consuming. Transgenic and genome editing approaches are being successfully employed for the creation of stress-tolerant cereals (Ansari et al., 2020). CRISPR, one of the genome editing tools, is now in use for the development of future-ready cereals. The challenge is to identify and use natural variations in wild plant populations for the development of outstanding stress-resilient cereal crops. Another challenge is to integrate all the information of pathways and candidate genes regulating abiotic stress tolerance, and diverse approaches for engineering climate-stable crops that can produce more yield under adverse conditions to feed the expanding society.

## References

AbdElgawad, H., Zinta, G., Hegab, M.M., Pandey, R., Asard, H. and Abuelsoud, W. (2016). High salinity induces different oxidative stress and antioxidant responses in maize seedlings organs. *Frontiers in Plant Science*, *7*, 276.

Abid, M., Shao, Y., Liu, S., Wang, F., Gao, J., Jiang, D., Tian, Z. and Dai, T. (2017). Pre-drought priming sustains grain development under post-anthesis drought stress by regulating the growth hormones in winter wheat (Triticum aestivum L.). *Planta*, *246*(3), 509–524. https://doi.org/10.1007/s00425-017-2698-4

Ahloowalia, B.S., Maluszynski, M. and Nichterlein, K. (2004). Global impact of mutation-derived varieties. *Euphytica*, *135*(2), 187–204. https://doi.org/10.1023/b:euph.0000014914.85465.4f

Ahmad, P., Rasool, S., Gul, A., Sheikh, S.A., Akram, N.A., Ashraf, M., Kazi, A.M. and Gucel, S. (2016). Jasmonates: Multifunctional Roles in Stress Tolerance. *Frontiers in Plant Science*, *7*, 813. PubMed. https://doi.org/10.3389/fpls.2016.00813

Ahmad, R., Kim, Y., Kim, M., Kwon, S., Cho, K., Lee, H. and Kwak, S. (2010). Simultaneous expression of choline oxidase, superoxide dismutase and ascorbate peroxidase in potato plant chloroplasts provides synergistically enhanced protection against various abiotic stresses. *Physiologia Plantarum*, *138*(4), 520–533. https://doi.org/10.1111/j.1399-3054.2010.01348.x

Alam, M.S., Kong, J., Tao, R., Ahmed, T., Alamin, M., Alotaibi, S.S., Abdelsalam, N.R. and Xu, J.-H. (2022). CRISPR/Cas9 Mediated Knockout of the OsbHLH024 Transcription Factor Improves Salt Stress Resistance in Rice (Oryza sativa L.). *Plants (Basel, Switzerland)*, *11*(9), 1184. PubMed. https://doi.org/10.3390/plants11091184

Ali, A., Raddatz, N., Pardo, J.M. and Yun, D.-J. (2021). HKT sodium and potassium transporters in Arabidopsis thaliana and related halophyte species. *Physiologia Plantarum*, *171*(4), 546–558. PubMed. https://doi.org/10.1111/ppl.13166

Almansouri, M., Kinet, J.M. and Lutts, S. (1999). Compared Effects of Sudden and Progressive Impositions of Salt Stress in Three Durum Wheat (Triticum durum Desf.) Cultivars. *Journal of Plant Physiology*, *154*(5–6), 743–752. https://doi.org/10.1016/s0176-1617(99)80253-3

Al-Tamimi, N., Langan, P., Bernád, V., Walsh, J., Mangina, E. and Negrão, S. (2022). Capturing crop adaptation to abiotic stress using image-based technologies. *Open Biology*, *12*(6), 210353. PubMed. https://doi.org/10.1098/rsob.210353

Amram, A., Fadida-Myers, A., Golan, G., Nashef, K., Ben-David, R. and Peleg, Z. (2015). Effect of GA-sensitivity on wheat early vigor and yield components under deep sowing. *Frontiers in Plant Science*, *6*, 487. PubMed. https://doi.org/10.3389/fpls.2015.00487

Ansari, W.A., Chandanshive, S.U., Bhatt, V., Nadaf, A.B., Vats, S., Katara, J.L., Sonah, H. and Deshmukh, R. (2020). Genome editing in cereals: Approaches, applications and challenges. *International Journal of Molecular Sciences*, *21*(11), 4040.

Apse, M.P., Aharon, G.S., Snedden, W.A. and Blumwald, E. (1999). Salt Tolerance Conferred by Overexpression of a Vacuolar Na + /H + Antiport in Arabidopsis. *Science*, *285*(5431), 1256–1258. https://doi.org/10.1126/science.285.5431.1256

Ashraf, M. and Hafeez, M. (2004). Thermotolerance of Pearl Millet and Maize at Early Growth Stages: Growth and Nutrient Relations. *Biologia Plantarum*, *48*(1), 81–86. https://doi.org/10.1023/b:biop.0000024279.44013.61

Assmann, S.M. and Jegla, T. (2016). Guard cell sensory systems: Recent insights on stomatal responses to light, abscisic acid, and CO2. *Current Opinion in Plant Biology*, *33*, 157–167. https://doi.org/10.1016/j.pbi.2016.07.003

Auler, P.A., Freire, F.B.S., Lima, V.F. and Daloso, D.M. (2022). On the role of guard cells in sensing environmental signals and memorising stress periods. *Theoretical and Experimental Plant Physiology*, *34*(3), 277–299. https://doi.org/10.1007/s40626-022-00250-4

Avalbaev, A.M., Yuldashev, R.A., Fatkhutdinova, R.A., Urusov, F.A., Safutdinova, Yu. V. and Shakirova, F.M. (2010). The influence of 24-epibrassidinolide on the hormonal status of wheat plants under sodium chloride. *Applied Biochemistry and Microbiology*, *46*(1), 99–102. https://doi.org/10.1134/s0003683810010175

Aziz, M.A. and Masmoudi, K. (2024). Molecular Breakthroughs in Modern Plant Breeding Techniques. *Horticultural Plant Journal*.

Bartels, D. and Sunkar, R. (2005). Drought and Salt Tolerance in Plants. *Critical Reviews in Plant Sciences*, *24*(1), 23–58. https://doi.org/10.1080/07352680590910410

Bhandawat, A., Jayaswall, K., Sharma, H. and Roy, J. (2020). Sound as a stimulus in associative learning for heat stress in Arabidopsis. *Communicative & Integrative Biology*, *13*(1), 1–5. https://doi.org/10.1080/19420889.2020.1713426

Bhattacharya, A. (2021). Effect of Soil Water Deficit on Growth and Development of Plants: A Review. *Soil Water Deficit and Physiological Issues in Plants*, 393–488. https://doi.org/10.1007/978-981-33-6276-5_5

Bhowmik, P., Ellison, E., Polley, B., Bollina, V., Kulkarni, M., Ghanbarnia, K., Song, H., Gao, C., Voytas, D.F. and Kagale, S. (2018). Targeted mutagenesis in wheat microspores using CRISPR/Cas9. *Scientific Reports*, *8*(1), 6502. PubMed. https://doi.org/10.1038/s41598-018-24690-8

Bittner, T., Nadler, S., Schulze, E. and Fischer-Iglesias, C. (2015). Two homolog wheat Glycogen Synthase Kinase 3/SHAGGY–like kinases are involved in brassinosteroid signaling. *BMC Plant Biology*, *15*, 247. PubMed. https://doi.org/10.1186/s12870-015-0617-z

Bokszczanin, K.L., Solanaceae Pollen Thermotolerance Initial Training Network (SPOT-ITN) Consortium, & Fragkostefanakis, S. (2013). Perspectives on deciphering mechanisms underlying plant heat stress response and thermotolerance. *Frontiers in Plant Science*, *4*, 315. PubMed. https://doi.org/10.3389/fpls.2013.00315

Brandt, K.M., Gunn, H., Moretti, N. and Zemetra, R.S. (2020). A Streamlined Protocol for Wheat (Triticum aestivum) Protoplast Isolation and Transformation With CRISPR–Cas Ribonucleoprotein Complexes. *Frontiers in Plant Science*, *11*, 769. PubMed. https://doi.org/10.3389/fpls.2020.00769

Briat, J.-F., Dubos, C. and Gaymard, F. (2015). Iron nutrition, biomass production, and plant product quality. *Trends in Plant Science*, *20*(1), 33–40. https://doi.org/10.1016/j.tplants.2014.07.005

Broadley, M.R., White, P.J., Hammond, J.P., Zelko, I. and Lux, A. (2007). Zinc in plants. *New Phytologist*, *173*(4), 677–702. https://doi.org/10.1111/j.1469-8137.2007.01996.x

Bukhov, N.G. and Carpentier, R. (2000). Heterogeneity of photosystem II reaction centers as influenced by heat treatment of barley leaves. *Physiologia Plantarum*, *110*(2), 279–285. https://doi.org/10.1034/j.1399-3054.2000.110219.x

Burzyński, M. and Kłobus, G. (2004). Changes of photosynthetic parameters in cucumber leaves under Cu, Cd, and Pb stress. *Photosynthetica*, *42*(4), 505–510. https://doi.org/10.1007/s11099-005-0005-2

Char, S.N. and Yang, B. (2019). Genome editing in grass plants. *aBIOTECH*, *1*(1), 41–57. PubMed. https://doi.org/10.1007/s42994-019-00005-x

Chaves, M.M., Flexas, J. and Pinheiro, C. (2008). Photosynthesis under drought and salt stress: Regulation mechanisms from whole plant to cell. *Annals of Botany*, *103*(4), 551–560. https://doi.org/10.1093/aob/mcn125

Che, P., Anand, A., Wu, E., Sander, J.D., Simon, M.K., Zhu, W., Sigmund, A.L., Zastrow-Hayes, G., Miller, M., Liu, D., Lawit, S.J., Zhao, Z.-Y., Albertsen, M.C. and Jones, T.J. (2018). Developing a flexible, high-efficiency Agrobacterium-mediated sorghum transformation system with broad application. *Plant Biotechnology Journal*, *16*(7), 1388–1395. PubMed. https://doi.org/10.1111/pbi.12879

Chen, H., An, R., Tang, J.-H., Cui, X.-H., Hao, F.-S., Chen, J. and Wang, X.-C. (2007). Over-expression of a vacuolar Na+/H+ antiporter gene improves salt tolerance in an upland rice. *Molecular Breeding*, *19*(3), 215–225. https://doi.org/10.1007/s11032-006-9048-8

Chen, Y.E., Cui, J.M., Li, G.X., Yuan, M., Zhang, Z.W., Yuan, S. and Zhang, H.Y. (2016). Effect of salicylic acid on the antioxidant system and photosystem II in wheat seedlings. *Biologia Plantarum*, *60*(1), 139–147. https://doi.org/10.1007/s10535-015-0564-4

Chinnusamy, V., Zhu, J. and Zhu, J.-K. (2007). Cold stress regulation of gene expression in plants. *Trends in Plant Science*, *12*(10), 444–451. https://doi.org/10.1016/j.tplants.2007.07.002

Choudhary, P., Pramitha, L., Rana, S., Verma, S., Aggarwal, P.R. and Muthamilarasan, M. (2021). Hormonal crosstalk in regulating salinity stress tolerance in graminaceous crops. *Physiologia Plantarum*, *173*(4), 1587–1596. https://doi.org/10.1111/ppl.13558

Chung, B.Y.W., Balcerowicz, M., Di Antonio, M., Jaeger, K.E., Geng, F., Franaszek, K., Marriott, P., Brierley, I., Firth, A.E. and Wigge, P.A. (2020). An RNA thermoswitch regulates daytime growth in Arabidopsis. *Nature Plants*, *6*(5), 522–532. PubMed. https://doi.org/10.1038/s41477-020-0633-3

Chunthaburee, S., Sanitchon, J., Pattanagul, W. and Theerakulpisut, P. (2015). Effects of salt stress after late booting stage on yield and antioxidant capacity in pigmented rice grains and alleviation of the salt-induced yield reduction by exogenous spermidine. *Plant Production Science*, *18*(1), 32–42.

Clemens, S. (2006). Toxic metal accumulation, responses to exposure and mechanisms of tolerance in plants. *Biochimie*, *88*(11), 1707–1719. https://doi.org/10.1016/j.biochi.2006.07.003

Colebrook, E.H., Thomas, S.G., Phillips, A.L. and Hedden, P. (2014). The role of gibberellin signalling in plant responses to abiotic stress. *Journal of Experimental Biology*, *217*(1), 67–75.

Cominelli, E., Galbiati, M., Vavasseur, A., Conti, L., Sala, T., Vuylsteke, M., Leonhardt, N., Dellaporta, S.L. and Tonelli, C. (2005). A Guard-Cell-Specific MYB Transcription Factor Regulates Stomatal Movements and Plant Drought Tolerance. *Current Biology*, *15*(13), 1196–1200. https://doi.org/10.1016/j.cub.2005.05.048

Cui, M., Zhang, W., Zhang, Q., Xu, Z., Zhu, Z., Duan, F. and Wu, R. (2011). Induced over-expression of the transcription factor OsDREB2A improves drought tolerance in rice. *Plant Physiology and Biochemistry*, *49*(12), 1384–1391. https://doi.org/10.1016/j.plaphy.2011.09.012

Cui, X., Balcerzak, M., Schernthaner, J., Babic, V., Datla, R., Brauer, E.K., Labbé, N., Subramaniam, R. and Ouellet, T. (2019). An optimised CRISPR/Cas9 protocol to create targeted mutations in homoeologous genes and an efficient genotyping protocol to identify edited events in wheat. *Plant Methods*, *15*, 119. PubMed. https://doi.org/10.1186/s13007-019-0500-2

Das, D., Konwar, T., Sarma, S., Baldodiya, G.M., Chikkaputtaiah, C. and Singha, D.L. (2022). Transgenic Strategies to Develop Abiotic Stress Tolerance in Cereals. *Omics Approach to Manage Abiotic Stress in Cereals*, 179–229. https://doi.org/10.1007/978-981-19-0140-9_9

de Bang, T.C., Husted, S., Laursen, K.H., Persson, D.P. and Schjoerring, J.K. (2021). The molecular–physiological functions of mineral macronutrients and their consequences for deficiency symptoms in plants. *New Phytologist*, *229*(5), 2446–2469.

Dwivedi, S.L., Ceccarelli, S., Blair, M.W., Upadhyaya, H.D., Are, A.K. and Ortiz, R. (2016). Landrace Germplasm for Improving Yield and Abiotic Stress Adaptation. *Trends in Plant Science*, *21*(1), 31–42. https://doi.org/10.1016/j.tplants.2015.10.012

Egawa, C., Kobayashi, F., Ishibashi, M., Nakamura, T., Nakamura, C. and Takumi, S. (2006). Differential regulation of transcript accumulation and alternative splicing of a DREB2 homolog under abiotic stress conditions in common wheat. *Genes & Genetic Systems*, *81*(2), 77–91. https://doi.org/10.1266/ggs.81.77

Elsayed, S., Elhoweity, M., Ibrahim, H.H., Dewir, Y.H., Migdadi, H.M. and Schmidhalter, U. (2017). Thermal imaging and passive reflectance sensing to estimate the water status and grain yield of wheat under different irrigation regimes. *Agricultural Water Management*, *189*, 98–110. https://doi.org/10.1016/j.agwat.2017.05.001

Feki, K., Quintero, F.J., Khoudi, H., Leidi, E.O., Masmoudi, K., Pardo, J.M. and Brini, F. (2013). A constitutively active form of a durum wheat Na+/H+ antiporter SOS1 confers high salt tolerance to transgenic Arabidopsis. *Plant Cell Reports*, *33*(2), 277–288. https://doi.org/10.1007/s00299-013-1528-9

Ferreyra, M.L.F., Serra, P. and Casati, P. (2021). Recent advances on the roles of flavonoids as plant protective molecules after UV and high light exposure. *Physiologia Plantarum*, *173*(3), 736–749.

Finkelstein, R. (2013). Abscisic Acid Synthesis and Response. *The Arabidopsis Book*, *11*, e0166. https://doi.org/10.1199/tab.0166

Fiorucci, A.-S., Galvão, V.C., Ince, Y.Ç., Boccaccini, A., Goyal, A., Allenbach Petrolati, L., Trevisan, M. and Fankhauser, C. (2020). PHYTOCHROME INTERACTING FACTOR 7 is important for early responses to elevated temperature in Arabidopsis seedlings. *The New Phytologist*, *226*(1), 50–58. PubMed. https://doi.org/10.1111/nph.16316

Flexas, J., Bota, J., Loreto, F., Cornic, G. and Sharkey, T.D. (2004). Diffusive and Metabolic Limitations to Photosynthesis under Drought and Salinity in C3 Plants. *Plant Biology*, *6*(3), 269–279. https://doi.org/10.1055/s-2004-820867

Fu, L., Shen, Q., Kuang, L., Yu, J., Wu, D. and Zhang, G. (2018). Metabolite profiling and gene expression of Na/K transporter analyses reveal mechanisms of the difference in salt tolerance between barley and rice. *Plant Physiology and Biochemistry*, *130*, 248–257. https://doi.org/10.1016/j.plaphy.2018.07.013

Furihata, T., Maruyama, K., Fujita, Y., Umezawa, T., Yoshida, R., Shinozaki, K. and Yamaguchi-Shinozaki, K. (2006). Abscisic acid-dependent multisite phosphorylation regulates the activity of a transcription activator AREB1. *Proceedings of the National Academy of Sciences of the United States of America*, *103*(6), 1988–1993. PubMed. https://doi.org/10.1073/pnas.0505667103

Fürtauer, L., Weiszmann, J., Weckwerth, W. and Nägele, T. (2019). Dynamics of Plant Metabolism during Cold Acclimation. *International Journal of Molecular Sciences*, *20*(21), 5411. https://doi.org/10.3390/ijms20215411

Gan, L., Wu, X. and Zhong, Y. (2015). Exogenously Applied Nitric Oxide Enhances the Drought Tolerance in Hulless Barley. *Plant Production Science*, *18*(1), 52–56. https://doi.org/10.1626/pps.18.52

Gharbi, E., Martínez, J.-P., Benahmed, H., Lepoint, G., Vanpee, B., Quinet, M. and Lutts, S. (2017). Inhibition of ethylene synthesis reduces salt-tolerance in tomato wild relative species Solanum chilense. *Journal of Plant Physiology*, *210*, 24–37. https://doi.org/10.1016/j.jplph.2016.12.001

Gill, S.S., Anjum, N.A., Gill, R., Yadav, S., Hasanuzzaman, M., Fujita, M., Mishra, P., Sabat, S.C. and Tuteja, N. (2015). Superoxide dismutase—Mentor of abiotic stress tolerance in crop plants. *Environmental Science and Pollution Research*, *22*(14), 10375–10394. https://doi.org/10.1007/s11356-015-4532-5

Gill, T., Gill, S.K., Saini, D.K., Chopra, Y., de Koff, J.P. and Sandhu, K.S. (2022). A Comprehensive Review of High Throughput Phenotyping and Machine Learning for Plant Stress Phenotyping. *Phenomics (Cham, Switzerland)*, *2*(3), 156–183. PubMed. https://doi.org/10.1007/s43657-022-00048-z

Godara, M., Das, D., Roy, J. and Bhandawat, A. (2023). Genetic Engineering Methods for Wheat Improvement. In S. Tiwari & B. Koul (Eds.), *Genetic Engineering of Crop Plants for Food and Health Security: Volume 1* (pp. 421–440). Springer Nature. https://doi.org/10.1007/978-981-99-5034-8_21

Gondor, O.K., Szalai, G., Kovács, V., Janda, T. and Pál, M. (2016). Relationship between polyamines and other cold-induced response mechanisms in different cereal species. *Journal of Agronomy and Crop Science*, *202*(3), 217–230.

González, A., Martín, I. and Ayerbe, L. (2008). Yield and Osmotic Adjustment Capacity of Barley Under Terminal Water-Stress Conditions. *Journal of Agronomy and Crop Science*, *194*(2), 81–91. https://doi.org/10.1111/j.1439-037x.2007.00289.x

Goyal, S., Wani, M.R., Raina, A., Laskar, R.A. and Khan, S. (2021). Phenotypic diversity in mutagenized population of urdbean (Vigna mungo (L.) Hepper). *Heliyon*, *7*(5), e06356. PubMed. https://doi.org/10.1016/j.heliyon.2021.e06356

Gu, L., Liu, Y., Zong, X., Liu, L., Li, D.-P. and Li, D.-Q. (2010). Overexpression of maize mitogen-activated protein kinase gene, ZmSIMK1 in Arabidopsis increases tolerance to salt stress. *Molecular Biology Reports*, *37*(8), 4067–4073. https://doi.org/10.1007/s11033-010-0066-6

Guttieri, M.J., Stein, R.J. and Waters, B.M. (2013). Nutrient partitioning and grain yield of TaNAM-RNAi wheat under abiotic stress. *Plant and Soil*, *371*(1–2), 573–591. https://doi.org/10.1007/s11104-013-1713-1

Hadiarto, T. and Tran, L.-S. P. (2010). Progress studies of drought-responsive genes in rice. *Plant Cell Reports*, *30*(3), 297–310. https://doi.org/10.1007/s00299-010-0956-z

Hafeez, U., Ali, M., Hassan, S.M., Akram, M.A. and Zafar, A. (2023). Advances in breeding and engineering climate-resilient crops: A comprehensive review. *International Journal of Research and Advances in Agricultural Sciences*, *2*(2), 85–99.

Hall, R.D., D'Auria, J.C., Silva Ferreira, A.C., Gibon, Y., Kruszka, D., Mishra, P. and van de Zedde, R. (2022). High-throughput plant phenotyping: A role for metabolomics? *Trends in Plant Science*, *27*(6), 549–563. https://doi.org/10.1016/j.tplants.2022.02.001

Hamada, H., Liu, Y., Nagira, Y., Miki, R., Taoka, N. and Imai, R. (2018). Biolistic-delivery-based transient CRISPR/Cas9 expression enables in planta genome editing in wheat. *Scientific Reports*, *8*(1), 14422. PubMed. https://doi.org/10.1038/s41598-018-32714-6

Hand, S.C., Menze, M.A., Toner, M., Boswell, L. and Moore, D. (2011). LEA Proteins During Water Stress: Not Just for Plants Anymore. *Annual Review of Physiology*, *73*(1), 115–134. https://doi.org/10.1146/annurev-physiol-012110-142203

Haque, M., Martinek, P., Watanabe, N. and Kuboyama, T. (2011). Genetic mapping of gibberellic acid-sensitive genes for semi-dwarfism in durum wheat. *Cereal Research Communications*, *39*(2), 171–178. https://doi.org/10.1556/crc.39.2011.2.1

Hasanuzzaman, M. and Fujita, M. (2022). Plant Responses and Tolerance to Salt Stress: Physiological and Molecular Interventions. *International Journal of Molecular Sciences*, *23*(9), 4810. https://doi.org/10.3390/ijms23094810

Hasegawa, P.M., Bressan, R.A., Zhu, J.-K. and Bohnert, H.J. (2000). Plant cellular and molecular responses to high salinity. *Annual Review of Plant Physiology and Plant Molecular Biology*, *51*(1), 463–499. https://doi.org/10.1146/annurev.arplant.51.1.463

Hatfield, J.L. and Prueger, J.H. (2015). Temperature extremes: Effect on plant growth and development. *Weather and Climate Extremes*, *10*, 4–10. https://doi.org/10.1016/j.wace.2015.08.001

Hayat, S., Khalique, G., Wani, A.S., Alyemeni, M.N. and Ahmad, A. (2014). Protection of growth in response to 28-homobrassinolide under the stress of cadmium and salinity in wheat. *International Journal of Biological Macromolecules*, *64*, 130–136. https://doi.org/10.1016/j.ijbiomac.2013.11.021

Hayes, S., Schachtschabel, J., Mishkind, M., Munnik, T. and Arisz, S.A. (2021). Hot topic: Thermosensing in plants. *Plant, Cell & Environment*, *44*(7), 2018–2033. PubMed. https://doi.org/10.1111/pce.13979

Hays, D.B., Do, J.H., Mason, R.E., Morgan, G. and Finlayson, S.A. (2007). Heat stress induced ethylene production in developing wheat grains induces kernel abortion and increased maturation in a susceptible cultivar. *Plant Science*, *172*(6), 1113–1123. https://doi.org/10.1016/j.plantsci.2007.03.004

Hedrich, R. and Shabala, S. (2018). Stomata in a saline world. *Current Opinion in Plant Biology*, *46*, 87–95. https://doi.org/10.1016/j.pbi.2018.07.015

Herbette, S., Labrouhe, D.T. de, Drevet, J.R. and Roeckel-Drevet, P. (2011). Transgenic tomatoes showing higher glutathione peroxydase antioxidant activity are more resistant to an abiotic stress but more susceptible to biotic stresses. *Plant Science*, *180*(3), 548–553. https://doi.org/10.1016/j.plantsci.2010.12.002

Howells, R.M., Craze, M., Bowden, S. and Wallington, E.J. (2018). Efficient generation of stable, heritable gene edits in wheat using CRISPR/Cas9. *BMC Plant Biology*, *18*(1), 215. PubMed. https://doi.org/10.1186/s12870-018-1433-z

Hsiao, T.C., Acevedo, E., Fereres, E., Henderson, D.W., Monteith, J.L. and Weatherley, P.E. (1997). Water stress, growth and osmotic adjustment. *Philosophical Transactions of the Royal Society of London. B, Biological Sciences*, *273*(927), 479–500. https://doi.org/10.1098/rstb.1976.0026

Huang, H. and Nusinow, D.A. (2016). Into the Evening: Complex Interactions in the Arabidopsis Circadian Clock. *Trends in Genetics*, *32*(10), 674–686. https://doi.org/10.1016/j.tig.2016.08.002

Jan, R., Kim, N., Lee, S.-H., Khan, M.A., Asaf, S., Lubna, Park, J.-R., Asif, S., Lee, I.-J. and Kim, K.-M. (2021). Enhanced flavonoid accumulation reduces combined salt and heat stress through regulation of transcriptional and hormonal mechanisms. *Frontiers in Plant Science*, *12*, 796956.

Jiang, W., Pan, R., Buitrago, S., Wu, C., Abou-Elwafa, S.F., Xu, Y. and Zhang, W. (2021). Conservation and divergence of the TaSOS1 gene family in salt stress response in wheat (Triticum aestivum L.). *Physiology and Molecular Biology of Plants : An International Journal of Functional Plant Biology*, *27*(6), 1245–1260. PubMed. https://doi.org/10.1007/s12298-021-01009-y

Jing, S., Zhou, X., Song, Y. and Yu, D. (2009). Heterologous expression of OsWRKY23 gene enhances pathogen defense and dark-induced leaf senescence in Arabidopsis. *Plant Growth Regulation*, *58*(2), 181–190. https://doi.org/10.1007/s10725-009-9366-z

Kant, S., Bi, Y.-M. and Rothstein, S.J. (2010). Understanding plant response to nitrogen limitation for the improvement of crop nitrogen use efficiency. *Journal of Experimental Botany*, *62*(4), 1499–1509. https://doi.org/10.1093/jxb/erq297

Khan, M.I.R., Iqbal, N., Masood, A., Per, T.S. and Khan, N.A. (2013). Salicylic acid alleviates adverse effects of heat stress on photosynthesis through changes in proline production and ethylene formation. *Plant Signaling & Behavior*, *8*(11), e26374. PubMed. https://doi.org/10.4161/psb.26374

Khush, G.S. (1999). Green revolution: Preparing for the 21st century. *Genome*, *42*(4), 646–655. https://doi.org/10.1139/g99-044

Kobayashi, F., Maeta, E., Terashima, A., Kawaura, K., Ogihara, Y. and Takumi, S. (2008). Development of abiotic stress tolerance via bZIP-type transcription factor LIP19 in common wheat. *Journal of Experimental Botany*, *59*(4), 891–905. https://doi.org/10.1093/jxb/ern014

Kobayashi, N.I., Yamaji, N., Yamamoto, H., Okubo, K., Ueno, H., Costa, A., Tanoi, K., Matsumura, H., Fujii-Kashino, M., Horiuchi, T., Nayef, M.A., Shabala, S., An, G., Ma, J.F. and Horie, T. (2017). OsHKT1;5 mediates Na+ exclusion in the vasculature to protect leaf blades and reproductive tissues from salt toxicity in rice. *The Plant Journal*, *91*(4), 657–670. https://doi.org/10.1111/tpj.13595

Kocsy, G., Szalai, G. and Galiba, G. (2004). Effect of osmotic stress on glutathione and hydroxymethylglutathione accumulation in wheat. *Journal of Plant Physiology*, *161*(7), 785–794. https://doi.org/10.1016/j.jplph.2003.12.006

Kohli, S.K., Handa, N., Sharma, A., Kumar, V., Kaur, P. and Bhardwaj, R. (2017). Synergistic effect of 24-epibrassinolide and salicylic acid on photosynthetic efficiency and gene expression in Brassica juncea L. under Pb stress. *Turkish Journal of Biology = Turk Biyoloji Dergisi*, *41*(6), 943–953. PubMed. https://doi.org/10.3906/biy-1707-15

Kooyers, N.J. (2015). The evolution of drought escape and avoidance in natural herbaceous populations. *Plant Science*, *234*, 155–162. https://doi.org/10.1016/j.plantsci.2015.02.012

Kosová, K., Vítámvás, P., Planchon, S., Renaut, J., Vanková, R. and Prášil, I.T. (2013). Proteome Analysis of Cold Response in Spring and Winter Wheat (Triticum aestivum) Crowns Reveals Similarities in Stress Adaptation and Differences in Regulatory Processes between the Growth Habits. *Journal of Proteome Research*, *12*(11), 4830–4845. https://doi.org/10.1021/pr400600g

Kosová, K., Vítámvás, P. and Prášil, I.T. (2014). Wheat and barley dehydrins under cold, drought, and salinity—What can LEA-II proteins tell us about plant stress response? *Frontiers in Plant Science*, *5*, 343. PubMed. https://doi.org/10.3389/fpls.2014.00343

Kotak, S., Larkindale, J., Lee, U., von Koskull-Döring, P., Vierling, E. and Scharf, K.-D. (2007). Complexity of the heat stress response in plants. *Current Opinion in Plant Biology*, *10*(3), 310–316. https://doi.org/10.1016/j.pbi.2007.04.011

Krugman, T., Peleg, Z., Quansah, L., Chagué, V., Korol, A.B., Nevo, E., Saranga, Y., Fait, A., Chalhoub, B. and Fahima, T. (2011). Alteration in expression of hormone-related genes in wild emmer wheat roots associated with drought adaptation mechanisms. *Functional & Integrative Genomics*, *11*(4), 565–583. https://doi.org/10.1007/s10142-011-0231-6

Lebedev, V.G., Lebedeva, T.N., Chernodubov, A.I. and Shestibratov, K.A. (2020). Genomic Selection for Forest Tree Improvement: Methods, Achievements and Perspectives. *Forests*, *11*(11), 1190. https://doi.org/10.3390/f11111190

Lee, S.C., Lim, C.W., Lan, W., He, K. and Luan, S. (2013). ABA Signaling in Guard Cells Entails a Dynamic Protein–Protein Interaction Relay from the PYL-RCAR Family Receptors to Ion Channels. *Molecular Plant*, *6*(2), 528–538. https://doi.org/10.1093/mp/sss078

Lee, S.-K., Kim, B.-G., Kwon, T.-R., Jeong, M.-J., Park, S.-R., Lee, J.-W., Byun, M.-O., Kwon, H.-B., Matthews, B.F., Hong, C.-B. and Park, S.-C. (2011). Overexpression of the mitogen-activated protein kinase gene OsMAPK33 enhances sensitivity to salt stress in rice (Oryza sativa L.). *Journal of Biosciences*, *36*(1), 139–151. https://doi.org/10.1007/s12038-011-9002-8

Lekshmy Sathee, Sairam, R.K., Chinnusamy, V. and Jha, S.K. (2015). Differential transcript abundance of salt overly sensitive (SOS) pathway genes is a determinant of salinity stress tolerance of wheat. *Acta Physiologiae Plantarum*, *37*(8). https://doi.org/10.1007/s11738-015-1910-z

Lemcoff, J.H., Chimenti, C.A. and Davezac, T.A.E. (1998). Osmotic Adjustment in Maize (Zea mays L.): Changes with Ontogeny and its Relationship with Phenotypic Stability. *Journal of Agronomy and Crop Science*, *180*(4), 241–247. https://doi.org/10.1111/j.1439-037x.1998.tb00530.x

Lin, L., Wu, J., Jiang, M. and Wang, Y. (2021). Plant Mitogen-Activated Protein Kinase Cascades in Environmental Stresses. *International Journal of Molecular Sciences*, *22*(4), 1543. PubMed. https://doi.org/10.3390/ijms22041543

Liu, G., Li, J. and Godwin, I.D. (2019). Genome Editing by CRISPR/Cas9 in Sorghum Through Biolistic Bombardment. *Methods in Molecular Biology*, 169–183. https://doi.org/10.1007/978-1-4939-9039-9_12

Llanes, A., Andrade, A., Alemano, S. and Luna, V. (2016). Alterations of Endogenous Hormonal Levels in Plants under Drought and Salinity. *American Journal of Plant Sciences*, *07*(09), 1357–1371. https://doi.org/10.4236/ajps.2016.79129

Lo, S.-F., Ho, T.-H. D., Liu, Y.-L., Jiang, M.-J., Hsieh, K.-T., Chen, K.-T., Yu, L.-C., Lee, M.-H., Chen, C.-Y., Huang, T.-P., Kojima, M., Sakakibara, H., Chen, L.-J. and Yu, S.-M. (2017). Ectopic expression of specific GA2 oxidase mutants promotes yield and stress tolerance in rice. *Plant Biotechnology Journal*, *15*(7), 850–864. PubMed. https://doi.org/10.1111/pbi.12681

Lou, D., Wang, H., Liang, G. and Yu, D. (2017). OsSAPK2 Confers Abscisic Acid Sensitivity and Tolerance to Drought Stress in Rice. *Frontiers in Plant Science*, *8*, 993. PubMed. https://doi.org/10.3389/fpls.2017.00993

Ma, H., Chen, J., Zhang, Z., Ma, L., Yang, Z., Zhang, Q., Li, X., Xiao, J. and Wang, S. (2017). MAPK kinase 10.2 promotes disease resistance and drought tolerance by activating different MAPKs in rice. *The Plant Journal*, *92*(4), 557–570. https://doi.org/10.1111/tpj.13674

Ma, X., Xin, Z., Wang, Z., Yang, Q., Guo, S., Guo, X., Cao, L. and Lin, T. (2015). Identification and comparative analysis of differentially expressed miRNAs in leaves of two wheat (Triticum aestivum L.) genotypes during dehydration stress. *BMC Plant Biology*, *15*, 21. PubMed. https://doi.org/10.1186/s12870-015-0413-9

Ma, Y., Szostkiewicz, I., Korte, A., Moes, D., Yang, Y., Christmann, A. and Grill, E. (2009). Regulators of PP2C Phosphatase Activity Function as Abscisic Acid Sensors. *Science*, *324*(5930), 1064–1068. https://doi.org/10.1126/science.1172408

Maraveas, C. (2024). Image Analysis Artificial Intelligence Technologies for Plant Phenotyping: Current State of the Art. *AgriEngineering*, *6*(3), Article 3. https://doi.org/10.3390/agriengineering6030193

McIsaac, M.A., Sanders, E., Kuester, T., Aronson, K.J. and Kyba, C.C.M. (2021). The impact of image resolution on power, bias, and confounding: A simulation study of ambient light at night exposure. *Environmental Epidemiology (Philadelphia, Pa.)*, *5*(2), e145. PubMed. https://doi.org/10.1097/EE9.0000000000000145

Melotto, M., Zhang, L., Oblessuc, P.R. and He, S.Y. (2017). Stomatal Defense a Decade Later. *Plant Physiology*, *174*(2), 561–571. PubMed. https://doi.org/10.1104/pp.16.01853

Mickelbart, M.V., Hasegawa, P.M. and Bailey-Serres, J. (2015). Genetic mechanisms of abiotic stress tolerance that translate to crop yield stability. *Nature Reviews Genetics*, *16*(4), 237–251. https://doi.org/10.1038/nrg3901

Mittler, R. (2006). Abiotic stress, the field environment and stress combination. *Trends in Plant Science*, *11*(1), 15–19. https://doi.org/10.1016/j.tplants.2005.11.002

Mittler, R., Finka, A. and Goloubinoff, P. (2012). How do plants feel the heat? *Trends in Biochemical Sciences*, *37*(3), 118–125. https://doi.org/10.1016/j.tibs.2011.11.007

Mohammadi, M., Xavier, A., Beckett, T., Beyer, S., Chen, L., Chikssa, H., Cross, V., Freitas Moreira, F., French, E., Gaire, R., Griebel, S., Lopez, M.A., Prather, S., Russell, B. and Wang, W. (2020). Identification, deployment, and transferability of quantitative trait loci from genome-wide association studies in plants. *Current Plant Biology*, *24*, 100145. https://doi.org/10.1016/j.cpb.2020.100145

Mora, C., Caldwell, I.R., Caldwell, J.M., Fisher, M.R., Genco, B.M. and Running, S.W. (2015). Suitable Days for Plant Growth Disappear under Projected Climate Change: Potential Human and Biotic Vulnerability. *PLOS Biology*, *13*(6), e1002167. https://doi.org/10.1371/journal.pbio.1002167

Morris, P.C., Reuveni, M. and Ortiz, R. (2024). Impact and mitigation of abiotic stress in cereals. *Frontiers in Plant Science*, *15*, 1374631.

Muluneh, M.G. (2021). Impact of climate change on biodiversity and food security: A global perspective—a review article. *Agriculture & Food Security*, *10*(1), 36. https://doi.org/10.1186/s40066-021-00318-5

Munné-Bosch, S. (2008). Do perennials really senesce? *Trends in Plant Science*, *13*(5), 216–220. https://doi.org/10.1016/j.tplants.2008.02.002

Munns, R. and Tester, M. (2008). Mechanisms of Salinity Tolerance. *Annual Review of Plant Biology*, *59*(1), 651–681. https://doi.org/10.1146/annurev.arplant.59.032607.092911

Nishiyama, Y. and Murata, N. (2014). Revised scheme for the mechanism of photoinhibition and its application to enhance the abiotic stress tolerance of the photosynthetic machinery. *Applied Microbiology and Biotechnology*, *98*(21), 8777–8796. https://doi.org/10.1007/s00253-014-6020-0

Niu, C., Wei, W., Zhou, Q., Tian, A., Hao, Y., Zhang, W., Ma, B., Lin, Q., Zhang, Z., Zhang, J. and Chen, S. (2012). Wheat WRKY genes TaWRKY2 and TaWRKY19 regulate abiotic stress tolerance in transgenic Arabidopsis plants. *Plant, Cell & Environment*, *35*(6), 1156–1170. https://doi.org/10.1111/j.1365-3040.2012.02480.x

# Chapter 2

# Advancements in Genome Editing Technologies for Enhancing Crop Improvement

Suhas Gorakh Karkute[1], Mahantesha B.N. Naika[2*], Jae-Yean Kim[3,4,5*] and Rahul Mahadev Shelake[3*]

## 1. Introduction

Crop cultivation faces significant challenges in the context of extreme weather and climate change. Crops are often exposed to high and low temperatures, drought, salinity, elevated $CO_2$, and heavy metals, which drastically affect the yield and quality of the crops (Francini and Sebastiani, 2019). The reduction in crop yields due to climate-change-related factors is projected to reach 11%–25% over the next 5 to 7 decades (Wing et al., 2021). Reduced crop yields seriously threaten world food security, mainly as the global population is expected to cross the mark of 9.7 billion by 2050 (United Nations, 2019). This underscores the necessity to develop improved crop cultivars that can thrive under adverse climatic conditions. In addition to food security, hidden hunger (nutrient deficiency) remains a major global challenge, particularly severe in underdeveloped nations (Lowe, 2021). Furthermore, the incidences of insect pest and pathogen attacks are rising due to accelerated evolution, allowing these pests to expand their geographic range, alter interactions with host plants, and infest non-host crops, thereby threatening cropping systems (Skendžić et al., 2021).

Advancements in plant science research have significantly contributed to ensuring a continuous food supply. However, current crop varieties are unlikely to remain sustainable in the future. In recent years, crop improvement programs have integrated advanced technologies such as transgenics, speed breeding, and genome editing (GE), with clustered regularly interspaced short palindromic repeats (CRISPR)/Cas-based GE being particularly transformative. CRISPR/Cas technology has been extensively utilized to develop nutrient-rich crops, enhance resilience to both abiotic and biotic stresses, and improve crop yields (Gupta and Karkute, 2021).

1 Division of Vegetable Improvement, ICAR-Indian Institute of Vegetable Research, Varanasi, India.
2 Centre for Biotechnology Research, Department of Biotechnology and Crop Improvement, College of Horticulture, University of Horticultural Sciences, Bagalkot, Karnataka, India.
3 Division of Applied Life Science (BK21 Four Program), Plant Molecular Biology and Biotechnology Research Center, Gyeongsang National University, Jinju, South Korea.
4 Nulla Bio Inc., 501 Jinju-daero, Jinju 52828, South Korea.
5 Division of Life Science, Gyeongsang National University, 501 Jinju-daero, Jinju 52828, South Korea.
* Corresponding author: rahultnau@gmail.com; mahantesha.bnn@uhsbagalkot.edu.in; kimjy@gnu.ac.kr

DOI: 10.1201/9781003584100-2

Further innovations in CRISPR/Cas-based tools have expanded their applications beyond simple mutation-based knockout studies and include chromosomal rearrangements, RNA editing, DNA and RNA detection for diagnostic purposes, as well as activation and suppression of desired genes (Russa and Qi, 2015; Hu and Li, 2022; Pramanik et al., 2022a; Heidersbach et al., 2023). The versatile CRISPR/Cas-based tools hold immense potential in functional genomics, enabling the exploration and characterization of critical genes and their subsequent use in future crop improvement efforts. This work reviews the basics of CRISPR/Cas-based GE technologies, their applications in agriculture, and the potential use of recently available genomic datasets for different plant species.

## 2. Basics of CRISPR-Based GE Technology

CRISPR/Cas is a part of the bacterial adaptive immune system, and its use for genomic DNA editing was first demonstrated in 2012 (Jinek et al., 2012). Engineered CRISPR/Cas technology offers unprecedented precision and efficiency in altering genetic information in different organisms, including plants (Shelake et al., 2023b). The two primary CRISPR/Cas9 system components are RNA molecules and an endonuclease Cas enzyme. Details about these two components and related terms are summarized in the following subsections.

### 2.1 Single Guide RNA (sgRNA)

RNA molecule used in CRISPR-based GE is termed as sgRNA. The sgRNA is a short RNA sequence that directs the Cas protein to a specific location in the genome. It is a hybrid form of trans-activating RNA (i.e., tracrRNA, which forms a scaffold) and CRISPR RNA (crRNA, complementary to the target site). The sgRNA is complementary to a target DNA sequence, ensuring that the Cas protein induces cuts precisely at the desired genomic site. The design of sgRNA is critical for the specificity (avoiding off-target effects) and efficiency of CRISPR-based editing.

### 2.2 Cas Protein

The Cas protein, typically Cas9 or its variants, acts as molecular scissors that cleave the DNA at the location specified by the gRNA. In the updated classification system, CRISPR/Cas proteins are categorized into two primary classes: Class 1 and Class 2. These two classes are again divided into various types (I, II, III, IV, V, and VI) and sub-types (Bhatia et al., 2023). The effector proteins consist of single (Class 2) or a combination of multiple Cas proteins (Class 1). Class 1 includes Types I, III, and IV, characterized by multiprotein effector fusions. On the other hand, Class 2 has Types II, V, and VI that feature a singular effector protein. This classification is based on the specific Cas proteins present. The most used Cas enzyme for GE purposes in plants is Cas9 from *Streptococcus pyogenes* (SpCas9), from the Class 2-Type II group. However, other variants, such as Cas12a (Cpf1) from the Class 2-Type V group, are also being explored for their unique properties. For instance, SpCas9 recognizes Guanine-Cytosine (GC) rich protospacer adjacent motif (PAM) sites (e.g., NGG, where N is any nucleobase followed by two guanine bases), creating blunt-end DNA cuts, whereas Cpf1 recognizes T-rich PAM sites [e.g., TTTV, where three thymine (T) bases followed by V; V, stands for A, C, or G], producing staggered-end DNA cuts at targeted sites.

### 2.3 Protospacer Adjacent Motif (PAM)

PAM is a short nucleotide sequence (generally 3 to 5 nucleotide bases) present on targeted DNA site recognized by the Cas protein and is present immediately after the target DNA sequence. The presence of PAM is essential for the Cas protein to bind and cut the DNA. The sequence of the PAM varies depending on the Cas protein used, with Cas9 requiring a 5'-NGG-3' sequence.

## 2.4 *Mechanism of CRISPR/Cas-Mediated GE*

The guide sequence of sgRNA binds to the target DNA sequence after locating the PAM site. The Cas enzyme cuts both DNA strands at the targeted genomic site, creating a double-stranded break (DSB). The natural endogenous DNA repair mechanisms activated to repair the DSB ultimately decide the fate of the editing outcome, which mainly involves nonhomologous end joining (NHEJ) or homologous recombination (HR). Among them, NHEJ is a preferred pathway that often results in small insertions or deletions, while HR can introduce precise and predefined changes by providing a donor DNA or RNA template (Pramanik et al., 2021a).

# 3. CRISPR-Based GE Toolset for Plants

Several modified versions in the CRISPR toolbox allow for specific, efficient, and targeted modifications to plant genomes, enabling advancements in crop improvement, disease resistance, and understanding of fundamental or applied aspects of plant biology (Figure 1). A versatile CRISPR-based toolkit enables various modifications, including gene disruption, insertion, replacement, transcriptional regulation, epigenetic modulation, base substitution, base deletion, and many more (Pramanik et al., 2021a; Wang and Doudna, 2023; Shelake et al., 2024). This section explores the different CRISPR-based tools for GE in plants, highlighting their advantages, challenges, and future potential.

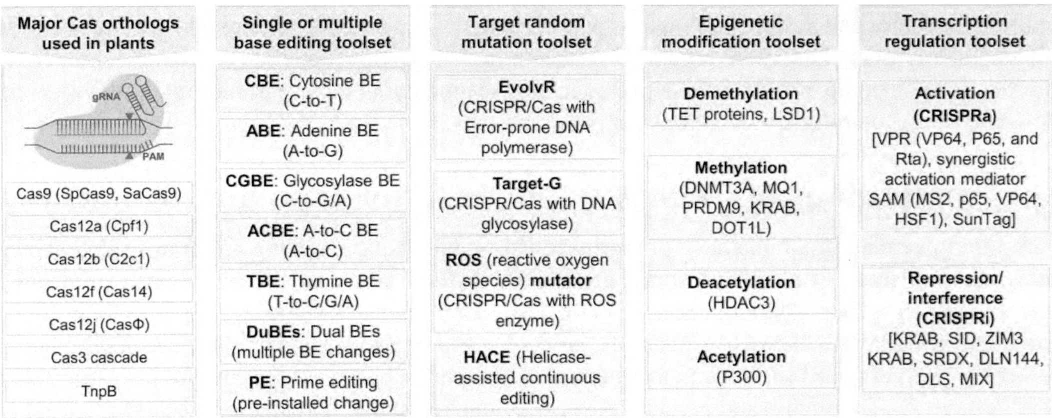


**Fig. 1** CRISPR toolset for plant genome editing. Details about the CRISPR toolset are discussed in the main text, along with relevant literature.

## 3.1 *SpCas9: A Foundation for CRISPR Tool Development*

The SpCas9 has been a cornerstone of CRISPR technology due to its broad applicability, well-characterized structure, and compatibility with fusion proteins. Its recognition of the 5'-NGG-3' PAM sequence is prevalent in many genomes, facilitating its use in diverse plant species. The fusion of SpCas9 with accessory proteins, for example, deaminases and enzymes involved in DNA repair mechanism, has expanded its capabilities without compromising its enzymatic activity.

## 3.2 *Expanding the CRISPR Toolset: Novel Cas Nucleases and Fusion Proteins*

Researchers have identified and characterized additional Cas nucleases to address the limitations of SpCas9 and explore new editing possibilities, including Cas12a, Cas12b, Cas12f, Cas12j, and

Cascade-Cas3 (Shelake et al., 2023a). Recent optimized transposon-associated TnpB in plant systems, which is about 400 amino acids (aa), provides an alternative to commonly used larger Cas versions, such as Cas9 (about 1000–1400 aa) and Cas12a (about 1300 aa) (Karmakar et al., 2024). These enzymes exhibit distinct editing properties, expanding the range of genetic modifications that can be achieved.

### 3.3 Advanced CRISPR Tools for Diverse Applications

Researchers have developed innovative tools with specialized editing functions by fusing accessory molecules with Cas variants. DNA base editors (BEs) allow precise base substitutions, while RNA base editors target RNA transcripts. Different BE and BE-based tools have recently been summarized for various applications in basic research, agriculture, and medical sciences (Shelake et al., 2024). DNA dual BEs (DuBEs) enable simultaneous editing of multiple DNA target sites, allowing the generation of mutant libraries for desired genetic loci in different organisms, including bacteria and plants.

Target-G and EvolvR are targeted random mutagenesis tools that can generate genetic diversity. Epigenetic modifiers can control gene expression by modifying epigenetic marks. Transcriptional regulators such as CRISPR activation (CRISPRa) and CRISPR interference (CRISPRi) tools precisely activate or suppress target genes, respectively. Prime editors (PEs) offer precise DNA rewriting capabilities, and ROS-mutator can introduce random mutations (summarized in Shelake et al., 2023a, 2024). While CRISPR-based tools have shown great promise, challenges such as inconsistent editing outcomes and delivery limitations persist. Ongoing research focuses on identifying new Cas effectors, improving DNA delivery systems, and optimizing CRISPR components to overcome these hurdles. For example, advancements in PE technology, including the use of multiple pegRNAs and site-specific recombinases, are addressing its limitations.

## 4. Applications of CRISPR-Based Tools for Crop Improvement

Crop improvement programs have primarily relied on molecular breeding and transgenics, which have enormously contributed to ensuring global food security. However, the limitations of molecular breeding, such as lack of genomic resources for desired traits and low public acceptance of transgenic products, require safe alternative methods, broadly accepted and sustainable, for accelerating crop breeding progress. CRISPR/Cas is the method of choice for crop breeding programs owing to its high efficiency, simplicity, and precision over previous GE tools. It has been successfully used to improve crops such as vegetables, fruits, flowers, pulses, oilseeds, cereals, etc. (Table 1) (Karkute et al., 2017; Zhang et al., 2018; Ku and Ha, 2020).

### 4.1 Crop Improvement for Biotic Stress Tolerance

Crop production is severely affected by biotic stressors such as plant pathogens, insects, nematodes, and weeds, leading to losses of up to 30% (Van Esse et al., 2020). Resistance breeding programs to develop crop tolerance against diseases and pests focus on using *Resistance* (*R*) genes. However, *R*-gene-mediated resistance is not durable due to the evolution of effector proteins in pathogens, which results in frequent resistance breakdown (Paineau et al., 2024). On the other hand, susceptibility genes required for successful pathogen infection are potential targets for modulation as their recessive alleles provide durable broad-spectrum resistance against pathogens in crops (Garcia-Ruiz et al., 2021).

The primary concern for targeting the susceptibility genes is the pleiotropic function of such genes, which, in turn, defines the strategy for targeting them through GE. For instance, if the gene is pleiotropic, such as sugar transporter genes (*SWEET*), which are targets of transcription activator-like effectors (TALE) of *Xanthomonas* bacteria, mutation needs to be carried out at the

**Table 1** Status of genome editing applications for crop improvement.

| *S. No.* | *Target trait* | *Target gene* | *Crop* | *Reference* |
|---|---|---|---|---|
| 1. | Bacterial blight resistance | *Promoter of SWEET11, SWEET13, SWEET14* | Rice | Oliva et al., 2019 |
| 2. | Bacterial blight resistance | Promoter of *OsSWEET14* | Basmati rice | Zafar et al., 2020 |
| 3. | Bacterial blight resistance | Promoter of *MeSWEET10a* | Cassava | Elliott et al., 2024 |
| 4. | Powdery mildew resistance | *SlMLO1* | Tomato | Nekrasov et al., 2017; Pramanik et al., 2021b |
| 5. | Powdery mildew resistance | *VvMLO3* | Grapevine | Wan et al., 2020 |
| 6. | Powdery mildew resistance | *MLO-B1* | Wheat | Li et al., 2022 |
| 7. | Powdery mildew resistance | *GmMLO02, GmMLO19, GmMLO20, GmMLO23* | Soybean | Bui et al., 2023 |
| 8. | Powdery mildew resistance | *CsaMLO8* | Cucumber | Shnaider et al., 2023 |
| 9. | Multiple biotic stress resistance | *DMR6-1* | Tomato | Thomazella et al., 2021 |
| 10. | Drought tolerance | *SlLBD40* | Tomato | Liu et al., 2020 |
| 11. | Heat stress tolerance | *SlMAPK3* | Tomato | Yu et al., 2019 |
| 12. | Drought and salt tolerance | *OsDST* | Rice | Santosh Kumar et al., 2020 |
| 13. | Salt tolerance | *OsPQT3* | Rice | Alfatih et al., 2020 |
| 14. | Cold tolerance | *OsPIN5b, GS3, OsMYB30* | Rice | Zeng et al., 2020 |
| 15. | Reduced enzymatic browning | *StPPO2* | Potato | Gonzalez et al., 2020 |
| 16. | Reduced enzymatic browning | *SmelPPO4, SmelPPO5, SmelPPO6* | Eggplant | Maioli et al., 2020 |
| 17. | Reduced enzymatic browning | *PPO1, PPO2* | Wheat | Wold-McGimsey et al., 2023 |
| 18. | Reduced enzymatic browning | *PPO* | Mushroom | Waltz, 2016 |
| 19. | Improved starch quality | *GBSSI* and *SBEII* | Sweet potato | Wang et al., 2019 |
| 20. | Increased amylose starch | *SBE1, SBE2* | Potato | Zhao et al., 2021 |
| 21. | Increased resistant starch | *SBE1, SBE2, SBE3, SBE4* | Rice | Biswas et al., 2023 |
| 22. | Increased amylose and resistant starch | *SBEI, SBEIIb* | Maize | Ma et al., 2024 |
| 23. | Increased resistant starch | *SSIIa, SSIIIa, SSIVa, SBEIIa, SBEIIb* | Barley | Yang et al., 2024 |
| 24. | Increased shelf life | *SlPG2a* and *SlPL* | Tomato | Ortega-Salazar et al., 2024 |
| 25. | Increased shelf life | *ACO1* | Muskmelon | Hu et al., 2021 |
| 26. | Increased oleic acid content in oil | *FAD2* | *Camelina sativa* | Jiang et al., 2017 |
| 27. | Increased oleic acid content in oil | *BnaA.FAD2.a* | *Brassica napus* | Okuzaki et al., 2018 |

*Contd.*

**Table 1** *Contd.*

| *S. No.* | *Target trait* | *Target gene* | *Crop* | *Reference* |
|---|---|---|---|---|
| 28. | Increased oleic acid content in oil | *AhFAD2* | Peanut | Yuan et al., 2019 |
| 29. | Low phytic acid content | *BnITPK* | *Brassica napus* | Sashidhar et al., 2020 |
| 30. | Low phytic acid content | *OsPLDα1* | Rice | Khan et al., 2019 |
| 31. | Low phytic acid content | *TaIPK1* | Wheat | Ibrahim et al., 2022 |
| 32. | Low phytic acid content | *GmIPK1* | Soybean | Song et al., 2022 |
| 33. | High level of γ-aminobutyric acid | *SlGAD2, SlGAD3* | Tomato | Nonaka et al., 2017 |
| 34. | Synthesis of 3-O-glucosides and 4-O-glucosides of gallic acid | *UGT84A23* and *UGT84A24* | Pomegranate | Chang et al., 2019 |
| 35. | High isoflavone content and resistance to soybean mosaic virus | *GmF3H1*, *GmF3H2*, and *GmFNSII-1* | Soybean | Zhang et al., 2020 |
| 36. | Natural rubber synthesis | *LsCPT3* | Lettuce | Kwon et al., 2023 |

TALE-binding site in the promoter region of these genes. Development of broad-spectrum resistance against bacterial leaf blight disease caused by *Xanthomonas oryzae* pv. oryzae in rice is one of the crucial achievements of CRISPR/Cas9-mediated GE technology using such a strategy (Oliva et al., 2019). A similar approach has been implemented in super basmati rice (Zafar et al., 2020) and recently in cassava (Elliott et al., 2024). A similar strategy can be adapted in other crops such as pomegranate, where bacterial blight caused by *Xanthomonas axonopodis* pv. punicae results in up to 80% yield loss annually.

Targeted mutagenesis of susceptible genes having pleiotropic functions could result in undesirable effects on normal plant development. For targeting trait-regulatory genetic loci, it is vital to create different alleles by targeting various positions in the coding region of the gene and then screening the mutant plants for desirable phenotypes without any adverse effect on the growth and yield of plants. One must consider all the factors, from seed germination to yield, and response to various stresses while screening mutant plants to ensure no side effects from editing the susceptibility genes.

Developing disease resistance by creating novel alleles of *Susceptibility* (*S*) genes is the most successful application of the GE tools. A well-known example of this strategy is the mutation of mildew resistance locus (*mlo*) governing resistance to powdery mildew. Since the identification of *mlo* mutation in barely (Jørgensen 1992), it has become the ideal gene target for powdery mildew resistance. Thus, powdery mildew resistance has been imparted in tomato (Nekrasov et al., 2017; Pramanik et al., 2021b), grapevine (Wan et al., 2020), wheat (Li et al., 2022), soybean (Bui et al., 2023), and cucumber (Shnaider et al., 2023). Besides the genes involved in pathogen infection and establishment, which usually provide resistance against specific pathogens, the genes involved in resistance response are also potential candidates for GE to develop broad-spectrum resistance against multiple pathogens. Salicylic acid 5-hydroxylase protein encoded by the *downy mildew resistance 6* (*DMR6*) gene in tomato catalyzes the hydroxylation of salicylic acid (SA) to produce dihydroxybenzoic acid (DHBA), which is an inactive form of SA. Mutation of *DMR6-1* homolog in tomato through CRISPR/Cas9 developed resistance against multiple diseases caused by

*Pseudomonas syringae*, *Pseudomonas capsici*, and *Xanthomonas* spp. without any adverse effects on plant growth and yield (Thomazella et al., 2021).

## 4.2 Crop Improvement for Abiotic Stress Tolerance

Different abiotic stresses such as high and low temperature, moisture, salinity, etc., drastically impact plant growth, causing a reduction in quality and yield (Shelake et al., 2022). Due to the complexity of abiotic stress response in plants, developing abiotic-stress-tolerant varieties is always challenging. However, there are multiple reports where a single gene has been identified to impart tolerance against abiotic stresses. Many of them are transcription factors such as dehydration-responsive element (DRE)-binding protein (DREB) (Yamaguchi-Shinozaki and Shinozaki 1994; Karkute et al., 2019) and NAC3 transcription factors providing tolerance to multiple abiotic stresses (Han et al., 2023). While all these genes must be either overexpressed through transgenics or transferred through breeding to elite cultivars with their limitations, GE allows the development of stress-tolerant crops with broad public acceptance.

CRISPR/Cas technology has assisted in both the identification of stress-responsive genes and the mutagenesis of negative regulators of abiotic stress response. Negative regulators of stress response are the prime targets for developing stress-tolerant plants using GE. The lateral organ boundaries domain 40 (*SlLBD40*) transcription factor in tomato is one such example, mutagenesis, which enhanced drought tolerance by improving water-holding capacity under water stress conditions (Liu et al., 2020). Similarly, the knockout of *drought and salt tolerance* (*OsDST*) gene in rice confers drought tolerance to edited lines through enhanced antioxidant activity and lowered $H_2O_2$ levels (Santosh Kumar et al., 2020). Several other examples include mutagenesis of *OsPQT3* (Alfatih et al., 2020), *SlMAPK3* (Yu et al., 2019), *OsPIN5b*, *GS3*, and *OsMYB30* (Zeng et al., 2020), which resulted in significant tolerance to various abiotic stresses.

## 4.3 Crop Improvement for Quality Traits

Enhancing quality and nutritional content is one of the most desired objectives of crop improvement programs. The lack of genetic resources for many such traits in germplasm is a significant limitation for the improvement of crops. On the other hand, targeted mutagenesis or overexpression of genes controlling biosynthesis of anticipated bioactive compounds needed for food quality and nutrition not only enhances the number of such compounds but also provides the opportunity to synthesize new compounds by engineering the pathways. To improve quality, nutrition, and shelf life, GE has been implemented in various crops, including cereals, oilseeds, pulses, and horticultural crops. At the same time, reduction of anti-nutritional compounds is also possible through GE.

Oxidation of phenolic compounds to form quinones results in flesh browning in cut fruits, lowering their quality for consumption. To avoid this enzymatic browning, the *polyphenol oxidase* (*PPO*) gene has been knocked out in potatoes (Gonzalez et al., 2020), eggplant (Maioli et al., 2020), wheat (Wold-McGimsey et al., 2023), and mushroom (Waltz 2016). Starch quality is essential in crops such as potato, sweet potato, maize, and rice for their utilization as food or in industries. CRISPR/Cas-based GE has been successfully used for modifying the amylose or amylopectin content by targeting *granule-bound starch synthase I* (*GBSSI*) gene and *starch branching enzyme II* (*SBEII*) gene, respectively, in sweet potato (Wang et al., 2019), potato (Zhao et al., 2021), rice (Biswas et al., 2023), maize (Ma et al., 2024), and barley (Yang et al., 2024). Shelf life or post-harvest life defines the quality of several horticultural crops. Shelf-life enhancement has been achieved by manipulating the genes involved in pectin metabolism, ripening, and ethylene synthesis. In a recent report, Ortega-Salazar et al. (2024) knocked out two genes, *polygalacturonase* (*SlPG2a*) and *pectate lyase* (*SlPL*), involved in pectin degradation to improve shelf life in tomato. Disruption of the ethylene production pathway by targeting *1-aminocyclopropane-1-carboxylic acid oxidase* (*ACO1*) gene by CRISPR/Cas9 showed enhanced shelf life in muskmelon (Nonaka et al., 2023) and banana fruits (Hu et al., 2021).

Oil composition has been improved to have reduced content of polyunsaturated fatty acid (PUFA) and a higher percentage of oleic acid by mutating the *FAD2* gene in *Camelina sativa* (Jiang et al., 2017), *Brassica napus* (Okuzaki et al., 2018), and peanut (Yuan et al., 2019). Phytic acid is an anti-nutrient that acts as a chelating agent for micronutrients, lowering their availability for humans. Thus, phytic acid is undesirable in consumable seeds. *Brassica napus* with low phytic acid content was produced by CRISPR-mediated mutation of the *BnITPK* gene (Sashidhar et al., 2020). Similarly, the low phytic acid trait has been achieved by targeting the *OsPLDα1* gene in rice (Khan et al., 2019), the *TaIPK1* gene in wheat (Ibrahim et al., 2022), and the *GmIPK1* gene in soybean (Song et al., 2022)

Metabolic engineering through CRISPR/Cas technology is an exciting field, especially in horticultural crops such as fruits, vegetables, flowers, and medicinal plants, where there is enormous scope to induce the synthesis of novel compounds beneficial for human health. Metabolic engineering is further empowered by the availability of multiplex editing tools that can target multiple genes simultaneously. There are several reports on using CRISPR tools to enhance the quantity of desirable nutrients in crops. For instance, in tomato, high levels of γ-aminobutyric acid (GABA) were produced by editing the *glutamate decarboxylase* (*GAD*) gene involved in GABA biosynthesis (Nonaka et al., 2017). In addition to GABA, improvement of several other nutrients such as total soluble solids (TSS), lycopene, carotenoids, and Vitamin D has been achieved in tomato (Gupta et al., 2024). Synthesis of novel compounds 3-O-glucosides and 4-O-glucosides of gallic acid was observed in pomegranate after knockout of *UGT84A24* and *UGT84A23* by CRISPR/Cas tool. Recently, multiplex editing approach using CRISPR has been implemented for metabolic engineering in soybeans by simultaneous mutagenesis of *GmF3H1*, *GmF3H2*, and *GmFNSII-1* genes, which resulted in increased isoflavone content as well as conferred resistance against soybean mosaic virus (Zhang et al., 2020). Kwon et al. (2023) knocked out the *LsCPT3* gene in lettuce and successfully showed the natural rubber synthesis of 1 million Daltons in molecular weight. Overall, metabolic engineering through CRISPR/Cas technology has enormous potential to produce novel metabolites in plants.

## 5. Advances in Next-Generation Sequencing and Their Uses in Plant GE

Next-generation sequencing (NGS) has revolutionized genomics research in plant biology. It can sequence millions of DNA and RNA molecules at once, allowing for rapid, cost-effective, and high-throughput sequencing and advances in data analysis, yielding extensive information about structure of genome, genetic variability, gene activity, gene expression pattern, and epigenetic modifications (Satam et al., 2023). The milestones of different generations of sequencing platforms (Gupta and Verma, 2019) are depicted in Figure 2. According to Marks et al. (2021), plant genome sequencing has expanded rapidly over the past two decades, resulting in more and higher-quality publicly available genomic information. Genomic information from different species provides new insights into the biology and evolution of plants.

Genome sequence information is critical for understanding the biological, physiological, and molecular mechanisms associated with plant genes. In plants, the first genome sequence published was of *Arabidopsis thaliana* in 2000, and 782 plant species have subsequently been sequenced over the last 20 years. During 2021–2023, 1031 plant species were sequenced. The “N3: plants, genomes, and technologies database” (http://ibi.zju.edu.cn/N3database/index.php) was created and maintains sequence information of 3517 genomes (Xie et al., 2024). Current status of sequenced genomes data available at National Center for Biotechnology Information (NCBI) as mentioned is: all genomes 2.47 million, 38.18 K reference, 2.02 M annotated; Eukaryota total 43.05 K, 17.77 K reference, 8.62K annotated; Archaea total 20.04 K, reference 704 K, annotated 14.14 K; bacteria total 2.19M, 19.70K reference, annotated 1.85M; and viruses total 192.60 K, reference 153.33 K, annotated 75745 information is as on date 6$^{th}$ September 2024 (NCBI database, 2024). We can also witness the

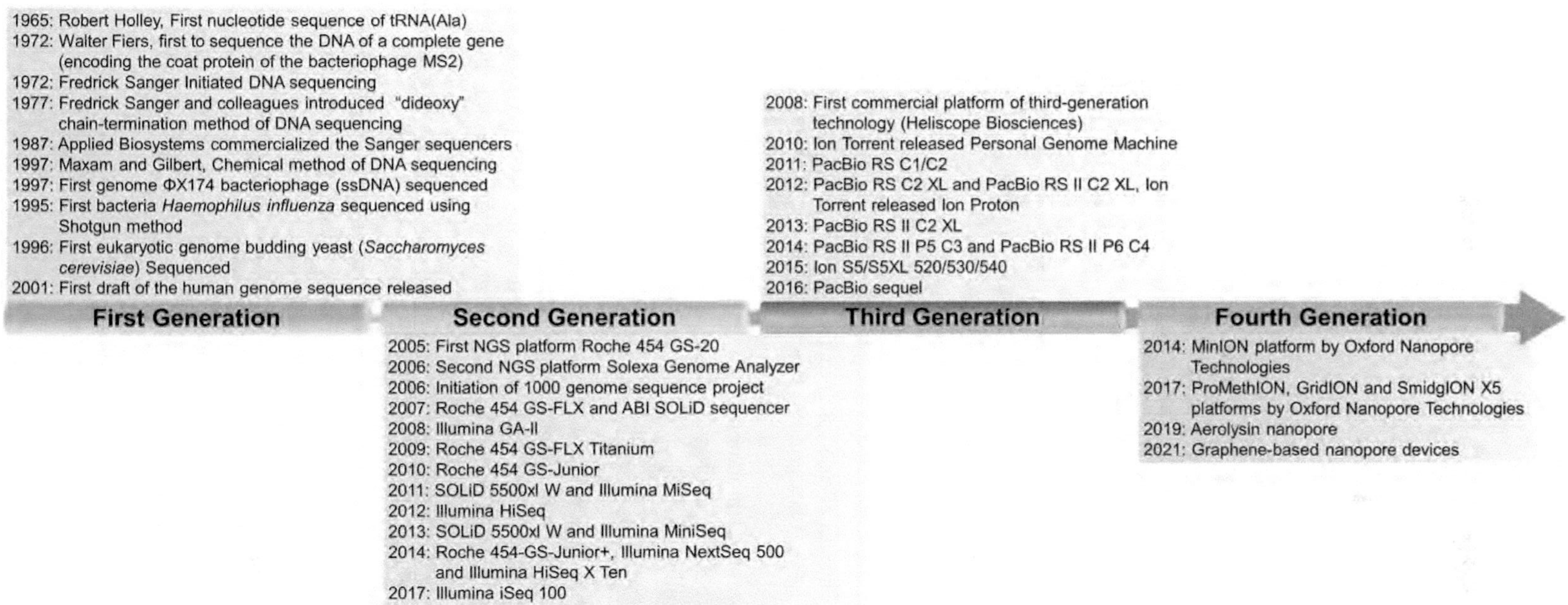


**Fig. 2** The milestones in the generations of different next-generation sequencing (NGS) platforms.

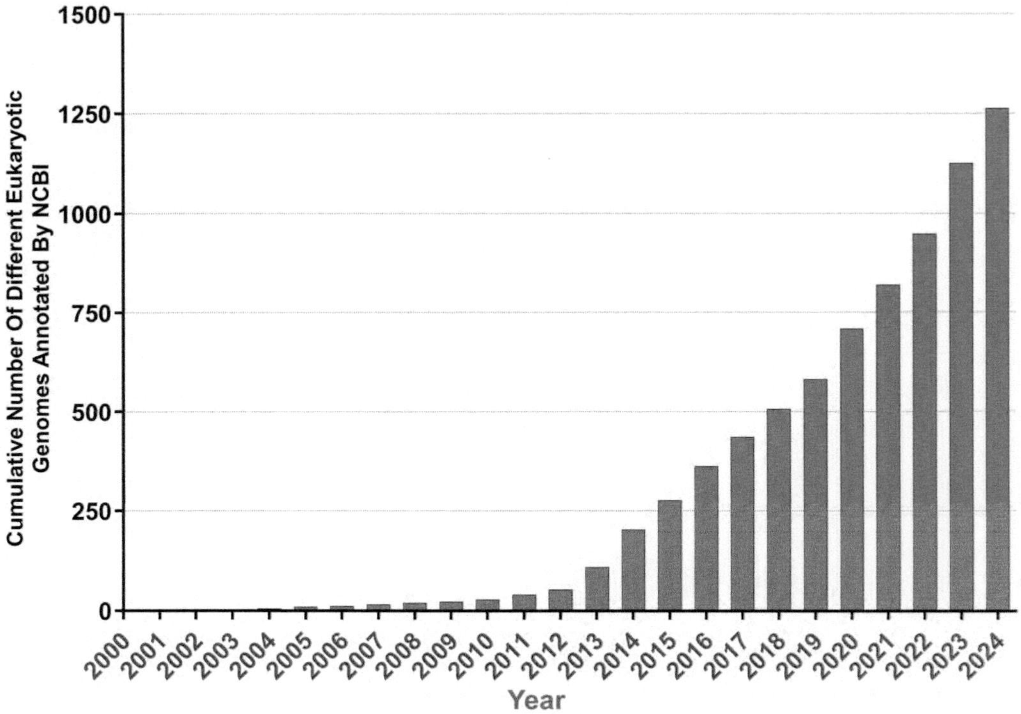


**Fig. 3** Cumulative number of eukaryotic genomes annotated on the NCBI webpages (data as of 19th October 2024).

growing information about genome annotation of the sequenced eukaryotic genomes accessible on the NCBI database site (Figure 3). Technological advances in NGS and various genomics tools such as single nucleotide polymorphism (SNP) chips and genomic selection have opened new avenues for agricultural genetic improvement programs. Bioinformatics tools effectively analyze omics data and help to discover novel genetic elements and candidate genes for GE. Similarly, gene editing tools have enabled precise modification of critical genes controlling agronomically important traits, resulting in novel varieties with desirable features (Shameer et al., 2019; Zhang et al., 2020).

NGS has played an important role in the fast adoption of CRISPR/Cas technology. NGS technique can help to identify target regions, confirm modifications, and discover off-target consequences. Gene expressions and functions can be analyzed using RNA-Seq and ChIP-Seq. Omics-based analytical approaches can help in optimizing the CRISPR/Cas system for higher efficacy and identify alternative GE methods (Illumina, Inc., 2017).

The significance of NGS technology in virus discovery and the effectiveness of CRISPR/Cas-based GE tools in engineering plants to develop virus resistance were reviewed and presented in recent works (Hadidi et al., 2016). NGS is a quick, efficient, and less expensive DNA/RNA high-throughput sequencing technology that is useful for sequencing the novel genomes of plant viruses and viroids. NGS is helpful in detecting and identifying known and new pathogens, genomic diversity, and evolution analysis, and studying the epidemiology of pathogens. The CRISPR/Cas system has developed resistance in plants against DNA Geminiviruses by incorporating mutations in different parts of the viral genome. Targeted DNA viruses include *Merremia mosaic virus*, *Tomato yellow leaf curl virus*, *Beet severe curly top virus*, *Beet curly top virus*, and *Bean yellow dwarf virus* (Khan et al., 2022). Targeted RNA viruses include *Turnip mosaic virus*, *Papaya ringspot virus*, *Zucchini yellow mosaic virus*, and *Cucumber vein yellowing virus* by targeting the endogenous plant genes such as *eIF4E* in cucumber or *Arabidopsis* plants (reviewed in Mustaq et al., 2021). Recent studies reported that the combined use of NGS and CRISPR-based GE enabled rapid resistance engineering by directly targeting desired genomic regions of plants, viruses, and viroids (Rajpal et al., 2023). Overall, undated genomic information provides a platform to study the evolution and mapping of trait-regulating genetic loci, ultimately enabling the use of GE technologies that focus on trait improvement and climate resilience in major food crops and underutilized dryland crops.

## 6. Regulatory Aspects of Genome-Edited Crop Varieties

Three main categories of GE techniques often discussed in terms of regulation differ based on the type of genetic modifications they introduce: site-directed nuclease-1 (SDN-1), SDN-2, and SDN-3 (Podevin et al., 2013). The mechanism of SDN-1 involves creating a DSB at a specific location in the genome using site-directed nucleases such as CRISPR/Cas system, and the changes are typically small (insertions or deletions, that is, indels at the target site) and do not involve the insertion of foreign DNA. SDN-2 also creates a DSB at a specific site. However, in this case, a repair template is provided along with the nuclease. This template guides the repair process to make precise changes in DNA sequence and introduces specific nucleotide changes or small insertions, but does not involve inserting foreign genes. The SDN-3 category involves inserting larger DNA sequences, which can include foreign genes. These distinctions are important for regulatory purposes, as SDN-1 and SDN-2 are often subject to less stringent regulations than SDN-3, which is treated similarly to traditional genetically modified organisms (GMOs).

The regulation of genome-edited crop varieties is complex and has significant deviations across different regions worldwide (Zaman et al., 2023). Some regulatory frameworks focus on the final product (product-based), while others emphasize the process of creating the GE crop variety (process-based). The distinct process- or product-based approaches can influence whether a genome-edited crop is subject to the same regulations as GMOs or is treated differently. For instance, USA regulatory authorities generally do not regulate GE crops as GMOs if they do not contain foreign DNA. Other relevant agencies such as the Food and Drug Administration (FDA) and Environmental Protection Agency (EPA) may also have regulatory roles depending on the crop's intended use. In the case of the European Union, stringent regulations have been sanctioned, and they treat GE crop varieties similarly to GMOs, requiring extensive safety assessments and approvals before they can be marketed. India and Japan have adopted a more lenient approach similar to the USA, exempting GE crop varieties of SDN-1 and SDN-2 categories from GMOs regulations if they do not contain foreign genes.

Regulatory bodies assess the safety of GE crops by evaluating potential risks such as unintended genetic changes, environmental impact, and food safety. This involves rigorous testing and data submission by developers. Public perception and scientific consensus play crucial roles in shaping regulatory policies. As GE technologies such as CRISPR continue to evolve, regulatory frameworks may adapt to reflect new scientific insights and societal attitudes. The regulatory landscape is expected to evolve as more countries develop specific guidelines for GE crops. Harmonizing international regulations could facilitate global trade and innovation in agricultural biotechnology.

## 7. Conclusions

Engineering of the CRISPR/Cas system for plant genome editing has unlocked multiple avenues to design crop varieties and tackle the issues of food and energy demand arising due to climate change and the increasing human population. The rapid advancements in CRISPR/Cas-based GE technologies have revolutionized crop improvement programs by allowing precise and efficient modifications of plant genomes. These technologies offer the potential to significantly increase crop yield, enhance resistance to pests and diseases, improve abiotic stress tolerance, and fortify crops with essential nutrients, contributing to global food security. The use of GE tools holds promises for reducing the use of chemical inputs such as pesticides and fertilizers, thereby promoting more sustainable and environmentally safe agricultural practices. Despite the potential, regulatory challenges, public acceptance issues, and ethical concerns still need to be addressed before widespread adoption of GE crops can be achieved. Continued research and development in GE technologies and supportive regulatory frameworks will further accelerate crop improvement efforts. Combining GE with other advanced breeding techniques and data-driven technologies such as genomics, phenomics, and artificial intelligence will create new opportunities for more targeted and efficient crop improvement strategies.

## Acknowledgments

The support provided by the National Research Foundation of Korea (grants NRF 2021R1I1A 3057067, 2021R1A5A8029490, and 2022R1A2C3010331) and research fund for the next generation of academics of Gyeongsang National University in 2024 are acknowledged.

## References

Alfatih, A., Wu, J., Jan, S.U., Zhang, Z.S., Xia, J.Q. and Xiang, C.B. (2020). Loss of rice PARAQUAT TOLERANCE 3 confers enhanced resistance to abiotic stresses and increases grain yield in field. *Plant, Cell & Environment*, 43(11), 2743–2754.

Bhatia, S. and Yadav, S.K. (2023). CRISPR-Cas for genome editing: classification, mechanism, designing and applications. *International Journal of Biological Macromolecules*, 238, 124054.

Biswas, S., Ibarra, O., Shaphek, M., Molina-Risco, M., Faion-Molina, M., Bellinatti-Della Gracia, M., Thomson, M.J. and Septiningsih, E.M. (2023). Increasing the level of resistant starch in 'Presidio' rice through multiplex CRISPR–Cas9 gene editing of starch branching enzyme genes. *The Plant Genome*, 16(2), e20225.

Bui, T.P., Le, H., Ta, D.T., Nguyen, C.X., Le, N.T., Tran, T.T., Van Nguyen, P., Stacey, G., Stacey, M.G., Pham, N.B. and Chu, H.H. (2023). Enhancing powdery mildew resistance in soybean by targeted mutation of MLO genes using the CRISPR/Cas9 system. *BMC Plant Biology*, 23(1), 533.

Chang, L., Wu, S. and Tian, L. (2019). Effective genome editing and identification of a regiospecific gallic acid 4-O-glycosyltransferase in pomegranate (Punica granatum L.). *Horticulture Research*, 6, 123.

Elliott, K., Veley, K.M., Jensen, G., Gilbert, K.B., Norton, J., Kambic, L., Yoder, M., Weil, A., Motomura-Wages, S. and Bart, R.S. (2024). CRISPR/Cas9-generated mutations in a sugar transporter gene reduce cassava susceptibility to bacterial blight. *Plant Physiology*, 195(4), 2566.

Francini, A. and Sebastiani, L. (2019). Abiotic stress effects on performance of horticultural crops. *Horticulturae* 5, 67–70. doi: 10.3390/horticulturae5040067

Garcia-Ruiz, H., Szurek, B. and Van den Ackerveken, G. (2021). Stop helping pathogens: engineering plant susceptibility genes for durable resistance. *Current Opinion in Biotechnology*, 70, 187–195.

Gonzalez, M.N., Massa, G.A., Andersson, M., Turesson, H., Olsson, N., Falt, A.S., Storani, L., Decima Oneto, C.A., Hofvander, P. and Feingold, S.E. (2020). Reduced enzymatic browning in potato tubers by specific editing of a polyphenol oxidase gene via ribonucleoprotein complexes delivery of the CRISPR/Cas9 system. *Frontiers in Plant Science* 10, 497481.

Gupta, E., Mishra, P., Singh, P. and Mishra, N. (2024). Evidence and prospects of lycopene as powerful red superfood: modern approach to food science. *Current Traditional Medicine*, 10(3), 50–60.

Gupta, N. and Verma, V. (2019). Next-generation sequencing and its application: Empowering in public health beyond reality. *Microbial Technology for the Welfare of Society*. 17, 313–341.

Gupta, O.P. and Karkute, S.G. (2021). *Genome Editing in Plants: Principles and Applications*. CRC Press.

Hadidi, A., Flores, R., Candresse, T. and Barba, M. (2016). Next-generation sequencing and genome editing in plant virology. *Frontiers in Microbiology*, 26(7), 1325.

Han, F., Wang, P., Chen, X., Zhao, H., Zhu, Q., Song, Y., Nie, Y., Li, Y., Guo, M. and Niu, S. (2023). An ethylene-induced NAC transcription factor acts as a multiple abiotic stress responsor in conifer. *Horticulture Research*, 10(8), 130.

Heidersbach, A.J., Dorighi, K.M., Gomez, J.A., Jacobi, A.M. and Haley, B. (2023). A versatile, high-efficiency platform for CRISPR-based gene activation. *Nature Communications*, 14(1), 902.

Hu, C., Sheng, O., Deng, G., He, W., Dong, T., Yang, Q., Dou, T., Li, C., Gao, H., Liu, S. and Yi, G. (2021). CRISPR/Cas9-mediated genome editing of MaACO1 (aminocyclopropane-1-carboxylate oxidase 1) promotes the shelf life of banana fruit. *Plant Biotechnology Journal*, 19(4), 654.

Hu, Y. and Li, W. (2022). Development and application of CRISPR-Cas based tools. *Frontiers in Cell and Developmental Biology*, 10, 834646.

Ibrahim, S., Saleem, B., Rehman, N., Zafar, S.A., Naeem, M.K. and Khan, M.R. (2022). CRISPR/Cas9 mediated disruption of Inositol Pentakisphosphate 2-Kinase 1 (TaIPK1) reduces phytic acid and improves iron and zinc accumulation in wheat grains. *Journal of Advanced Research*, 37, 33–41.

Illumina, Inc. (2017). Gene editing research review. https://sapac.illumina.com/science/publication-reviews.html

Jiang, W.Z., Henry, I.M., Lynagh, P.G., Comai, L., Cahoon, E.B. and Weeks, D.P. (2017). Significant enhancement of fatty acid composition in seeds of the allohexaploid, Camelina sativa, using CRISPR/Cas9 gene editing. *Plant Biotechnology Journal*, 15(5), 648–657.

Jinek, M., Chylinski, K., Fonfara, I., Hauer, M., Doudna, J.A. and Charpentier, E. (2012). A programmable dual-RNA–guided DNA endonuclease in adaptive bacterial immunity. *Science*, 337(6096), 816–821.

Jørgensen, I.H. (1992). Discovery, characterization and exploitation of Mlo powdery mildew resistance in barley. *Euphytica*, 63, 141–152.

Karkute, S.G., Krishna, R., Ansari, W.A., Singh, B., Singh, P.M., Singh, M. and Singh, A.K. (2019). Heterologous expression of the AtDREB1A gene in tomato confers tolerance to chilling stress. *Biologia Plantarum*, 63(1), 268–277.

Karkute, S.G., Singh, A.K., Gupta, O.P., Singh, P.M. and Singh, B. (2017). CRISPR/Cas9 mediated genome engineering for improvement of horticultural crops. *Frontiers in Plant Science*, 8, 1635.

Karmakar, S., Panda, D., Panda, S., Dash, M., Saha, R., Das, P., ... and Molla, K.A. (2024). A miniature alternative to Cas9 and Cas12: transposon-associated tnpb mediates targeted genome editing in plants. *Plant Biotechnology Journal*, https://doi.org/10.1111/pbi.14416

Khan, M.S.S., Basnet, R., Islam, S.A. and Shu, Q. (2019). Mutational analysis of OsPLDα1 reveals its involvement in phytic acid biosynthesis in rice grains. *Journal of Agricultural and Food Chemistry*, 67(41), 11436–11443.

Khan, Z.A., Kumar, R. and Dasgupta, I. (2022). CRISPR/Cas-mediated resistance against viruses in plants. *International Journal of Molecular Sciences*, 23(4), 2303.

Ku, H.K. and Ha, S.H. (2020). Improving nutritional and functional quality by genome editing of crops: status and perspectives. *Frontiers in Plant Science*, 11, 577313.

Kwon, M., Hodgins, C.L., Salama, E.M., Dias, K.R., Parikh, A., Mackey, A.V., Catenza, K.F., Vederas, J.C. and Ro, D.K. (2023). New insights into natural rubber biosynthesis from rubber-deficient lettuce mutants expressing goldenrod or guayule cis-prenyltransferase. *New Phytologist*, 239(3), 1098–1111.

La Russa, M.F. and Qi, L.S. (2015). The new state of the art: Cas9 for gene activation and repression. *Molecular and Cellular Biology*, 35(22), 3800–3809.

Li, S., Lin, D., Zhang, Y., Deng, M., Chen, Y., Lv, B., Li, B., Lei, Y., Wang, Y., Zhao, L. and Liang, Y. (2022). Genome-edited powdery mildew resistance in wheat without growth penalties. *Nature*, 602(7897), 455–460.

Liu, L., Zhang, J., Xu, J., Li, Y., Guo, L., Wang, Z., Zhang, X., Zhao, B., Guo, Y.D. and Zhang, N. (2020). CRISPR/Cas9 targeted mutagenesis of SlLBD40, a lateral organ boundaries domain transcription factor, enhances drought tolerance in tomato. *Plant Science*, 301, 110683.

Lowe, N.M. (2021). The global challenge of hidden hunger: perspectives from the field. *Proceedings of the Nutrition Society*, 80(3), 283–289.

Ma, M., Sun, S., Zhu, J., Qi, X., Li, G., Hu, J., Xie, C. and Liu, C. (2024). Engineering high amylose and resistant starch in maize by CRISPR/Cas9-mediated editing of starch branching enzymes. *The Crop Journal*, 12(4), 1252–1258.

Maioli, A., Gianoglio, S., Moglia, A., Acquadro, A., Valentino, D., Milani, A.M., Prohens, J., Orzaez, D., Granell, A., Lanteri, S. and Comino, C. (2020). Simultaneous CRISPR/Cas9 editing of three PPO genes reduces fruit flesh browning in *Solanum melongena* L. *Frontiers in plant science*, 11, 607161.

Marks, R.A., Hotaling, S., Frandsen, P.B. and VanBuren, R. (2021). Representation and participation across 20 years of plant genome sequencing. *Nature Plants*, 7(12), 1571–1578.

Mushtaq, M., Dar, A.A., Basu, U., Bhat, B.A., Mir, R.A., Vats, S., ... and Wani, S.H. (2021). Integrating CRISPR-Cas and next generation sequencing in plant virology. *Frontiers in Genetics*, 12, 735489.

NCBI database. (2024). https://www.ncbi.nlm.nih.gov/genome/browse#!/eukaryotes/

Nekrasov, V., Wang, C., Win, J., Lanz, C., Weigel, D. and Kamoun, S. (2017). Rapid generation of a transgene-free powdery mildew resistant tomato by genome deletion. *Scientific Reports*, 7(1), 482.

Nonaka, S., Arai, C., Takayama, M., Matsukura, C. and Ezura, H. (2017). Efficient increase of γ-aminobutyric acid (GABA) content in tomato fruits by targeted mutagenesis. *Scientific Reports*, 7(1), 7057.

Nonaka, S., Ito, M. and Ezura, H. (2023). Targeted modification of CmACO1 by CRISPR/Cas9 extends the shelf-life of Cucumis melo var. reticulatus melon. *Frontiers in Genome Editing*, 5, 1176125.

Okuzaki, A., Ogawa, T., Koizuka, C., Kaneko, K., Inaba, M., Imamura, J. and Koizuka, N. (2018). CRISPR/Cas9-mediated genome editing of the fatty acid desaturase 2 gene in Brassica napus. *Plant Physiology and Biochemistry*, 131, 63–69.

Oliva, R., Ji, C., Atienza-Grande, G., Huguet-Tapia, J.C., Perez-Quintero, A., Li, T., Eom, J.S., Li, C., Nguyen, H., Liu, B. and Auguy, F. (2019). Broad-spectrum resistance to bacterial blight in rice using genome editing. *Nature Biotechnology*, 37(11), 1344–1350.

Ortega-Salazar, I., Crum, D., Sbodio, A.O., Sugiyama, Y., Adaskaveg, A., Wang, D., Seymour, G.B., Li, X., Wang, S.C. and Blanco-Ulate, B. (2024). Double CRISPR knockout of pectin degrading enzymes improves tomato shelf-life while ensuring fruit quality. *Plants, People, Planet*, 6(2), 330–340.

Paineau, M., Minio, A., Mestre, P., Fabre, F., Mazet, I.D., Couture, C., Legeai, F., Dumartinet, T., Cantu, D. and Delmotte, F. (2024). Multiple deletions of candidate effector genes lead to the breakdown of partial grapevine resistance to downy mildew. *New Phytologist*, 243(4), 1490–1505.

Podevin, N., Davies, H.V., Hartung, F., Nogué, F. and Casacuberta, J.M. (2013) Site-directed nucleases: a paradigm shift in predictable, knowledge-based plant breeding. *Trends in Biotechnology*, 31, 375–383.

Pramanik, D., Shelake, R.M., Kim, M.J. and Kim, J.-Y. (2021a). CRISPR-mediated engineering across the central dogma in plant biology for basic research and crop improvement. *Molecular Plant*, 14, 127–150.

Pramanik, D., Shelake, R.M., Park, J., Kim, M.J., Hwang, I., Park, Y. and Kim, J.Y. (2021b). CRISPR/Cas9-mediated generation of pathogen-resistant tomato against tomato yellow leaf curl virus and powdery mildew. *International Journal of Molecular Sciences*, 22(4), 1878.

Rajpal, V.R., Sehgal, D., Valluru, R. and Singh, S. (2023). Current advances in genomics and gene editing tools for crop improvement in a changing climate scenario. *Frontiers in Genetics*, 14, 1214679.

Santosh Kumar, V.V., Verma, R.K., Yadav, S.K., Yadav, P., Watts, A., Rao, M.V. and Chinnusamy, V. (2020). CRISPR-Cas9 mediated genome editing of drought and salt tolerance (OsDST) gene in indica mega rice cultivar MTU1010. *Physiology and Molecular Biology of Plants*, 26, 1099–1110.

Sashidhar, N., Harloff, H.J., Potgieter, L. and Jung, C. (2020). Gene editing of three BnITPK genes in tetraploid oilseed rape leads to significant reduction of phytic acid in seeds. *Plant Biotechnology Journal*, 18(11), 2241–2250.

Satam, H., Joshi, K., Mangrolia, U., Waghoo, S., Zaidi, G., Rawool, S., ... and Malonia, S.K. (2023). Next-generation sequencing technology: current trends and advancements. *Biology*, 12(7), 997.

Shameer, K., Naika, M.B., Shafi, K.M. and Sowdhamini, R. (2019). Decoding systems biology of plant stress for sustainable agriculture development and optimized food production. *Progress in Biophysics and Molecular Biology*, 145,19–39.

Shelake, R.M., Jadhav, A.M., Bhosale, P.B. and Kim, J.Y. (2023a). Unlocking secrets of nature's chemists: Potential of CRISPR/Cas-based tools in plant metabolic engineering for customized nutraceutical and medicinal profiles. *Plant Physiology and Biochemistry*, 108070.

Shelake, R.M., Kadam, U.S., Kumar, R., Pramanik, D., Singh, A.K. and Kim, J.Y. (2022). Engineering drought and salinity tolerance traits in crops through CRISPR-mediated genome editing: Targets, tools, challenges, and perspectives. *Plant Communications*, 3(6).

Shelake, R.M., Pramanik, D. and Kim, J.Y. (2024). CRISPR base editor-based targeted random mutagenesis (BE-TRM) toolbox for directed evolution. *BMB Reports*, 57(1), 30.

Shelake, R.M., Pramanik, D., Waghunde, R.R. and Kim, J.Y. (2023b). Plant Mutagenesis Tools for Precision Breeding: Conventional CRISPR/Cas9 Tools and Beyond. In: Penna, S., Jain, S.M. (eds.). *Mutation Breeding for Sustainable Food Production and Climate Resilience*. Springer, Singapore.

Shnaider, Y., Elad, Y., Rav-David, D., Pashkovsky, E., Leibman, D., Kravchik, M., Shtarkman-Cohen, M., Gal-On, A. and Spiegelman, Z. (2023). Development of powdery mildew resistance in cucumber using CRISPR/Cas9-mediated mutagenesis of CsaMLO8. *Phytopathology®*, 113(5), 786–790.

Skendžić, S., Zovko, M., Živković, I.P., Lešić, V. and Lemić, D. (2021). The impact of climate change on agricultural insect pests. *Insects*, 12(5), 440.

Song, J.H., Shin, G., Kim, H.J., Lee, S.B., Moon, J.Y., Jeong, J.C., Choi, H.K., Kim, I.A., Song, H.J., Kim, C.Y. and Chung, Y. S. (2022). Mutation of GmIPK1 gene using CRISPR/Cas9 reduced phytic acid content in soybean seeds. *International Journal of Molecular Sciences*, 23(18), 10583.

Thomazella, D.P.D.T., Seong, K., Mackelprang, R., Dahlbeck, D., Geng, Y., Gill, U.S., ... and Staskawicz, B. (2021). Loss of function of a DMR6 ortholog in tomato confers broad-spectrum disease resistance. *Proceedings of the National Academy of Sciences*, *118*(27), e2026152118.

United Nations. (2019). World Population Prospects: The 2019 Revision. New York: United Nations.

Van Esse, H.P., Reuber, T.L. and van der Does, D. (2020). Genetic modification to improve disease resistance in crops. *New Phytologist*, 225(1), 70–86.

Waltz, E. (2016). Gene-edited CRISPR mushroom escapes US regulation. *Nature,* 532, 293.

Wan, D.Y., Guo, Y., Cheng, Y., Hu, Y., Xiao, S., Wang, Y. and Wen, Y.Q. (2020). CRISPR/Cas9-mediated mutagenesis of VvMLO3 results in enhanced resistance to powdery mildew in grapevine (Vitis vinifera). *Horticulture Research*, 7.

Wang, H., Wu, Y., Zhang, Y., Yang, J., Fan, W., Zhang, H., Zhao, S., Yuan, L. and Zhang, P. (2019). CRISPR/Cas9-based mutagenesis of starch biosynthetic genes in sweet potato (Ipomoea Batatas) for the improvement of starch quality. *International Journal of Molecular Sciences*, 20(19), 4702.

Wang, J.Y. and Doudna, J.A. (2023). CRISPR technology: a decade of genome editing is only the beginning. *Science*, 379, eadd8643.

Wing, I.S., De Cian, E. and Mistry, M.N. (2021). Global vulnerability of crop yields to climate change. *Journal of Environmental Economics and Management*, 109, 102462.

Wold-McGimsey, F., Krosch, C., Alarcón-Reverte, R., Ravet, K., Katz, A., Stromberger, J., Mason, R.E. and Pearce, S. (2023). Multi-target genome editing reduces polyphenol oxidase activity in wheat (Triticum aestivum L.) grains. *Frontiers in Plant Science*, 14, 1247680.

Xie, L., Gong, X., Yang, K., Huang, Y., Zhang, S., Shen, L., ... and Fan, L. (2024). Technology-enabled great leap in deciphering plant genomes. *Nature Plants*, 10(4), 551–566.

Yamaguchi-Shinozaki, K. and Shinozaki, K. (1994). A novel cis-acting element in an Arabidopsis gene is involved in responsiveness to drought, low-temperature, or high-salt stress. *The Plant Cell*, 6(2), 251–264.

Yang, Q., Ral, J.P., Wei, Y., Zheng, Y., Li, Z. and Jiang, Q. (2024). Genome editing of five starch synthesis genes produces highly resistant starch and dietary fibre in barley grains. *Plant Biotechnology Journal*, 22(7), 2051.

Yu, W., Wang, L., Zhao, R., Sheng, J., Zhang, S., Li, R. and Shen, L. (2019). Knockout of SlMAPK3 enhances tolerance to heat stress involving ROS homeostasis in tomato plants. *BMC Plant Biology*, 19, 1–13.

Yuan, M., Zhu, J., Gong, L., He, L., Lee, C., Han, S., Chen, C. and He, G. (2019). Mutagenesis of FAD2 genes in peanut with CRISPR/Cas9 based gene editing. *BMC Biotechnology*, 19, 1–7.

Zafar, K., Khan, M.Z., Amin, I., Mukhtar, Z., Yasmin, S., Arif, M., Ejaz, K. and Mansoor, S. (2020). Precise CRISPR-Cas9 mediated genome editing in super basmati rice for resistance against bacterial blight by targeting the major susceptibility gene. *Frontiers in Plant Science*, 11, 575.

Zaman, Q.U., Raza, A., Lozano-Juste, J., Chao, L., Jones, M.G., Wang, H.F. and Varshney, R.K. (2023). Engineering plants using diverse CRISPR-associated proteins and deregulation of genome-edited crops. *Trends in Biotechnology*, 42, 560–574.

Zeng, Y., Wen, J., Zhao, W., Wang, Q. and Huang, W. (2020). Rational improvement of rice yield and cold tolerance by editing the three genes OsPIN5b, GS3, and OsMYB30 with the CRISPR–Cas9 system. *Frontiers in Plant Science*, 10, 1663.

Zhang, P., Du, H., Wang, J., Pu, Y., Yang, C., Yan, R., Yang, H., Cheng, H. and Yu, D. (2020). Multiplex CRISPR/Cas9-mediated metabolic engineering increases soya bean isoflavone content and resistance to soya bean mosaic virus. *Plant Biotechnology Journal*, 18(6), 1384–1395.

Zhang, Y., Massel, K., Godwin, I.D. and Gao, C. (2018). Applications and potential of genome editing in crop improvement. *Genome Biology*, 19(1), 210.

Zhao, X., Jayarathna, S., Turesson, H., Fält, A.S., Nestor, G., González, M.N., Olsson, N., Beganovic, M., Hofvander, P., Andersson, R. and Andersson, M. (2021). Amylose starch with no detectable branching developed through DNA-free CRISPR-Cas9 mediated mutagenesis of two starch branching enzymes in potato. *Scientific Reports*, 11(1), 4311.

# CHAPTER 3

# Accelerated Breeding Approaches for Enabling Sustainable Rice Farming

Ashvinkumar Katral[1], Anilkumar C.[2,5*] Lokeshkumar B.M.[3], Jeevan B.[2], Sunitha N.C.[2,6], Basavaraj P.S.[4], Basana Gowda G.[2,7], Raghu S.[2], Shridhar Ragi[1], Vishvjeet S. Patel[3] and Rameswar Prasad Sah[2]

## 1. Introduction

Rice (*Oryza sativa* L.) is one of the most vital staple crops worldwide, providing sustenance for over half of the global population. It serves as a primary source of nutrition and income for people in many Asian and African countries (Khush and Jena, 2009; Basavaraj et al., 2021). The increase in global population, coupled with diminishing agricultural land and other resources, imposes more demand on rice production (Patra et al., 2020). It was predicted that rice production should be enhanced to 771 million tonnes to meet the demand of the rice-consuming population in 2030 (Khush and Jena, 2009). Ensuring food security through rice production is becoming increasingly important as the world's population increases and environmental problems worsen. Meeting the target of rice production is possible only when the production constraints are mitigated by adopting modern breeding and production technologies (Chakraborti et al., 2021). With the efforts of traditional breeding approaches, rice yields were stabilized with improving high-yielding varieties in many rice-growing parts of the world. Of late, erratic changes in climate have played a major role in the resurgence of many pests and diseases that infest rice plants leading to unpredictable and unprecedented yield declines (Hussain et al., 2020). Furthermore, the change in climate is challenging rice crops with sudden, unpredicted drought conditions or flash floods. These climatic factors have a major impact on sustainable rice improvement and production. Therefore, exploring new breeding approaches with the use of modern genomic tools will help improve rice for sustainable production.

Pre-genomic or traditional breeding methods contributed a significant number of improved rice varieties with high yielding ability in the past century. Traditional breeding techniques,


1 ICAR-Indian Agricultural Research Institute, New Delhi, India.
2 ICAR-Central Rice Research Institute, Cuttack, India.
3 ICAR-Central Soil Salinity Research Institute, Karnal, India.
4 ICAR-National Institute of Abiotic Stress Management, Baramati, India.
5 Department of Agronomy and Plant Genetics, University of Minnesota, MN, USA.
6 University of Agricultural Sciences, Bengaluru, India.
7 ICAR-National Research Center for Pomegranate, Solapur, India.
* Corresponding author: anilcgpb@gmail.com


DOI: 10.1201/9781003584100-3

including hybridization, selection, and acclimation of desired characters, have historically been used to improve rice (Bradshaw, 2017). This process is long and frequently imprecise because of intricate genetic and environmental interactions. Traditional selection techniques have given way to sophisticated molecular and biotechnological strategies in rice breeding over time. However, advances in genomics offer a variety of tools and methods to accelerate rice breeding by combining genetic determinants for many complex traits (Katara et al., 2021). The genomic-based approaches, including molecular markers, genome or gene editing tools, genome-wide prediction, and selection approaches, and integrating data science with breeding methods reduce the difficulties in breeding for genetically complex traits (Kumar et al., 2024). Taking advantage of genomic tools and data science, the focus of rice breeding is being transitioned from breeding for high yield or resistance to a specific disease pest toward improving rice varieties for sustainable production. Of late, rice breeding objectives are oriented to develop multiple stress-tolerant, high-yielding, and nutrient-dense rice varieties.

Rice, being researchers' favorite crop, has vast genomic resources developed over decades. Rice has a relatively small genome, and the availability of exhaustive genomic resources helps breeders to utilize in adopting new breeding methods (Brar and Khush, 2018). Early development of DNA-based markers opened new path for dissecting the genetics of many complex characters and utilizing them for breeding through approaches such as marker-assisted selection (MAS). Similarly, some of the key genomic resources, including the availability of whole genome sequences and databases, such as the Rice Annotation Project (RAPdb) (https://rapdb.dna.affrc.go.jp/), the Rice SNP-Seek Database (https://snp-seek.irri.org/), and the Rice Expression Profile Database (RiceXPro) (https://ricexpro.dna.affrc.go.jp/) for retrieving genomic information, impacted positively on precision rice breeding goals. Some of the genomic resources helped in genetic understanding of complex characters before breeding, and some directly helped in updating breeding methodologies to improve genetic gains for complex traits. A quick understanding of the role of various genomic resources in rice helps strategic planning of rice-breeding activities.

## 2. Molecular Markers and Marker-Aided Rice Breeding

In contemporary rice breeding, molecular markers have emerged as crucial tools that allow for a more accurate and effective selection of desired traits, especially those that are governed by intricate genetic factors or challenging to evaluate using conventional phenotyping techniques. Different DNA markers, starting from restriction fragment length polymorphic (RFLP) markers to recent single-nucleotide polymorphism (SNP) markers, have been utilized extensively in rice breeding. These molecular markers were utilized in identification of quantitative trait loci (QTL) regulating complex traits. In general, QTL in rice for complex traits are located through recombination-based linkage mapping and/or association mapping. Genome-wide association study (GWAS) is also an approach that helps identify major and minor QTL controlling the target trait. Some of the major achievements in QTL identification in rice include the identification of the *Sub1* locus for submergence tolerance (Xu and Mackill, 1996), several loci for blast resistance (Fang et al., 2016, Devanna et al., 2022), QTL for grain weight (Anilkumar et al., 2022a), yield-related QTL (Bai et al., 2012), and for many other traits.

Some of the QTLs identified with high confidence and high phenotypic variation explained were exploited in rice breeding programs using a marker-assisted selection approach where QTL-associated markers were used as surrogates to select the trait. MAS is believed to be more precise than phenotype-based selection owing to the climate and crop stage neutrality of DNA markers. Some of the characteristics improved through MAS in rice breeding are highlighted in Table 1. MAS helps introgression of important genomic regions into the elite background through a modified approach called marker-assisted backcross breeding (MABB), where the backcross progeny with the target genomic region is identified in the early crop stage and only positive progenies will be used in further backcrossing. Similarly, MAS can also help select progeny in the early crop stage, in a forward breeding program. MAS in rice breeding has been particularly efficient in improving abiotic and biotic stresses. MAS helps improve selection efficiency by eliminating the environmental influence and enhancing the precision of selection. While breeding for improved cultivar development, MAS helps in improving genetic gains for the target trait. Even though the MAS is an effective approach, its effectiveness for complex traits is comparatively low as large number of genes control these traits. Linkage drag also plays a negative role in the success of the MAS in rice breeding.

**Table 1** Examples of improvement of traits through MAS.

| *S.No* | *Trait* | *Genes/QTL involved* | *Objective* | *Reference* |
|---|---|---|---|---|
| 01 | Rice blast | *Pi-1, Pi 2, Pi 4(t), Pi 5, Piz-5, Pi 5(t), Pi 7(t), Pi 10(t), Pi-b, Pi54, Pi21, Pia* | Resistance to *Magnaporthe oryzae* | Cho et al. (2007); Singh et al. (2011a) |
| 02 | Bacterial blight | *Xa21, xa13, Xa10, xa5, Xa4, Xa3, Xa1, Xa7* | Resistance to *Xanthomonas oryzae* | Ma et al. (1999); Porter et al. (2003); Shin et al. (2007); Singh et al. (2011b) |
| 03 | Rice sheath blight | *qShB1, qShB2-1, qSB5, qShB6, qShB9-2* | Resistance to *Rhizoctonia solani* | Liu et al. (2009) |
| 04 | Submergence | *Sub1* | Tolerance to submergence | Das and Rao (2015) |
| 05 | Salinity | *Saltol* | Tolerance to salinity | Das and Rao (2015) |
| 06 | Brown plant hopper | *Bph-1, Bph-10(t), Bph-3, Bph-17, Bph-18, Bph-20, Bph21, Bph25, Bph26* | Resistance to *Nilaparvata lugens* | Singh et al. (2011b) |
| 07 | Cold | *qPSST-3, qPSST-7, qPSST-9, qSCT1a, qSCT2* | Cold tolerance | Jena et al. (2010); Kim et al. (2014) |

## 3. Genomewide Selection for Rice Breeding

The MAS approach in rice breeding has primarily been effective in improving oligogenic traits or QTLs with major effects. However, its application for improving quantitative traits controlled by multiple minor-effect genes located at unknown positions across the rice genome has been limited. This restriction hindered progress in enhancing yield-related traits, resulting in stagnated genetic gains (Muralidharan et al., 2019). Over the past three decades, the advancement of genomic selection (GS) in various crops has demonstrated its potential to improve genetic gains for quantitative traits. Genomic selection is similar to MAS, but instead of relying on QTL flanking markers, it uses marker information from all over the genome. Inspired by its success in livestock breeding, plant breeders have increasingly adopted genomic selection for crop improvement (Bernardo, 1994; Meuwissen et al., 2001; Smith et al., 2008). The GS approach consists of two populations: a training population, which is precisely phenotyped for the target quantitative trait and assayed with markers distributed on all chromosomes, and a testing population, which is genotyped but not evaluated for the target trait. The relationship between these populations is vital for the success of genomic selection. Additional factors such as trait heritability, marker density, and prediction models also play significant roles in determining the accuracy of genomic selection predictions (Anilkumar et al., 2022b).

Genomic resources in rice have been developed over several decades, making them highly suitable for GS breeding. The abundance availability of molecular markers, extensive genomic information, affordable genotyping facilities, and the relatively small genome size of rice all contribute to its potential for effective genomic selection. Despite these advantages, GS has not been as widely adopted in rice breeding as it has been in other crops such as maize and wheat (Anilkumar et al., 2022c). Recently, however, efforts have been made to integrate genomic selection with other breeding methods to accelerate genetic gains in rice. Table 2 provides recent examples where GS has been applied in rice breeding. Although many of these GS studies were focused on optimizing methodology and improving prediction accuracy, they have nonetheless laid the foundation for the broader implementation of genomic selection in rice breeding.

**Table 2** Recent examples of genomic selection used in rice improvement.

| *S.No.* | *Trait* | *Populations* | *Statistical models* | *Reference* |
|---|---|---|---|---|
| 01 | Grain yield and plant height | Diversity panel (280 individuals) | GBLUP<br>RKHS | Bhandari et al. (2019) |
| 02 | Arsenic content in flag leaf | A diversity panel (225 individuals) and elite lines (95) | GBLUP<br>RKHS | Frouin et al. (2019) |
| 03 | Grain yield and contributing traits | 1495 rice hybrids | GBLUP | Yanru et al. (2020) |
| 04 | Grain yield and grain number | Recombinant inbred lines (210) | 2D BLUP | Wang et al. (2021) |
| 05 | Grain yield | 10,716 single-cross hybrids | GBLUP | Labroo et al. (2021) |
| 06 | Agronomic traits | Germplasm accessions (469) | rrBLUP<br>GBLUP<br>RKHS | Zhang et al. (2023) |
| 07 | Grain weight | Elite breeding lines, germplasm accession (189 accessions) | GBLUP<br>RKHS<br>RFR | Anilkumar et al. (2023a) |
| 08 | Grain yield and plant height | Elite breeding lines (111 lines) | GBLUP | Nguyen et al. (2023) |
| 09 | Grain yield | Breeding lines (1445 lines) | rrBLUP | Biswas et al. (2023) |
| 10 | Salinity tolerance | Reference panel of *Oryza sativa* cv *japonica* (241 lines) | GBLUP<br>RKHS | Bartholomé et al. (2023) |

GBLUP: genomic best linear unbiased prediction; rrBLUP: ridge regression best linear unbiased prediction; RFR: random forest regression; RKHS: reproducing kernel Hilbert space regression; 2D BLUP: two-dimensional best linear unbiased predictor.

The GS approach can be implemented for the improvement of inbred rice varieties through a pure-line selection-based program where potential lines are chosen based on predicted genetic values or additive genetic effects estimated based on marker information. Best performing elite breeding lines predicted through a genomic selection approach may be evaluated before nominating for national testing. Currently, most studies focus on the optimization of populations and markers for implementing GS with high prediction accuracy in rice breeding. Spindel et al. (2015) reported predictive abilities of 0.63 and 0.31 for flowering time and grain yield, respectively, in a panel of 363 elite breeding lines. Similarly, Hassen et al. (2018) reported the predictive ability of 0.35 and 0.38 for flowering date and 100 panicle weight, respectively, in the biparental population. On the other hand, genomic selection can also be used for the development of hybrids. The performance of hybrid combinations can be predicted using an appropriate prediction model and marker information of inbred parental lines (Yanru et al., 2020). A large number of hybrid performances can be predicted, and the best hybrid can be selected based on predicted performance. Furthermore, multi-trait GS models have demonstrated their effectiveness in predicting the best line or hybrid, considering the information from correlated traits. The predictive abilities of primary traits are reported to increase upon considering correlated trait information in prediction (Wang et al., 2021; Anilkumar et al., 2025). To optimize the breeding efficiency and achieve accelerated genetic improvement in rice breeding, it is important to integrate genomic selection with other modern technologies, including data science and high-throughput plant phenotyping. Capturing environmental variables as factors of prediction can play an important role in understanding the influence of the environment on plant phenotype expression and identifying the best candidate as a variety or hybrid for commercial release.

## 4. Speed Breeding in Rice

The traditional methods of rice varietal breeding programs are time-consuming. Approximately, the varietal development process in rice takes around 8–9 years for breeding and 3 or more years for the dissemination of the variety. Considering the rapid climate change, the products of breeding programs are becoming obsolete before reaching the target stakeholder, farmers. Therefore, it is important to accelerate the breeding programs to the demands of changing climate. The most important opportunity the breeder has to accelerate the breeding program is through reducing the breeding cycle duration (Sah et al., 2021). To reach this target, maize breeders use double haploid technology (Thomas et al., 2003), where homozygosity is achieved through chromosomal doubling in the first filial generation through colchicine treatment under laboratory conditions. However, the doubled haploid technology in rice is still in a nascent stage and not available at the commercial scale. The use of molecular markers, genetic engineering tools (Araki and Ishii, 2015), and growing in off-season nurseries following the shuttle breeding approach are some approaches in rice to speed up breeding activities. However, these methods have their own limitations. Alternatively, accelerated generation advancement is possible through modification of the growing environment. The approach involves the use of artificial light and nutritional conditions to encourage plants to accelerate their growth and flowering in a shorter time than their regular flowering duration (Ghosh et al., 2018; Alahmad et al., 2018). This process of inducing plants to flower and complete the life cycle within a short time is called speed breeding (Watson et al., 2018). Speed breeding accelerates the physiological development of a plant and helps to complete several generations within a short period (Sysoeva et al., 2010). This process helps the breeder by speeding up the process of completing each generation in a breeding cycle.

Rice is a short-day plant that shows a positive response to variations in light duration and intensity during its early growth and maturation phases. However, the application of speed breeding in rice has been restricted until recently. In recent years, speed breeding techniques have been effectively integrated into rice breeding methods to accelerate the breeding process. On a global scale, the International Rice Research Institute (IRRI) has established speed breeding centers at its main stations in the Philippines and India to improve the efficacy of breeding projects. When combined with molecular markers, speed breeding ensures both precision and rapid progress in the breeding process. Marker-assisted backcross breeding, along with speed breeding, made the trait introgression process faster and more precise for many quantitative traits in rice (Gregorio et al., 2013; Sandhu and Kumar, 2017; Pradhan et al., 2019). Similarly, the efficient application of genomic selection along with the speed breeding approach reduces the cost of the breeding program and saves the time required for breeding product development (Hickey et al., 2019). Rana et al. (2019) accelerated the generation cycle and increased the rate of genetic gain in rice breeding by combining speed breeding and genomic selection. According to Heffner et al. (2010), genomic selection enables precise selection for complex traits such as grain yield and shortens the breeding cycle, both of which enhance the rate of genetic gain per unit time. By using a multivariate genomic selection model and comparing speed breeding features with field-evaluated traits, Watson et al. (2019) demonstrated that speed breeding resulted in greater selection intensity and genetic gain compared to direct field phenotyping. Therefore, the ideas behind speed breeding, its uses, and its contribution to GS for enhancing genetic gain in rice breeding programs prompted breeders to incorporate this technique as a novel approach.

# 5. Genome Editing in Rice Breeding

Genome editing is a precise approach that modifies the phenotype of a targeted trait by directly altering the DNA bases of the gene responsible for it. Recently, CRISPR/Cas9 (clustered regularly interspaced short palindromic repeats/CRISPR-associated protein 9) and its variants have become widely used for gene editing in rice and other crops. The invention of base editors has further enhanced the precision of genome editing, making it easier to edit genes in various crops. These tools help breeders create genetic variation for specific traits. Genome editing is particularly effective for modifying oligogenic traits, where a small number of genes significantly impact trait expression. As a result, genome editing is frequently employed to alter genes associated with oligogenic traits such as quality and disease tolerance in rice.

Precise information on thousands of gene positions and sequences is publicly available for rice. Important yield-contributing genes such as *Gn1a*, *GS3*, *DEP1*, and *IPA* were more extensively used in demonstrating genome editing in rice (Li et al., 2016). A list of genes targeted in editing and improvement of rice can be seen in Mishra et al. (2018). Editing of grain-weight genes such as *GW2*, *GW5*, and *TGW6* using the CRISPR/Cas9 system increases grain weight by 29% (Xu et al., 2016). In the nutritional rice improvement, CRISPR/Cas9 technology was successfully used to edit *SBEI* and *SBEIIb* genes which led to the development of high amylose rice variants (Sun et al., 2017). Similarly, the metal transporter gene *OsNramp5* was edited to develop rice variants with low cadmium accumulation (Tang et al., 2017).

Genome editing was found promising in editing genes responsible for biotic stress tolerance in rice. The sucrose transporter gene, *SWEET14*, has been edited to create a new rice variant tolerant against bacterial blight disease (Li et al., 2012). It has been shown that targeted deletion of the ethylene response factor (ERF) transcription factor gene *OsERF922* using the CRISPR system increases resistance to rice blast (Wang et al., 2016a). Similarly, the genes identified and characterized for abiotic stress tolerance are also being considered for editing. However, reports are very limited in the case of abiotic stress tolerance. A transcription factor known to impart cold tolerance, *TIFY1b*, and its homologous gene, *TIFY1a*, were edited to develop a cold-tolerant rice line (Huang et al., 2017). Several variants created through genome editing have been utilized as new sources of variation in rice breeding. Overall, genome editing techniques hold a greater significance in rice research and improvement.

# 6. Breeding for Climatic Resilience

## *6.1 Genetic Improvement for Biotic Stress Tolerance*

Rice production can be impacted by several factors, but diseases such as blast, sheath blight, brown spot, bakanae, bacterial blight, and false smut are the most significant ones (Jeevan et al., 2023; Li et al., 2023; Patil et al., 2024). Depending on their severity, these diseases can result in yield losses ranging from 20% to 100% (Shivappa et al., 2022). Chemical fungicides have helped manage the disease, but they come at a high cost (Panda et al., 2017), are unsuccessful in high-disease-pressure situations, and may even contribute to pathogen resistance (Yamaguchi, 2004). Therefore, the use of host resistance (R genes) in developing resistant rice cultivars is the most efficient and affordable way of rice disease management (Jeevan et al., 2020).

A key step in successfully introducing R genes into resistance breeding programs is identifying efficient sources of resistance and conducting fine-mapping of R genes within the rice genome (Tan et al., 2022). Rice blast resistance is either mediated by R genes or regulated by multiple genes as a quantitative trait. Since quantitative resistance is race-neutral and provides long-lasting protection against the blast pathogen, mapping the genomic regions responsible for this resistance has gained increasing importance (Lopez-Gerena, 2006). Since the first rice blast genetic study in 1922, 118 R genes have been mapped, with 35 genes cloned and characterized for their role in rice blast resistance

(Wang and Valent, 2017; Kalia and Rathour, 2019). However, only a few R genes, including *Pi25(t)* (Zhuang et al., 2002), *Pb1* (Hayashi et al., 2010), *Pi64* (Ma et al., 2015), *Pi-jnw1* (Wang et al., 2016b), and *Pi68(t)* (Devi et al., 2020), as well as 14 QTLs (Kalia and Rathour, 2019), have been identified for neck blast resistance. Most of the cloned blast R genes, such as *Pi37*, *Pit*, *Pi-sh*, *Pi64*, *Pi-b*, *Pi63*, *Pi9*, *Piz-t*, *Pid3*, *Pigm*, *Pi25*, *Pi36*, *Pi5*, *Pi-54*, *Pik-m*, *Pik*, *Pik-p*, *Pik-e*, *Pi-a*, *Pi1*, and *Pita*, share common nucleotide-binding site (NBS) and leucine-rich repeat (LRR) domains (Kalia and Rathour, 2019). In contrast, *Pitr* encodes a putative E3 ligase with four armadillo repeats (Zhao et al., 2018). The protein *Pi21* is a heavy-metal-binding protein with proline-rich domains (Fukuoka et al., 2009), while the R gene *pid2* encodes a serine/threonine kinase membrane-spanning protein (Chen et al., 2006). Most identified R genes are located on chromosomes 6, 11, and 12, with cloned R genes found specifically on chromosomes 6 and 11 (Devanna et al., 2022). Similarly, the majority of QTLs are present in chromosomes 6, 11, and 12, containing 21, 27, and 27 genes, respectively. The fewest R genes are located on chromosomes 3 and 7 (Kalia and Rathour, 2019; Tan et al., 2022).

Since the first publication of QTL mapping for sheath blight resistance using the RFLP marker, various genetic approaches and mapping populations have been developed and employed in sheath-blight-resistant breeding programs (Li et al., 1995). To date, over 200 QTLs have been identified using different mapping populations, including $F_2$s (Zuo et al., 2014), recombinant inbred lines (RILs) (Liu et al., 2009), backcross inbred lines (BILs) (Taguchi-Shiobara et al., 2013), doubled haploids (DH) (Liu et al., 2009; Xu et al., 2011), and chromosomal segment substitution lines (CSSLs) (Zuo et al., 2014). Most of the identified QTLs are located on chromosomes 2, 3, and 9, with each chromosome having more than 30 QTLs. These are followed by chromosomes 7, 1, and 11, which have 26, 24, and 21 QTLs, respectively. The fewest QTLs are found on chromosomes 4 and 10, with 10 and 2 QTLs, respectively (Zarbafi and Ham, 2019).

Rice plants are more vulnerable to brown spot disease during the booting and flowering stages as well as the seedling stage (Chakrabarti, 2001). The first investigation to indicate varietal differences in brown spot resistance levels was conducted in 1930 when field observation revealed that "Kutto-urupe," a Korean cultivar, was more sensitive than other cultivars under natural infection (Nagai and Hara, 1930). In the very first study to identify QTLs for rice brown spot resistance, populations of RILs were developed by crossing "Tadukan" (resistant) with "Hinohikari" (susceptible) rice cultivars. Three QTLs including *qBS2*, *qBS9*, and *qBS11* that show resistance in seedlings are located on chromosomes 2, 9, and 11, respectively (Sato et al., 2008). To confirm the impact of these QTLs, the parents and advanced generations of the RIL populations were evaluated for brown spot resistance in the field conditions. Subsequently, three additional QTLs that confer resistance at the field level were found to be *qBSfR1*, *qBSfR4*, and *qBSfR11* (Sato et al., 2015). The QTL, *qBSfR11*, was found to coincide with *qBS11*, which was identified for resistance at the seedling stage. As a result, the resistance associated with *qBSfR11* (*qBS11*) remained intact irrespective of the growth stages (Sato et al., 2015). Eleven QTLs were identified using DH lines obtained from a cross involving the resistant "CT 9993-5-10-1-M" and the susceptible "IR 62266-42-6-2." Four of these QTLs (*BSq2.1v&i*, *BSq2.2v&i*, *BSq4.1v&i*, and *BSq11.1v&i*) were identified in trials conducted on both vertisol and inceptisol soil types (Katara et al., 2010).

Bacterial blight, a severe disease in rice caused by *Xanthomonas oryzae* pv. *oryzae* (*Xoo*), is widespread in many rice-producing regions globally and can lead to yield losses of up to 50% (Singh et al., 2011b). Like other traits in rice, this disease is reported to be controlled by many genes. As of now, approximately 46 resistance (R) genes that provide resistance to various strains of *Xoo* have been identified from diverse rice sources. While many of these genes have been linked with molecular markers, 11 have been cloned and characterized, showcasing different strategies of R-gene-mediated resistance against *Xoo* (Kumar et al., 2020a; Wang et al., 2020). The dominant cloned R genes include *Xa1*, *Xa3*/*Xa26*, *Xa4*, *Xa10*, *Xa21*, *Xa23*, and *Xa27(t)*, while the recessive genes include *xa5*, *xa13*, *xa25(t)*, and *xa41(t)* (Fiyaz et al., 2022). Among the markers, Lj138 and A83B4 on chromosome 11 were used to identify the *Xa23* gene from the wild rice species

*Oryza rufipogon* (Zhang et al., 1998). A recessive R gene, *xa25(t)*, was discovered in Minghui 63 and provides resistance during both the seedling and maturity stages (Lee et al., 2003). Another recessive gene, *xa26(t)*, was also identified from Minghui 63 (Lee et al., 2003). In rice breeding programs, *Xa21*, a member of the receptor kinase (RLK) family with broad-spectrum resistance, has been widely used to confer durable resistance against bacterial blight (Sridhar et al., 2001). The Chinese landrace "Zhachanglong" carries *Xa22(t)*, a gene that provides strong resistance to 16 different *Xoo* strains (Lin et al., 1996). Additionally, genes such as *Xa1*, *Xa4*, *Xa5*, *Xa21*, *Xa26*, and *Xa27* have been frequently employed in pyramiding strategies for various susceptible rice cultivars (Fiyaz et al., 2022). The identification of multiple genes controlling rice resistance to bacterial blight has significantly advanced rice breeding, particularly through marker-aided breeding for gene pyramiding. Consequently, resistance to bacterial blight has been enhanced in more than 70 rice varieties and parental lines of hybrid rice, either individually or through the combination of genes and QTLs.

The bakanae disease has increased in prevalence over the past few years in most of the world's rice-growing regions (Raghu et al., 2023). Through extensive screening of rice accessions, researchers worldwide have made significant progress toward identifying resistant sources. Further, these resistant sources have been used in the breeding program for mapping QTLs for bakanae resistance (Lee et al., 2022; Chen et al., 2023; Kushwaha et al., 2023). To date, around 29 QTLs that confer resistance to bakanae across 12 chromosomes have been mapped using linkage mapping and GWAS analysis. Thirteen QTLs including *qBK1*, *qBK1.1*, *qBK1Z*, *qBK1.2*, *qBK1.3*, *qBK1WD*, *qFfR1*, *qBK1*, *qB1*, *qBK1.4*, *qBK1.5*, *qBK1.6*, and *qBK1.7* have been reported on chromosome 1. Other QTLs reported are *qBK3.1* on chromosome 3, *qBK4_31750955* and *qBK4T* on chromosome 4, *qBK4_628091* on chromosome 6, *qFfR9* on chromosome 9, and *qB10* on chromosome 10 (Yang et al., 2006; Fiyaz et al., 2016; Kang et al., 2019; Lee et al., 2022). One main effect QTL, *qBK5.1*, was identified on chromosome 5 in non-basmati rice (Khan et al., 2024). The only QTL that has been fine-mapped to date is *qBK1*, which is located in a 35 kb genomic area and is flanked by the InDel makers InDel 18 and InDel 19–14 (Lee et al., 2019). Four putative candidate R genes, *LOC_Os01g41770*, *LOC_Os01g4180*, *LOC_Os01g41790*, and *LOC_Os01g41800*, have been found to confer resistance to bakanae disease, according to a transcriptome study. The molecular mechanisms controlling resistance to bakanae disease remain unclear. However, transcriptome analyses have revealed that the resistant Selenio variety contains genes involved in the jasmonic acid biosynthesis pathway (Matić et al., 2016). The detection of numerous genes and QTLs-linked biotic stress tolerance has enabled breeders to incorporate them into marker-aided breeding and GS strategies, aiding the improvement of sustainable rice varieties.

## 6.2 Genetic Improvement for Abiotic Stress Tolerance

Climate change has led to more frequent and intense hot and dry spells (Gornall et al., 2010). By the end of 21st century, global average temperatures are projected to increase by 7°C (Kalkuhl and Wenz, 2020), which will significantly impact the productivity of key crops such as rice and disrupt normal ecosystem functions. The growing demand for food highlights the urgent need for advanced crop improvement research to address the challenges posed by climate change (Dhankher and Foyer, 2018; Yang et al., 2022). Understanding how plants respond to abiotic stress is important in developing stress-resilient crops. While traditional breeding methods have greatly improved crop yields and resistance to both biotic and abiotic stresses (Takeda and Matsuoka, 2008), recent advancements have made significant contributions to understanding the complex quantitative traits involved. This includes identifying key genes, QTLs, and molecular pathways that regulate plant responses to drought, heat, salinity, and other abiotic stresses (Krasensky and Jonak, 2012; Esmaeili et al., 2022).

Water shortages, exacerbated by industrialization, agricultural expansion, population growth, and rising global temperatures, are projected to become more frequent and severe, significantly limiting rice production (Aide, 2019). Rice has different water needs at various growth stages, and its response to drought stress is influenced by the timing of the stress. Drought during critical stages can severely affect growth, development, and physiological functions, resulting in suboptimal yield levels and product quality. Therefore, exploring the mechanisms of drought tolerance in rice is essential for optimizing water use and enhancing rice production. Drought poses a major threat to future global food production, with forecasts suggesting that more than 50% of the world's arable land will be affected by 2050 (Katral et al., 2022; Pixley et al., 2023). Rice, in particular, is highly susceptible to drought, with potential yield losses of up to 100% under severe conditions (Chengqi et al., 2024). Therefore, understanding plant reaction to drought stress is essential for improving drought-resistant rice varieties. Numerous QTLs and genes associated with drought resistance traits—such as maintaining leaf water status, regulating root architecture, and controlling stomatal closure—have been identified (Courtois et al., 2000; Yue et al., 2006; Baldoni, 2022). Genes involved in key plant processes, including preserving cell membrane integrity (e.g., Late Embryogenesis Abundant (LEA) proteins) (Xu et al., 1996), signaling pathways (e.g., kinases) (Saijo et al., 2001; Liu et al., 2003), water uptake and transport (e.g., aquaporins) (Martre et al., 2002), carbohydrate metabolism (Jang et al., 2003), and oxidative stress regulation (Prakash et al., 2016), all play significant roles in drought tolerance.

Key root traits, including root length, growth angle, volume, and thickness, play a important role in drought tolerance. Among these, the *Deeper Rooting 1* (*DRO1*) QTL is particularly significant and has been extensively studied in rice for rapid generation advancement (Uga et al., 2013, 2015). *DRO1* has been shown to interact with other QTLs like *DRO2* and *DRO3*. The early auxin-responsive factor underlying the *DRO1* QTL has been cloned using deep-rooted (Kinandang Patong) and shallow-rooted (IR64) cultivars. Singh et al. (2021) performed allele mining of the 3058 bp *DRO1* gene using 24 prominent Indian upland and lowland rice genotypes, as well as drought-yield QTL (DTY) donors. Based on SNPs and the resulting amino acid changes, the genotypes were classified into 16 primary structure variants, ranging from fully functional alleles (Satti, IR36, and DTY 3.1 donor IR81896-B-B-195) to truncated, nonfunctional alleles (PMK2, IR64, IR20, and Swarna). In reaction to drought stress, plants express a range of proteins, including transcription factors, which are critical regulatory elements. Drought stress activates transcription factor families such as MYB/MYC, AP2/ERF, NAC, bZIP, HD-ZIP, and NF-Y, which influence drought tolerance by regulating gene expression (Hu et al., 2006; Li et al., 2008). Dehydration-responsive element-binding (DREB) transcription factors play a vital role in controlling drought-induced gene expression, mainly through an ABA-independent pathway. They interact with Dehydration-responsive element/C-repeat (DRE/CRT) cis-elements in the promoter regions of abiotic-stress-responsive genes (Dubouzet et al., 2003; Lata et al., 2011). Wang et al. (2008) cloned and characterized the *OsDREB1F* gene, a DREB transcription factor from upland rice (Nipponbare). Expression analysis revealed that *OsDREB1F* is activated by drought, salt, and cold stresses, as well as by abscisic acid (ABA) application. The protein specifically binds to DRE/CRT elements, but not to ABA-responsive elements (ABRE) elements. Singh et al. (2015) analyzed natural allelic diversity in the *OsDREB1F* gene across Indian wild rice germplasm, identifying six protein variants associated with drought resistance. Promoters of eight drought-resistance genes were studied across five cultivars, revealing significant allelic variation, including genes such as *OsALF11*, *OsAE128*, and *OsZIM14*, from families such as Alfin-like, AP2/EREBP, CCAAT-HAP3, GRF, and Tify (Park et al., 2004; Svensson et al., 2006). The allelic variations that confer diverse drought tolerance mechanisms are preserved in germplasm accessions and can be utilized in marker-assisted breeding programs. Rice also contains numerous genes activated by low-temperature stress (Fowler & Thomashow, 2002; Seki et al., 2002; Rabbani et al., 2003), including regulatory proteins such as transcription factors and protein kinases. Among these, *DREB1A* and *DREB1B* transcription factors, encoded by the *DREB1*-type genes in the rice genome, regulate low-temperature tolerance (Dubouzet et al., 2003). Challam et al. (2015) reported allelic variations in these genes, leading to breeding lines with improved cold tolerance. Several major and consistent QTLs for rice grain yield under reproductive stage drought stress have been identified, including *qDTY1.1* (chromosome 1), *qDTY2.1*, *qDTY2.2*, *qDTY2.3* (chromosome 2), *qDTY3.1*, *qDTY3.2* (chromosome 3), *qDTY4.1* (chromosome 4), *qDTY6.1*, *qDTY6.2* (chromosome 6), *qDTY9.1* (chromosome 9), *qDTY10.1* (chromosome 10), and *qDTY12.1* (chromosome 12) (Anilkumar et al., 2023b). These QTLs have been validated across different rice genetic backgrounds, such as Swarna, IR64, MTU1010, TDK1-Sub1, MRQ74, MR219, Savitri, and Vandana, using various molecular marker technologies (Kumar et al., 2020b). Additionally, two major QTLs for grain yield under drought, *qDTY3*.2 and *qDTY12*.1, were introgressed into the popular Nepalese variety Sabitri to create a drought-tolerant version of the variety (Dixit et al., 2017).

Rice has a distinctive ability among cereals to adapt to water stagnation. The Sub1 QTL, a crucial locus for submergence tolerance, is located in chromosome 9. It contains a group of ERF genes: *SUB1A*, *SUB1B*, and *SUB1C*. The *SUB1A-1* allele, which confers tolerance and is derived from the a subgroup of rice under oryza species subgroup of indica rice, was identified by Xu et al. (2006). Niroula et al. (2012) tested 109 accessions of cultivated and wild rice species, including *Oryza rufipogon* and *Oryza nivara*, and found submergence-tolerant accessions carrying the *SUB1A-1* allele, confirming the role of the *SUB1* locus in tolerance. Additionally, rice plants adapt to deep water stagnation (greater than 1 meter) by elongating their stems, a process controlled by the *Snorkel1* and *Snorkel2* genes (Hattori et al., 2009). For anaerobic germination, Angaji et al. (2010) identified five QTLs—*qAG*-1-2, *qAG*-3-1, *qAG*-7-2, *qAG*-9-1, and *qAG*-9-2—from Khao Hlan On, which explained 17.9%–33.5% of the phenotypic variance. Later, Septiningsih et al. (2013) reported six QTLs on chromosomes 2, 5, 6, and 7 from the landrace *Ma-Zhan Red*. The submergence-tolerant *Sub1A* gene has been introduced into many high-yielding rice cultivars suited for rainfed lowland flood-prone areas. This development helps mitigate the adverse effects of flash flooding, ensuring yield stability and providing farmers with more reliable benefits compared to conventional high-yielding but flood-susceptible cultivars (Kuanar et al., 2019, 2023). Similarly, the *Saltol* QTL, which is associated with salt tolerance during the seedling stage, is being integrated into varieties such as ADT 45, Savitri, Gayatri, MTU 1010, PR 114, Pusa 44, and Sarjoo 52 (Singh et al., 2016; Krishnamurthy et al., 2020). To develop multi-gene combinations through marker-aided breeding, various alleles from different germplasm sources must be incorporated into elite breeding lines. This strategy helps minimize linkage drag during marker-assisted backcross breeding.

## 7. Breeding for Resource-Use Efficiency

Rice cultivation is resource-intensive, requiring approximately 3500 liters of water to produce 1 kg of rice. Over the decades, the use of nitrogen (N) and phosphorus (P) fertilizers in India has surged significantly, with N consumption increasing from 0.06 million tonnes in 1950–1951 to 10.80 million tonnes in 2000–2001 and P use growing from 0.01 to 1.80 million tonnes during the same period. Further, N and P fertilizers contribute approximately 64% and 78% to Indian agriculture, respectively, but their use has been inefficient (Pathak et al., 2010). Continuous flooding of paddy fields exacerbates nitrogen losses through leaching, with cereals—especially rice—accounting for two thirds of urea consumption in India. Poor nitrogen-use efficiency (NUE) in rice cultivation (20%–40%) leads to substantial economic and environmental losses, including nitrogen waste worth ₹1 trillion annually in India and over $170 billion globally (Mohanty et al., 2023; Alam et al., 2023; Tyagi et al., 2024). Similarly, approximately 80%–90% of applied phosphorus is unavailable to plants, often contributing to environmental issues such as eutrophication (Schindler et al., 2016). These inefficiencies result in greenhouse gas emissions, water pollution, and diminished returns for farmers. Resources such as land, water, nutrients, and radiation are increasingly limited and degraded due to global warming and climate change, all of which are essential for agriculture. Enhancing the efficiency of water, nitrogen, and phosphorus use is crucial for sustainable rice farming. Strategies should focus on genetic improvements, breeding nutrient-efficient rice varieties, and adopting better management practices to reduce input costs, increase yields, and protect soil and water quality. Developing rice cultivars that perform efficiently under low-input conditions is particularly important for marginal lands with limited nutrient availability. Given the current challenges of resource scarcity and unpredictability, there is an urgent need to develop crop varieties that can maximize resource-use efficiency while sustaining productivity.

## 7.1 Water-Use Efficiency (WUE)

With 2.4% of the world's land area and 4% of its water resources, India faces significant distress to maintain the livelihoods of around 18% of the world's population and a developing economy (Pathak et al., 2022). This creates overstrain on the utilization of uses of natural resources for agriculture such as water. Rice is an extremely significant food crop in India, where it provides sustenance for almost 60% of the population. The area under rice farming increased from 30.81 million hectares in 1950–1951 to 43.86 million hectares in 2020–2021, a 142% increase (Das et al., 2023). For production of 1 kg rice, it takes 3000 liters of water (Seeli et al., 2024). Around 34%–43% of the total irrigation water is solely exploited for the cultivation of rice. Thus, genetic improvement of rice for higher WUE is an alarming demand of time amid water scarcity arising due to the climate change and groundwater deprivation for different reasons (Edwards et al., 2024). WUE is a cornerstone in economic yield, especially under stress circumstances such as drought and heat.

WUE is commonly defined as the amount of dry matter or grain yield produced per unit of water consumed. Alternatively, it can be expressed as the ratio of carbon assimilation via photosynthesis to the water lost through transpiration. In this context, total biomass production is directly influenced by both overall water usage and WUE. To improve both total water use and WUE in crop plants, four primary strategies can be employed: (1) optimizing water uptake from the soil to support maximal photosynthetic efficiency, (2) reducing water loss by modifying plant architecture, such as altering stomatal behavior and the cuticle, without compromising photosynthetic activity, (3) enhancing productivity per unit of water used, and (4) increasing cellular tolerance to water scarcity.

Both agronomic and physiological perspectives can be used to assess plants in context of WUE. Physiologically, WUEp is the correlation between the plant's transpiration rate and the $CO_2$ photosynthetic assimilation rate (A) (Bänziger et al., 2004):

$$WUEp = \frac{(PA-PI)}{1.6(VPI-VPA)},$$

WUEp: is the water-use efficiency, physiologically,
PA: partial pressure of $CO_2$ in the air,
PI: partial pressure of $CO_2$ inside the leaf,
VPI: vapor pressure of the water inside the leaf, and
VPA: vapor pressure of the water in the air.

In the context of agronomy, WUEa is the correlation between the dry mass generated and the amount of water utilized in the cultivated area's cycle (irrigation plus precipitation) (Jones, 1993):

$$WUEa = \frac{GY}{V},$$

WUE: water-use efficiency in agronomic terms,
GY: grain yield or dry mass yield, and
V: total water volume used in the cycle by the crop.

Roots have an important role in nutrient uptake, while also serving as sensors for soil water and nutrient availability. Root morphology and associated traits are key factors in improving water uptake from soil during drought conditions. Several QTLs related to root traits have been identified and tagged with molecular markers, allowing for the screening and selection of desirable plant types from segregating populations (Babu et al., 2003; Courtois et al., 2003; Zheng et al., 2003). Current research efforts are focused on using MAS to introgress these QTLs for root traits into breeding programs, thereby improving drought tolerance and water-use efficiency (Steele et al., 2002). Stomatal pores control the uptake of $CO_2$ for photosynthesis as well as the loss of water through transpiration. Genetic studies in Arabidopsis have identified genes that control stomatal responses to water deficits (Schroeder et al., 2001). In rice, QTLs associated with traits that minimize water loss such as stomatal regulation (Price et al., 1997), leaf ABA accumulation (Quarrie et al., 1997), and leaf rolling (Courtois et al., 2000) have been mapped using molecular markers. These QTLs and genes present opportunities for genetically improving stomatal regulation and boosting WUE. WUE can also be further enhanced by selecting for traits such as higher harvest index, spikelet fertility, stay-green characteristics, seedling vigor, shorter growth duration, and reduced photorespiration. Additionally, plants with high WUE tend to show lower discrimination against the heavier carbon isotope ($^{13}CO_2$) during photosynthesis, a trait that is positively correlated with WUE in rice (Peng et al., 1998).

Selection for higher WUE without improvement of photosynthetic potential will lead to decline in plant water use, transpiration, total plant biomass, and grain yield (Blum, 2009). Grain yield under drought conditions was demonstrated to be improved by the allele of qtl12.1, mainly through a 7% increase in plant water uptake. This ostensibly slight increase in water absorption may play a major role in the greater yield effects seen in field conditions (Bernier et al., 2009). Mutant rice lines with large size (LS) stomata showed improved WUE (iWUE) as compared to higher density (HD) of stomata due to the decline in stomatal conductance. The molecular approaches used to manipulate two orthologs of EPF genes, that is, overexpression of EPF1 gene (negative regulator of stomatal density) (Caine et al., 2019) and knockdown of EPFL9 gene using CRISPR/Cas9/CpF1 approach (Yin et al., 2017), cause significant reduction in stomatal density and ultimately result in iWUE. Overexpression of PsbS genes causes increase in non-photochemical quenching (NPQ), which is an important mechanism to disperse excess radiation and protect plants from photodamage and enhance photosynthetic efficiency and iWUE. The reason for this iWUE cannot be attributed to reduced stomatal density, but perhaps to the more complex interwoven control of PsbS overexpression over NPQ, photosynthetic efficiency, and iWUE (Patel-Tupper et al., 2024). Due to its lower stomatal density, the drought-insensitive tilling line-4 (ditl4) mutant demonstrated higher WUE while displaying decreased stomatal conductance, transpiration, and $CO_2$ assimilation. A SNP in *OsDIRH2* (*LOC_Os11g39640*), a gene encoding a RING-H2 type E3 ligase, was discovered by whole-genome resequencing of ditl4, which resulted in an early stop codon. When compared to control plants, CRISPR/Cas9-mediated knockout mutants showed improved drought tolerance through decreased stomatal density (Kim et al., 2024). Marker-assisted breeding is an effective tool to develop crop genotypes with a specific desirable feature by introgression identified and validated major QTL governing that particular trait. While retaining more than 97% of the recurrent parent's genome, five of the $BC_3F_3$ generation's best trait introgressed lines (TILs) showed yield increases of 52% and 63% above IR-64 under 100% and 60% field capacity, respectively. In comparison to recurrent parent, TILs showed enhanced WUE with a 28.5% decrease in $\Delta^{13}C$, a 53% increase in root weight, and a 23.5% increase in root length (Dharmappa et al., 2019). Osckx2 lines which are mutated for gene *OsCKX2* (cytokinin oxidase 2) using CRISPR/Cas9 had significant higher WUE under drought conditions, and led to greater photochemical performance and increased carbon assimilation (Rashid et al., 2024). A single gene on chromosome 6, *BLANKET LEAF* (BKL), regulates leaf pubescence. When exposed to direct sunlight, the presence of leaf hairs seems to increase the leaf's resistance to the boundary layer, making the leaf surface warmer. As a result, under intermediate to high light levels, chromosomal segment introgression lines (IL-hairy) showed a reduced transpiration rate, which enhanced the efficiency of photosynthetic water usage. This demonstrates how the *BKL* gene could be potentially used for breeding rice cultivars with improved photosynthetic and WUE (Hamaoka et al., 2017).

## 7.2 Nutrient-Use Efficiency

Nutrient-use efficiency (NUE) refers to a plant's ability to absorb and effectively utilize available nutrients for growth and development throughout its life cycle. In sustainable agriculture, plants with high NUE can produce greater biomass or grain while requiring less fertilizer input. Currently, India is the world's second-largest producer and consumer of fertilizers, following China. While field crop varieties that are highly responsive to fertilizers and irrigation have significantly boosted food production, they have also created challenges in soil and water management. Much scientific attention has been focused on the rising use of fertilizers in the country, the large amount of nitrogen released into the environment as a result, and the resulting environmental consequences (Abrol et al., 2007).

Higher use of chemical fertilizers without improving the fertilizer uptake, fertilizer responsiveness, and nutrient-use efficiency does not confer yield advancement. Rather, it generates several environmental issues such as water, air, and soil pollution. Considering this, improving nutrient-use efficiency is in great demand due to agronomical and environmental reasons. Nutrient-use efficiency is expressed as uptake, utilization, and remobilization of efficiencies, but ideally defined as grain yield per unit of nutrient uptake.

Physiological nutrient-use efficiency (PNUE) is the degree to which a plant effectively transforms the nutrients it has taken into biomass or yield. At the cellular and metabolic levels, it emphasizes the internal mechanisms and the efficiency with which the plant consumes the nutrients (Reich et al., 2014):

$$\text{PNUE} = \frac{\text{Dry matter generated (G)}}{\text{Nutirent uptake (G)}}.$$

Agronomical nutrient-use efficiency (ANUE) quantifies the amount of agricultural production generated per unit of nutrient application (Pathak, 1999). Basically, it gives the measure of the efficiency of fertilization in raising crop output:

$$\text{ANUE} = \frac{\text{Yield}\left(\frac{\text{kg}}{\text{ha}}\right)}{\text{Fertillizers applied}\left(\frac{\text{kg}}{\text{ha}}\right)}.$$

Effective nutrient uptake depends on the depth and structure of root systems. In order to obtain nutrients from a greater volume of soil, varieties with deeper and more widespread root systems can achieve higher NUE (Lynch, 2011; Mann et al., 2023). The efficiency with which plants take up nutrients from the soil is influenced by genetic differences across crop varieties. Certain cultivars have been engineered to increase root exudation, which can boost the uptake and solubilization of nutrients (Baligar et al., 2001). The development of leaf area, photosynthetic capability, and nutrient remobilization during reproductive growth are examples of physiological responses to nutrition availability that can be influenced by genetic variations (Nguyen, et al., 2016; Nguyen, 2018). Because these features maximize biomass production per unit of nutrient taken, they can improve NUE. The goal of breeding efforts aimed at NUE is to find and choose cultivars with the best root characteristics, improved nutrient uptake, and effective nutrient utilization in metabolic processes.

## 7.3 Nitrogen-Use Efficiency

By 2050, it is predicted that fertilizer use in India will double (Calicioglu et al., 2019), as fertilizers are essential for modern crop cultivation to maintain high productivity to meet the rapidly growing population's needs. The consumption of nitrogen fertilizer has risen dramatically, from 0.2 million tonnes (1.4 kg/ha of gross cropped area) in 1960 to 17.6 million tonnes (86 kg/ha of gross cropped area) in 2018. Among cereals, rice has the lowest nitrogen-use efficiency, resulting in the highest nitrogen consumption. Given the vast area and production scale of rice, it accounts for the majority of nitrogen fertilizer use in food crops. Strategic breeding of rice to improve both grain yield and nitrogen-use efficiency could lead to significant advancements in these traits. Rice has considerable germplasm diversity, which holds great potential for improving nitrogen-use efficiency (Haque et al., 2016; Li et al., 2024), and with the availability of a large genomic toolbox in the post-genomic era of crop improvement (Yu et al., 2002), this potential can be fully realized.

Genetic variation in NUE has been extensively studied in rice, with landraces identified as valuable genetic resources for improving NUE (Broadbent et al., 1987; Tirol-Padre et al., 1996; Zhang et al., 2009). Research conducted at Indian Institute of Rice Research on approximately 800 rice genotypes under both low and recommended nitrogen conditions revealed significant genetic variation in yield and nitrogen-related traits. Low nitrogen availability led to reduced plant height, fewer tillers, and impaired chloroplast function, resulting in a decrease in panicle number (approximately 20%), panicle weight (approximately 36%), and grain yield (30%–50%). These results underscore the detrimental effects of nitrogen deficiency and the need to identify genotypes with efficient NUE for improved productivity.

Nutrient-use efficiency is a highly complex trait that varies significantly due to the extensive genomic variation among different rice varieties. QTL analysis is the most commonly used method for studying genetic variation in NUE (Takeda and Matsuoka, 2008). Several studies have identified various QTLs associated with NUE in rice, using RIL populations and diverse phenotyping methods. For example, Wei et al. (2012) identified five NUE-related QTLs by analyzing grain yield response to nitrogen (GR) and combining grain yield with nitrogen uptake into a physiological NUE trait (PE), isolating four additional QTLs. Bai et al. (2021) focused on NUE traits such as grain nitrogen increase rate (GIR) and agronomic NUE (agNUE), discovering one GIR QTL and four agNUE QTLs. These studies highlight useful phenotyping strategies and genetic loci for improving NUE, though challenges persist in fine-mapping and identifying causal genes. Further validation of these QTLs could facilitate the development of high-NUE rice varieties. Rice's vast genetic diversity, including wild rice, landraces, and modern cultivars, offers an excellent resource for NUE improvement. GWAS have proven effective in uncovering genetic variations within rice germplasm populations, leading to the identification of NUE-related genes such as *OsNAC42*, *OsNPF6*.1, *OsNLP4*, and *OsNiR* (Huang and Han, 2014; Gupta et al., 2019). In addition to GWAS, identifying and manipulating nitrogen signaling regulators at the system level offer great potential for enhancing NUE. While genes like *NRT1.1B* have been explored, other components such as *SPX4* and *NBIP1* remain underutilized for NUE improvement. Signaling pathways, including the LBDs-TCP19-DLT (nitrogen-BR signaling) and GID1-NGR5/DELLA-GRF4-GIF (nitrogen-GA signaling), offer frameworks for discovering novel regulatory elements. However, genes such as *NRT1.1A*, *NRT2*.3b, and *DREB1C* likely lack sufficient natural variation, limiting their direct use for breeding elite high-NUE variants (Fan et al., 2016; Wei et al., 2022).

The crucial functions that *OsAAP1* and *OsLHT1* play in nitrogen transport and uptake are highlighted by their increased expression, especially in lower nitrogen environments. *OsLHT1* assists in the uptake of nitrogen from the soil, whereas *OsAAP1* enhances the transport of amino acids (Zhu et al., 2020). Increased expression of these important genes and a higher amino acid content are two characteristics that lead to improved NUE. This demonstrates an appealing approach for selecting the crop types that will maximize the use of nitrogen in low-input farming systems. NIL-dep1 carrying the dep1-1 gene from Qianchonglang 2 showed significantly higher grain yield, 6.7%–23.7% higher nitrogen-use efficiency for grain production (NUEg) and 11.7%–21.4% higher grain N content/total N uptake (NHI) at same nitrogen level as compared to control (Li et al., 2024). The discovery of 62 potential NUE genes through the combination of QTL co-localization and functional assessments, with chromosome 1 identified as a key hotspot, underscores the complex regulation of NUE evidenced by tissue- and stage-specific expression patterns, while experimental validation demonstrates that high NUE genotypes exhibit improved photosynthetic performance and water-use efficiency, suggesting these genes are viable targets for agricultural enhancement (Kumari et al., 2021). G-quadraplex (G4) sequences are associated with nitrogen uptake, hence play key role in nitrogen availability. Targeting G4 sequences may offer new approaches for genetic engineering or breeding programs aimed at improving NUE, and knowledge of their role in nitrogen transport may result in the development of more resilient and effective crop varieties because approximately 17% of nitrogen-responsive genes with G4 sequences are associated with yield, suggesting their potential role in enhancing NUE through epigenetic regulation. However, 83% of these genes may be influenced by traditional genetic mechanisms or other epigenetic factors, indicating the complexity of NUE regulation (Sharma et al., 2023).

## 7.4 Phosphorous-Use Efficiency (PUE)

Phosphorus (P) is a crucial nutrient for crop yields, but it is frequently limited in agricultural systems. Rock phosphate, the primary source of phosphorus fertilizer, is a non-renewable resource that is expected to become increasingly scarce. Moreover, the overuse of phosphorus in agriculture has resulted in significant environmental challenges. Enhancing PUE in crops through breeding offers a promising solution to address both nutrient scarcity and environmental concerns. Plants have evolved intricate regulatory mechanisms to adapt to phosphorus-deficient environments, optimizing their root architecture to improve phosphorus acquisition. These adaptations include the development of shallow roots, longer primary roots, root hairs, and lateral roots. Additionally, plants secrete acid phosphatase (APase) to release inorganic phosphorus from organic compounds in the soil (Wasaki et al., 2003). The Pup1 QTL, with its causal gene *PSTOL1* (*PHOSPHORUS STARVATION TOLERANCE 1*), is known to improve root growth and surface area, significantly enhancing phosphorus uptake and yield in phosphorus-deficient soils. Overexpression of *OsPTF1* has been shown to improve phosphorus starvation tolerance by altering root architecture, while proteins such as RMD and transcription factors such as *OsMYB1*, *OsMYB2P*-1, *OsMYB4P*, and *OsMYB5P* further regulate root responses under phosphorus deficiency (Gu et al., 2023). Auxin signaling also plays a key role in root adaptation. For instance, *OsAUX1* promotes root hair elongation under low phosphorus conditions, and auxin response factors *OsARF12* and *OsARF16* regulate phosphorus acquisition. Mutations in these genes result in altered root morphology and phosphorus uptake efficiency (Giri et al., 2018).

Plants employ various strategies to enhance phosphorus (P) acquisition under P-limited conditions. One such strategy involves the secretion of APase such as *OsPAP10a* and *OsPAP10c*, which are upregulated during P deficiency and regulated by the *Pi* signaling regulator *OsPHR2* to release inorganic phosphorus from organic compounds (Lu et al., 2023). Overexpression of genes such as *OsPAP21b* and *OsHAD1* has been shown to improve phosphorus uptake and plant performance under phosphorus-deficient conditions (Pandey et al., 2017). Transcription factors such as *OsWRKY74* further promote APase activity, facilitating phosphorus acquisition (Dai et al., 2016). Phosphorus is absorbed by roots through low- and high-affinity Pi transporters encoded by the *PHT1* gene family. In rice, ten *PHT1* genes, including *OsPT1*, *OsPT4*, and *OsPT8*, are crucial for phosphorus uptake and allocation. While *OsPT1* and *OsPT8* function independently of phosphorus availability, *OsPT4* is prominent in roots and embryos. Genes such as *OsPT3*, *OsPT6*, *OsPT9*, and *OsPT10* are strongly induced under phosphorus starvation, while *OsPT2* is mainly expressed in root steles during phosphorus deficiency. Collectively, these genes enhance phosphorus acquisition and utilization under varying phosphorus conditions.

Efficient phosphorus transport from roots to shoots is vital for plant growth, and proteins from the SPX-EXS subfamily play a key role in this process. In rice, *OsPHO1*;2, a member of the SPX-EXS subfamily, is primarily expressed in roots and mediates phosphorus translocation from root to shoot. Mutations in *OsPHO1*;2 result in reduced phosphorus transfer to shoots, causing phosphorus accumulation in roots and deficiency in shoots (Secco et al., 2010). *OsPHO2*, an E2 ubiquitin-conjugase, negatively regulates this process, and the *Ospho2* mutant exhibits excessive phosphorus in shoots. The regulation of *OsPHO2* is mediated by *OsmiR399* and *OsIPS1* in response to cellular phosphorus levels (Wang et al., 2018). Additionally, *OsPHO2* interacts with *OsPHO1*, although the mechanism remains unclear (Ying et al., 2017). *PHT1* family phosphorus transporters, such as *OsPT1*, *OsPT4*, and *OsPT8*, also contribute to phosphorus root-to-shoot translocation, with *OsPT2* enhancing translocation to shoots upon overexpression. Similarly, *OsPT3* plays a role in phosphorus transport under extremely low phosphorus levels, regulated by *OsPHR2*. These findings suggest a coordinated role for SPX-EXS proteins and *PHT1* transporters in regulating phosphorus translocation.

Studies have demonstrated that modifying phosphorus metabolism genes can enhance plant PUE without adversely affecting yield or other agronomic traits. Silencing the *IPK1* gene, which encodes the enzyme responsible for the final step in phytic acid (PA) biosynthesis, reduces phosphorus and phytic acid levels while increasing inorganic phosphorus levels in rice seeds, with no negative impact on germination or agronomic traits (Ali et al., 2013). The *OsSULTR3*;3 gene, identified in the low-phytic acid (*lpa*) mutant, regulates phytic acid and total phosphorus accumulation in grains. It is expressed in vascular bundles, stems, and seeds, and its protein localizes to the ER, though it does not function as a phosphorus transporter (Zhao et al., 2016). Another important gene, *SPDT* (SULTR-like phosphorus distribution transporter), encodes a plasma membrane phosphorus transporter crucial for phosphorus allocation to grains. Mutants lacking *SPDT* show a 20%–30% reduction in total phosphorus and phytic acid in brown rice, while phosphorus levels increase in leaves. Notably, yield, germination, and seedling vigor remain unaffected under field conditions (Yamaji et al., 2017). These findings highlight the potential for enhancing PUE by reducing phosphorus allocation to grains without compromising yield or plant performance.

Differential gene expression analyses have also shown an inverse association between harvest index and phosphorus levels across various rice genotypes, revealing significant relationships between these genes and varying nutrient doses. Key genes with unique regulation patterns, such as *OsFER2*, *OsSULTR3*;4, and *OsCYP75B4*, show promise as targets for enhancing NUE and PUE (Madan and Raghuram, 2024). The divergent expression responses between genotypes suggest that specific genetic processes govern nutrient efficiency, providing a foundation for future genetic modification efforts aimed at maximizing rice's nutrient utilization.

## *7.5 Radiation-Use Efficiency (RUE)*

Solar radiation is one of the key drivers of plant survival and economics as it is the ultimate source of energy in plant system by means of the photosynthesis. Plants utilize photosynthetically active radiation (PAR) spectrum of visible light with the wavelength range 400–700 nanometers (nm) for the process of photosynthesis (Krause and Weis, 1991; Long and Bernacchi, 2003). Within the PAR spectrum, photosynthesis reaches its maximum at particular wavelengths, particularly in the red (about 640–680 nm) and blue (approximately 430–450 nm) regions (Farquhar et al., 1980). The degree with which plants transform absorbed PAR into biomass is known as radiation-use efficiency, or RUE (Monteith, 1977). It is a crucial metric for comprehending how plants respond to their surroundings and how productive they are.

Agronomically, the quantity of biomass generated per unit of PAR received by the plant is known as RUE. It aids in determining how productive crops are in different light levels:

$$\text{RUEa} = \frac{\text{Biomass produced}}{\text{PAR absorbed}},$$

usually stated as g $MJ^{-1}$, or grams of biomass per megajoule of PAR.

The physiological definition of RUE is the ratio of net primary production (NPP) to the quantity of PAR that the plant intercepts (Monteith, 1972). This metric shows how well a plant uses the light radiation that is available for development and photosynthesis:

$$RUEp = \frac{NPP}{PAR\ intercepted},$$

where NPP can be calculated as: NPP = GPP – Respiration.

GPP: gross primary production (total photosynthesis).
Respiration: total respiration by the plant.

The photosynthetic efficiency of C3 and C4 plants varies greatly. C3 plants use the enzyme RuBisCO to fix carbon directly from the Calvin cycle. This can result in inefficiencies, especially when temperatures are high and $CO_2$ levels are low because of photorespiration, when RuBisCO fixes oxygen rather than carbon dioxide, causing a net loss of energy and carbon. C4 plants, on the other hand, use a different series of reactions to fix $CO_2$ first into oxaloacetate, a four-carbon molecule, before going into the Calvin cycle (Brown and Hattersley, 1989). This adaptation improves light-use efficiency (LUE), particularly in warm, bright conditions, by reducing photorespiration and enabling C4 plants to concentrate $CO_2$ around RuBisCO. As a result, C3 plants are more susceptible to environmental factors that worsen these inefficiencies and typically show lower LUE.

The evaluation of high-yielding rice cultivars developed during time period of 1936 to 2005 for RUE and NUE suggested that both of these traits have a significant correlation with grain yield. Thus, the gain in yield is not only due to effortful selection for harvest index, biomass, and sink size, particularly the number of spikelets per panicle, but it is also closely attributed to the improvement of RUE and NUE (Zhu et al., 2016). The enhanced NPQ and reduced photo-inhibition in PsbS overexpression lines are likely contributors to their higher biomass and grain yield, particularly under fluctuating light conditions due to higher RUE (Hubbart et al., 2018). On chromosome 1, a significant QTL affecting panicle counts and yield under LL was discovered (*qPNLL1*.1, *qGYLL1*.1). The QTL on chromosome 1 contained the shade-responsive gene monoculm 2 (MOC2), which was upregulated in the tolerant but repressed insusceptible lines (Ganguly et al., 2024). There is potential for direct use in MAS to create shade-tolerant rice because of the QTL's large additive effect on chromosome 1, which has been validated across multiple genotypes.

The combined overexpression of three essential C4 enzymes—pyruvate orthophosphate dikinase (PPDK), phosphoenol pyruvate carboxylase (PEPC), and carbonic anhydrase (CA)—offers a viable approach to enhancing the physiological and morphological characteristics of C3 rice, which may result in increased yield (Sen et al., 2017). In light of climate change and food security, traditional rice varieties can profit by utilizing C4 pathway mechanisms. A potential field of study targeted at improving photosynthetic efficiency and productivity is the conversion of rice (a C3 plant) to a C4 plant. To insert genes for the C4 photosynthesis pathway into rice, researchers have used genetic engineering. This involves working with important enzymes and structural elements that are part of the C4 cycle. Enhanced yields due to a possible improvement in photosynthetic efficiency in high light and temperature environments can be achieved by this approach (Cui, 2021; Swain, et al., 2021; Lin et al., 2016; Nomura et al., 2005). Whereas, intricate genetic relationships and difficulties in obtaining consistent and effective gene expression are the major limitations in this mission (Pradhan et al., 2022). A C4-like pathway in rice has been built using synthetic biology techniques, which also include the development of artificial organelles for $CO_2$ concentration (Sahoo et al., 2024). It makes it possible to regulate metabolic processes more precisely and to become more adaptive to changing environmental circumstances. However, this approach is still untested in terms of commercial benefits and is currently a hot topic for the scientific community due to ethical concerns over synthetic organisms and potential ecological repercussions (Ermakova et al., 2020). In an effort to produce a C4-like phenotype, rice genes involved in photosynthesis have been edited using CRISPR/Cas9 technology (Jethva et al., 2024; Zheng et al., 2021; Karki et al., 2013). Without adding extraneous DNA, targeted changes can result in particular improvements in photosynthesis and productivity, but off-target consequences and regulatory issues with crops that have had their genomes altered are one of the constraints for this approach (Ermakova et al., 2020).

# 8. Breeding for Nutrient-Dense Rice

Nutrition is fundamental for all living organisms, supporting essential functions such as growth, development, and productivity. Rice is a major food source for over three billion population globally, accounting for 20% of the world's nutritional intake. In recent years, agricultural efforts have increasingly prioritized the development of nutritionally enhanced varieties of staple crops such as rice, wheat, maize, and legumes to meet human dietary needs (Ricachenevsky et al., 2019). This focus is particularly prominent in developing and underdeveloped countries, where improving crop yield and accessibility is a key priority. While these staples provide the majority of caloric intake, they often lack essential nutrients, contributing to widespread health problems such as anemia (iron deficiency), night blindness (vitamin A deficiency), stunted growth, and impaired mental health (Oakley et al., 2004; Darnton-Hill et al., 2005). This nutritional gap is often referred to as "hidden hunger," which poses significant socioeconomic challenges (Hossain et al., 2022; Katral et al., 2024). The situation is further worsened by the limited consumption of balanced diets that include diverse food sources. In 2021, 828 million people globally faced hunger. In recognition of the importance of a healthy diet, the United Nations has included it in 12 of the 17 Sustainable Development Goals (FAO et al., 2022). Supplementing diets with foods rich in carotenoids, proteins, iron, and zinc can help address these nutrient deficiencies (Senguttuvel et al., 2023). However, for economically disadvantaged populations, incorporating fortified or nutrient-dense foods may not always be financially feasible. Enhancing the nutritional quality of rice, such as increasing protein content and micronutrient levels, offers a practical and sustainable solution. This process, known as crop biofortification, is vital for addressing deficiencies in iron (Fe), zinc (Zn), provitamin A, and protein (Bouis and Saltzman, 2017; Bin Rahman and Zhang, 2023; Senguttuvel et al., 2023). Consequently, extensive research is underway to enhance rice's nutritional content, targeting both micronutrients and macronutrients through various genomic approaches, including conventional breeding and transgenic techniques (Figure 1).

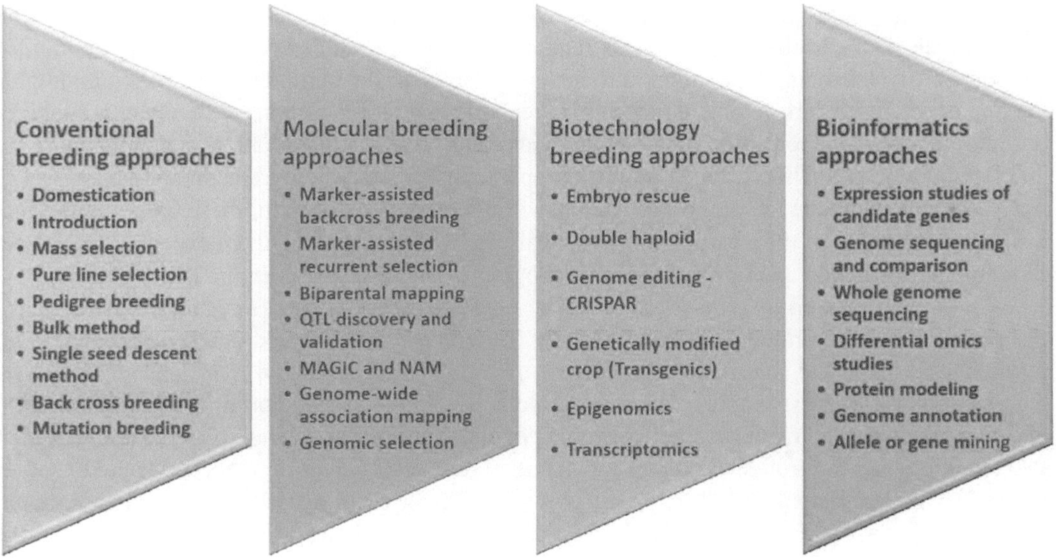


**Fig. 1** Schematic representation of breeding for biofortified rice crop.

A total of 14 QTLs linked with grain Fe and Zn contents have been identified, with several candidate genes located near these QTLs. For Fe content, genes such as *OsYSL1* and *OsMTP1* have been highlighted, while *OsARD2*, *OsIRT1*, *OsNAS1*, and *OsNAS2* are linked to Zn content.

Additionally, genes such as *OsNAS3*, *OsNRAMP1*, and *APRT* have been associated with both Fe and Zn contents (Anuradha et al., 2012). GWAS, coupled with grain Fe content data, have revealed key genes involved in Fe homeostasis, including *OsNAS3* (nicotianamine synthase), *OsVIT1* (vacuolar Fe transport), *OsMTP6*, *OsMT2D*, and *OsNRAMP7*, which colocate with QTLs for both Fe and Zn (Descalsota et al., 2018). Utilizing multi-season genotypic and phenotypic data, Dixit et al. (2019) identified two significant QTLs for Fe and three for Zn, with candidate genes such as *OsPOT*, *OsZIP4*, *OsFDR3*, and *OsIAA5*, which play a role in regulating grain Fe and Zn content. A seasonal comparison revealed two QTLs each for Fe and Zn content, with two QTLs on chromosome 2 showing consistent results. These included *qZnPR*.2.1 (Phenotypic variance explained 19.42%) for polished rice and *qZnBR*.2.1 (PVE 20.80%) for brown rice (Uttam et al., 2022).

In most cereal grains, around 80% of the overall phytic acid is concentrated in the aleurone layer, with maize being the exception. Phytic acid predominantly exists as mixed salts, known as phytates (Rojaria et al., 2024). Due to its six negatively charged ions, phytate acts as a strong chelator of essential divalent cations such as $Fe^{2+}$, $Zn^{2+}$, $Ca^{2+}$, and $Mg^{2+}$, thereby reducing the bioavailability of these vital minerals (Majumder et al., 2019). Myo-inositol-3-phosphate synthase (MIPS) is an enzyme responsible for catalyzing the conversion of D-glucose-6-phosphate to 1D-myo-inositol-3-phosphate [Ins(3)P1], a key step in the biosynthesis of phytic acid. Mutations in the MIPS gene have led to the development of low phytic acid mutants in rice, such as lpa1, with its locus identified on chromosome 3S (Larson et al., 2000). In rice mutants Os-lpa-XS110-1 and Os-lpa-XS110-2, created through gamma irradiation, the mutations responsible for the reduced phytic acid content remain unidentified (Liu et al., 2007). Kim et al. (2008) developed a novel low phytic acid (lpa) rice line, lpa N15-186, through chemical mutagenesis. This line is orthologous to the maize myo-inositol kinase (MIK) gene on chromosome 3, which encodes MIK, an enzyme involved in the synthesis of Ins(3)P1. Furthermore, the SPDT was identified as a rice transporter responsible for the delivery of phosphorus to the grain. A single nucleotide change in the SPDT gene resulted in reduced phytic acid content and enhanced mineral bioavailability in rice grains (Kumar et al., 2021). Furthermore, allelic variations in the SPDT gene were found to regulate phytic acid content across different rice genotypes (Azharudheen et al., 2022). The positive association between grain zinc (Zn) and iron (Fe) content, with correlation coefficients ranging between 0.2 and 0.5, indicates the potential for simultaneously enhancing these micronutrients in rice through breeding. This correlation highlights the importance of understanding heritability in genomic prediction, as traits with higher heritability are more accurately predicted using GS methods (Hariprasanna et al., 2014; Senguttuvel et al., 2023). Genomic prediction offers a promising approach for estimating grain Zn and Fe content in rice. Additionally, the moderate-to-high, broad-sense heritability values for these traits suggest that a significant portion of their variation is genetically determined, making them excellent candidates for genomic prediction and targeted breeding programs (Dhanyalakshmi et al., 2024).

For protein content in brown rice, three key QTLs—*qPC1*.1, *qPC11.1*, and *qPC11.2*—have been associated with protein quantity, with *qPC11.1* and *qPC11.2* located on chromosome 11. Yu et al. (2009) identified five additional QTLs—*qPC-3*, *qPC-4*, *qPC-5*, *qPC-6*, and *qPC-10*—on chromosomes 3, 4, 5, 6, and 10, respectively. Notably, *qPC-6* on the short arm of chromosome 6 was mapped near the Wx marker between RM190 and RZ516. Zhong et al. (2011) later identified two reliable QTLs for protein concentration in milled rice, *qPr1* and *qPr7*, located on chromosomes 1 and 7, with consistent expression over two years. Yang et al. (2019) discovered two major, consistent QTLs—*qGPC-1* and *qGPC-10*—with the latter harboring the *OsGluA2* gene, which influences protein variation in both japonica and indica rice. In an association mapping study, Pradhan et al. (2019) identified three novel QTLs—*qPC3.1*, *qPC5.1*, and *qPC9*. Additionally, three QTLs—*qPC1.1*, *qPC1.2*, and *qPC7.1*—were found to colocalize with QTLs for the amino acid concentrations of serine, histidine, and threonine in rice (Jang et al., 2020).

While several crops, such as maize, cassava, and sweet potato, have successfully been biofortified with elevated levels of provitamin A, rice germplasm lacks naturally occurring variation for this trait (De Moura et al., 2014). As a result, achieving provitamin A biofortification in rice has largely relied on genetic engineering. This method entailed introducing two genes into the temperate japonica rice variety, Kaybonnet, from the USA. The genes included *phytoene synthase* (*Zmpsy1*) from *Zea mays* and *carotene desaturase* (*crtI*) from the soil bacterium *Pantoea ananatis* (formerly known as *Erwinia uredovora*). These modifications completed the carotenoid biosynthesis pathway in the rice grains, leading to the accumulation of β-carotene (provitamin A) in the endosperm (Paine et al., 2005). For the broad adoption of golden rice in Asia, it is essential to transfer the provitamin A trait from the temperate japonica variety, Kaybonnet, into locally adapted and widely cultivated rice varieties. Among the six second-generation golden rice (GR2) events developed by the IRRI, the GR2 event stood out for containing a single, intact copy of the inserted DNA at one specific site in the rice genome. This event produced agronomically superior progeny with optimal grain carotenoid levels. The use of marker-assisted backcrossing (MABC) successfully facilitated the transfer of the GR2E trait into Asian rice varieties (Biswas et al., 2021). MABC has also proven effective in introducing high-value genes and QTLs related to disease resistance, submergence tolerance, and drought tolerance into popular rice varieties, without compromising their essential agronomic traits (Swamy et al., 2013). Thus, a comprehensive understanding of the genetic and molecular processes driving the biofortification of both micronutrients and macronutrients in rice, combined with advancements in breeding technologies, holds great promise for improving global nutritional stability and addressing the issues of hidden hunger and malnutrition.

## References

Abrol, Y.P., Raghuram, N. and Sachdev, M.S. (Eds.). (2007). *Agricultural nitrogen use and its environmental implications.* IK International Pvt. Ltd.

Aide, M.T. (2019). Rice production with restricted water usage: A global perspective. *Egyptian Journal of Agronomy*, *41*(3), 197–206.

Alahmad, S., Dinglasan, E., Leung, K.M., Riaz, A., Derbal, N., Voss-Fels, K.P. and Hickey, L.T. (2018). Speed breeding for multiple quantitative traits in durum wheat. *Plant Methods*, *14*, 1–15.

Alam, M.S., Khanam, M., and Rahman. M.M. (2023). Environment-friendly nitrogen management practices in wetland paddy cultivation. *Frontiers in Sustainable Food Systems 7*, 1020570. doi: 10.3389/fsufs.2023.1020570.

Ali, N., Paul, S., Gayen, D., Sarkar, S.N., Datta, K. and Datta, S.K. (2013). Development of low phytate rice by RNAi mediated seed-specific silencing of inositol 1, 3, 4, 5, 6-pentakisphosphate 2-kinase gene (IPK1). *PloS One*, *8*(7), e68161.

Angaji, S.A., Septiningsih, E.M., Mackill, D.J., Ismail, A.M. (2010). QTLs associated with tolerance of flooding during germination in rice (Oryza sativa L.). *Euphytica*, 172, 159–168, doi: 10.1007/s10681-009-0014-5

Anilkumar, C., Muhammed Azharudheen, T.P., Sah, R.P., Sunitha, N.C., Devanna, B.N., Marndi, B.C. and Patra, B.C. (2023a). Gene based markers improve precision of genome-wide association studies and accuracy of genomic predictions in rice breeding. *Heredity*, *130*(5), 335–345.

Anilkumar, C., Sah, R.P., Beena, R., Azharudheen TP, M., Kumar, A., Behera, S., Sunitha, N.C., Pradhan, S.K., Reshmi Raj, K.R., Parameswaran, C., Marndi, B.C. and Singh, A.K. (2023b). Conventional and contemporary approaches for drought tolerance rice breeding: Progress and prospects. *Plant Breeding, 142*(4), 418–438. https://doi.org/10.1111/PBR.13119

Anilkumar, C., Sah, R.P., Muhammed Azharudheen, T.P., Behera, S., Mohanty, S.P., Anandan, A. and Samantaray, S. (2025). Integrating multi-trait genomic selection with simulation strategies to improve grain yield and parental line selection in rice. *Annals of Applied Biology. 186 (2)*, 217–227. https://doi.org/10.1111/aab.12964

Anilkumar, C., Sah, R.P., Muhammed Azharudheen, T.P., Behera, S., Singh, N., Prakash, N.R. and Nair, S.K. (2022a). Understanding complex genetic architecture of rice grain weight through QTL-meta analysis and candidate gene identification. *Scientific Reports*, *12*(1), 13832.

Anilkumar, C., Sah, R.P., Muhammed, A.T., Sunitha, N.C., Behera, S., Marndi, B.C. and Singh, A.K. (2022c). Genomic selection in rice: current status and future prospects. In: *Genomic Selection in Plants*, 68–82, CRC Press, Boca Raton.

Anilkumar, C., Sunitha, N.C., Harikrishna, Devate, N.B. and Ramesh, S. (2022b). Advances in integrated genomic selection for rapid genetic gain in crop improvement: a review. *Planta*, *256*(5), 87.

Anuradha, K., Agarwal, S., Rao, Y.V., Rao, K.V., Viraktamath, B.C., Sarla, N. (2012). Mapping QTLs and candidate genes for iron and zinc concentrations in unpolished rice of Madhukar× Swarna RILs. *Gene*, *508*(2), 233–240. https://doi.org/10.1016/j.gene.2012.07.054

Araki, M. and Ishii, T. (2015). Towards social acceptance of plant breeding by genome editing. *Trends in Plant Science*, *20*(3), 145–149.

Azharudheen, M.T.P., Kumar, A., Anilkumar, C., Sah, R.P., Behera, S. and Marndi, B.C. (2022). Understanding natural genetic variation for grain phytic acid content and functional marker development for phytic acid-related genes in rice. *BMC Plant Biology*, *22*(1), 446.

Babu, R.C., Nguyen, B.D., Chamarerk, V., Shanmugasundaram, P., Chezhian, P., Jeyaprakash, P. and Nguyen, H.T. (2003). Genetic analysis of drought resistance in rice by molecular markers: Association between secondary traits and field performance. *Crop Science*, *43*(5), 1457–1469.

Bai, J., Piao, Z., Wan, C., Lee, G., Ruan, X., Luo, Z., Shi, F., Cong, X. and Yang, R. (2021). SLAF-based linkage map construction and QTL mapping of nitrogen use efficiency in rice (Oryza sativa L.). *Plant Molecular Biology Reports*, *39*, 727–738.

Bai, X., Wu, B. and Xing, Y. (2012). Yield-related QTLs and their applications in rice genetic improvement. *Journal of Integrative Plant Biology*, *54*(5), 300–311.

Baldoni, E. (2022). Improving drought tolerance: can comparative transcriptomics support strategic rice breeding? *Plant Stress*, *3*, 100058.

Baligar, V.C., Fageria, N.K. and He, Z.L. (2001). Nutrient use efficiency in plants. *Communications in Soil Science and Plant Analysis*, *32*, 921–950. https://doi.org/10.1081/CSS-100104098

Bänziger, M., Setimela, P.S., Hodson, D. and Vivek, B. (2004). Breeding for improved abiotic stress tolerance in maize adapted to Southern Africa. *Agricultural Water Management*, *80*, 212–224. http://dx.doi.org/10.1016/j.agwat.2005.07.014.

Bartholomé, J., Frouin, J., Brottier, L., Cao, T.V., Boisnard, A., Ahmadi, N. and Courtois, B. (2023). Genomic selection for salinity tolerance in japonica rice. *PloS One*, *18*(9), e0291833.

Basavaraj, P.S., Muralidhara, B., Manoj, C.A., Anantha, M.S., Rathod, S., Raju, C.D. and Gireesh, C. (2021). Identification and molecular characterization of high-yielding, blast resistant lines derived from Oryza rufipogon Griff in the background of 'Samba Mahsuri' rice. *Genetic Resources and Crop Evolution*, *68*, 1905–1921.

Bernardo, R. (1994). Prediction of maize single-cross performance using RFLPs and information from related hybrids. *Crop Science*, *34*(1), 20–25.

Bernier, J., Serraj, R., Kumar, A., Venuprasad, R., Somayanda, I., Veeresh, R.P. and Atlin, G. (2009). The large-effect drought-resistance QTL qtl12.1 increases water uptake in upland rice. *Field Crops Research*, *113*(2), 139–146. https://doi.org/10.1016/j.fcr.2008.07.010

Bhandari, A., Bartholomé, J., Cao-Hamadoun, T.V., Kumari, N., Frouin, J., Kumar, A. and Ahmadi, N. (2019). Selection of trait-specific markers and multi-environment models improve genomic predictive ability in rice. *PLoS One*, *14*(5), e0208871.

Bin Rahman, A.N.M.R. and Zhang, J. (2023). Trends in rice research: 2030 and beyond. *Food Energy Security, 12,* e390.

Biswas, P.S., Ahmed, M.E., Afrin, W., Rahman, A., Shalahuddin, A.K.M., Islam, R. and Islam, M.R. (2023). Enhancing genetic gain through the application of genomic selection in developing irrigated rice for the favorable ecosystem in Bangladesh. *Frontiers in Genetics, 14,* 1083221.

Biswas, P.S., Swamy, B.M., Kader, M.A., Hossain, M.A., Boncodin, R., Samia, M. and Reinke, R. (2021). Development and field evaluation of near-isogenic lines of GR2-EBRRI dhan29 golden rice. *Frontiers in Plant Science*, 12, 619739.

Blum, A. (2009). Effective use of water (EUW) and not water-use efficiency (WUE) is the target of crop yield improvement under drought stress. *Field Crops Research*, *112*(2–3), 119–123. https://doi.org/10.1016/J.FCR.2009.03.009

Bouis, H.E. and Saltzman, A. (2017). Improving nutrition through biofortification: a review of evidence from HarvestPlus, 2003 through 2016. *Glob Food Security, 12,* 49–58.

Bradshaw, J.E. (2017). Plant breeding: past, present and future. *Euphytica*, *213*, 1–12.

Brar, D.S., and Khush, G.S. (2018). Wild relatives of rice: a valuable genetic resource for genomics and breeding research. *The wild Oryza genomes*, 1–25. DOI https://doi.org/10.1007/978-3-319-71997-9_1

Broadbent, F.E., De Datta, S.K. and Laureles, E.V. (1987). Measurement of nitrogen utilization efficiency in rice genotypes. *Agronomy Journal*, *79*(5), 786–791.

Brown, R.H. and Hattersley, P.W. (1989). Leaf anatomy of C3–C4 species as related to evolution of C4 photosynthesis. *Plant Physiology*, *91*(4), 1543–1550. https://doi.org/10.1104/PP.91.4.1543

Caine, R.S., Yin, X., Sloan, J., Harrison, E.L., Mohammed, U., Fulton, T., Biswal, A.K., Dionora, J., Chater, C.C., Coe, R.A., Bandyopadhyay, A., Murchie, E.H., Swarup, R., Quick, W.P. and Gray, J.E. (2019). Rice with reduced stomatal density conserves water and has improved drought tolerance under future climate conditions. *New Phytologist*, *221*(1), 371–384. https://doi.org/10.1111/NPH.15344

Calicioglu, O., Flammini, A., Bracco, S., Bellù, L. and Sims, R. (2019). The future challenges of food and agriculture: An integrated analysis of trends and solutions. *Sustainability*, *11*(1), 222. https://doi.org/10.3390/su11010222

Chakrabarti, N.K. (2001). Epidemiology and disease management of brown spot of rice in India. In: Major Fungal Diseases of Rice Recent Advances. Kluwer Academic Publishers. The Netherlands. 293–306.

Chakraborti, M., Anilkumar, C., Verma, R.L., Fiyaz, R.A., Raj, K.R., Patra, B.C. and Rao, L.S. (2021). Rice breeding in India: eight decades of journey towards enhancing the genetic gain for yield, nutritional quality, and commodity value. *ORYZA-An International Journal of Rice*, *58*(Special Issue), 69–88. https://doi.org/10. 35709/ory.2021.58.spl.

Challam, C., Ghosh, T., Rai, M. and Tyagi, W. (2015). Allele mining across DREB1A and DREB1B in diverse rice genotypes suggests a highly conserved pathway inducible by low temperature. *Journal of Genetics*, *94*, 231–238.

Chen, S.Y., Lai, M.H., Chu, Y.L., Wu, D.H., Tung, C.W., Chen, Y.J. and Chung, C.L. (2023). Identification of qBK2.1, a novel QTL controlling rice resistance against *Fusarium fujikuroi*. *Botanical Studies*, *64*(1), 11. doi: 10.1186/s40529-023-00375-y

Chen, X.W., Shang, J.J., Chen, D.X., Lei, C.L., Zou, Y., Zhai, W.X. and Zhu, L.H. (2006). A B-lectin receptor kinase gene conferring rice blast resistance. *The Plant Journal*, *46*(5), 794–804.

Chengqi, Z., Yuxuan, Y., Tian, Q., Yafan, H., Jifeng, Y. and Zhicheng, S. (2024). Drought-tolerant rice at molecular breeding eras: an emerging reality. *Rice Science*, *31*(2), 179–189. https://doi.org/10.1016/j.rsci.2023.11.005.

Cho, Y.C., Kwon, S.J., Choi, I.S., Lee, S.K., Jeon, J.S., Oh, M.K., et al. (2007). Identification of major blast resistance genes in Korean rice varieties (*Oryza sativa* L.) using molecular markers. *Journal of Crop Science and Biotechnology*, *10*(4), 265–276.

Courtois, B., McLaren, G., Sinha, P.K., Prasad, K., Yadav, R. and Shen, L. (2000). Mapping QTL associated with drought avoidance in upland rice. *Molecular Breeding*, *6*(1), 55–66.

Courtois, B., Shen, L., Petalcorin, W., Carandang, S., Mauleon, R. and Li, Z. (2003). Locating QTLs controlling constitutive root traits in the rice population IAC 165 × Co39. *Euphytica*, *134*(3), 335–345.

Cui, H. (2021). Challenges and approaches to crop improvement through C3-to-C4 engineering. *Frontiers in Plant Science*, *12*, 715391. https://doi.org/10.3389/FPLS.2021.715391/BIBTEX

Dai, X., Wang, Y. and Zhang, W.H. (2016). OsWRKY74, a WRKY transcription factor, modulates tolerance to phosphate starvation in rice. *Journal of Experimental Botany*, *67*(3), 947–960.

Darnton-Hill, I., Webb, P., Harvey, P.W., Hunt, J.M., Dalmiya, N., Chopra, M., Ball, M.J., Bloem, M.W. and deBenoist, B. (2005). Micronutrient deficiencies and gender: Social and economic costs. *American Journal of Clinical Nutrition*, *81*(5): 1198S–1205S.

Das, G., and Rao, G.J.N. (2015). Molecular marker assisted gene stacking for biotic and abiotic stress resistance genes in an elite rice cultivar. *Frontiers in Plant Science,* 6, 698. doi: 10.3389/fpls.2015.00698

Das, S.R., Nayak, B.K., Dey, S., Sarkar, S., Chatterjee, D., Saha, S., Sarkar, D., Pradhan, A., Saha, S. and Nayak, A.K. (2023). Potential soil organic carbon sequestration vis-a-vis methane emission in lowland rice agroecosystem. *Environmental Monitoring and Assessment*, *195*(9), 1–29. https://doi.org/10.1007/S10661-023-11673-0/TABLES/3

De Moura, F.F., Palmer, A.C., Finkelstein, J.L., Haas, J.D., Murray-Kolb, L.E., Wenger, M.J. and Peña-Rosas, J.P. (2014). Are biofortified staple food crops improving vitamin A and iron status in women and children? New evidence from efficacy trials. *Advances in Nutrition*, *5*(5), 568–570.

Descalsota, G.I., Swamy, B.M., Zaw, H., Inabangan-Asilo, M.A., Amparado, A., Mauleon, R., Chadha-Mohanty, P., Arocena, E.C., Raghavan, C., Leung, H. and Hernandez, J.E. (2018). Genome-wide association mapping in a rice magic plus population detects QTLs and genes useful for biofortification. *Frontiers in Plant Science, 9*, 1347.

Devanna, B.N., Jain, P., Solanke, A.U., Das, A., Thakur, S., Singh, P.K., Kumari, M., Dubey, H., Jaswal, R., Pawar, D., et al. (2022). Understanding the dynamics of blast resistance in rice-magnaporthe oryzae interactions. *Journal of Fungi*, *8*, 584. https://doi.org/10.3390/jof8060584

Devi, S.J.S.R., Singh, K., Umakanth, B., Vishalakshi, B., Rao, K.V.S., Suneel, B., et al. (2020). Identification and characterization of a large-effect QTL from *Oryza glumaepatula* revealed Pi68(t) as a putative candidate gene for rice blast resistance. *Rice*, *13*(1), 1–13.

Dhankher, O.P. and Foyer, C.H. (2018). Climate resilient crops for improving global food security and safety. *Plant, Cell & Environment*, *41*(5), 877–884.

Dhanyalakshmi, K.H., Mohan, R., Behera, S., Jha, U.C., Moharana, D., Behera, A. and Beena, R. (2024). Next generation nutrition: Genomic and molecular breeding innovations for iron and zinc biofortification in rice. *Rice Science*, *31*(5), 526–544.

Dharmappa, P.M., Doddaraju, P., Malagondanahalli, M.V, Rangappa, R.B., Mallikarjuna, N.M., Rajendrareddy, S.H., Ramanjinappa, R., Mavinahalli, R.P., Prasad, T.G., Udayakumar, M. and Sheshshayee, S.M. (2019). Introgression of root and water use efficiency traits enhances water productivity: An evidence for physiological breeding in rice (*Oryza sativa* L.). *Rice*, 12, 1–14. https://doi.org/10.1186/s12284-019-0268-z

Dixit, S., Singh, U.M., Abbai, R., Ram, T., Singh, V.K., Paul, A., Virk, P.S., Kumar, A. (2019). Identification of genomic region(s) responsible for high iron and zinc content in rice. *Scientific Reports, 9*, 8136. https://doi.org/10.1038/s41598-019-43888-y

Dixit, S., Yadaw, R.B., Mishra, K.K. and Kumar, A. (2017). Marker-assisted breeding to develop the drought-tolerant version of Sabitri, a popular variety from Nepal. *Euphytica*, *213*, 1–16.

Dubouzet, J.G., Sakuma, Y., Ito, Y., Kasuga, M., Dubouzet, E.G., Miura, S., Seki, M., Shinozaki, K., Yamaguchi-Shinozaki, K. (2003). OsDREB genes in rice, Oryza sativa L., encode transcription activators that function in drought-, high-salt-, and cold-responsive gene expression. *Plant Journals*, *33*(4): 751–763.

Edwards, G.I., Nelson, K.M., Le Clec'H, S., Luu, T., Coast, O., Futakuchi, K. and Kok, K. (2024). Twenty-five rice research priorities for sustainable rice systems by 2050. *Global Sustainability*, *7*, e23. https://doi.org/10.1017/SUS.2024.17

Ermakova, M., Danila, F.R., Furbank, R.T. and von Caemmerer, S. (2020). On the road to C4 rice: advances and perspectives. *The Plant Journal*, *101*(4), 940–950. https://doi.org/10.1111/TPJ.14562

Esmaeili, N., Shen, G. and Zhang, H. (2022). Genetic manipulation for abiotic stress resistance traits in crops. *Frontiers in Plant Science*, *13*, 1011985.

Fan, X., Tang, Z., Tan, Y., Zhang, Y., Luo, B., Yang, M. and Xu, G. (2016). Overexpression of a pH-sensitive nitrate transporter in rice increases crop yields. *Proceedings of the National Academy of Sciences of the United States of America*, *113*(26), 7118–7123.

Fang, N., Wang, R., He, W., Yin, C., Guan, C., Chen, H. and Zhang, H. (2016). QTL mapping of panicle blast resistance in japonica landrace Heikezijing and its application in rice breeding. *Molecular Breeding*, *36*(12), 1–8. https://doi.org/10.1007/s11032-016-0603-7

FAO, IFAD, UNICEF, WFP and WHO. (2022). The State of Food Security and Nutrition in the World 2022. Repurposing food and agricultural policies to make healthy diets more affordable. Rome, FAO. https://doi.org/10.4060/cc0639en

Farquhar, G.D., von Caemmerer, S. and Berry, J. A. (1980). A biochemical model of photosynthetic CO2 assimilation in leaves of C3 species. *Planta*, *149*(1), 78–90. https://doi.org/10.1007/BF00386231

Fiyaz, R.A., Shivani, D., Chaithanya, K., Mounika, K., Chiranjeevi, M., Laha, G.S. and Sundaram, R.M. (2022). Genetic improvement of rice for bacterial blight resistance: Present status and future prospects. *Rice Science*, *29*(2), 118–132.

Fiyaz, R.A., Yadav, A.K., Krishnan, S.G., Ellur, R.K., Bashyal, B.M., Grover, N. and Singh, A.K. (2016). Mapping quantitative trait loci responsible for resistance to bakanae disease in rice. *Rice*, *9*(1), 45. https://doi.org/10.1186/s12284-016-0117-2

Fowler, S. and Thomashow, M.F. (2002). Arabidopsis transcriptome profiling indicates that multiple regulatory pathways are activated during cold acclimation in addition to the CBF cold response pathway. *The Plant Cell*, *14*(8), 1675–1690.

Frouin, J., Labeyrie, A., Boisnard, A., Sacchi, G.A., Ahmadi, N., et al. (2019). Genomic prediction offers the most effective marker assisted breeding approach for ability to prevent arsenic accumulation in rice grains. *PLoS ONE, 14*(6), e0217516. DOI: 10.1371/journal.pone.0217516

Fukuoka, S., Saka, N., Koga, H., Ono, K., Shimizu, T., Ebana, K. and Yano, M. (2009). Loss of function of a proline-containing protein confers durable disease resistance in rice. *Science*, *325*(5943), 998–1001.

Ganguly, S., Nimitha, K., Saha, S., Sinha Mahapatra, N., Bhattacharya, K., Kundu, R. and Bhattacharyya, S. (2024). Identification and analysis of low light responsive yield enhancing QTLs in rice. *Scientific Reports*, *14*, 21011. https://doi.org/10.1038/s41598-024-71593-y

Ghosh, S., Watson, A., Gonzalez-Navarro, O.E., Ramirez-Gonzalez, R.H., Yanes, L., Mendoza-Suárez, M. and Hafeez, A. (2018). Speed breeding in growth chambers and glasshouses for crop breeding and model plant research. *Nature Protocols*, *13*(12), 2944–2963.

Giri, J., Bhosale, R., Huang, G., Pandey, B.K., Parker, H., Zappala, S. and Bennett, M.J. (2018). Rice auxin influx carrier OsAUX1 facilitates root hair elongation in response to low external phosphate. *Nature Communications*, *9*(1), 1408.

Gornall, J., Betts, R., Burke, E., Clark, R., Camp, J., Willett, K. and Wiltshire, A. (2010). Implications of climate change for agricultural productivity in the early twenty-first century. *Philosophical Transactions of the Royal Society B: Biological Sciences*, *365*(1554), 2973–2989.

Gregorio, G.B., Islam, M.R., Vergara, G.V. and Thirumeni, S. (2013). Recent advances in rice science to design salinity and other abiotic stress-tolerant rice varieties. *Sabrao Journal of Breeding and Genetics*, *45*, 31–41.

Gu, P., Tao, W., Tao, J., Sun, H., Hu, R., Wang, D. and Zhang, Y. (2023). The D14-SDEL1-SPX4 cascade integrates the strigolactone and phosphate signalling networks in rice. *New Phytologist*, 239(2), 673–686.

Gupta, P.K., Kulwal, P.L. and Jaiswal, V. (2019). Association mapping in plants in the post-GWAS genomics era. *Advances in Genetics*, *104*, 75–154.

Hamaoka, N., Yasui, H., Yamagata, Y., Inoue, Y., Furuya, N., Araki, T., Ueno, O. and Yoshimura, A. (2017). A hairy-leaf gene, BLANKET LEAF, of wild *Oryza nivara* increases photosynthetic water use efficiency in rice. *Rice*, *10*(1), 1–8. https://doi.org/10.1186/S12284-017-0158-1

Haque, Md Anamul and Haque, M. (2016). Growth, yield and nitrogen use efficiency of new rice variety under variable nitrogen rates. *American Journal of Plant Sciences*, 7, 612–622. 10.4236/ajps.2016.73054.

Hariprasanna, K., Agte, V., Elangovan, M. and Patil, J.V. (2014). Genetic variability for grain iron and zinc content in cultivars, breeding lines and selected germplasm accessions of Sorghum [Sorghum bicolor (L.) Moench]. *Indian Journal of Genetics and Plant Breeding*, *74*(1), 42–49.

Hassen, B.M., Cao, T.V., Bartholome, J., Orasen, G., Colombi, C., Rakotomalala, J., Razafinimpiasa, L., Bertone, C., Biselli, C., Volante, A. and Desiderio, F. (2018). Rice diversity panel provides accurate genomic predictions for complex traits in the progenies of biparental crosses involving members of the panel. *Theoretical and Applied Genetics*, *131*, 417–435.

Hattori, Y., Nagai, K., Furukawa, S., Song, X.J., Kawano, R., Sakakibara, H. and Ashikari, M. (2009). The ethylene response factors SNORKEL1 and SNORKEL2 allow rice to adapt to deep water. *Nature*, *460*(7258), 1026–1030.

Hayashi, N., Inoue, H., Kato, T., Funao, T., Shirota, M., Shimizu, T. and Shimizu, T. (2010). Durable panicle blast-resistance gene Pb1 encodes an atypical CC-NBS-LRR protein and was generated by acquiring a promoter through local genome duplication. *Plant Journal*, *64*(3), 498–510.

Heffner, E.L., Lorenz, A.J., Jannink, J.L. and Sorrells, M.E. (2010). Plant breeding with genomic selection: gain per unit time and cost. *Crop Science*, *50*(5), 1681–1690. https://doi.org/10.2135/cropsci2009.11.0662

Hickey, L.T., Amber, N.H., Robinson, H., Jackson, S.A., Leal-Bertioli, S.C.M., Tester, M., Gao, C., Godwin, I.D., Hayes, B.J. and Wulff, B.B.H. (2019). Breeding crops to feed 10 billion. *Nature Biotechnology*, *37*(7), 744–754. https://doi.org/10.1038/s41587-019-0152-6.

Hossain, F., Zunjare, R.U., Muthusamy, V., Bhat, J.S., Mehta, B.K., Sharma, D., et al., (2022). Biofortification of maize for nutritional security. In *Biofortification of staple crops* (pp. 147–174). Singapore: Springer Singapore. https://doi.org/10.1007/978-981-16-3280-8_6

Hu, H., Daim, M., Yao, J., Xiao, B., Li, X., Zhang, Q. and Xiong, L. (2006). Overexpressing a NAM, ATAF, and CUC (NAC) transcription factor enhances drought resistance and salt tolerance in rice. *Proceedings of the National Academy of Sciences of the United States of America*, *103*(35), 12987–12992.

Huang, X.Z., Zeng, X.F., Li, J.R. and Zhao, D.G. (2017). Construction and analysis of tify1a and tify1b mutants in rice (*Oryza sativa*) based on CRISPR/Cas9 technology. *Journal of Agricultural Biotechnology*, *25*, 1003–1012. https://doi.org/10.3969/j.issn.1674-7968.2017.06.015.

Huang, X. and Han, B. (2014). Natural variations and genome-wide association studies in crop plants. *Annual Review of Plant Biology*, *65*, 531–551.

Hubbart, S., Smillie, I.R.A., Heatley, M., Swarup, R., Foo, C.C., Zhao, L. and Murchie, E.H. (2018). Enhanced thylakoid photoprotection can increase yield and canopy radiation use efficiency in rice. *Communications Biology*, *1*(1), 26. https://doi.org/10.1038/s42003-018-0026-6.

Hussain, S., Huang, J., Huang, J., Ahmad, S., Nanda, S., Anwar, S., Shakoor, A., Zhu, C., Zhu, L., Cao, X. and Jin, Q. (2020). Rice production under climate change: adaptations and mitigating strategies. In *Environment, Climate, Plant, and Vegetation Growth*, 659–686.

Jang, I.C., Oh, S.J., Seo, J.S., Choi, W.B., Song, S.I., Kim, C.H., Kim, Y.S., Seo, H.S., DoChoi, Y., Nahm, B.H. and Kim, J.K. (2003). Expression of a bifunctional fusion of the Escherichia coli genes for trehalose-6-phosphatesynthase and trehalose-6-phosphate phosphatase in transgenic rice plants increases trehalose accumulation and abiotic stress tolerance without stunting growth. *Plant Physiology*, *131*, 516–524.

Jang, S., Han, J.-H., Lee, Y.K., Shin, N.-H., Kang, Y.J., Kim, C.-K. and Chin, J.H. (2020). Mapping and validation of QTLs for the amino acid and total protein content in brown rice. *Frontiers in Genetics*, *11*, 240. https://doi.org/10.3389/fgene.2020.00240.

Jeevan, B., Gogoi, R., Sharma, D., et al. (2020). Genetic analysis of maydis leaf blight resistance in subtropical maize (Zea mays L.) germplasm. *Journal of Genetics*, *99*(1), 89. https://doi.org/10.1007/s12041-020-01245-3.

Jeevan, B., Hosahatti, R., Koti, P.S., Devappa, V.H., Ngangkham, U., Devanna, P., et al. (2023). Phenotypic and genotypic screening of fifty-two rice (*Oryza sativa* L.) genotypes for desirable cultivars against blast disease. *PLoS One*, *18*(3), e0280762. https://doi.org/10.1371/journal.pone.0280762.

Jena, K.K., Kim, S.M., Suh, J.P. and Kim, Y.G. (2010). Development of cold-tolerant breeding lines using QTL analysis in rice. In *Second Africa Rice Congress: Innovation and Partnerships to Realize Africa's Rice Potential*.

Jethva, J., Hahn, F., Giuliani, R., Peeters, N., Cousins, A. and Kelly, S. (2024). Realisation of a key step in the evolution of C4 photosynthesis in rice by genome editing. *BioRxiv*. https://doi.org/10.1101/2024.05.21.595093

Jones, H. (1993). Drought tolerance and water-use efficiency. *CAB International*. Retrieved from https://www.cabidigitallibrary.org.

Kalia, S. and Rathour, R. (2019). Current status on mapping of genes for resistance to leaf- and neck-blast disease in rice. *3 Biotech, 9*(1), 1–14.

Kalkuhl, M. and Wenz, L. (2020). The impact of climate conditions on economic production. Evidence from a global panel of regions. *Journal of Environmental Economics and Management*, *103*, 102360.

Kang, D.Y., Cheon, K.S., Oh, J., Oh, H., Kim, S.K., Kim, N., Lee, E., Choi, I., Baek, J., Kim, K.H., Chung, N.J. and Ji, H. (2019). Rice genome resequencing reveals a major quantitative trait locus for resistance to bakanae disease caused by *Fusarium fujikuroi*. *International Journal of Molecular Sciences*, *20*(10), 2598. https://doi.org/10.3390/ijms20102598.

Karki, S., Rizal, G. and Quick, W.P. (2013). Improvement of photosynthesis in rice (Oryza sativa L.) by inserting the C4 pathway. *Rice, 6*(1), 1–8. https://doi.org/10.1186/1939-8433-6-28/TABLES/1.

Katara, J.L., Parameswaran, C., Devanna, B.N., Verma, R.L., Anilkumar, C., Patra, B.C. and Samantaray, S. (2021). Genomics assisted breeding: The need and current perspective for rice improvement in India. *ORYZA – An International Journal of Rice, 58*(1), 61–68.

Katara, J.L., Sonah, H., Deshmukh, R.K., Chaurasia, R. and Kotasthane, A.S. (2010). Molecular analysis of QTLs associated with resistance to brown spot in rice (Oryza sativa L.). *Indian Journal of Genetics and Plant Breeding, 70*(1), 17–21.

Katral, A., Biradar, H., Harijan, Y., Aruna, Y.R., Hadimani, J. and Hittalmani, S. (2022). Genetic analysis and traits association study in marker-assisted multi-drought-traits pyramided genotypes under reproductive-stage moisture stress in rice (Oryza sativa L.). *Euphytica, 218*(3), 21.

Katral, A., Hossain, F., Zunjare, R.U., Mishra, S.J., Ragi, S., Kasana, R., Chhabra, R., Thimmegowda, V., Vasudev, S., Kumar, S., Bhat, J.S., Neeraja, C.N., Yadava, D.K. and Muthusamy, V. (2024). Enrichment of kernel oil and fatty acid composition through genomics-assisted breeding for *dgat1-2* and *fatb* genes in multi-nutrient rich maize. *The Plant Journal, 119*, 2402–2422. https://doi.org/10.1111/tpj.16926.

Khan, M., Raghu, S., Sah, R.P., Azharudheen, T.M., Moharana, D., Behera, S., Mohanty, P.S., Jeevan, B., Behera, L., Marndi, B.C., Samantaray, S. and Anilkumar, C. (2024). Identification of novel QTL for bakanae disease resistance in non-basmati indica rice. *Cereal Research Communications*, 1–10.

Khush, G.S. and Jena, K.K. (2009). Current status and future prospects for research on blast resistance in rice (Oryza sativa L.). In *Advances in genetics, genomics, and control of rice blast disease*, 1–10. Springer, Dordrecht. https://doi.org/10.1007/978-1-4020-9500-9_1.

Kim, J.H., Cho, A.Y., Lim, S.D., Jang, C.S. and Jang, C.S. (2024). Mutation of a RING E3 ligase, OsDIRH2, enhances drought tolerance in rice with low stomata density. *Physiologia Plantarum, 176*(5), e14565. https://doi.org/10.1111/ppl.14565.

Kim, S.I., Andaya, C.B., Newman, J.W., Goyal, S.S. and Tai, T.H. (2008). Isolation and characterization of a low phytic acid rice mutant reveals a mutation in the rice orthologue of maize MIK. *Theoretical and Applied Genetics, 117*(8), 1291–1301.

Kim, Y.S., Kim, S., Shin, S.Y., Park, T.H., Park, H.M., Kim, Y.H., et al. (2014). Overexpression of dehydroascorbate reductase confers enhanced tolerance to salt stress in rice plants (Oryza sativa L. japonica). *Journal of Agronomy and Crop Science, 200*, 444–456. https://doi.org/10.1111/jac.12078.

Krasensky, J. and Jonak, C. (2012). Drought, salt, and temperature stress-induced metabolic rearrangements and regulatory networks. *Journal of Experimental Botany, 63*(4), 1593–1608.

Krause, G. and Weis, E. (1991). Chlorophyll fluorescence and photosynthesis: the basics. Retrieved from https://www.cabidigitallibrary.org.

Krishnamurthy, S.L., Pundir, P., Warraich, A.S., Rathor, S., Lokeshkumar, B.M., Singh, N. K. and Sharma, P.C. (2020). Introgressed saltol QTL lines improve the salinity tolerance in rice at seedling stage. *Frontiers in Plant Science, 11*, 833.

Kuanar, S.R., Molla, K.A., Chattopadhyay, K., Sarkar, R.K. and Mohapatra, P.K. (2019). Introgression of Sub1 (SUB1) QTL in mega rice cultivars increases ethylene production to the detriment of grain-filling under stagnant flooding. *Scientific Reports, 9*(1), 18567.

Kuanar, S.R., Sarkar, R.K., Panigrahi, R. and Mohapatra, P.K. (2023). Introgression of SUB1 aggravates the susceptibility of the popular rice cultivars Swarna and Savitri to stagnant flooding. *Scientific Reports, 13*(1), 9032.

Kumar, A., Kumar, R., Sengupta, D., Das, S.N., Pandey, M.K., Bohra, A., Sharma, N.K., Sinha, P., Sk H., Ghazi, I.A., Laha, G.S. and Sundaram, R. M. (2020a). Deployment of genetic and genomic tools toward gaining a better understanding of rice-*Xanthomonas oryzae* pv. *oryzae* interactions for development of durable bacterial blight resistant rice. *Frontiers in Plant Science, 11*, 1152.

Kumar, A., Nayak, S., Ngangkham, U., Sah, R.P., Lal, M.K., Tp, A., Behera, S., Swain, P., Behera, L. and Sharma, S. (2021). A single nucleotide substitution in the SPDT transporter gene reduced phytic acid and increased mineral bioavailability from rice grain (*Oryza sativa* L.). *Journal of Food Biochemistry, 45*(7):e13822. https://doi.org/10.1111/jfbc.13822

Kumar, A., Sandhu, N., Venkateshwarlu, C., Priyadarshi, R., Yadav, S., Majumder, R.R. and Singh, V.K. (2020b). Development of introgression lines in high yielding, semi-dwarf genetic backgrounds to enable improvement of modern rice varieties for tolerance to multiple abiotic stresses free from undesirable linkage drag. *Scientific Reports, 10*(1), 13073.

Kumar, R., Das, S.P., Choudhury, B.U., Kumar, A., Prakash, N.R., Verma, R., Chakraborti, M., Devi, A.G., Bhattacharjee, B., Das, R. and Das, B. (2024). Advances in genomic tools for plant breeding: harnessing DNA molecular markers, genomic selection, and genome editing. *Biological Research, 57*, 80.

Kumari, S., Sharma, N. and Raghuram, N. (2021). Meta-analysis of yield-related and N-responsive genes reveals chromosomal hotspots, key processes, and candidate genes for nitrogen-use efficiency in rice. *Frontiers in Plant Science, 12*, 627955. https://doi.org/10.3389/fpls.2021.627955.

Kushwaha, A.K., Ellur, R.K., Maurya, S.K., Krishnan, S.G., Bashyal, B.M., Bhowmick, P.K., Vinod, K.K., Bollinedi, H., Singh, N.K. and Singh, A.K. (2023). Fine mapping of qBK1.2, a major QTL governing resistance to bakanae disease in rice. *Frontiers in Plant Science*, *14*, 1265176. https://doi.org/10.3389/fpls.2023.1265176.

Labroo, M.R., Ali, J., Aslam, M.U., de Asis, E.J., dela Paz, M.A., Sevilla, M.A., Lipka, A.E., Studer, A.J. and Rutkoski, J.E. (2021). Genomic prediction of yield traits in single-cross hybrid rice (Oryza sativa L.). *Frontiers in Genetics*, *12*, p. 692870.

Larson, S.R., Rutger, J.N., Young, K.A. and Raboy, V. (2000). Isolation and genetic mapping of a non-lethal rice (*Oryza sativa* L.) *low phytic acid 1* mutation. *Crop Science*, *40*(5):1397–1405.

Lata, C., Bhutty, S., Bahadur, R.P., Majee, M. and Prasad, M. (2011). Association of an SNP in a novel DREB2-like gene SiDREB2 with stress tolerance in foxtail millet [Setaria italica (L.)]. *Journal of Experimental Botany*, *62*(10):3387–3401. https://doi.org/10.1093/jxb/err016.

Lee, K.S., Rasabandith, S., Angeles, E.R. and Khush, G.S. (2003). Inheritance of resistance to bacterial blight in 21 cultivars of rice. *Phytopathology*, *93*(2), 147–152.

Lee, S.B., Kim, N., Hur, Y.J., Cho, S.M., Kim, T.H., Lee, J.Y., Cho, J.H., Lee, J.H., Song, Y.C., Seo, Y.S., Ko, J.M. and Park, D. S. (2019). Fine mapping of qBK1, a major QTL for bakanae disease resistance in rice. *Rice*, *12*, 36. https://doi.org/10.1186/s12284-019-0295-9.

Lee, S.B., Lee, J.Y., Kang, J. W., Mang, H., Kabange, N.R., Seong, G.U., Kwon, Y., Lee, S.M., Shin, D., Lee, J.H., Cho, J.H., Oh, K.W. and Park, D.S. (2022). A novel locus for bakanae disease resistance, qBK4T, identified in rice. *Agronomy*, *12*(10), 2567. https://doi.org/10.3390/agronomy12102567.

Li, G.H., Zhang, Y., Zhou, C., Xu, J. W., Zhu, C.J., Ni, C., Huo, Z.Y., Dai, Q.G. and Xu, K. (2024). Agronomic and physiological characteristics of high yield and nitrogen use efficient varieties of rice: Comparison between two near-isogenic lines. *Food and Energy Security*, *13*(2). https://doi.org/10.1002/FES3.539.

Li, M., Li, X., Zhou, Z., Wu, P., Fang, M., Pan, X., et al. (2016). Reassessment of the four yield-related genes Gn1a, DEP1, GS3, and IPA1 in rice using a CRISPR/Cas9 system. *Frontiers in Plant Science*, *7*, 377. https://doi.org/10.3389/fpls.2016.00377.

Li, R., Chen, S., Matsumoto, H., Gouda, M., Gafforov, Y., Wang, M. and Liu, Y. (2023). Predicting rice diseases using advanced technologies at different scales: present status and future perspectives. *aBIOTECH*, *4*(4), 359–371. https://doi.org/10.1007/s42994-023-00126-4

Li, T., Liu, B., Spalding, M.H., Weeks, D.P. and Yang, B. (2012). High-efficiency TALEN-based gene editing produces disease-resistant rice. *Nature Biotechnology*, *30*, 390–392. https://doi.org/10.1038/nbt.2199.

Li, W.X., Oono, Y., Zhu, J., He, X.J., Wu, J.M., et al. (2008). The Arabidopsis NFYA5 transcription factor is regulated transcriptionally and post-transcriptionally to promote drought resistance. *Plant Cell*, *20*(8):2238–2251. https://doi.org/10.1105/tpc.108.059444.

Li, Z., Pinson, S.R.M., Marchetti, M.A., Stansel, J.W. and Park, W.D. (1995). Characterization of quantitative trait loci (QTLs) in cultivated rice contributing to field resistance to sheath blight (Rhizoctonia solani). *Theoretical and Applied Genetics*, *91*, 382–388.

Lin, H., Karki, S., Coe, R.A., Bagha, S., Khoshravesh, R., Balahadia, C.P., Sagun, J., Tapia, R., Israel, W.K., Montecillo, F., De Luna, A., Danila, F.R., Lazaro, A., Realubit, C.M., Acoba, M.G., Sage, T.L., Von Caemmerer, S., Furbank, R.T., Cousins, A.B. and Covshoff, S. (2016). Targeted knockdown of GDCH in rice leads to a photorespiratory-deficient phenotype useful as a building block for C4 rice. *Plant and Cell Physiology*, *57*(5), 919–932. https://doi.org/10.1093/pcp/pcw033.

Lin, X.H., Zhang, D.P., Xie, Y.F., Gao, H.P. and Zhang, Q.F. (1996). Identifying and mapping a new gene for bacterial blight resistance in rice based on RFLP markers. *Phytopathology*, *86*, 1156–1159.

Liu, G., Jia, Y., Correa-Victoria, F.J., Prado, G.A., Yeater, K.M., McClung, A., et al. (2009). Mapping quantitative trait loci responsible for resistance to sheath blight in rice. *Phytopathology*, *99*, 1078–1084. https://doi.org/10.1094/PHYTO-99-9-1078.

Liu, Q.L, Xu, X.H., Ren, X.L., Fu, H.W., Wu, D.X., and Shu, Q.Y. (2007). Generation and characterization of low phytic acid germplasm in rice (*Oryza sativa* L.). *Theoretical and Applied Genetics*, *114*(5), 803–814. https://doi.org/10.1007/s00122-006-0478-9.

Liu, W., Xu, Z.H., Luo, D. and Xue, H.W. (2003). Roles of OsCKI1, a rice case in kinase i, in root development and plant hormone sensitivity. *Plant Journal*, *36*, 189–202. https://doi.org/10.1046/j.1365-313x.2003.01866.x.

Long, S. and Bernacchi, C.J. (2003). Gas exchange measurements, what can they tell us about the underlying limitations to photosynthesis? *Journal of Experimental Botany*, *54* (392), 2393–2401. https://doi.org/10.1093/jxb/erg262.

Lopez-Gerena, J. (2006). Mapping QTL controlling durable resistance to rice blast in the cultivar Oryzica Llanos 5 (Ph.D. thesis). Plant Pathology, College of Agriculture, Universidad del Valle, Cali, Colombia, Kansas State University, Manhattan, KS, USA.

Lu, H., Wang, F., Wang, Y., Lin, R., Wang, Z. and Mao, C. (2023). Molecular mechanisms and genetic improvement of low-phosphorus tolerance in rice. *Plant, Cell & Environment*, 46(4), 1104–1119.

Lynch, J.P. (2011). Root phenes for enhanced soil exploration and phosphorus acquisition: tools for future crops. *Plant Physiology*, *156*(3), 1041–1049. https://doi.org/10.1104/PP.111.175414

Ma, B.J., Wang, W.M., Zhao, B., Zhou, Y.L., Zhu, L. and Zhai, W. (1999). Studies of PCR marker for the rice bacterial blight resistance gene Xa-4. *Hereditas*, *21*, 9–12.

Ma, J., Lei, C., Xu, X., Hao, K., Wang, J., Cheng, Z. and Wan, J. (2015). Pi64, encoding a novel CC-NBS-LRR protein, confers resistance to leaf and neck blast in rice. *Molecular Plant-Microbe Interactions*, 28(5), 558–568.

Madan, B. and Raghuram, N. (2024). Phenotypic, physiological, and gene expression analysis for nitrogen and phosphorus use efficiencies in three popular genotypes of rice (*Oryza sativa*). *MDPI, Plants, 13*(18), 2567. https://doi.org/10.3390/2223-7747/13/18/2567.

Majumder, S., Datta, K. and Datta, S.K. (2019). Rice biofortification: high iron, zinc, and vitamin-A to fight against "hidden hunger". *Agronomy*, *9*(12), 803.

Mann, A., Mirza, S., Chandra, P., Kumar, A., et al. (2023). Root system architecture and phenotyping for improved resource use efficiency in crops. *Springer*, 229–255. https://doi.org/10.1007/978-981-19-7498-4_11.

Martre, P., Morillon, R., Barrieu, F., North, G.B., Nobel, P.S. and Chrispeels, M.J. (2002). Plasma membrane aquaporins play a significant role during recovery from water deficit. *Plant Physiology*, *130*(4), 2101–2110. https://doi.org/10.1104/pp.009019

Matić, S., Bagnaresi, P., Biselli, C., et al. (2016). Comparative transcriptome profiling of resistant and susceptible rice genotypes in response to the seedborne pathogen *Fusarium fujikuroi*. *BMC Genomics*, *17*, 608. https://doi.org/10.1186/s12864-016-2925-6.

Meuwissen, T.H.E., Hayes, B.J. and Goddard, M.E. (2001). Prediction of total genetic value using genome-wide dense marker maps. *Genetics*, *157*, 1819–1829.

Mishra, R., Joshi, R.K. and Zhao, K. (2018). Genome editing in rice: Recent advances, challenges, and future implications. *Frontiers in Plant Science*, *9*, 1361.

Mohanty, S., Nayak, A.K., Tripathi, R., Bhaduri, D., Chatterjee, D., Kumar, A., Shahid, M., Kumar, U., Munda, S., Mandi, G. and Pathak, H. (2023). Nitrogen use efficiency of rice in India: A regional analysis. *International Journal of Sustainable Development & World Ecology*, *30*(8), 869–882.

Monteith, J.L. (1972). Solar radiation and productivity in tropical ecosystems. *Journal of Applied Ecology*, *9*(3), 747–766.

Monteith, J.L. (1977). Climate and the efficiency of crop production in Britain. *Philosophical Transactions of the Royal Society of London. B, Biological Sciences*, *281*(980), 277–294. https://doi.org/10.1098/rstb.1977.0140

Muralidharan, K., Prasad, G.S., Rao, C.S. and Siddiq, E.A. (2019). Genetic gain for yield in rice breeding and rice production in India to meet with the demand from increased human population. *Current Science*, *116*(4), 544.

Nagai, I. and Hara, S. (1930). On the inheritance of variegation disease in a strain of rice plant. *Japanese Journal of Genetics*, *5*, 140–144.

Nguyen, G. (2018). Improving nitrogen use efficiency in plants: Effective phenotyping in conjunction with agronomic and genetic approaches. CSIRO Publishing.

Nguyen, G., Panozzo, J., et al. (2016). Phenotyping approaches to evaluate nitrogen-use efficiency related traits of diverse wheat varieties under field conditions. CSIRO Publishing. https://www.publish.csiro.au/CP/CP16173.

Nguyen, V.H., Morantte, R.I.Z., Lopena, V., Verdeprado, H., Murori, R., Ndayiragije, A., Katiyar, S.K., Islam, M.R., Juma, R.U., Flandez-Galvez, H. and Glaszmann, J. C. (2023). Multi-environment genomic selection in rice elite breeding lines. *Rice*, *16*(1), 7.

Niroula, R.K., Pucciariello, C., Ho, V.T., Novi, G., Fukao, T. and Perata, P. (2012). SUB1A-dependent and independent mechanisms are involved in the flooding tolerance of wild rice species. *Plant Journal*, *72*(2), 282–293.

Nomura, M., Higuchi, T., Ishida, Y., et al. (2005). Differential expression pattern of C4 bundle sheath expression genes in rice, a C3 plant. *Plant and Cell Physiology*, *46*(5), 754–760.

Oakley, Jr., G.P., Bell, K.N. and Weber, M.B. (2004). Recommendations for accelerating global action to prevent folic acid–preventable birth defects and other folate-deficiency diseases: Meeting of experts on preventing folic acid–preventable neural tube defects. *Birth Defects Research Part A: Clinical and Molecular Teratology*, *70*(11), 835–837.

Paine, J.A., Shipton, C.A., Chaggar, S., Howells, R.M., Kennedy, M.J., Vernon, G. and Drake, R. (2005). Improving the nutritional value of Golden Rice through increased pro-vitamin A content. *Nature Biotechnology*, *23*(4), 482–487.

Panda, G., Sahu, C., Yadav, M.K., Aravindan, S., Umakanta, N., Raghu, S., et al. (2017). Morphological and molecular characterization of *Magnaporthe oryzae* from Chhattisgarh. *ORYZA – An International Journal on Rice*, *54*, 330–336.

Pandey, B.K., Mehra, P., Verma, L., Bhadouria, J. and Giri, J. (2017). OsHAD1, a haloacid dehalogenase-like APase, enhances phosphate accumulation. *Plant Physiology*, *174*(4), 2316–2332.

Park, H.C., Kim, M.L., Kang, Y.H., Jeon, J.M., Yoo, J.H., Kim, M.C. and Yoon, H.W. (2004). Pathogen- and NaCl-induced expression of the SCaM-4 promoter is mediated in part by a GT-1 box that interacts with a GT-1-like transcription factor. *Plant Physiology*, *135*(4), 2150–2161. https://doi.org/10.1104/pp.104.041442.

Patel-Tupper, D., Kelikian, A., Leipertz, A., Maryn, N., Tjahjadi, M., Karavolias, N.G., Cho, M.J. and Niyogi, K.K. (2024). Multiplexed CRISPR-Cas9 mutagenesis of rice PsbS1 noncoding sequences for transgene-free overexpression. *Science Advances*, *10*(23). https://doi.org/10.1126/SCIADV.ADM7452

Pathak, H. (1999). Fertilizer and Nitrate Pollution in India. Managing Risks of Nitrates to Humans and the Environment, 228–239. https://doi.org/10.1533/9781845693206.228

Pathak, H., Mishra, J.P. and Mohapatra, T. (2022). Indian agriculture after independence. *Indian Council of Agricultural Research, New Delhi*, *110*(001), 426.

Pathak, H., Mohanty, S., Jain, N. and Bhatia, A. (2010). Nitrogen, phosphorus, and potassium budgets in Indian agriculture. *Nutrient Cycling in Agroecosystems*, *86*, 287–299.

Patil, N., Raghu, S., Mohanty, L., et al. (2024). Rhizosphere bacteria isolated from medicinal plants improve rice growth and induce systemic resistance in host against pathogenic fungus. *Journal of Plant Growth Regulation*, *43*, 770–786. https://doi.org/10.1007/s00344-023-11137-2.

Patra, B.C., Anilkumar, C. and Chakraborti, M. (2020). Rice breeding in India: A journey from phenotype-based pure-line selection to genomics-assisted breeding. *Agricultural Research Journal*, *57*(6), 816–825. https://doi.org/10.5958/2395-146X.2020.00120.9.

Peng, S., Laza, R.C., Khush, G.S., Sanico, A.L., Visperas, R.M. and Garcia, F.V. (1998). Transpiration efficiencies of indica and improved tropical japonica rice grown under irrigated conditions. *Euphytica*, *103*, 103–108.

Pixley, K.V., Cairns, J.E., Lopez-Ridaura, S., Ojiewo, C.O., Dawud, M.A., Drabo, I. and Zepeda-Villarreal, E.A. (2023). Redesigning crop varieties to win the race between climate change and food security. *Molecular Plant*, *16*(8), 1590–1611.

Porter, B.W., Chittoor, J.M., Yano, M., Sasaki, T. and White, F.F. (2003). Development and mapping of markers linked to the rice bacterial blight resistance gene, Xa7. *Crop Science*, *43*, 1484–1492. https://doi.org/10.2135/cropsci2003.1484.

Pradhan, B., Panda, D., Bishi, S.K., Chakraborty, K., Muthusamy, S.K. and Lenka, S.K. (2022). Progress and prospects of C4 trait engineering in plants. *Plant Biology*, *24*(6), 920–931. https://doi.org/10.1111/PLB.13446

Pradhan, S.K., Pandit, E., Pawar, S., Baksh, S.Y., Mukherjee, A.K. and Mohanty, S.P. (2019). Development of flash-flood tolerant and durable bacterial blight resistant versions of mega rice variety 'Swarna' through marker-assisted backcross breeding. *Scientific Reports*, *9*, 12810. https://doi.org/10.1038/s41598-019-49176-z.

Prakash, C., Mithra, S.V.A., Singh, P.K., Mohapatra, T. and Singh, N.K. (2016). Unraveling the molecular basis of oxidative stress management in a drought-tolerant rice genotype Nagina 22. *BMC Genomics*, *17*, 774–787. https://doi.org/10.1186/s12864-016-3131-2

Price, A.H., Young, E.M. and Tomos, A.D. (1997). Quantitative trait loci associated with stomatal conductance, leaf rolling, and heading date mapped in upland rice (*Oryza sativa*). *New Phytologist*, 137, 83–91.

Quarrie, S.A., Laurie, D.A., Zhu, J., Lebreton, C., Semikhodskii, A., Steed, A., Witsenboer, A. and Calestani, C. (1997). QTL analysis to study the association between leaf size and abscisic acid accumulation in droughted rice leaves and comparisons across cereals. *Plant Molecular Biology*, 35, 155–165.

Rabbani, M.A., Maruyama, K., Abe, H., Khan, M.A., Katsura, K., Ito, Y. and Yamaguchi-Shinozaki, K. (2003). Monitoring expression profiles of rice genes under cold, drought, and high-salinity stresses and abscisic acid application using cDNA microarray and RNA gel-blot analyses. *Plant Physiology*, *133*(4), 1755–1767.

Raghu, S., Baite, M.S., Yadav, M.K., Prabhukarthikeyan, S.R., et al. (2023). Population structure, genetic diversity, and bakanae disease resistance among rice varieties. *Plant Genetic Resources: Characterization and Utilization*, 1–9. https://doi.org/10.1017/S1479262123000199

Rana, M.M., Takamatsu, T., Baslam, M., Kaneko, K., Itoh, K., Harada, N., Sugiyama, T., Ohnishi, T., Kinoshita, T., Takagi, H. and Mitsui, T. (2019). Salt tolerance improvement in rice through efficient SNP marker-assisted selection coupled with speed-breeding. *International Journal of Molecular Sciences*, *20*(10), 2585.

Rashid, A., Achary, V.M.M., Abdin, M.Z., Karippadakam, S., Parmar, H., Panditi, V., Prakash, G., Bhatnagar-Mathur, P. and Reddy, M.K. (2024). Cytokinin oxidase2-deficient mutants improve panicle and grain architecture through cytokinin accumulation and enhance drought tolerance in indica rice. *Plant Cell Reports*, *43*(8). https://doi.org/10.1007/S00299-024-03289-6

Reich, M., Aghajanzadeh, T. and De Kok, L.J. (2014). Physiological Basis of Plant Nutrient Use Efficiency – Concepts, Opportunities and Challenges for Its Improvement. In: Hawkesford, M., Kopriva, S., De Kok, L. (eds) Nutrient Use Efficiency in Plants. Plant Ecophysiology, vol 10. Springer, Cham. https://doi.org/10.1007/978-3-319-10635-9_1

Ricachenevsky, F.K., Vasconcelos, M.W., Shou, H.X., Johnson, A.A.T. and Sperotto, R.A. (2019). Editorial: Improving the nutritional content and quality of crops: Promises, achievements, and future challenges. *Frontiers in Plant Science*, *10*, 738.

Rojaria, V., Hossain, F., Zunjare, R.U., Bhatt, V., Katral, A., Kasana, R.K., et al. and Muthusamy, V. (2024). Development and characterization of lpa1 and lpa2-based low phytate double mutants in maize for enhancing the nutritional quality of food and feed. *Journal of Food Composition and Analysis*, *136*, 106771.

Sah, R.P., Kumar, A., Dash, S.K., Azharudheen, M.T.P and Pradhan, S.K. (2021). Speed breeding–A unique tool for higher genetic gain in rice. *Advances in Rice Breeding: Stress Tolerance*, 316.

Sahoo, J., Mahapatra, D., Mahapatra, M., Dweh, T.J., Kayastha, S., Pradhan, P., Tripathy, S.K., Samal, K., Mishra, A., Dash, M. and Nanda, S. (2024). Understanding the biochemical, physiological, molecular, and synthetic biology approaches towards the development of C4 rice (Oryza sativa L.). *Cereal Research Communications*, *52*, 1459–1471. https://doi.org/10.1007/s42976-024-00489-4.

Saijo, Y., Kinoshita, N., Ishiyama, K., Hata, S., Kyozuka, J., Hayakawa, T. and Yamaya, T.I.K. (2001). A $Ca^{2+}$-dependent protein kinase that endows rice plants with cold and salt stress tolerance functions in vascular bundles. *Plant Cell Physiology*, *42*(11), 1228–1233. https://doi.org/10.1093/pcp/pce158

Sandhu, N. and Kumar, A. (2017). Bridging the rice yield gaps under drought: QTL, genes, and their use in breeding programs. *Agronomy, 7*(27). https://doi.org/10.3390/agronomy7020027.

Sato, H., Ando, I., Hirabayashi, H., Takeuchi, Y., Arase, S., Kihara, J., Kato, H., Imbe, T. and Nemoto, H. (2008). QTL analysis of brown spot resistance in rice (*Oryza sativa* L.). *Breeding Science*, *58*(1), 93–96.

Sato, H., Matsumoto, K., Ota, C., Yamakawa, T., Kihara, J. and Mizobuchi, R. (2015). Confirming a major QTL and finding additional loci responsible for field resistance to brown spot (Bipolaris oryzae) in rice. *Breeding Science*, *65*(2), 170–175.

Schindler, D.W., Carpenter, S.R., Chapra, S.C., Hecky, R.E. and Orihel, D.M. (2016). Reducing phosphorus to curb lake eutrophication is a success. *Environmental Science & Technology*, *50*, 8923–8929.

Schroeder, J.I., Kwak, J.M. and Allen, G.J. (2001). Guard cell abscisic acid signalling and engineering drought hardiness in plants. *Nature*, *410*, 327–330.

Secco, D., Baumann, A. and Poirier, Y. (2010). Characterization of the rice PHO1 gene family reveals a key role for OsPHO1; 2 in phosphate homeostasis and the evolution of a distinct clade in dicotyledons. *Plant Physiology*, *152*(3), 1693–1704.

Seeli, F.D.P., Manoharan, M., Ayyenar, B., Kambale, R., Mohanavel, V., Rajagopalan, V.R., Manickam, S., Muthurajan, R. and Swaminathan, M. (2024). Genetic improvement of drought tolerance in a mega-rice variety improved White Ponni through marker-assisted backcross breeding. *Agriculture*, *14*(3), 431. https://doi.org/10.3390/agriculture14030431.

Seki, M., Narusaka, M., Ishida, J., Nanjo, T., Fujita, M., Oono, Y. and Shinozaki, K. (2002). Monitoring the expression profiles of 7000 Arabidopsis genes under drought, cold and high-salinity stresses using a full-length cDNA microarray. *The Plant Journal*, *31*(3), 279–292.

Sen, P., Ghosh, S., Sarkar, S.N., Chanda, P., Mukherjee, A., Datta, S.K. and Datta, K. (2017). Pyramiding of three C4 specific genes towards yield enhancement in rice. *Plant Cell, Tissue and Organ Culture*, *128*(1), 145–160. https://doi.org/10.1007/S11240-016-1094-2.

Seguttuvel, P., Padmavathi, G., Jasmine, C., Sanjeeva Rao, D., Neeraja, C.N., Jaldhani, V., Beulah, P., Gobinath, R., Aravindkumar, J., Sai Prasad, S.V., Subba Rao, L.V., Hariprasad, A.S., Sruthi, K., Shivani, D., Sundaram, R.M. and Govindaraj, M. (2023). Rice biofortification: breeding and genomic approaches for genetic enhancement of grain zinc and iron contents. *Frontiers in Plant Science*, *14*, 1138408.

Septiningsih, E.M., Ignacio, J.C.I., Sendon, P.M.D., Sanchez, D.L., Ismail, A.M. and Mackill, D.J. (2013). QTL mapping and confirmation for tolerance of anaerobic conditions during germination derived from the rice landrace. *Theoretical and Applied Genetics*, *126*(6), 1357–1366.

Sharma, N., Madan, B., Khan, M.S., Sandhu, K.S. and Raghuram, N. (2023). Weighted gene co-expression network analysis of nitrogen (N)-responsive genes and the putative role of G-quadruplexes in N use efficiency (NUE) in rice. *Frontiers in Plant Science, 14*, 1135675. https://doi.org/10.3389/FPLS.2023.1135675/BIBTEX.

Shin, S.H., Kim, K.Y., Shin, M.S., Ko, J.K. and Kim, C.K. (2007). Identification of rice varieties containing bacterial blight resistance gene Xa1 by SNP marker (16PFXa). *Korean Journal of Breed Science*, *39*, 63–69.

Shivappa, R., B. Navadagi, D., Seikholen Baite, M., Kumar Yadav, M., S. Rathinam, P., Umapathy, K. and Chandra Rath, P. (2022). Emerging Minor Diseases of Rice in India: Losses and Management Strategies. *IntechOpen*. doi: 10.5772/intechopen.99898

Singh, A.K., Gopalakrishnan, S., Singh, V.P., Prabhu, K.V., Mohapatra, T., Singh, N.K., et al. (2011a). Marker assisted selection: a paradigm shift in Basmati breeding. *Indian Journal of Genetics*, *71*, 120–128.

Singh, B.K., Ramkumar, M.K., Dalal, M., Singh, A., Solanke, A.U., Singh, N.K. and Sevanthi, A.M. (2021). Allele mining for a drought responsive gene DRO1 determining root growth angle in donors of drought tolerance in rice (Oryza sativa L.). *Physiology and Molecular Biology of Plants*, *27*, 523–534.

Singh, B.P., Jayaswal, P.K., Singh, B., Singh, P.K., Kumar, V., Mishra, S. and Singh, N.K. (2015). Natural allelic diversity in OsDREB1F gene in the Indian wild rice germplasm led to ascertain its association with drought tolerance. *Plant Cell Reports*, *34*(6), 993–1004. https://doi.org/10.1007/s00299-015-1760-6.

Singh, R., Singh, Y., Xalaxo, S., Verulkar, S., Yadav, N., Singh, S. and Rao, P.V.R. (2016). From QTL to variety—harnessing the benefits of QTLs for drought, flood, and salt tolerance in mega rice varieties of India through a multi-institutional network. *Plant Science*, *242*, 278–287. doi: 10.1016/j.plantsci.2015.08.008.

Singh, U.D., Gogoi, R. and Mondal, K.K. (2011b). Molecular approach for augmenting disease resistance in cereals: Rice and maize. *Plant Pathology in India*, *20*, 21–30.

Smith, J.S., Hussain, T., Jones, E.S., Graham, G., Podlich, D., et al. (2008). Use of doubled haploids in maize breeding: implications for intellectual property protection and genetic diversity in hybrid crops. *Molecular Breeding 22*(1), 51–59.

Spindel, J., Begum, H., Akdemir, D., Virk, P., Collard, B., Redona, E., Atlin, G., Jannink, J.L., and McCouch, S.R., (2015). Genomic selection and association mapping in rice (*Oryza sativa*): effect of trait genetic architecture, training population composition, marker number and statistical model on accuracy of rice genomic selection in elite, tropical rice breeding lines. *PLoS Genetics*, *11*(2), e1004982.

Sridhar, R., Redd, R.N., Shanti, M.L. and Singh, U.D. (2001). Characterization of rice bacterial blight pathogen isolates from eastern India and breeding for bacterial resistance by molecular tools. In S. Nagarajan & D. P. Singh (Eds.), *Role of Resistance in Intensive Agriculture*, 183–194, Kalyani, New Delhi.

Steele, K.A., Singh, D.N., Kumar, R., Prasad, S.C., Virk, D.S., Gangwar, J.S. and Witcombe, J.R. (2002). Combining molecular marker technology and participatory techniques: A case study for drought-tolerant rice in eastern India II: Farmer evaluation of SLSMAS bulks in participatory plant breeding. In: Breeding rainfed rice for droughtprone environments: integrating conventional and participatory plant breeding in South and Southeast Asia. (Eds.) J. R. Witcombe, L. B. Parr, and G. N. Atlin. Proceedings of a DFID Plant Sciences Research Programme/IRRI Conference, 12–15 March 2002, Philippines: IRRI, Los Baños, Laguna, pp: 29–31.

Sun, Y., Jiao, G., Liu, Z., Zhang, X., Li, J., Guo, X., et al. (2017). Generation of high-amylose rice through CRISPR/Cas9-mediated targeted mutagenesis of starch branching enzymes. *Frontiers in Plant Science*, *8*, 298. https://doi.org/10.3389/fpls.2017.00298

Svensson, J.T., Crosatti, C., Campoli, C., Bassi, R., Stanca, A.M., Close, T.J. and Cattivelli, L. (2006). Transcriptome analysis of cold acclimation in barley Albina and Xantha mutants. *Plant Physiology*, *141*(1), 257–270. https://doi.org/10.1104/pp.105.072645

Swain, A., Behera, D., Karmakar, S., Dash, M., Dash, B.P., Swain, P., Molla, K.A. and Baig, M.J. (2021). Morphophysiological alterations in transgenic rice lines expressing PPDK and ME genes from the C4 model Setaria italica. *Journal of Plant Physiology*, *264*, 153482. https://doi.org/10.1016/J.JPLPH.2021.153482.

Swamy, B.P.M., Ahmed, H.U., Henry, A., Mauleon, R., Dixit, S., Vikram, P. and Kumar, A. (2013). Genetic, physiological, and gene expression analyses reveal that multiple QTL enhance yield of rice mega-variety IR64 under drought. *PLoS One*, *8*(5), e62795.

Sysoeva, M.I., Markovskaya, E.F. and Shibaeva, T.G. (2010). Plants under continuous light: A review. *Plant Stress*, *4*, 5–17.

Taguchi-Shiobara, F., Ozaki, H., Sato, H., Maeda, H., Kojima, Y., Ebitani, T. and Yano, M. (2013). Mapping and validation of QTLs for rice sheath blight resistance. *Breeding Science*, *63*, 301.

Takeda, S. and Matsuoka, M. (2008). Genetic approaches to crop improvement: responding to environmental and population changes. *Nature Reviews Genetics*, *9*(6), 444–457.

Tan, Q., He, H., Chen, W., Huang, L., Zhao, D., Chen, X., Li, J. and Yang, X. (2022). Integrated genetic analysis of leaf blast resistance in upland rice: QTL mapping, bulked segregant analysis and transcriptome sequencing. *AoB Plants*, *14*(6), plac047. https://doi.org/10.1093/aobpla/plac047

Tang, X., Lowder, L.G., Zhang, T., Malzahn, A.A., Zheng, X., Voytas, D.F., et al. (2017). A CRISPR-Cpf1 system for efficient genome editing and transcriptional repression in plants. *Nature Plants*, *3*, 17018. https://doi.org/10.1038/nplants.2017.18

Thomas, W.T.B., Forster, B.P. and Gertsson, B. (2003). Doubled haploids inbreeding. In: Doubled haploid production in crop plants. *Springer*, Dordrecht, 337–349.

Tirol-Padre, A., Ladha J.K., Singh U., Laureles E., Punzalan G. and Akita S. (1996). Grain yield performance of rice genotypes at suboptimal levels of soil N as affected by N uptake and utilization efficiency. *Field Crops Research*, *46*, 127–143.

Tyagi, A., Chakraborty, N. and Raghuram, N. (2024). Morphophysiological evaluation of indica rice genotypes with contrasting crop duration for nitrogen use efficiency under graded urea doses. *Journal of Plant Growth Regulation*, *43*, 4826–4847. https://doi.org/10.1007/s00344-024-11437-1

Uga, Y., Kitomi, Y., Yamamoto, E., Kanno, N., Kawai, S., Mizubayashi, T. and Fukuoka, S. (2015). A QTL for root growth angle on rice chromosome 7 is involved in the genetic pathway of DEEPER ROOTING 1. *Rice*, *8*, 1–8.

Uga, Y., Sugimoto, K., Ogawa, S., Rane, J., Ishitani, M., Hara, N. and Yano, M. (2013). Control of root system architecture by DEEPER ROOTING 1 increases rice yield under drought conditions. *Nature Genetics*, *45*(9), 1097–1102.

Uttam, G.A., Suman, K., Jaldhani, V., Babu, P.M., Rao, D.S., Sundaram, R.M. and Neeraja, C.N. (2022). Identification of genomic regions associated with high grain Zn content in polished rice using genotyping-by-sequencing (GBS). *Plants*, *12*(1), 144. https://doi.org/10.3390/plants12010144

Wang, E., Wang, J., Zhu, X., Hao, W., Wang, L., Li, Q. and Ma, H. (2008). Control of rice grain-filling and yield by a gene with a potential signature of domestication. *Nature Genetics*, *40*(11), 1370–1374. https://doi.org/10.1038/ng.220

Wang, F., Deng, M., Xu, J., Zhu, X. and Mao, C. (2018, February). Molecular mechanisms of phosphate transport and signaling in higher plants. In Seminars in cell & developmental biology (Vol. 74, pp. 114–122). Academic Press.

Wang, F., Wang, C., Liu, P., Lei, P., Hao, W., Gao, Y., et al. (2016a). Enhanced rice blast resistance by CRISPR/Cas9-Targeted Mutagenesis of the ERF Transcription Factor Gene OsERF922. *PLoS One*, *11*, e0154027. https://doi.org/10.1371/journal.pone.0154027

Wang, G.L. and Valent, B. (2017). Durable resistance to rice blast. *Science*, *355*, 906–907. https://doi.org/10.1126/science.aam9517

Wang, R., Fang, W., Guan, C., He, W., Bao, Y. and Zhang, H. (2016b). Characterization and fine mapping of a blast resistant gene Pi-jnw1 from the japonica rice landrace Jiangnanwan. *PLoS One*, *11*, 1–12. https://doi.org/10.1371/journal.pone.0169417

Wang, S., Liu, W., Lu, D., Lu, Z., Wang, X., Xue, J. and He, X. (2020). Distribution of bacterial blight resistance genes in the main cultivars and application of Xa23 in rice breeding. *Frontiers in Plant Science*, *11*, 555228. https://doi.org/10.3389/fpls.2020.555228

Wang, S., Xu, Y., Qu, H., Cui, Y., Li, R., Chater, J.M., Yu, L., Zhou, R., Ma, R., Huang, Y. and Qiao, Y. (2021). Boosting predictabilities of agronomic traits in rice using bivariate genomic selection. *Briefings in Bioinformatics, 22*(3), bbaa103.

Wasaki, J., Yamamura, T., Shinano, T. and Osaki, M. (2003). Secreted acid phosphatase is expressed in cluster roots of lupin in response to phosphorus deficiency. *Plant and Soil*, 248, 129–136.

Watson, A., Ghosh, S., Williams, M.J., Cuddy, W.S., Simmonds, J., Rey, M.D., Hatta, M.A.M., Hinchliffe, A., Steed, A., Reynolds, D. and Adamski, N. M. (2018). Speed breeding is a powerful tool to accelerate crop research and breeding. *Nature Plants, 4*(1), 23–29.

Watson, A., Hickey, L.T., Christopher, J., Rutkoski, J., Poland, J. and Hayes, B.J. (2019). Multivariate genomic selection and potential of rapid indirect selection with speed breeding in spring wheat. *Crop Science*, *59*(5), 1945–1959.

Wei, D., Cui, K., Ye, G., Pan, J., Xiang, J., Huang, J. and Nie, L. (2012). QTL mapping for nitrogen-use efficiency and nitrogen-deficiency tolerance traits in rice. *Plant Soil*, *359*, 281–295.

Wei, S., Li, X., Lu, Z., Zhang, H., Ye, X., Zhou, Y., Li, J., Yan, Y., Pei, H., Duan, F., et al. (2022). A transcriptional regulator that boosts grain yields and shortens the growth duration of rice. *Science*, *377*, eabi8455.

Xu, D., Duan, X., Wang, B., Hong, B., Ho, T.H.D. and Wu, R. (1996). Expression of a late embryogenesis abundant protein gene, HVA1, from barley confers tolerance to water deficit and salt stress in transgenic rice. *Plant Physiology*, *110*(1), 249–257.

Xu, K. and Mackill, D.J. (1996). A major locus for submergence tolerance mapped on rice chromosome 9. *Molecular Breeding*, *2*(3), 219–224. https://doi.org/10.1007/BF00564199

Xu, K., Xu, X., Fukao, T., Canlas, P., Maghirang-Rodriguez, R., Heuer, S. and Mackill, D.J. (2006). Sub1A is an ethylene-response-factor-like gene that confers submergence tolerance to rice. *Nature*, *442*(7103), 705–708.

Xu, Q., Yuan, X., Yu, H., Wang, Y., Tang, S. and Wei, X. (2011). Mapping quantitative trait loci for sheath blight resistance in rice using double haploid population. *Plant Breeding, 130*, 404–406. https://doi.org/10.1111/j.1439-0523.2010.01806.x

Xu, R., Yang, Y., Qin, R., Li, H., Qiu, C., Li, L., et al. (2016). Rapid improvement of grain weight via highly efficient CRISPR/Cas9-mediated multiplex genome editing in rice. *Journal of Genetics and Genomics*, *43*, 529–532. https://doi.org/10.1016/j.jgg.2016.07.003

Yamaguchi, I. (2004). Overview on the Chemical Control of Rice Blast Disease. *Kluwer Academic Publishers*. 1–13.

Yamaji, N., Takemoto, Y., Miyaji, T., Mitani-Ueno, N., Yoshida, K.T. and Ma, J.F. (2017). Reducing phosphorus accumulation in rice grains with an impaired transporter in the node. *Nature*, 541(7635), 92–95.

Yang, C.D., Kuo, L.B., Ji, Z.J., et al. (2006). Analysis of QTLs for resistance to rice bakanae disease. *Chinese Journal of Rice Science, 20*, 657–659.

Yang, Y., Guo, M., Sun, S., Zou, Y., Yin, S., Liu, Y. and Yan, C. (2019). Natural variation of OsGluA2 is involved in grain protein content regulation in rice. *Nature Communications*, *10*(1), 1–2. https://doi.org/10.1038/s41467-019-09919-y.

Yang, Y., Yu, J., Qian, Q. and Shang, L. (2022). Enhancement of heat and drought stress tolerance in rice by genetic manipulation: A systematic review. *Rice*, *15*(1), 67.

Yanru, C., Li, R., Li, G., Zhang, F., Zhu, T., et al. (2020). Hybrid breeding of rice via genomic selection. *Plant Biotechnology Journal, 18*(1), 57–67.

Yin, X., Biswal, A.K., Dionora, J., Perdigon, K.M., Balahadia, C.P., Mazumdar, S., Chater, C., Lin, H.C., Coe, R.A., Kretzschmar, T., Gray, J.E., Quick, P.W. and Bandyopadhyay, A. (2017). CRISPR-Cas9 and CRISPR-Cpf1 mediated targeting of a stomatal developmental gene EPFL9 in rice. *Plant Cell Reports, 36*(5), 745–757. https://doi.org/10.1007/S00299-017-2118-Z

Ying, Y., Yue, W., Wang, S., Li, S., Wang, M., Zhao, Y. and Shou, H. (2017). Two h-type thioredoxins interact with the E2 ubiquitin conjugase PHO2 to fine-tune phosphate homeostasis in rice. *Plant Physiology*, *173*(1), 812–824.

Yu, J., Hu, S., Wang, J., Wong, G.K.S., Li, S., Liu, B., Deng, Y., Dai, L., Zhou, Y., Zhang, X., Cao, M., Liu, J., Sun, J., Tang, J., Chen, Y., Huang, X., Lin, W., Ye, C., Tong, W., Yang, H. (2002). A draft sequence of the rice genome (*Oryza sativa* L. ssp. indica). *Science*, *296*(5565), 79–92. https://doi.org/10.1126/SCIENCE.1068037

Yu, Y.H., Li, G., Fan, Y.Y., Zhang, K.Q., Min, J., Zhu, Z.W. and Zhuang, J.Y. (2009). Genetic relationship between grain yield and the contents of protein and fat in a recombinant inbred population of rice. *Journal of Cereal Science*, *50*(1), 121–125. https://doi.org/10.1016/j.jcs.2009.03.008

Yue, B., Xue, W., Xiong, L., Yu, X., Luo, L., Cui, K. and Zhang, Q. (2006). Genetic basis of drought resistance at reproductive stage in rice: separation of drought tolerance from drought avoidance. *Genetics*, *172*(2), 1213–1228.

Zarbafi, S.S. and Ham, J.H. (2019). An overview of rice QTLs associated with disease resistance to three major rice diseases: Blast, sheath blight, and bacterial panicle blight. *Agronomy*, *9*(4), 177.

Zhang, L., Lin, S., Bouman, B.A.M., Xue, C., Wei, F., Tao, H., Yang, X., Wang, H., Zhao, D. and Dittrt, K. (2009). Response of aerobic rice growth and grain yield to N fertilizer at low contrasting sites near Beijing, China. *Field Crops Research*, *114*, 45–53.

Zhang, Q., Lin, S.C., Zhao, B.Y., Wang, C.L., Yang, W.C., Zhou, Y.L., Li, D.Y., Chen, C.B. and Zhu, L.H. (1998). Identification and tagging a new gene for resistance to bacterial blight (*Xanthomonas oryzae* pv. *oryzae*) from *O. rufipogon*. *Rice Genetics Newsletter*, *15*, 138–142.

Zhang, Y., Zhang, M., Ye, J., Xu, Q., Feng, Y., Xu, S., Hu, D., Wei, X., Hu, P. and Yang, Y. (2023). Integrating genome-wide association study into genomic selection for the prediction of agronomic traits in rice (*Oryza sativa* L.). *Molecular Breeding*, *43*(11), 81.

Zhao, H., Frank, T., Tan, Y., Zhou, C., Jabnoune, M., Arpat, A.B. and Shu, Q. (2016). Disruption of Os SULTR 3; 3 reduces phytate and phosphorus concentrations and alters the metabolite profile in rice grains. *New Phytologist*, 211(3), 926–939.

Zhao, H., Wang, X., Jia, Y., Minkenberg, B., Wheatley, M., Fan, J., Jia, M.H., Famoso, A., Edwards, D., Wamishe, Y., Valent, B., Wang, G. L. and Yang, Y. (2018). The rice blast resistance gene Ptr encodes an atypical protein required for broad-spectrum disease resistance. *Nature Communications*, *9*(1), 2039. https://doi.org/10.1038/s41467-018-04369-4

Zheng, B.S., Yang, L., Zhang, W.P., Mao, C.Z., Wu, Y.R., Yi, K.K., Liu, F.Y. and Wu, P. (2003). Mapping QTLs and candidate genes for rice root traits under different water-supply conditions and comparative analysis across three populations. *Theoretical and Applied Genetics*, *107*, 1505–1515.

Zheng, S., Ye, C., Lu, J., Liufu, J., Lin, L., Dong, Z., Li, J. and Zhuang, C. (2021). Improving the rice photosynthetic efficiency and yield by editing oshxk1 via CRISPR/Cas9 system. *International Journal of Molecular Sciences*, *22*(17), 9554. https://doi.org/10.3390/IJMS22179554/S1

Zhong, M., Wang, L., Yuan, J., Luo, L., Xu, C. and He, Y.Q. (2011). Identification of QTL affecting protein and amino acid contents in rice. *Rice Science*, *18*(3), 187–195. https://doi.org/10.1016/S1672-6308(11)60026-7

Zhu, G., Peng, S., Huang, J., Cui, K., Nie, L. and Wang, F. (2016). Genetic improvements in rice yield and concomitant increases in radiation- and nitrogen-use efficiency in middle reaches of Yangtze river. *Scientific Reports*, *6*, 1–12. https://doi.org/10.1038/srep21049

Zhu, K., Zhou, Q., Shen, Y., Yan, J., Xu, Y., Wang, Z. and Yang, J. (2020). Agronomic and physiological performance of an indica–japonica rice variety with a high yield and high nitrogen use efficiency. *Crop Science*, *60*(3), 1556–1568. https://doi.org/10.1002/CSC2.20150

Zhuang, J.Y., Ma, W.B., Wu, J.L., Chai, R.Y., Lu, J., Fan, Y.Y., et al. (2002). Mapping of leaf and neck blast resistance genes with resistance gene analog, RAPD and RFLP in rice. *Euphytica*, *128*, 363–370.

Zuo, S., Zhang, Y., Yin, Y., Li, G., Zhang, G., Wang, H., Chen, Z., et al. (2014). Fine-mapping of qSB-9 TQ, a gene conferring major quantitative resistance to rice sheath blight. *Molecular Breeding*, *34*, 2191–2203.

# Chapter 4

# Navigating the Development of Next-Generation Crop Varieties Using Genomic Selection for Food and Nutritional Security

Suma Krishnappa[1,2], Likhithashree TR[1], Sharon Roja Kavuri[3], Mahesh Pujar[2], Naresh Bomma[2], Shruthi HB[2] and Prakash I Gangashetty[2*]

## 1. Introduction

### *1.1 Next-Generation Varieties*

The green revolution phenomenon greatly contributed toward developing novel high-yielding varieties, enabling India to attain food security. However, in the face of future climate change, the current varieties so developed may not be sustainable, making it essential to focus breeding efforts on developing varieties that can withstand the adverse conditions expected in future times. Even though it significantly increased productivity, Green Revolution failed to address nutritional security in general. Nutritional security primarily deals with giving access to inexpensive, secure, and healthy foods to common people that support wellness and fend against disease (FAOSTAT, 2012). Considering these facts, it is therefore crucial to bring changes which improve the newly produced varieties' nutritional content and resilience to biotic and abiotic challenges, particularly with regard to staple crops (Anilkumar et al., 2022). Thereby, next-generation varieties can be defined as advanced cultivars developed via modern breeding techniques including molecular methods and genetic engineering, which is the combination of more than two desirable traits. These varieties are designed to offer enhanced characteristics such as increased yield, improved stress resistance, higher nutritional value, with greater environmental resilience. Classical plant breeding, also titled conventional plant breeding, emphasizes crop development by phenotypic selection, which includes basic selection in the form of pure lines, mass selection, or hybridization, followed

[1] Department of Genetics and Plant Breeding, University of Agricultural Sciences, GKVK, Bengaluru 560065, Karnataka, India.

[2] Pigeonpea Breeding, Accelerated Crop Improvement, International Crops Research Institute for Semi-Arid Tropics, Patancheru, Hyderabad 502324, Telangana, India.

[3] Department of Genetics and Plant Breeding, Acharya N.G. Ranga Agricultural University, Lam, Guntur 522034, Andhra Pradesh, India.

* Corresponding author: prakash.gangashetty@icrisat.org

DOI: 10.1201/9781003584100-4

by selection using pedigree breeding and bulk breeding procedures. Phenotypic selection also encompasses other traditional breeding techniques, such as population improvement by recurrent selection, among others. However, the fact that conventional breeding strategies only consider phenotype poses numerous questions about the reliability in relation to variables controlled by quantitative genes, such as grain yield and several nutritional traits. Furthermore, it is hard to meet the steadily rising demands due to the slow genetic progress achieved by traditional breeding. As a result, new methodologies such as 'genomic selection (GS)' that facilitate accurate and efficient variety manipulation—which results in the generation of Next Generation varieties (NGVs)—need to be developed and put into practice.

## *1.2 Overview of Genomic Selection*

Since 1980s, the introduction of molecular marker technology has helped geneticists and plant breeders to utilize markers for detecting quantitative trait loci (QTLs) (Geldermann, 1975) to perform marker-assisted selection (MAS), and also helped in the transfer of desired traits (Lande and Thompson, 1990). Marker-aided selection has greatly complimented classical plant breeding to bring about desirable changes in a trait per se more precisely and swiftly to realize rapid genetic gain than conventional plant breeding alone. In addition, MAS is becoming the first choice approach for improvement of specific traits largely due to the swift decline in marker costs and wider coverage throughout the genome, made possible by advanced genotyping techniques. However, MAS is effective for QTLs with substantial genetic effects, while markers pose challenges in handling quantitative traits governed by multiple genes with minor effects.

Around 2.4% genetic increase is needed to provide food security for the world's population, which is predicted to reach a record high by 2050 (Hickey et al., 2017). Continuous decrease in arable land along with reduced cultivated land largely due to low productivity and climate change makes it difficult to meet the growing demand. In addition, genetic gain achieved by the intervention of MAS is very narrow and largely restricted to specific trait. Therefore, with the exponential increase in the demand for food, it is vital to develop varieties with multifaceted utilization like NGVs. This can be acclaimed through advanced breeding methodologies. GS as proposed by Meuwissen et al. (2001) is a novel technique that predicts an individual's breeding value using greater number of markers spread throughout the complete genome. Instead of concentrating on particular markers associated with qualities, it makes predictions about complex traits—including those regulated by several genes—using genome-wide data. This method is also known as genome-wide selection or genomic prediction. GS is extensively employed in animal breeding for the improvement of traits (Hayes et al., 2009). Recently, GS has been employed in plant breeding for the improvement of complex traits in crops such as maize (Shikha et al., 2017; Beyene et al., 2021; Gunundu et al., 2023), rice (Grenier et al., 2015; Xu et al., 2018; Cui et al., 2020), wheat (Michel et al., 2016; Belamkar et al., 2018), lentil (Gebremedhin et al., 2024), and chickpea (Li et al., 2018).

Genomic selection consists of the training and breeding phases. Genome-estimated breeding values (GEBVs) are estimated from the training population using robust genotyping and phenotyping in conjunction with the development of a prediction model. The estimated GEBVs are then used to perform the selection process in the breeding population, where higher GEBVs are used to select individuals. Figure 1 depicts the overview of GS method used for selection of superior varieties. Superior performing varieties are selected solely based on GEBVs, devoid of phenotyping in breeding population. Thereby, it reduces the cost of phenotyping and increases the genetic gain that can be obtained per selection.

## *1.3 Prospects and Potential of Genomic Selection Approach*

Genomic selection increases the genetic gain through a rapid breeding cycle by estimating GEBVs at an early stage of the plant growth (seedling stage) with the ability to detect both major and minor

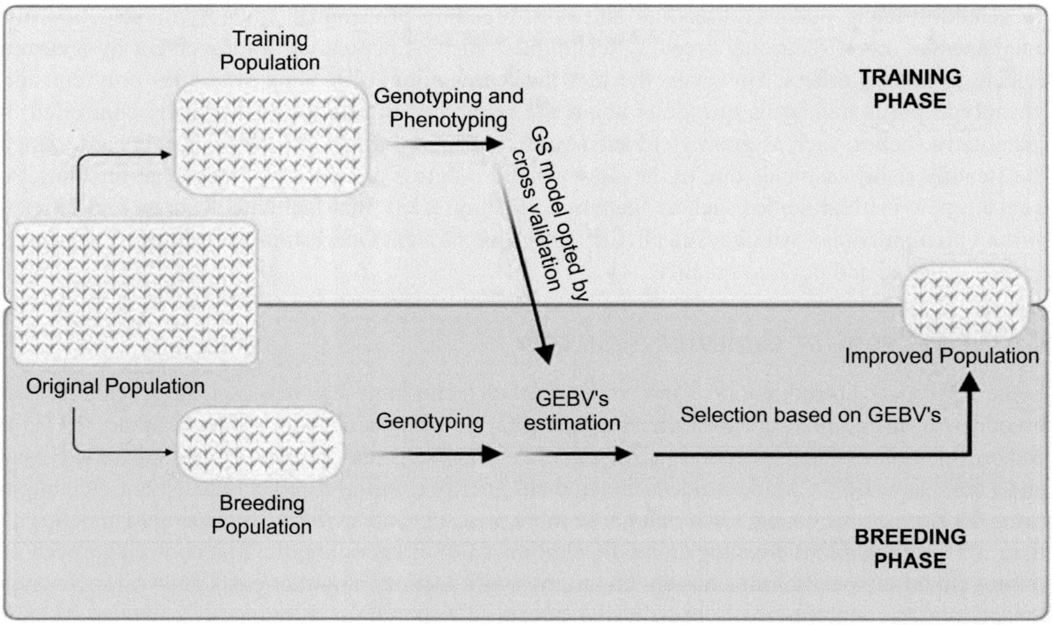


**Fig. 1** An overview of genomic selection procedure in plants.

gene effects, thus improving precision of the calculated breeding values (Crossa et al., 2017; Zhang et al., 2018) and enhancing exploitation of existing genetic resources via genome-informed mate selection (Akdemir and Sánchez, 2016). It is an ideal way to meet the food demand by improving the genetic gain of the food crops. Desirable traits such as grain yield, tolerance to heat and drought, few diseases, and pest resistance that are governed by the many genes with small/negligeble effects can be improved simultaneously through GS. Large-scale field trials, encompassing multiple locations and seasons, are typically carried out to get reliable phenotypic data. This is more expensive and time-consuming, but GS can help by minimizing the requirement for extensive phenotyping over several generations, which lowers the overall cost of breeding operations over time. This is accomplished by initially investing in genotyping. Decisions in GS are data driven and thus help in selection with more precision that can be used across all the crops including both major and orphan crops. The GS offers a better prospect of accurately predicting hybrid performance since it captures the genetic potential of parents employed in the crosses.

Gaynor et al. (2017) suggested splitting traditional breeding programs into two segments: a population enhancement phase employing recurrent GS, and a product development (PD) phase resembling traditional breeding approaches. The PD pipelines are incorporated into prevailing breeding frameworks, facilitating extensive exploration for GS implementation (Bassi et al., 2016; Hoefler, 2022). Many factors affect the accuracy of GS, which needs to be optimized. In this chapter, an effort is made to review the principles of GS, the optimization of factors affecting GS, and the role of advanced genotyping and phenotyping facilities that guide the development of varieties to feed the ever-growing population.

## 2. Principles of Genomic Selection

### *2.1 SNP Genotyping, Linkage Disequilibrium, and Marker-Trait Associations (MTAs)*

Single-nucleotide polymorphisms (SNPs) are the common type of genetic variation and are widely used in genomic selection due to their abundance, stability, and ease of detection. Hence, SNP genotyping has a crucial role in providing the genetic data needed to predict breeding value of individuals based on genome-wide marker information. SNP markers rarely show direct effect on phenotypic traits but are in linkage disequilibrium (LD) with causal genes. SNP genotyping is a process of identifying and analyzing the polymorphisms through various methodologies. The automation and miniaturization of SNP genotyping methods have increased marker density while reducing the genotyping costs and time in comparison to earlier approaches. Among traditional methods, Sanger sequencing identifies SNPs in an organism or population which is preferred for detailed study on specific variation. Next-generation sequencing (NGS) techniques such as Illumina sequencing, 454 pyrosequencing, and Ion Torrent sequencing are high throughput and efficient methods to identify the SNP polymorphisms. The recent advancements in sequencing methodologies are third-generation sequencing platforms such as PacBio sequencing, Oxford Nanopore, and 10x Genomics. These enable analysis of complex genomes with greater accuracy, making the methods more expensive. Ultimately, the choice of method should align with the research objectives, budget, and resources available for analysis.

LD is a nonrandom association of alleles with different loci (Jennings, 1917). It is the correlation of polymorphisms caused due to their shared history of mutation with recombination (Sherry et al., 2003). The LD in GS is essential to optimize the training population to improve the selection efficiency. By enabling markers all around the genome to function as stand-ins for genes governing major attributes, linkage disequilibrium improves genomic selection. In addition, it makes sure that the adjacent markers and genes are inherited together allowing GS to capture the major and minor genes' effects, and thus helps to increase prediction accuracy of breeding values, accelerating genetic gain, and enabling the selection of individuals at early stages, even for complex, polygenic traits. Among the different metrices of estimation of LD, $r^2$, $D'$, $d$, $\delta$, and Q are different standardized versions of the D estimate, wherein $D'$ and $r^2$ are the most relevant estimates of LD for plant species.

Marker trait association refers to the relationship between genetic markers and specific phenotypic traits in an organism. Marker-trait association is a powerful tool in modern genetics and plant breeding, enhancing the ability to improve crops by leveraging genetic markers linked to economically important traits. Various statistical procedures are adopted to decipher the markers linked to superior traits, enabling targeted breeding and genetic improvement. MTAs are identified through genome-wide association studies (GWAS) and QTL mapping. Further identified MTA is beneficial for marker-aided selection and also to understand genetic mechanism of the complex trait.

## 2.2 Genetic Gain: MAS vs GS

Response to selection following any breeding method is quantified using genetic gain, which is a reward offered to selection. It is defined as improved performance gained through genetic improvement breeding procedures with consecutive selection cycles. Genetic gain attained by appropriate selection techniques, such as GS, can be used to assess response to selection. The key target of implementing GS is to obtain more substantial genetic gains at faster and cheaper rates in comparison with standard pedigree breeding techniques. This is also known as genetic response (R), which can be derived by the breeder's equation, that is,

$$\frac{ir\sigma_A}{t},$$

where $i$ is the selection intensity, $\sigma_A$ is the square root of the additive genetic variance, $r$ is the selection accuracy, which is equivalent to narrow-sense heritability ($h^2$), and $t$ is the cycle time (Falconer, 1996).

In essence, the way that genomic selection uses genotypic information dispersed throughout the genome sets it apart from marker-assisted selection. While GS uses several markers (ideally SNPs spread over the entire genome to assist in forecasting an individual's breeding value), MAS considers genetic markers which are intimately linked to attributes under study. Unlike GS, which is more robust and better suited for grain yield (which are complex attributes impacted by more genes with tiny effects), MAS is more effective at selecting basic attributes conferred by one/few genes. A significant distinction is in selection methodology employed in MAS. Specifically, judgments are made based on presence/absence of a particular marker, taking into account the correlation between a lesser number of genes and the underlying trait. GS, on the other hand, is a statistical model that aggregates all of the available marker data from across the genome to forecast an individual's total genetic worth. When it comes to traits with major gene inheritance, MAS is a cost-effective method of applying selection; however, its effectiveness decreases when the trait has polygenic inheritance. Compared to MAS, GS involves more genotyping, which can increase initial costs. However, the cost of genotyping would go down, and the prediction model would get better, making GS economically achievable, mainly for qualities that are influenced by multiple genes.

### *2.3 Increasing Genetic Gain Using GS*

Several elements of the response to selection equation are considered to move up the genetic gain. One can increase genetic gain by increasing selection precision, breeding population variation, and selection intensity. The heritability of a trait is influenced by its relationship to its target environment, which in turn affects the selection accuracy process. Therefore, more genetic gain and improved heritability result from less environmental effect. The expense of breeding programs, which influences the strength of selection, is another factor impacting genetic gain (Hickey et al., 2017). Deployment of conventional phenotypic systems paves to increased duration to breed cultivars as the time required to collect sufficient amount of data increases. Higher replicates and precision of the phenotypic data are required to improve the genetic gain (Stock and Reents, 2013). The precision of phenotypic data relies on the cost allocated to phenotyping. As the expense of phenotyping increases, the population size managed with the stipulated amount decreases which causes increase in selection intensity.

Phenotypic precision is affected by the cost available for breeding program. A higher cost will reduce the population size with which breeders can work by increasing the selection intensity. Greater genetic gain results from shorter recycling times because the length of a breeding cycle has an inverse effect on genetic gain (Heffner et al., 2010). Conversely, the programs without GS would lose genetic gain quickly each year as compared to the early adopters, to the point that their germplasm could soon become obsolete (Bassi et al., 2024). Heffner et al. (2010) stated that GS provides threefold and twofold genetic gain for a year in comparison with MAS for maize and winter wheat when costs were equivalent, respectively. Furthermore, its application in maize breeding has led to 30–50% decrease in whole breeding expenditures (Beyene et al., 2015).

## 3. Key Players in Effective Implementation of GS

### *3.1 Training Population*

Numerous population types with large and diverse sets of germplasm including wild relatives, elite lines, landraces, and cultivars are employed to conduct GS analysis. GS has been applied to populations with broad base including multi-lines of barley, wheat, and maize (Zhong et al., 2009, Crossa et al., 2017, Heffner et al., 2011), and also to populations with narrow base, that is, biparental populations (Lorenzana and Bernardo, 2009). Along with this, natural populations can be deployed to exploit historical LD. Despite the population type, the fundamental step in implementing GS is developing training populations with comprehensive phenotypic and genotypic data.

Training population often consists of multiple individuals related to one another or whose family ancestry is known. The main objective of Training Population (TP) development is to assess marker effects by employing statistical techniques. According to several simulation experiments, prediction accuracy will be improved with increasing TP size (Poland et al., 2012; Asoro et al., 2011). Hence, panel must comprise of diverse genotypes. The best possible TP size for crops such as barley, maize, and rice was 50, 250, and 100–500, respectively, which turns into an accurate prediction (Bernardo and Yu, 2007, Hamblin et al., 2010, Frouin et al., 2019). A key challenge is enhancing the diversity of TP without increasing the diversity of TP and Breeding population (BP) (Crossa et al., 2014). Developing a good TP considering the abovementioned features is crucial for the successful GS.

## 3.2 Breeding Population

The pool of individuals (plants, animals, etc.) that breeders try to enhance through selective breeding is known as the breeding population. According to genomic selection models, these individuals are usually candidates for parents to be employed in a crossing program because of their genetic value. In general, members of BP are either direct descendants of TP or similarly related individuals. Lines that are developed through natural or artificial processes, such as $F_2$, $F_3$, DH, and/or Recombinant Inbred Lines (RIL), can be used as BP. Based on allelic similarity that pertains to a trait phenotype in the TP, breeding values of individuals within a BP are predicted (Krishnappa et al., 2021). The selection in breeding population is performed through genotyping of individuals from breeding population, and thereby, estimating the GEBVs based on their genomic information.

## 3.3 Relationship between Training and Breeding Population

The relationship between TP and BP influences the prediction accuracy of GS; closely related TP and BP will give more accurate prediction values and vice versa (Hickey et al., 2014; Meuwissen et al., 2016). Full-sib families generally provide greater accuracy than half-sib families due to their higher relatedness (Riedelsheimer and Melchinger, 2013). Increasing relatedness between TP and BP is more critical for accuracy than simply enlarging the TP with unrelated crosses (Brandariz and Bernardo, 2019). Accuracy improves when TP and BP consist of segregating lines of same cross (Combs and Bernardo, 2013). The TP which includes individuals from both related and unrelated groups can be advantageous for breeders. Combining progenies from different families with similar ancestry but varying degrees of relatedness can enhance the TP's ability to effectively execute GS to diverse populations. However, this approach is not widely used in GS because its accuracy can significantly diminish when the TP is not related closely to BP (Anilkumar et al., 2022).

## 3.4 Marker Selection

In general, GS utilizes all the genetic variation with the help of markers that cover the complete genome. For complete genome coverage, a high number of markers are required, regardless of whether they are SNPs or SSRs. Even if Simple sequence repeats (SSRs) are more informative, SNPs are preferred for broad genome coverage and wider availability. With genome sequencing, SNPs work more effectively for comprehensive genome coverage in GS. Furthermore, a lot of GS statistical approaches are built to function better with SNP data (Anilkumar et al., 2022).

The number of markers essential for GS depends mostly on how often markers are in LD, influenced by population structure and effective population size (Thistlethwaite et al., 2020). Marker density is vital for achieving higher accuracy in GS. In comparison with marker density, wider coverage of markers is important. Depending on the structure and degree of LD of each population, different populations have different optimal marker densities. Because of their relatively large LD and population structure, self-pollinated species and biparental crosses require less marker density than wild populations (Hao et al., 2019). According to Smith et al. (2008), biparental populations require a reduced marker density than natural populations because of less recombination and maintain a clearly identifiable structure.

A high enough marker density ensures that at least one marker and each QTL have nearly ideal LD, which boosts prediction accuracy. In conceptual terms, an increase in number of markers leads to more accurate prediction. In practical circumstances, accuracy is hard to significantly increase once the density reaches a certain point. One such example is the study conducted by Su et al. (2012), which revealed that prediction accuracy increased only 0.5–1.0% when marker density was switched from 54 K (medium density) to 777 K (high density). Other studies have demonstrated a similar tendency. Once marker density reaches an extent, it is no longer pertinent to genomic prediction. To sum up briefly, the accuracy of GS does not increase in direct proportion to the cost as marker density increases (Wang et al., 2018).

## 3.5 Statistical Analysis

Selection of a statistical model to find the mathematical relationship between phenotypic and genotypic data is one of the important steps in GS. To develop the model, evaluated markers are provided with specific weights to their phenotypic effect that allow the GEBVs of candidate individuals in the BP to be estimated (Alemu et al., 2024). Several statistical machine-learning techniques are proposed for genomic prediction (GP) where "*p*" denote the predictors, that is, large SNP markers, and "*n*" is number of observations. The different mechanisms for normalization applied to estimate genome-wide SNP marker effects from a regression with large "*p*" and small "*n*" has led to the emergence of numerous statistical machine-learning approaches (Montesinos-López et al., 2022). These statistical models are grouped into parametric, semi-parametric, and non-parametric groups which are briefly explained.

### 3.5.1 Parametric Models

The basic principle of parametric models is the assumption that the response or dependent variables and the independent or predictor variables follow a predefined function. They are characterized by a certain number of parameters calculated from the data. It solves the dilemma of small '*p*" and large "*n*" using various methods such as shrinkage procedure. Few models are: linear least square regression (LLSR) model; BLUP models [ridge regression (RR), ridge regression BLUP (rrBLUP), gBLUP, compressed BLUP (cBLUP), andsuper BLUP (sBLUP)]; and Bayesian prediction models (BayesA, BayesB, BayesCπ, Bayesian LASSO, and Bayesian ridge regression). Each of the statistical model used for GS analysis is briefly explained in the following sections.

A. **Linear least square regression model:** It is the simplest parametric model. However, it is not easy to perform the estimation when "*p*" > "*n*" (Lohithaswa et al., 2022). Although modifications have been done to the model by utilizing the subsets of markers, still the accuracy could not reach the standard. Therefore, Meuwissen et al. (2001) suggested eliminating the independent variables (predictors) than dependent variables (observations). However, this modification fails to take the best advantage of genome-wide markers.
B. **Ridge regression:** When $p >> n$, multicollinearity is the problem which can be overcome by using RR model (Howard et al., 2014). With the assumption of random marker effects ($m_{j's}$ $j = 1...p$), RR in GS is implemented. Markers were taken from a group with a normal distribution and Var(mj) = $\sigma^2_m$, where $\sigma^2_m = \sigma^2_a/n_k$ $\sigma^2_a$ represents the additive component of genetic variance expressed among individuals and $n_k$ represents number of the markers (Habier et al., 2007).
C. **Best linear unbiased prediction (BLUP):** It is utilized for handling unbalanced datasets like multi-locational dataset (Bernardo, 2010). Genomic BLUP takes advantage of genomic relationship matrix between the individuals to know about the genetic relationship, calculated based on SNPs data whereas traditional pedigree BLUP uses a pedigree relationship matrix (Henderson, 1975, Habier et al., 2013).
D. **Least absolute shrinkage and selection operator (LASSO):** LASSO is developed to prevail the shortcomings of LLSR model. It is the compelled form of ordinary least square (Tibshirani, 1996). LASSO picks only one among the highly correlated marker and ignores the

remaining markers. Hence, it is indifferent to correlated markers (Wang et al., 2018). Due to its penalized regression-based methodology, LASSO gives accurate estimates with more markers than individuals used ($p > n$) (Budhlakoti et al., 2020).

**E. Bayesian prediction models:** BayesA assumes a prior distribution of genetic effects, whereas BayesB is devoid of these assumptions. But, in both the models, the data and variances of marker positions are required to be modeled (Lohithaswa et al., 2022). Bayesian B employs ridge penalty to handle the multicollinearity among predictors, but has the problem of regression coefficient shrinking toward zero. The BayesCπ gives a more sensible formulation of the mixture. However, it poses the same objectives and challenges as BayesB (Gianola, 2013). LASSO results in the sparse model, whereas Bayesian LASSO yields an effectively sparse specification like BayesB (Meuwissen et al., 2001), since it combines LASSO's ability to enforce sparsity (i.e., setting some coefficients to zero) with Bayesian inference to estimate the coefficients and their uncertainties. Initially, these models started as BayesA and BayesB models (Meuwissen et al., 2001). But later, several modifications such as BayesCπ and BayesDπ (Habier et al., 2011), fast EM-BayesA (Sun et al., 2012), fast BayesB (Meuwissen et al., 2009), BRR (Bayesian ridge regression on markers) (VanRaden, 2008), Bayesian LASSO (Park and Casella, 2008), etc., were evolved.

In general terms, additive genetic effect at each marker loci in a homozygous population can be accurately estimated and captured using nearly all statistical models. When TPs and BPs are made up of heterozygous lines and are phenotyped in different places and years (environments), they are less effective in assessing nonadditive or nongenetic effects, such as dominance, epistatic, and Genotype x environment interaction (GEI) effects.

### *3.5.2 Nonparametric Models*

Nonparametric models do not clearly assume anything regarding the distribution or functional form of the underlying data. In these models, predictors are not predefined; instead they are developed on the grounds of insights drawn from the data. Hence, these models propose higher flexibility for determining the structure of model (Alemu et al., 2024). Nonparametric models are beneficial while dealing with complex or unknown relationships since they adopt to various data patterns without predefined assumptions. The models are developed to account for dominance, epistasis, and genotype by environment interaction effects in prediction models.

**A. Nadaraya–Watson estimator:** This is one of the most used nonparametric models for GS. It utilizes a nonparametric kernel estimator (Silverman, 2018), which is used in the estimation of $p(x)$. The estimator is just a weighted sum of observations *yi*, *i* ¼ 1, 2, 3...*n*, and is called Nadaraya–Watson's equation. This model accounts for the epistatic interaction when additive effects are present and its prediction ability is less in comparison to other nonparametric models (Howard et al., 2014).

**B. Support vector machine (SVM) regression:** In this model, the training dataset is utilized in creation of maximum marginal classifier that results in the biggest possible separation between the comparing classes of observations, proposed by Vapnik (1995) and Cortes and Vapnik (1995). The link between a marker's genotypes and phenotypes in plant breeding is explained by SVM regression. It is described with a nonlinear or linear mapping function, which transfers data to an abstract from a predictor space, that is, multidimensional feature space (Hastie et al., 2009; Long et al., 2011). Unlike conventional regression approaches which focus on minimizing the error between predicted and actual values, SVR aims to fit a function within a specified margin of tolerance while minimizing the model complexity.

**C. Neural networks (NNs):** NN techniques are familiar prediction approaches in machine learning. NN can quantify both additivity and epistasis interaction because of its nonparametric nature, which can also simulate both linear and complex nonlinear functions (Howard et al., 2014).

### 3.5.3 Semi-Parametric Models

Gianola et al. (2006) proposed this semi-parametric model which combines the elements of both parametric and nonparametric approaches. They incorporate a parametric component to capture the main effects and a nonparametric component to model complex, nonlinear relationships or interactions by considering some of the predetermined statistical assumptions.

A. **Reproducing kernel Hilbert space (RKHS):** The RKHS model combines a matrix of genomic relationship (G) and relationship matrix from pedigree-based numerator (A) in kernel matrix while considering the weaker assumptions on the interaction of G and A (Rodríguez-Ramilo et al., 2014).

## 3.6 Validation Methods

Various validation methods are utilized to know the GS accuracy wherein two sets are formed from the total individuals; one to train the model and a separate set for prediction. The widely used methods are k-fold cross-validation, forward validation, and resampling-based cross-validation.

K-fold cross-validation method divides the individuals into *k* sets, where *k* times prediction models are applied. Each time prediction model is fit to the sets leaving the $k^{th}$ set, value of the $k^{th}$ set is predicted, which serves as the validation set. Historical data of previous years are used in forward validation method. The model is trained by using the data from experiments of previous years, and individuals of the current experiment are used for prediction. It mimics the idea of general breeding program. Resampling-based cross-validation involves selection of individuals in a random manner used for training the model, and prediction model is used for remaining individuals. Depending on the genetic resources and objective of the study, any of the abovementioned validation methods can be used. Predictive ability of different validation methods is the criteria to narrow down on the validation method. Pearson correlation is calculated between the estimated breeding values using phenotype and breeding values predicted from GS, which in turn gives the predictive ability of validation methods. Further, predictive ability is divided by the square root of the line mean heritability (r/h) in the validation set to determine the prediction accuracy. The obtained prediction accuracy can be directly used in the selection formula replacing genetic gain.

# 4. Genetic and Genomic Resources that Enroute the Development of NGV

## 4.1 Overview of Genetic/Genomic Resources Available for Developing Next-Generation Varieties

Plant breeders intend to alter the genetic makeup of plants to produce superior cultivars with desirable traits. Efficient usage of genetic and genomic resources will fasten the development of cultivars that cater to the future needs of the population by developing next-generation varieties. Genomic selection relies heavily on genetic resources that provide the foundational data for predicting performance of individuals depending on their genetic makeup. Genetic diversity available in panel largely shapes the genetic gain that can be accomplished via GS. Here are some key genetic resources crucial for genomic selection.

It might include genetically diverse and sufficiently varied populations, such as elite inbred lines, biparental populations, multi-parental populations, or naturally occurring populations such as germplasm accessions. Plant genetic resources (PGR), which are the storehouse of huge amount of genetic variability, form an efficient tool for the development of superior varieties (Singh et al., 2019). Highly valuable PGRs comprise landraces and wild relatives of crop species, which are a valuable source for extending the genetic base of contemporary cultivars and for creating cultivars with increased productivity that can tolerate shifting environmental conditions, having high nutritional value and stress resistance. Apart from natural populations, advanced generations such as RILs can be utilized to simulate the models in GS. These are collections of genetically diverse individuals within a species. They represent a broad spectrum of genetic variation found in the population and are utilized to capture the full spectrum of genetic factors influencing desirable traits. Advanced breeding lines can be employed to estimate marker effects (Berro et al., 2019). Large population of elite lines relatively increases the predictive abilities of GS (Allier et al., 2020). Any of the abovementioned panel can be utilized to follow GS keeping in mind its genetic diversity and population size.

## 4.2 High-Throughput Genotyping Technologies and Their Role in GS Models

Genetic gains can be hastened with time through cost-effective high-throughput genotyping, efficient phenotyping (phenomics), bioinformatics, and statistical tools (Varshney et al., 2016). Genotyping of large populations with genome-wide markers, a crucial aspect of GS studies, is evolving to reduce the genotyping cost of individual samples (Ronne et al., 2023). The decreasing expense of sequencing technologies is the primary catalyst for the growing number of fully assembled genomes across various crops and their wild counterparts (Brozynska et al., 2016). However, sequencing a single reference genome isn't enough to encompass the entire genetic variability within a species. Therefore, additional wild accessions, cultivars, and landraces are resequenced to capture the broader genomic landscape of each species. Various resequencing methodologies prove invaluable in achieving this goal.

The initial drafts of the human genome in 2000 marked the beginning of the modern era of DNA sequencing. This milestone spurred subsequent innovations and advancements, leading to the development of high-throughput DNA sequencing strategies known as high-throughput NGS. Genotyping by sequencing (GBS) employs NGS technology to genotype many individuals, with the potential to lower the cost per sample to under \$10 through extensive multiplexing (Elshire et al., 2011; Poland et al., 2012). These advanced high-throughput NGS techniques were specifically designed to meet the increasing demands for sequencing throughput and cost-effectiveness, making them versatile for a broad spectrum of current and future applications in plant breeding (Parekh et al., 2016). Next-generation platforms produce shorter reads of lesser quality. This decrease in read length and quality requires the advancement of bioinformatics tools to aid in either mapping these shorter reads to reference sequences or performing de novo assembly (Deschamps and Campbell, 2010). The reduced representation sequencing (RRS) approach like RAD-sequencing focuses on a fraction of the genome. The fraction to be selected for sequencing depends on the objective of the project. The most widely used RRS approach is GBS, which is flexible, cost-effective, and is faster in genotyping (He et al., 2014). The idea of GBS used for multiplexed samples will reduce the genotyping cost. The appeal of GBS has resulted in numerous optimizations, including the selection of enzymes, protocols for SNP calling, enhancements in marker density (such as double-digest GBS and high-density GBS), and improvements in library preparation methods (Sonah et al., 2013; Wang et al., 2018; Torkamaneh et al., 2021). Even though the cost is reduced, the genotyping cost will become higher to cater large number of samples. This needs further reduction in the fraction of the genome to genotype larger samples via multiplexing (Pértille et al., 2016). Meanwhile, extreme reduction of the genome will reduce the number of markers, which is rectified by selecting uniform markers all through the genome (Ronne et al., 2023).

These advanced technologies in the field of genomics provide new tools to generate and utilize genetic variability. The genomic tools developed in the recent past accelerated the genetic gain that can be achieved in a stipulated time. The ease and access to high-throughput genotyping technologies such as low-density, mid-density, and high-density platforms increase genetic gain that can be achieved via GS.

## 4.3 High-Throughput Phenotyping for Effective Varietal Development

Phenotyping is an important step in GS, as the breeding values of BP are predicted based on the phenotype of TP. Accurate genotypic data when integrated with precise phenotypic data will be advantageous and increases the genetic gain to be achieved. As the GS process involves phenotyping of only the TP, the efficacy of phenotyping plays an important part in accurate estimation of GEBVs. Hence, it is vital to obtain high-quality phenotypic data. The prediction accuracy of model depends highly on the effectiveness of phenotyping. Basic and largely followed method of phenotyping is field phenotyping, which involves manually investigating each individual to obtain the visual data points. This method is prone to human error and consumes more time for phenotyping of huge population involved in the GS. Thereby, field phenotyping becomes a bottleneck for the genetic improvement of crops. In recent times, various phenotyping platforms with improved efficiency have been employed to get accurate data for different quantitative and qualitative traits. Agronomically important traits such as grain yield, nutrient content, and stress tolerance are mostly complex. These traits with intricate characteristics are influenced by numerous genes, typically with minor individual effects and significant epistatic interactions (Li et al., 1997, 2001). High-speed analysis of traits and using genetic selection for complex characteristics hold the potential to transform breeding (Cabrera-Bosquet et al., 2012). As phenotype is the product of genotype and environment, the role of genotype × environment (G×E) interaction effects should be dissected to effectively predict GEBVs. While phenotyping, G×E interaction should be included and high-throughput phenotyping (HTPP) needs to be modeled to consider G×E interaction effects.

The advanced phenomic technologies, such as those highlighted by Finkle (2009), play a crucial role by providing detailed measurements of plant characteristics that collectively yield accurate estimates of phenotypic traits. The HTPPs used for phenotyping various nutritional traits are enlisted in Table 1. These platforms are also instrumental in modeling and predicting genotypic performance under diverse climate conditions. HTPPs operate with high data recording and processing capacities, rapid data collection, and employ non-destructive, noninvasive, and remote-sensing methods (Prasanna et al., 2013). While these platforms enhance understanding of crop physiology, their

**Table 1** High-throughput phenotyping platforms used in different crops.

| *Phenotyping platform* | *Applications* | *Crops* | *References* |
|---|---|---|---|
| Phenopsis | Plant responses to soil water | | Granier et al., 2006 |
| HyperART | Quantification of leaf chlorophyll content | Barley, maize, rapeseed, tomato | Bergsträsser et al., 2015 |
| Radiation detection and ranging (RADAR) and light detection and ranging (LiDAR) | | | Pineda et al., 2020 |
| LemnaTec Scanalyzer 3D | Nitrogen content, salt and nutrient stress | Chickpea, wheat | Bruning et al., 2019 |
| Glyph | Water-use efficiency | Soybean | Peirone et al., 2018 |
| Hyperspectral imaging | Protein content | | Sun et al., 2019 |
| X-ray fluorescence (XRF) | Elemental nutrient concentrations (Ca, K, P, S, and Zn) | Various crops | Singh et al., 2022; Montanha et al., 2020 |
| LeasyScan | Leaf area, transpiration, and leaf area index | Several semi-arid crops | Vadez et al., 2015 |
| Lysimeter | Evapotranspiration | | Liu et al., 2002 |

most effective techniques for high-throughput phenotyping, akin to GS, are empirical and rely on extensive data acquisition and statistical processing rather than deep physiological understanding. Examples include near-infrared spectroscopy (NIRS), spectral reflectance analysis, and other methods such as photography and sonar, already integrated into existing HTPPs primarily designed for controlled plant environments. These techniques are anticipated to see increased deployment in future field-based HTPPs for assessing complex traits under varying environmental stresses (Montes et al., 2007 and Montes et al., 2011).

The advanced techniques lend hand in estimation of important traits such as major and minor nutrients, which paves way to develop biofortification of crops. Specifically, spectral techniques such as NIRS and plant spectral reflectance have emerged as promising tools for field-based high-throughput phenotyping of crops subjected to both abiotic and biotic stresses, as indicated by studies from Araus et al. (2002), Araus et al. (2008), Lopes et al. (2011), and Prasanna et al. (2012).

# 5. Considerations for Optimum Implementation of GS Models in Various Crops

## 5.1 *Nature of Pollination in Crops*

Breeding methods mostly depend on the mode of pollination of a crop since it influences marker density optimization and TP optimization. In case of self-pollinated (SP) crops, LD decay is slow because of minimum recombination events. Hence, number of markers needed to achieve better GP accuracy is less. That is, SP obtains good results with a less-dense genetic map than cross-pollinated (CP) crops where the latter needs a dense genetic map to achieve the same. Self-pollinated crops often have a more structured genetic makeup compared to CP crops, because genetic variation offers to be more localized in a population because of reduced gene flow from exterior sources, thereby influencing the relatedness between the TP and BP. The SP crops have population structure compared to CP crops, hence the bias will be more in SP crops than the CP crops. Hence, SP can have small-sized TP. GS is used to develop a homozygous line in SP crop, whereas it is used to develop a hybrid in CP crops.

## 5.2 *Trait of Interest*

The architecture of the trait influences the density of the genetic map and type of model to be implemented. Complex traits influenced by many genes (quantitative traits) such as nutrient content and grain yield may benefit more from GS than qualitative traits such as biotic stress resistance since it can capture polygenic effects. The architecture of the trait also influences the type of model to be implemented for GS. It is noted that parametric models were only known to be effective for traits with additive genetic architectures, and that using parametric models became difficult for characters which are strongly influenced by nonadditive or epistatic interactions (Moore & Williams, 2009). Whereas, nonparametric and semi-parametric models account for both additive and nonadditive (epistatic) interactions.

## 5.3 *Heritability of a Trait*

Heritability ($h^2$) of the target trait directly influences the accuracy of GP. A trait with low $h^2$ should be compensated by increasing the TP size (N) to achieve an optimum GP accuracy, since $Nh^2$ determines the power of GP models (Bernardo, 2016). Furthermore, machine-learning models consider epistatic interactions with the ability to accelerate the prediction accuracy when epistatic interaction largely or partially renders to the true architecture of a trait (de Los Campos et al., 2010; Morgante et al., 2018). Several empirical investigations and simulation research have demonstrated that the $r_{MG}$ generally rises as QTL number decreases and trait heritability increases.

## 5.4 Genotype by Environment Interaction

Multi-environment trials for assessing the stable phenotypic performance of the genotypes are a mandatory step in plant breeding. The experiments account for G×E interactions, which constitute a basic difficulty in plant breeding and are becoming better acknowledged as a significant intricacy in GS models (Crossa et al., 2017). The G×E interaction can complicate the prediction of plant performance based solely on genomic data because the same genotype can perform differently in different environments. However, incorporating G×E data into GS models will potentially improve prediction accuracy by accounting for environmental effects. Hence, tackling this complicacy of multi-trait genomics and varied environments needs a theoretical framework which considers complex interactions such as Bayesian multi-trait, multi-environment genomic model.

For training the simulation model, molecular data and accurate phenotyping data will be employed. A model yielding higher accuracy of prediction can be developed by combining data with an accurate prediction model. Line breeding will so employ the model and data that were acquired. The genotypes with high GEBVs will be chosen for selection and crossing program will be implemented. The progenies from $F_1$s will be forwarded into advanced generations to select superior performing varieties. The overall procedure of model training integrated with line development program is graphically depicted in Figure 2.

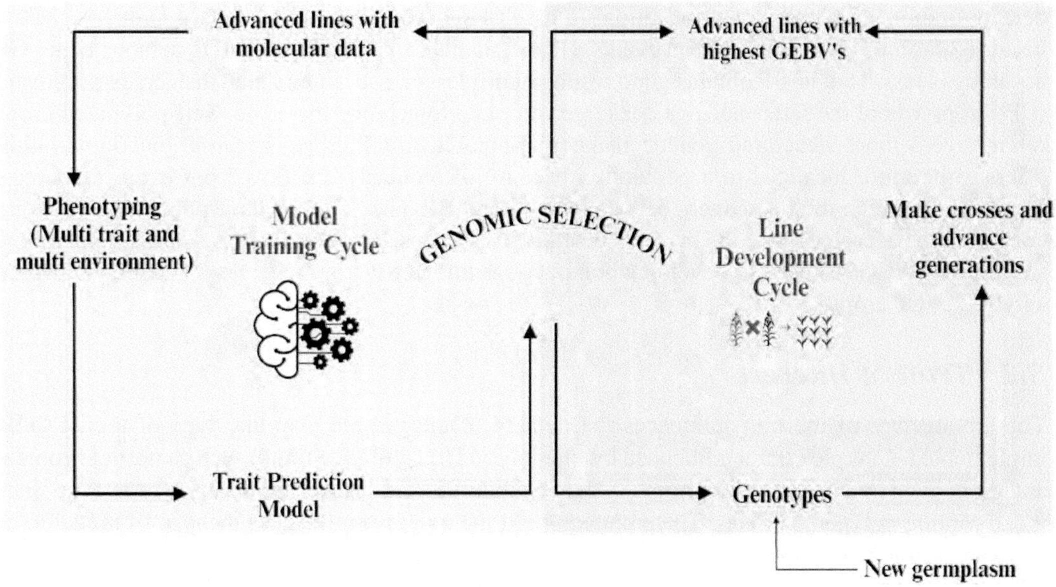


**Fig. 2** Integrating genomic selection procedure with line development program.

## 5.5 Linkage Disequilibrium

Linkage disequilibrium is the key factor that is driving the GP process. To estimate the LD, microsatellite markers and SNPs are more known along with the varying degree of the map density. The SNP markers require three times higher density in comparison to using microsatellites to accomplish similar accuracy (Solberg et al., 2014). Types of markers and their density of the linkage map utilized to calculate the LD play a major role in optimizing the TP. Since most of the markers are phenotypically neutral and small proportions of them are relevant to a specific trait, it is beneficial to optimize the marker density.

### *5.6 Marker Density*

To optimize the marker density, fixed effects were fitted as sub-setting depending on prior association studies where the significant markers were identified via GWAS. Another optimization approach is applying haplotype blocks based on marker LD in prediction models. Predictions based on haplotype blocks, rather than single SNP markers, could efficiently capture local epistasis and better account for LD to QTLs leading to improved GP accuracy (Weber et al., 2023). In comparison to marker density, size of TP had a greater impact on prediction accuracy (Brown et al., 2019).

## 6. Crop-Specific Examples of Utilizing GS Models for Variety Development

Genomic selection is potentially used in prediction of performance of many hybrid combinations (VanRaden, 2008; Crossa et al., 2017) and also filters out the important breeding material for traits such as yield, quality parameters, and biotic and abiotic stress resistance. To combat crop loss due to biotic stress, identification of resistance genes in the germplasm and their incorporation into the breeding procedures are the most important steps. Furthermore, the prediction models are utilized to select candidate genotypes or parents for biofortification of cereals. The GS is employed in various crops to get higher prediction accuracy and successful examples of them are presented in Table 2.

**Table 2** Successful results of genomic selection in plants.

| *Crops* | *Population* | *Traits* | *Genetic gain/ prediction accuracy* | *References* |
|---|---|---|---|---|
| **Cereals** | | | | |
| 1. Maize | Maize yellow synthetic 1 | Grain yield with drought tolerance | 0.110 and 0.135 t ha$^{-1}$ | Das et al., 2020 |
| | Maize yellow synthetic population | Grain yield with drought and water-logging tolerance | 0.038 and 0.113 t ha$^{-1}$ | |
| | Test crosses | Grain yield | 0.78 | Rio et al., 2019 |
| | $F_1$ from test crosses, North Carolina design II and full diallel | Grain yield | 0.10, 0.58, and 0.60 | Fristche-Neto et al., 2018 |
| | Biparental population | Grain yield with drought tolerance | 0.086 t ha$^{-1}$ | Beyene et al., 2015 |
| 2. Wheat | Spring wheat lines | Grain yield and mineral content | 0.55 | Joukhadar et al., 2021 |
| | Winter wheat lines | Grain yield | −0.21 to 0.42 | Lozada et al., 2020 |
| | Winter wheat breeding lines | Powdery mildew resistance<br>Grain yield | 0.60<br>0.64 | Sarinelli et al., 2019 |
| 3. Rice | 3K rice genome | Zinc concentration | 17.1 to 40.2 ppm 0.51 | Rakotondramanana et al., 2022 |
| | Hybrid population | Grain yield | 0.54 | Cui et al., 2020 |
| | Germplasm lines | Drought tolerance | 0.228 to 0.809 | Bhandari et al., 2019 |
| | Rice population | Blast resistance | 0.17 to 0.73 | Huang et al., 2019 |

*Contd.*

**Table 2** *Contd.*

| *Crops* | *Population* | *Traits* | *Genetic gain/ prediction accuracy* | *References* |
|---|---|---|---|---|
| 4. Barley | | Malting quality | 0.4–0.8 | Schmidt et al., 2016 |
| **Pulses** | | | | |
| 1. Lentils | Lentil breeding lines | Yield, biotic and abiotic stress | 0.34 to 0.83<br>0.22 to 0.54 | Gebremedhin et al., 2024 |
| | Lentil diversity panel, RILs | Maturity duration | 0.58 to 0.84 | Haile et al., 2020 |
| 2. Common bean | RILs, MAGIC population | Cooking time | 0.22 to 0.55 | Diaz et al., 2021 |
| 3. Chickpea | Breeding lines | Drought tolerance | 0.56 to 0.61 | Li et al., 2018 |
| **Oil seeds** | | | | |
| 1. *Brasssicanapus* | Double haploids | Oil content<br>Single-plant yield | 0.90<br>0.55 | Xiong et al., 2020 |
| | Spring canola lines | Yield and oil content | 0.69 and 0.64, respectively | Fikere et al., 2020 |
| | Test cross F1s | Oil content and oil yield | 0.81 and 0.75, respectively | Jan et al., 2016 |
| 2. Sunflower | Hybrids | Oil content | 0.78 | Mangin et al., 2017 |
| 3. Soybean | Nested association mapping (NAM) population | Grain yield | 0.68 | Beche et al., 2021 |
| 4. Groundnut | Breeding lines | Yield<br>Protein<br>Rust resistance<br>Late leaf-spot resistance | 0.49 to 0.60<br>0.41 to 0.46<br>0.74 to 0.75<br>0.57 to 0.65 | Pandey et al., 2020 |

In lentil, Haile et al. (2020) compared different GS models and prediction scenarios and reported that if large-effect QTLs were present in the population, multi-trait-based BayesB was found as the best GS model. They suggested it for low heritable traits with G×E interactions, as multi-trait GS improves the predictability. Diaz et al. (2021) evaluated GS using different populations, that is, recombinant inbred lines, multi-parent advanced-generation inter–crosses (MAGIC), and Andean and Mesoamerican breeding lines, in *Phaseolus* for cooking time. They inferred that in comparison to Mesoamerican genotypes (0.22), genomic prediction accuracy for cooking time using the MAGIC population was high (0.55) and showed promise. To promote drought-tolerant breeding lines, Li et al. (2018) integrated GWAS results into GS model involving 132 chickpea varieties. The observed prediction accuracy when genome-wide SNPs were considered was 0.148–0.186, whereas when filtered-yield-related causal SNPs were used prediction accuracy increased to 0.56–0.61. This improved prediction accuracy further provides another factor to improve prediction accuracy with limited resources. In oilseeds, the most important attributes are oil quality along with yield and stresses tolerance. GS accounts for G×E interactions as oil quality and yield traits are affected by environment (Patil et al., 2020). Hence, it is vital to use the appropriate GS models along with better phenotyping tools and techniques, to account for the G×E effects for accurate selection.

## 7. Challenges in Applying GS Models to Develop Next-Generation Varieties Across Diverse Crops

Along with humungous advantages, it renders certain challenges while applying GS in plant breeding. To implement the GS as a blanket approach for developing NGVs in different crops, many areas need attention, as outlined in the following text.

1. GS is mostly adopted to extend genetic gain by qualitative traits on the architecture, of which our knowledge is limited. Moreover, the considered traits are affected by G×E interactions which pose a difficulty in decision-making.
2. GS is not well accepted by plant breeding community because of lack of sufficient evidences for its practical utility, since many evaluations are conducted as simulation studies.
3. The estimation of GEBVs from marker effect by using simulation models considers additive genetic variance and seems to ignore the epistatic effects. Nevertheless, this is good only for self-pollinated crop species, since homozygous lines are used as parents and varieties. Whereas in other species, progeny performance depends on epistatic and dominant effect.
4. The selection response declines at faster rate in GS than the pedigree selection. Hence, increasing number of markers to estimate GEBVs can minimize decline. And placing the higher weights on low frequent alleles can yield the long-term selection response under GS (Nakaya & Isobe, 2012).
5. Implementation of GS requires a considerable infrastructure and other resources which add to the cost for breeding program. This makes the GS beyond the reach of moderate-size to small-size public sector breeding program particularly in developing countries.
6. Genotyping of large population for thousands of marker loci adds a huge cost to plant breeding program. Since the future depends on marker data for selection of parents and candidate genotypes, it is the need of the hour to reduce the cost per unit marker.
7. Planning and execution of GS are quite skill-demanding. Hence, plant breeders should reorient their approach to adopt the GS.

## 8. Integration with Breeding Programs and Future Perspectives

To meet the future food and nutrition demands of the country, it is necessary to effectively increase the genetic gain of staple crops exponentially. Expansion of GS from livestock toward plants will successfully achieve the goal of genetic gain. Furthermore, the development of next-generation sequencing tools that reduces the cost of genotyping while increasing its efficiency and high-throughput phenotyping technique which provides accurate phenotypic data for complex nutritional traits will fasten the deployment of GS in plants. The amalgamation of different disciplines with the new breeding pipeline is necessary and the success of such breeding pipelines depends on the way transdisciplinary approach is combined with GS. Approaches such as gene editing, speed breeding, omics technology, high-end genotyping, and phenotyping will improve genetic gain of GS (Figure 3). Speed breeding reduces the breeding cycle that yields more number of generations per year.

The closely linked genes/QTLs identified through GWAS or traditional QTL mapping need to be further pyramided to a single variety to develop a multipurpose NGV. Tightly linked genes that inherit together termed as haplotype often cover a region on the chromosome. This idea should be extended to entire genome using functional markers, such that it makes easier to transfer the desirable traits to the offspring. GS aids in using this wider haplotype in developing an NGV with desirable genes for agronomically and nutritionally important traits. By considering effect of these haplotypes, prediction of GS will be optimized and favorable genes can be introgressed into a single line. This line is superior in performance and utilized to develop varieties such as NGVs. In comparison to phenotypic selection for longer duration for traits such as oil and protein content, it is evident that GS offers more advantages and is faster in achieving the said amount of genetic gain

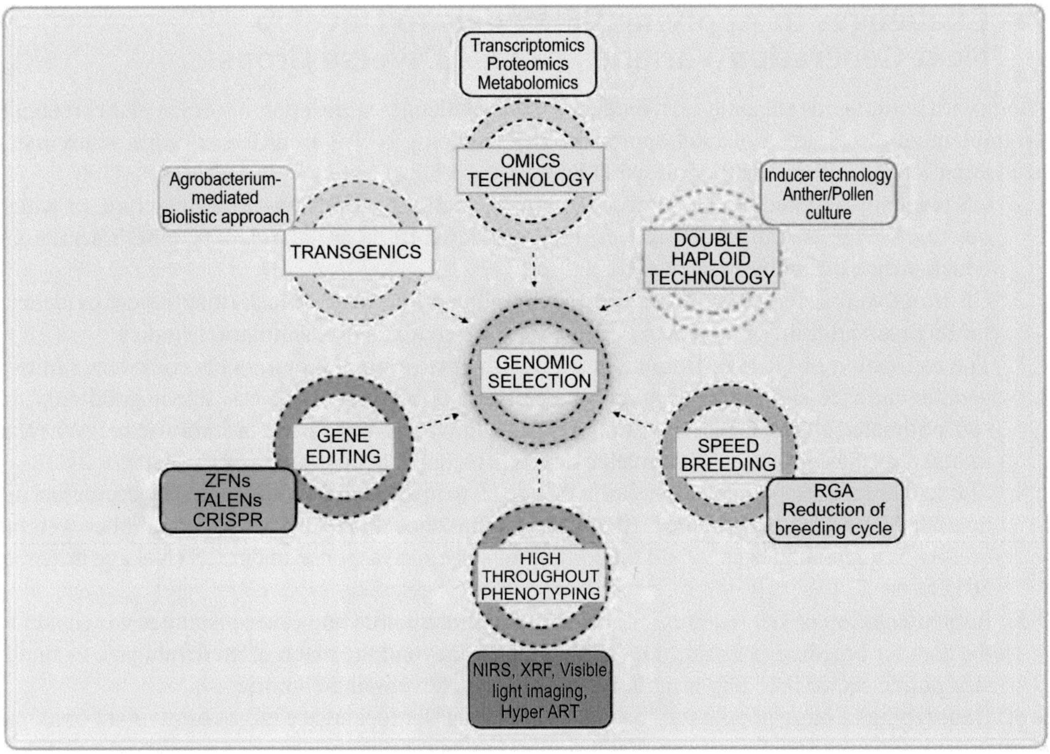


**Fig. 3** Integrating genomic selection with other breeding methods.

(Goldman et al., 1993). Hereby, to increase the deployment of GS in future breeding pipelines, high-end genotyping platforms with low cost should be developed.

Alongside knowledge on GS, advantages and simulation models should be disseminated with much efficiency, as concept and statistical analysis of GS is complex. Furthermore, the effect of G×E interaction on the trait inheritance should be considered for shooting up prediction accuracy by conducting evaluation over environments and seasons/years. Effective statistical model such as GBLUP when combined with G×E data for testing of early generations increases the prediction accuracy. These characteristics further increase GS adoption and lead to its large-scale implementation. Current breeding practices should normalize using GS to obtain better genetic gain. And institutions with the facilities and advanced technologies should focus efforts on GS, which makes use of historical data obtained through long-standing process. Integrated and open platform which provides access to breeders globally will further strengthen the process of GS with transdisciplinary approaches. GS framework should be flexible enough to effectively combine the results of research conducted all over the world. These developments will enable breeders to achieve the food demands of the world in the near future via GS.

## 9. Concluding Remarks and Future Prospects

The advent of new genotyping technologies with lower cost and ease of application has amplified utilization of markers in improvement of crops. These features have led to the surge of new techniques such as genomic selection, utilizing the genome-wide marker data to predict the breeding values. The association between TP and BP determines prediction accuracy of the model. Accurate prediction of breeding values of unobserved individuals will increase the efficiency of GS. Distant individuals and untested individuals in the population lead to lower prediction accuracy. High-throughput genotyping and phenomics widen the scope of GS toward the development of next-generation varieties, which are varieties with combination of highly desirable traits such as quantity and quality of nutrients, higher grain yield, and stress tolerance. High-end phenotyping and genotyping methodologies would enable us to accurately estimate breeding values and utilize genome-wide marker data. Such usage of genome-wide data will aid in improving complex attributes controlled by many genes/QTLs with small effects. Thereby, it aids in tailoring the varieties depending on the future needs to combine many traits. GS leads to development of superior varieties, which is otherwise a long and tedious process. Careful considerations should be made to include GS into breeding programs focused on line or/and hybrid development for combining agronomic traits such as grain yield, nutritional quality, and stress tolerance. GS can be integrated with conventional and modern breeding technologies along with allied disciplines to improve the efficiency and genetic gain. Ancestral data obtained in pedigree breeding will enhance the prediction accuracy when coupled with GS and statistical models such as GBLUP and rrBLUP integrated with machine learning. When combined with other interdisciplinary approaches, GS yields better results. Multi-year/-season data will enable the inclusion of G×E interaction effects, according to which models can be developed. There is need to overcome obstacles such as genotyping larger samples, complex statistical analysis, and integration into breeding cycles to further deploying GS in plant breeding. Futuristic ideas such as multi-trait, multi-environment models will increase the genetic gain attained per breeding cycle. Diverse genetic resources harboring superior alleles will be deployed in GS to develop lines for further development of next-generation varieties. Consequently, the new-age technologies can be used efficiently with GS technology to feed the world population and meet the nutritional demands of people.

## References

Akdemir, D. and Sánchez, J.I. (2016). Efficient breeding by genomic mating. *Frontiers in Genetics*, 7, 210.

Alemu, A., Åstrand, J., Montesinos-López, O.A., Sanchez, J.I., Fernandez-Gonzalez, J., Tadesse, W. and Chawade, A. (2024). Genomic selection in plant breeding: key factors shaping two decades of progress. *Molecular Plant*, *17*(4), 552–578.

Allier, A., Teyssèdre, S., Lehermeier, C., Moreau, L. and Charcosset, A. (2020). Optimised breeding strategies to harness genetic resource with different performance levels. *BMC Genomics*, 21, 1–16.

Anilkumar, C., Sunitha, N.C., Harikrishna, Devate, N.B. and Ramesh, S. (2022). Advances in integrated genomic selection for rapid genetic gain in crop improvement: a review. *Planta*, *256*(5), 87.

Araus, J.L., Slafer, G.A., Reynolds, M.P. and Royo, C. (2002). Plant breeding and drought in C3 cereals: what should we breed for?. *Annals of Botany*, *89*(7), 925–940.

Araus, J.L., Slafer, G.A., Royo, C. and Serret, M.D. (2008). Breeding for yield potential and stress adaptation in cereals. *Critical Reviews in Plant Science*, *27*(6), 377–412.

Asoro, F.G., Newell, M.A., Beavis, W.D., Scott, M.P. and Jannink, J.L. (2011). Accuracy & TP design for GS on quantitative traits in elite North American oats. *The Plant Genome*, *4*(2), 132–144.

Bassi, F.M., Bentley, A.R., Charmet, G., Ortiz, R. and Crossa, J. (2016). Breeding schemes for the implementation of GS in wheat (*Triticum* spp.). *Plant Science*, 242, 23–36.

Bassi, F.M., Sanchez-Garcia, M. and Ortiz, R., (2024). What plant breeding may (and may not) look like in 2050?. *The Plant Genome*, *17*(1), e20368.

Beche, E., Gillman, J.D., Song, Q., Nelson, R., Beissinger, T., Decker, J. and Scaboo, A.M. (2021). Genomic prediction using TP design in interspecific soybean populations. *Molecular Breeding*, 41, 1–15.

Belamkar, V., Guttieri, M.J., Hussain, W., Jarquín, D., El-Basyoni, I., Poland, J. and Baenziger, P.S. (2018). Genomic selection in PYTs in a winterwheat breeding program. *G3: Genes, Genomes, Genetics*, *8*(8), 2735–2747.

Bergsträsser, S., Fanourakis, D., Schmittgen, S., Cendrero-Mateo, M.P., Jansen, M., Scharr, H. and Rascher, U. (2015). HyperART: non-invasive quantification of leaf traits using hyperspectral absorption-reflectance-transmittance imaging. *Plant Methods*, 11, 1–17.

Bernardo, A.B. (2010). Extending hope theory: internal and external locus of trait hope. *Personality and Individual Differences*, 49, 944–949.

Bernardo, R. (2016). Bandwagons I, too, have known. *Theoretical and Applied Genetics*, 129, 2323–2332.

Bernardo, R. and Yu, J. (2007). Prospects for genome-wide selection for quantitative traits in maize. *Crop Science*, *47*(3), 1082–1090.

Berro, I., Lado, B., Nalin, R.S., Quincke, M. and Gutiérrez, L. (2019). TP optimization for GS. *The Plant Genome*, *12*(3), 190028.

Beyene, Y., Gowda, M., Pérez-Rodríguez, P., Olsen, M., Robbins, K.R., Burgueño, J. and Crossa, J. (2021). Application of GS at the early stage of breeding pipeline in tropical maize. *Frontiers in Plant Science*, 12, 685488.

Beyene, Y., Semagn, K., Mugo, S., Tarekegne, A., Babu, R., Meisel, B. and Crossa, J. (2015). Genetic gains in grain yield through genomic selection in eight bi-parental maize populations under drought stress. *Crop Science*, *55*(1), 154–163.

Bhandari, A., Bartholomé, J., Cao-Hamadoun, T.V., Kumari, N., Frouin, J., Kumar, A. and Ahmadi, N. (2019). Selection of trait-specific markers & multi-environment models improve genomic predictive ability in rice. *PLoS One*, *14*(5), e0208871.

Brandariz, S.P. and Bernardo, R. (2019). Small ad hoc versus large general training populations for genome-wide selection in maize biparental crosses. *Theoretical and Applied Genetics*, 132, 347–353.

Brown Stewart, B.B., Song, Q., Vaughn, J.N. and Li, Z. (2019). Genomic selection for yield & seed composition traits within an applied soybean breeding program. *G3: Genes, Genomes, Genetics*, *9*(7), 2253–2265.

Brozynska, M., Furtado, A. and Henry, R.J. (2016). Genomics of crop wild relatives: expanding the gene pool for crop improvement. *Plant Biotechnology Journal*, *14*(4), 1070–1085.

Bruning, B., Liu, H., Brien, C., Berger, B., Lewis, M. and Garnett, T. (2019). The development of hyperspectral distribution maps to predict the content, & distribution of nitrogen and water in wheat (*Triticum aestivum*). *Frontiers in Plant Science*, 10, 1380.

Budhlakoti, N., Rai, A. and Mishra, D.C. (2020). Statistical approach for improving genomic prediction accuracy through efficient diagnostic measure of influential observation. *Scientific Reports*, *10*(1), 8408.

Cabrera-Bosquet, L., Crossa, J., von Zitzewitz, J., Serret, M.D. and Luis Araus, J. (2012). High-throughput phenotyping and genomic selection: the frontiers of crop breeding converge F. *Journal of Integrative Plant Biology*, *54*(5), 312–320.

Combs, E. and Bernardo, R. (2013). Accuracy of genome-wide selection for different traits with constant population size, heritability, and number of markers. *The Plant Genome*, *6*(1), 1–7.

Cortes C. and Vapnik V. (1995). Support-vector networks. *Machine Learning 20*(3): 273–297.

Crossa, J., Perez, P., Hickey, J., Burgueno, J., Ornella, L., Cerón-Rojas, J. and Mathews, K. (2014). Genomic prediction in CIMMYT maize & wheat breeding programs. *Heredity*, *112*(1), 48–60.

Crossa, J., Pérez-Rodríguez, P., Cuevas, J., Montesinos-López, O., Jarquín, D., de Los Campos, G. and Varshney, R.K. (2017). Genomic selection in plant breeding: methods, models, and perspectives. *Trends in Plant Science*, *22*(11), 961–975.

Cui, Y., Li, R., Li, G., Zhang, F., Zhu, T., Zhang, Q. and Xu, S. (2020). Hybrid breeding of rice via genomic selection. *Plant Biotechnology Journal*, *18*(1), 57–67.

Das, R.R., Vinayan, M.T., Patel, M.B., Phagna, R.K., Singh, S.B., Shahi, J.P. and Zaidi, P.H. (2020). Genetic gain with rapid-cycle genomic selection for combined drought & waterlogging tolerance in tropical maize (*Zea mays* L.). *The Plant Genome*, *13*(3), e20035.

de Los Campos, G.U.S.T.A.V.O., Gianola, D., Rosa, G.J., Weigel, K.A. and Crossa, J. (2010). Semi-parametric genomic-enabled prediction of genetic values using reproducing kernel Hilbert spaces methods. *Genetics Research*, *92*(4), 295–308.

Deschamps, S. and Campbell, M.A. (2010). Utilization of next-generation sequencing platforms in plant genomics and genetic variant discovery. *Molecular Breeding*, 25, 553–570.

Diaz, S., Ariza-Suarez, D., Ramdeen, R., Aparicio, J., Arunachalam, N., Hernandez, C. and Raatz, B. (2021). Genetic architecture and genomic prediction of cooking time in common bean (*Phaseolus vulgaris* L.). *Frontiers in Plant Science*, 11, 622213.

Elshire, R.J., Glaubitz, J.C., Sun, Q., Poland, J.A., Kawamoto, K., Buckler, E.S. and Mitchell, S.E. (2011). A robust, simple GBS approach for high diversity species. *PLoS One*, *6*(5), e19379.

Falconer, D.S. (1996). *Introduction to Quantitative Genetics*. Pearson Education India.

Fikere, M., Barbulescu, D.M., Malmberg, M.M., Maharjan, P., Salisbury, P.A., Kant, S. and Daetwyler, H.D. (2020). Genomic prediction and genetic correlation of agronomic, blackleg disease, and seed quality traits in canola (*Brassica napus* L.). *Plants*, *9*(6), 719.

Finkle, E. (2009). With "phenomics" plant scientists hope to shift breeding into overdrive. *Science*, 325, 380–381.

Food and Agriculture Organization (FAOSTAT), 2012, https://www.fao.org/faostat/en/#data/QCL

Fristche-Neto, R., Akdemir, D. and Jannink, J.L. (2018). Accuracy of genomic selection to predict maize single-crosses obtained through different mating designs. *Theoretical and Applied Genetics*, 131, 1153–1162.

Frouin, J., Labeyrie, A., Boisnard, A., Sacchi, G.A. and Ahmadi, N. (2019). Genomic prediction offers the most effective marker assisted breeding approach for ability to prevent arsenic accumulation in rice grains. *PLoS One*, *14*(6), e0217516.

Gaynor, R.C., Gorjanc, G., Bentley, A.R., Ober, E.S., Howell, P., Jackson, R. and Hickey, J.M. (2017). A two-part strategy for using genomic selection to develop inbred lines. *Crop Science*, *57*(5), 2372–2386.

Gebremedhin, A., Li, Y., Shunmugam, A.S., Sudheesh, S., Valipour-Kahrood, H., Hayden, M.J. and Kaur, S. (2024). Genomic selection for target traits in the Australian lentil breeding program. *Frontiers in Plant Science*, 14, 1284781.

Geldermann, H. (1975). Investigations on inheritance of quantitative characters in animals by gene markers I. Methods. *Theoretical and Applied Genetics*, *46*(7), 319–330.

Gianola, D. (2013). Priors in whole-genome regression: the Bayesian alphabet returns. *Genetics*, *194*(3), 573–596.

Gianola, D., Fernando, R.L. and Stella, A. (2006). Genomic-assisted prediction of genetic value with semiparametric procedures. *Genetics*, *173*(3), 1761–1776.

Goldman, I.L., Rocheford, T.R. and Dudley, J.W. (1993). Quantitative trait loci influencing protein & starch concentration in the Illinois long term selection maize strains. *Theoretical and Applied Genetics*, 87, 217–224.

Granier, C., Aguirrezabal, L., Chenu, K., Cookson, S.J., Dauzat, M., Hamard, P. and Tardieu, F. (2006). PHENOPSIS is an automated platform for reproducible phenotyping of plant responses to soil water deficit in *Arabidopsis thaliana* permitted the identification of an accession with low sensitivity to soil water deficit. *New Phytologist*, *169*(3), 623–635.

Grenier, C., Cao, T.V., Ospina, Y., Quintero, C., Châtel, M.H., Tohme, J. and Ahmadi, N. (2015). Accuracy of genomic selection in a rice synthetic population developed for recurrent selection breeding. *PLoS One*, *10*(8), e0136594.

Gunundu, R., Shimelis, H. and Mashilo, J. (2023). Genomic selection and enablers for agronomic traits in maize (*Zea mays*): a review. *Plant Breeding*, *142*(5), 573–593.

Habier, D., Fernando, R.L. and Dekkers, J. (2007). The impact of genetic relationship information on genome-assisted breeding values. *Genetics*, *177*(4), 2389–2397.

Habier, D., Fernando, R.L. and Garrick, D.J. (2013). Genomic BLUP decoded: a look into the black box of genomic prediction. *Genetics*, *194*(3), 597–607.

Habier, D., Fernando, R.L., Kizilkaya, K. and Garrick, D.J. (2011). Extension of the Bayesian alphabet for genomic selection. *BMC Bioinformatics*, 12, 1–12.

Haile, T.A., Heidecker, T., Wright, D., Neupane, S., Ramsay, L., Vandenberg, A. and Bett, K.E. (2020). Genomic selection for lentil breeding: empirical evidence. *The Plant Genome*, *13*(1), e20002.

Hamblin, M.T., Close, T.J., Bhat, P.R., Chao, S., Kling, J.G., Abraham, K.J. and Jannink, J.L. (2010). Population structure and linkage disequilibrium in US barley germplasm: implications for association mapping. *Crop Science*, *50*(2), 556–566.

Hao, Y., Wang, H., Yang, X., Zhang, H., He, C., Li, D. and Fu, J. (2019). Genomic prediction using existing historical data contributing to selection in biparental populations: a study of kernel oil in maize. *The Plant Genome*, *12*(1), 180025.

Hastie, T., Tibshirani, R. and Friedman. J. (2009). The elements of statistical learning: data mining, inference, and prediction. Springer, New York.

Hayes, B.J., Bowman, P.J., Chamberlain, A.J. and Goddard, M.E. (2009). Invited review: Genomic selection in dairy cattle: progress & challenges. *Journal of Dairy Science*, *92*(2), 433–443.

He, J., Zhao, X., Laroche, A., Lu, Z.X., Liu, H. and Li, Z. (2014). Genotyping-by-sequencing (GBS), an ultimate marker-assisted selection (MAS) tool to accelerate plant breeding. *Frontiers in Plant Science*, 5, 484.

Heffner, E.L., Jannink, J.L. and Sorrells, M.E. (2011). Genomic selection accuracy using multifamily prediction models in a wheat breeding program. *The Plant Genome*, *4*(1), 65–75

Heffner, E.L., Lorenz, A.J., Jannink, J.L. and Sorrells, M.E. (2010). Plant breeding with genomic selection: gain per unit time and cost. *Crop Science*, *50*(5), 1681–1690.

Henderson, C.R. (1975). Best linear unbiased estimation & prediction under a selection model. *Biometrics*, 30, 423–447.

Hickey, J.M., Chiurugwi, T., Mackay, I. and Powell, W. (2017). Genomic prediction unifies animal & plant breeding programs to form platforms for biological discovery. *Nature Genetics*, *49*(9), 1297–1303.

Hickey, J.M., Dreisigacker, S., Crossa, J., Hearne, S., Babu, R., Prasanna, B.M. and Gorjanc, G. (2014). Evaluation of genomic selection training population designs & genotyping strategies in plant breeding programs using simulation. *Crop Science*, *54*(4), 1476–1488.

Hoefler, R. (2022). *Increasing the Efficiency of Cereal Breeding Programs Through Resource Allocation Optimization*. The University of Wisconsin-Madison.

Howard, R., Carriquiry, A.L. and Beavis, W.D. (2014). Parametric & nonparametric statistical methods for genomic selection of traits with additive & epistatic genetic architectures. *G3: Genes, Genomes, Genetics*, *4*(6), 1027–1046.

Huang, M., Balimponya, E.G., Mgonja, E.M., McHale, L.K., Luzi-Kihupi, A., Wang, G.L. and Sneller, C.H. (2019). Use of genomic selection in breeding rice (*Oryza sativa* L.) for resistance to rice blast (*Magnaporthe oryzae*). *Molecular Breeding*, 39, 1–16.

Jan, H.U., Abbadi, A., Lücke, S., Nichols, R.A. and Snowdon, R.J. (2016). Genomic prediction of testcross performance in canola (*Brassica napus*). *PLoS One*, *11*(1), e0147769.

Jennings, H.S. (1917). The numerical results of diverse systems of breeding with respect to two pairs of characters, linked or independent, with special relation to the effects of linkage. *Genetics*, *2*(2), 97.

Joukhadar, R., Thistlethwaite, R., Trethowan, R.M., Hayden, M.J., Stangoulis, J., Cu, S. and Daetwyler, H.D. (2021). Genomic selection can accelerate the biofortification of spring wheat. *Theoretical and Applied Genetics*, 134, 3339–3350.

Krishnappa, G., Savadi, S., Tyagi, B.S., Singh, S.K., Mamrutha, H.M., Kumar, S. and Singh, G.P. (2021). Integrated genomic selection for rapid improvement of crops. *Genomics*, *113*(3), 1070–1086.

Lande, R. and Thompson, R. (1990). Efficiency of marker-assisted selection in the improvement of quantitative traits. *Genetics*, *124*(3), 743–756.

Li, Y., Ruperao, P., Batley, J., Edwards, D., Khan, T., Colmer, T.D. and Sutton, T. (2018). Investigating drought tolerance in chickpea using genome-wide association mapping & genomic selection based on whole-genome resequencing data. *Frontiers in Plant Science*, 9, 190.

Li, Z.K., Luo, L.J., Mei, H.W., Wang, D.L., Shu, Q.Y., Tabien, R. and Paterson, A. (2001). Over-dominant epistatic loci are the primary genetic basis of inbreeding depression & heterosis in rice. I. Biomass and grain yield. *Genetics*, 158(4), 1737–1753.

Li, Z., Pinson, S.R., Park, W.D., Paterson, A.H. and Stansell, J.W. (1997). Epistasis for three grain yield components in rice (*Oryza sativa* L.). *Genetics*, *145*(2), 453–465.

Lohithaswa, H.C., Shreekanth, S.M., Banakara, S.K., Sripathy, K.V. and Mallikarjuna, M.G. (2022). Genomic selection for enhanced stress tolerance in maize. In *Next-Generation Plant Breeding Approaches for Stress Resilience in Cereal Crops* (pp. 121–160). Singapore: Springer Nature Singapore.

Long, N., Gianola, D., Rosa, G.J. and Weigel, K.A. (2011). Application of support vector regression to genome-assisted prediction of quantitative traits. *Theoretical and Applied Genetics*, 123, 1065–1074.

Lopes, M.S., Araus, J.L., Van Heerden, P.D. and Foyer, C.H. (2011). Enhancing drought tolerance in C4 crops. *Journal of Experimental Botany*, *62*(9), 3135–3153.

Lorenzana, R.E. and Bernardo, R. (2009). Accuracy of genotypic value predictions for marker-based selection in biparental plant populations. *Theoretical and Applied Genetics*, 120, 151–161.

Lozada, D.N., Ward, B.P. and Carter, A.H. (2020). Gains through selection for grain yield in a winter wheat breeding program. *PLoS One*, *15*(4), e0221603.

Mangin, B., Bonnafous, F., Blanchet, N., Boniface, M.C., Bret-Mestries, E., Carrère, S. and Langlade, N.B. (2017). Genomic prediction of sunflower hybrids oil content. *Frontiers in Plant Science*, 8, 1633.

Meuwissen, T.H., Hayes, B.J. and Goddard, M. (2001). Prediction of total genetic value using genome-wide dense marker maps. *Genetics*, *157*(4), 1819–1829.

Meuwissen, T.H., Solberg, T.R., Shepherd, R. and Woolliams, J.A. (2009). A fast algorithm for BayesB type of prediction of genome-wide estimates of genetic value. *Genetics Selection Evolution*, 41, 1–10.

Meuwissen, T., Hayes, B. and Goddard, M. (2016). Genomic selection: a paradigm shift in animal breeding. *Animal Frontiers*, *6*(1), 6–14.

Michel, S., Ametz, C., Gungor, H., Epure, D., Grausgruber, H., Löschenberger, F. and Buerstmayr, H. (2016). Genomic selection across multiple breeding cycles in applied bread wheat breeding. *Theoretical and Applied Genetics*, 129, 1179–1189.

Montes, J.M., Melchinger, A.E. and Reif, J.C. (2007). Novel throughput phenotyping platforms in plant genetic studies. *Trends in Plant Science*, *12*(10), 433–436.

Montes, J.M., Technow, F., Dhillon, B.S., Mauch, F. and Melchinger, A.E. (2011). High-throughput non-destructive biomass determination during early plant development in maize under field conditions. *Field Crops Research*, *121*(2), 268–273.

Montesinos-López, O.A., Montesinos-López, A., Bernal Sandoval, D.A., Mosqueda-Gonzalez, B.A., Valenzo-Jiménez, M.A. and Crossa, J. (2022). Multi-trait genome prediction of new environment with partial least squares. *Frontiers in Genetics*, 13, 966775.

Morgante, F., Huang, W., Maltecca, C. and Mackay, T.F. (2018). Effect of genetic architecture on the prediction accuracy of quantitative traits in samples of unrelated individuals. *Heredity*, *120*(6), 500–514.

Nakaya, A. and Isobe, S.N. (2012). Will genomic selection be a practical method for plant breeding? *Annals of Botany*, *110*(6), 1303–1316.

Pandey, M.K., Chaudhari, S., Jarquin, D., Janila, P., Crossa, J., Patil, S.C. and Varshney, R.K. (2020). Genome-based trait prediction in multi-environment breeding trials in groundnut. *Theoretical and Applied Genetics*, 133, 3101–3117.

Parekh, M.J., Kumar, S., Zala, H.N., Fougat, R.S., Patel, C.B., Bosamia, T.C. and Parihar, A. (2016). Development and validation of novel fiber relevant dbEST–SSR markers and their utility in revealing genetic diversity in diploid cotton (*Gossypium herbaceum* and *G. arboreum*). *Industrial Crops and Products*, 83, 620–629.

Park, T. and Casella, G. (2008). The bayesian lasso. *Journal of the American Statistical Association*, *103*(482), 681–686.

Patil, R., Viswanatha, K.P., Upadhyaya, H.D., Lokesha, R., Khan, H. and Gururaj, S. (2020). Genetic diversity, association & principle component analyses for agronomical & quality traits in genomic selection training population of groundnut (*Arachis hypogaea* L.*). Indian Journal of Genetics and Plant Breeding*, *80*(03), 282–290.

Peirone, L.S., Pereyra Irujo, G.A., Bolton, A., Erreguerena, I. and Aguirrezábal, L.A. (2018). Assessing the efficiency of phenotyping early traits in a greenhouse automated platform for predicting drought tolerance of soybeon in the field. *Frontiers in Plant Science*, 9, 587.

Pértille, F., Guerrero-Bosagna, C., Silva, V.H.D., Boschiero, C., Nunes, J.D.R.D.S., Ledur, M.C. and Coutinho, L.L. (2016). High-throughput and cost-effective chicken genotyping using next-generation sequencing. *Scientific Reports*, *6*(1), 26929.

Pineda, M., Barón, M. and Pérez-Bueno, M.L. (2020). Thermal imaging for plant stress detection & phenotyping. *Remote Sensing*, *13*(1), 68.

Poland, J., Endelman, J., Dawson, J., Rutkoski, J., Wu, S., Manes, Y. and Jannink, J.L. (2012). Genomic selection in wheat breeding using genotyping-by-sequencing. *The Plant Genome*, *5*(3), 103–113

Prasanna, B.M., Araus, J.L., Crossa, J., Cairns, J.E., Palacios, N., Das, B. and Magorokosho, C. (2013). High-throughput & precision phenotyping for cereal breeding programs. *Cereal Genomics II*, 2, 341–374.

Prasanna, B.M., Chaikam, V. and Mahuku, G. (2012). *Doubled Haploid Technology in Maize Breeding: Theory & Practice*. Cimmyt.

Rakotondramanana, M., Tanaka, R., Pariasca-Tanaka, J., Stangoulis, J., Grenier, C. and Wissuwa, M. (2022). Genomic prediction of zinc-biofortification potential in rice gene bank accessions. *Theoretical and Applied Genetics*, *135*(7), 2265–2278.

Riedelsheimer, C. and Melchinger, A.E. (2013). Optimizing the allocation of resources for genomic selection in one breeding cycle. *Theoretical and Applied Genetics*, 126, 2835–2848.

Rio, S., Mary-Huard, T., Moreau, L. and Charcosset, A. (2019). Genomic selection efficiency & a priori estimation of accuracy in a structured dent maize panel. *Theoretical and Applied Genetics*, 132, 81–96.

Rodríguez-Ramilo, S.T., García-Cortés, L.A. and González-Recio, Ó. (2014). Combining genomic & genealogical information in a reproducing kernel Hilbert spaces regression model for genome-enabled predictions in dairy cattle. *PLoS One*, *9*(3), e93424.

Ronne De, M., Légaré, G., Belzile, F., Boyle, B. and Torkamaneh, D. (2023). 3D-GBS: a universal genotyping-by-sequencing approach for genomic selection and other high-throughput low-cost applications in species with small to medium-sized genomes. *Plant Methods*, *19*(1), 13.

Sarinelli, J.M., Murphy, J.P., Tyagi, P., Holland, J.B., Johnson, J.W., Mergoum, M. and Brown-Guedira, G. (2019). TP selection & use of fixed effects to optimize genomic predictions in a historical USA winter wheat panel. *Theoretical and Applied Genetics*, 132, 1247–1261.

Schmidt, M., Kollers, S., Maasberg-Prelle, A., Großer, J., Schinkel, B., Tomerius, A. and Korzun, V. (2016). Prediction of malting quaity traits in barley based on genome-wide marker data to assess the potential of genomic selection. *Theoretical and Applied Genetics*, 129, 203–213.

Sherry Flint-Garcia, A., Thornsberry, J.M. and Buckler IV, E.S. (2003). Structure of linkage disequilibrium in plants. *Annual Review of Plant Biology*, *54*(1), 357–374.

Shikha, M., Kanika, A., Rao, A.R., Mallikarjuna., M.G., Gupta., H.S. and Nepolean, T. (2017). Genomic selection for drought tolerance using genome-wide SNPs in maize. *Frontiers in Plant Science*, 8, 550.

Silverman, B.W. (2018). *Density Estimation for Statistics and Data Analysis*. Routledge.

Singh, K., Kumar, S., Kumar, S.R., Singh, M. and Gupta, K. (2019). Plant genetic resources management & pre-breeding in genomics era. *Indian Journal of Genetics and Plant Breeding*, 79(Sup-01), 117–130.

Smith, J.S.C., Hussain, T., Jones, E.S., Graham, G., Podlich, D., Wall, S. and Williams, M. (2008). Use of DH in maize breeding: implications for intellectual property protection & genetic diversity in hybrid crops. *Molecular Breeding*, 22, 51–59.

Solberg Woods, L.C. (2014). QTL mapping in outbred populations: successes and challenges. *Physiological genomics*, *46*(3), 81–90.

Sonah, H., Bastien, M., Iquira, E., Tardivel, A., Legare, G., Boyle, B. and Belzile, F. (2013). An improved GBS approach offering increased versatility & efficiency of SNP discovery and genotyping. *PLoS One*, *8*(1), e54603.

Stock, K.F. and Reents, R. (2013). Genomic selection: status in different species and challenges for breeding. *Reproduction in Domestic Animals*, 48, 2–10.

Su, G., Brøndum, R.F., Ma, P., Guldbrandtsen, B., Aamand, G. P. and Lund, M. S. (2012). Comparison of genomic predictions using medium-density (~54,000) and high-density (~777,000) SNP marker panels in Nordic Holstein and Red Dairy Cattle populations. *Journal of Dairy Science*, *95*(8), 4657–4665.

Sun, H., Zheng, X., Lu, X. and Wu, S. (2019). Spectral–spatial attention network for hyperspectrol image classification. *IEEE Transactions on Geoscience and Remote Sensing*, *58*(5), 3232–3245.

Sun, X., Qu, L., Garrick, D.J., Dekkers, J.C. and Fernando, R.L. (2012). A fast EM algorithm for BayesA-like prediction of genomic breeding values. *PLoS One*, *7*(11), e49157.

Thistlethwaite, F.R., Gamal El-Dien, O., Ratcliffe, B., Klápště, J., Porth, I., Chen, C. and El-Kassaby, Y.A. (2020). Linkage disequilibrium vs. pedigree: genomic selection prediction accuracy in conifer species. *PLoS One*, *15*(6), e0232201.

Tibshirani, R. (1996). Regression shrinkage and selection via the lasso. *Journal of the Royal Statistical Society Series B: Statistical Methodology*, *58*(1), 267–288.

Torkamaneh, D., Laroche, J., Valliyodan, B., O'Donoughue, L., Cober, E., Rajcan, I. and Belzile, F. (2021). Soybean (*Glycine max*) Haplatype Map (GmHapMap): a universal resource for soybean translational & functional genomics. *Plant Biotechnology Journal*, *19*(2), 324–334.

VanRaden, P.M. (2008). Efficient methods to compute genomic predictions. *Journal of Dairy Science*, 91(11), 4414–4423.

Vapnik V. (1995). The nature of statistical learning theory, 2nd edn. Springer, New York.

Varshney, R.K., Singh, V.K., Hickey, J.M., Xun, X., Marshall, D.F., Wang, J. and Ribaut, J.M. (2016). Analytical & decision support tools for genomics-assisted breeding. *Trends in Plant Science*, *21*(4), 354–363.

Vadez, V., Kholova, J., Hummel, G., Gupta, S.K. and Hash, C.T. (2015). LeasyScan: a novel concept combining 3D imaging and lysimetry for high-throughput phenotyping of traits controlling plant water budget. *Journal of Experimental Biology*, *66*(18), 5581–5593.

Wang, X, Xu, Y., Hu, Z. and Xu, C. (2018). Genomic selection methods for crop improvement: current status & prospects. *The Crop Journal*, *6*(4), 330–340.

Weber, S.E., Frisch, M., Snowdan, R.J. and Voss-Fels, K.P. (2023). Haplotype blocks for genomic prediction: a comparative evaluation in multiple crop datasets. *Frontiers in Plant Science*, 14, 1217589.

Xiong, S., Wang, M., Zou, J., Meng, J. and Liu, Y. (2020). A two-stage method for improving the prediction accuracy of complex traits by incorparating genotype × environment interactions in *Brassica napus*. *Discrete Dynamics in Nature and Society*, 2020(1), 7959508.

Xu, Y., Wang, X., Ding, X., Zheng, X., Yong, Z., Xu, C. and Hu, Z. (2018). Genomic selection of agronomic traits in hybrid rice using an NCII population. *Rice*, 11, 1–10.

Zhang, Y., Massel, K., Godwin, I.D. and Gao, C. (2018). Applications and potential of genome editing in crop improvement. *Genome Biology*, *19*(1), 210.

Zhong, S., Dekkers, J.C., Fernondo, R.L. and Jannink, J.L. (2009). Factors affecting accuracy from genomic selection in populations derived from multiple inbred lines: a barley case study. *Genetics*, *182*(1), 355–364.

CHAPTER 5

# Genetic Innovations in Wheat for Sustainable Agriculture and Global Food Security

Shoeb Ahmed[1], Jitendra Kumar[2*] Sangeeta Dayal[3], Pradeep Singh[2], Rahul Kumar[1], Pradeep Kumar Sharma[1] and Joy Roy[2*]

## 1. Introduction

Cereals are important to the human population for food security and are the sources of essential nutrients and calories. Their contribution is approximately 56% of global energy needs and 50% of protein requirements (Borrill, 2020; Muitire et al., 2021). Cereals, as staple crops, are important in many underdeveloped and developing nations (in Africa and Asia), where they are the dominant source of dietary energy and essential nutrients (Shahane and Shivay, 2024; Ibrahim et al., 2024). Wheat, in particular, is a leading player in global food security due to its sustainability and high demand in these regions (Dixit et al., 2023). The cereals are nutrient-dense, which contain mineral elements, protein, and fat, and a few anti-nutritional factors underscore their importance in addressing food security challenges (Shahane and Shivay, 2024).

Cereals are not only a major source of calories and protein but also provide genetic resources for desirable traits, contribute to agricultural and dietary diversification, and are essential to achieving global security of food and nutrition (Ibrahim et al., 2024). In the face of challenges such as micronutrient deficiencies, vitamin deficiencies, and protein-energy malnutrition, the development of nutrient-dense cereal varieties has become increasingly important. The significant prevalence of iron (Fe) and zinc (Zn) deficiencies, particularly in Asia and Africa, highlights the critical role of cereals in combating hidden hunger (Shahane and Shivay, 2024). Goal 2 for Sustainable Development under the agenda of the United Nations 2030 underscores the significance of enhancing the nutritional status of cereals. The objective is to eradicate hunger, ensure food security, enhance nutrition, and support sustainable agricultural practices (Raheem et al., 2021).

As the world population is rapidly expanding and the climate continues to change, cereal agriculture confronts enormous challenges in meeting the increasing requirements of cereals as a staple food source (Boruah et al., 2023). Nevertheless, due to climate change, cereal grain production and quality are at risk, which affects critical organic components (e.g., carbohydrates, protein, and

1 Ch. Charan Singh University, Meerut, U.P., India.
2 BRIC-National Agri-Food & Biomanufacturing Institute (NABI), Sector 81, Mohali, Punjab, India.
3 KVSCOS, Subharti University, Meerut, U.P., India.
* Corresponding author: joykroy@nabi.res.in; jitendra_biotech09@rediffmail.com

DOI: 10.1201/9781003584100-5

lipids) in grains (Divte et al., 2022). The Intergovernmental Panel on Climate Change (IPCC) has reported a substantial increase in global cereal production over the years, leading to heightened environmental pressures and emissions of greenhouse gases from the agriculture and food sectors (Vinci et al., 2022). The uncertainty in climate sensitivity of agriculture, coupled with regional variations in weather patterns, presents challenges in adapting management practices for optimal harvests (Bhat et al., 2024).

Genetic approaches offer promising strategies to overcome the challenges in cereal agriculture. These approaches aim to improve nitrogen-use efficiency, increase crop diversity, and develop nutrient-dense varieties that can withstand environmental stresses (Dixit et al., 2023; Kumar et al., 2024; Kumar et al., 2024). By leveraging genetic insights from plant systems biology, researchers aim to improve cereal crops' resilience, productivity, and nutritional value while reducing reliance on fertilizers and diminishing environmental impacts.

To ensure food security with the increasing population growth and climate change, researchers are developing resilient cereal crops using advanced technologies of genome editing such as CRISPR/Cas (Basu et al., 2023). Additionally, the combination of marker-assisted selection (MAS) and genome selection (GS) has transformed plant breeding, allowing the creation of improved traits in desirable crop varieties (Dreisigacker et al., 2016; Pandurangan et al., 2021; Verma et al., 2024).

Omics technologies play a significant role in advancing the improvement of crops by identifying and mapping genes associated with important traits in major cereal crops (Faryad et al., 2021; Kaur et al., 2021; Yang et al., 2021; Naqvi et al., 2024). These methodologies have significantly accelerated breeding efforts, offering both time and cost efficiencies. However, challenges persist in effectively integrating complex omics datasets into practical crop enhancement strategies (Yang et al., 2021; Faryad et al., 2021).

Biofortifying wheat with increased Fe and Zn content presents a promising genetic strategy to address widespread micronutrient deficiencies (Roy et al., 2022; Borrill et al., 2014). The development of nutritionally enhanced crop varieties relies on genetic variability, which can be sourced from wild relatives or generated through mutagenesis. Both forward and reverse genetics play key roles in crop improvement—forward genetics identifies genes linked to inheritable traits, while reverse genetics modifies gene activity to validate predicted gene functions (Kumar et al., 2024). Combining genetic approaches with advanced breeding techniques and effective management strategies can significantly enhance the sustainability of agriculture and the security of food in the face of global challenges. Fig. 1 illustrates various genetic methods used for wheat improvement. This chapter examines various methodologies, including GS, MAS, and omics technologies, utilized for crop improvement for global food demand.

## 2. Marker-Assisted Breeding

MAS has become an essential technique in cereal breeding, greatly improving the identification and integration of favorable alleles for key agronomic traits. It is widely employed to enhance various characteristics in cereals, such as stress tolerance, disease resistance, and yield-related traits (Liu et al., 2007; Kamboj et al., 2020). The development of molecular markers, such as Random Amplified Polymorphic DNAs (RAPDs), Restriction Fragment Length Polymorphisms (RFLPs), Amplified Fragment Length Polymorphisms (AFLPs), and Simple Sequence Repeats (SSRs), has significantly advanced the construction of genetic maps and the identification of quantitative trait loci (QTLs) of desirable traits (Idrees and Irshad, 2014).

By expediting breeding cycles and improving variety development, marker-assisted breeding has revolutionized cereal crop enhancement. DNA markers enable precise trait selection, reducing the influence of environmental variables on phenotypic selection (Vagndorf et al., 2018). This technology has been widely adopted in wheat breeding initiatives across the globe, including in Canada and India, to enhance disease resistance, cereal quality, and adaptation to specific market requirements (Randhawa et al., 2013; Rai et al., 2019). MAS has now become fundamental to plant

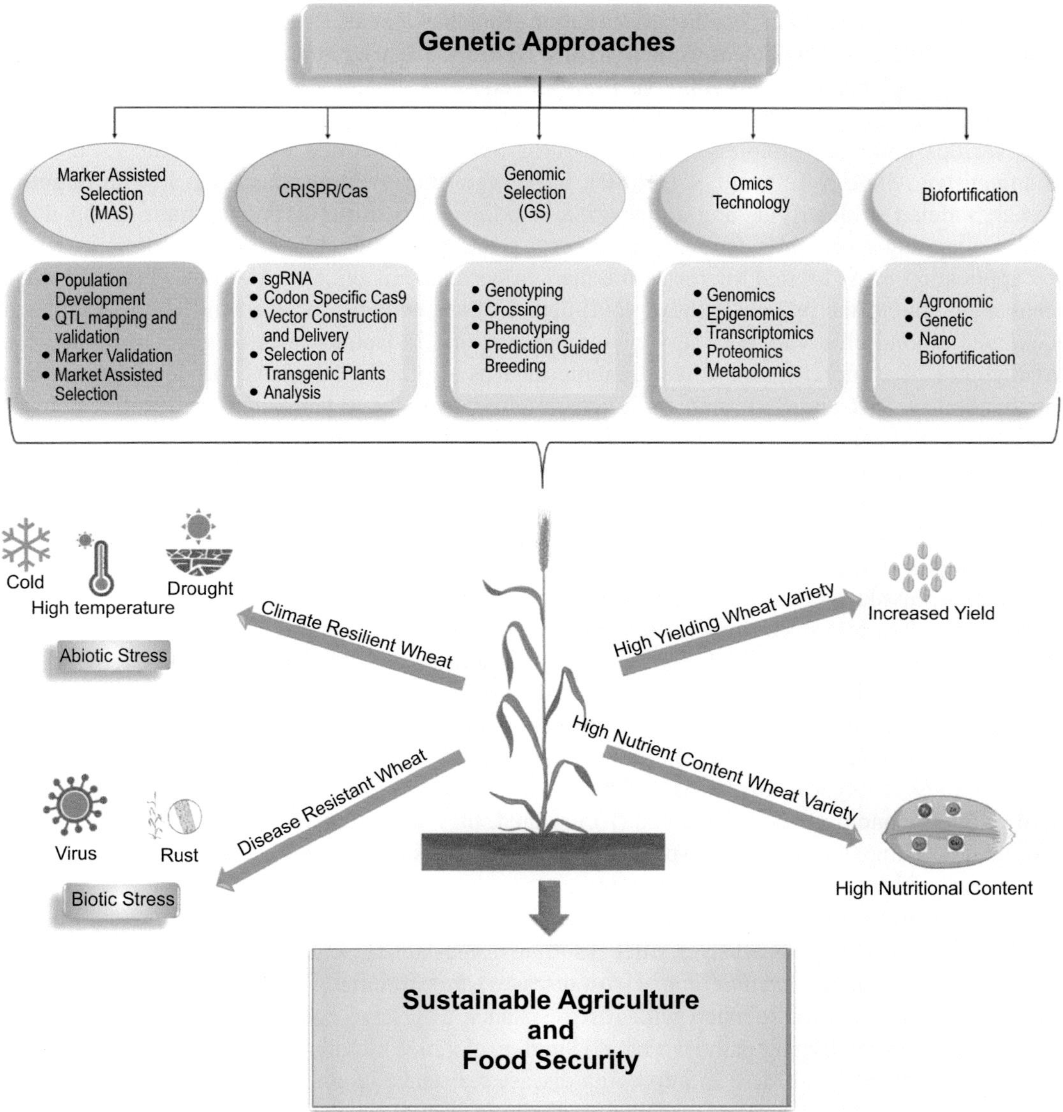


**Fig. 1** Different genetic approaches used to improve wheat for food security.

breeding, significantly reducing the required time for the development and release of new crop varieties suited to challenging environments (Baloch et al., 2023).

Furthermore, MAS has supported the introgression of genes for improvement in quality, boosted yield, and enhanced resistance to different stresses (Kamboj et al., 2020). The integration of MAS with phenotypic selection is important in developing high-yielding varieties of wheat with improved drought tolerance (Gautam et al., 2020). By characterizing numerous genes and QTLs for trait improvement, MAS supports the development of crops with enhanced different traits (Lateef, 2015; Kamboj et al., 2020; Gautam et al., 2021).

MAS is a highly valuable tool in wheat breeding, providing substantial benefits in terms of time and cost efficiency. It has been extensively utilized for the incorporation of genes that are resistant to disease and other traits. For example, MAS supports the introduction of rust-resistance genes such as *Lr34* and *Yr36*, the *Pch1* gene for eyespot resistance, and QTLs associated with Fusarium head blight (FHB) resistance (Miedaner and Korzun, 2012; Song et al., 2023). In Canada, wheat breeding programs routinely utilize MAS for traits such as rust resistance, orange wheat blossom midge resistance, and grain protein concentration (Randhawa et al., 2013). Similarly, International Maize and Wheat Improvement Center (CIMMYT for Centro Internacional de Mejoramiento de Maíz y Trigo) global wheat program has incorporated MAS strategies with recent advancements in genome-wide prediction methods for complex traits (Dreisigacker et al., 2016).

Various molecular markers, including SSR, Single Nucleotide Polymorphism (SNP) and Kompetitive Allele Specific PCR (KASP), have been employed in wheat MAS, with ongoing research aimed at identifying and utilizing markers linked to critical agronomic traits (Song et al., 2023). Ongoing research is focused on inspecting the genetic foundation of key traits in wheat by application of AFLP and microsatellite marker systems. Historically, MAS in wheat has faced challenges due to the polyploid (hexaploid) nature of the genome and the limited polymorphism observed among elite varieties. However, advancements in technologies of genotyping and the availability of comprehensive wheat genome data have greatly enhanced the development and application of SSR and SNP markers. These improvements have facilitated the identification of agriculturally important candidate genes, contributing to the refinement of wheat breeding strategies (Srivastava, 2019).

In India, wheat breeding programs have primarily aimed to improve disease resistance and yield to fulfill the requirements of a growing population. A significant milestone was achieved with the development of PBW723 (first Indian wheat variety produced via MAS). This variety possesses genes that provide resistance to both stripe and leaf rust (Sharma et al., 2021). Additionally, five wheat lines with improved drought tolerance were developed by introducing three linked QTLs. These lines, evaluated under both rainfed and arid conditions, exhibited enhanced productivity (Rai et al., 2018; Mujtaba et al., 2018). MAS has also been effectively utilized to improve wheat varieties for higher yield and better drought tolerance. For example, a yield QTL (Qyld.csdh.7AL) was introduced by Gautam et al. (2020) into four wheat cultivars, which gave rise to drought-tolerant, high-yielding genotypes. After these advanced lines (55 lines) were identified and evaluated under two different conditions (i) rainfed and (ii) irrigated, they were subjected to phenotypic selection. The advanced lines exhibited a low-pressure sensitivity index and a higher yield, which suggests that they are capable of withstanding water stress.

Wheat crops are increasingly vulnerable to abiotic and biotic factors, with climate change further intensifying these challenges, often resulting in substantial yield losses. Among biotic factors, diseases caused by the *Puccinia* fungus (rust diseases) are particularly problematic. The integration and stacking of genes associated with rust resistance also have been vital in ensuring durable resistance against emerging pathogen races (Singh et al., 2015). Different studies demonstrated the successful application of MAS in improving disease resistance in wheat. For instance, Gupta et al. (2005) identified RAPD markers associated with the *Lr9* gene, which imparts resistance to leaf rust. This led to the development of a species-specific sequence-characterized amplified region (SCAR) marker for *Lr9*, which was later validated across diverse wheat genetic backgrounds. Similarly, Singh et al. (2009) designed a polymerase chain reaction (PCR)-based assay to facilitate the selection of the *Yr10* gene, which provides resistance to stripe rust. Another significant achievement was the development of a genotype resistant to Pre-Harvest Sprouting (PHS) and leaf rust. This was accomplished by introducing the PHS-associated QTL *QPhs.ccsu3A.1* alongside the *Lr24* and *Lr28* genes into the elite PHS-susceptible wheat variety HD2329 (Kumar et al., 2010).

MAS has also been instrumental in improving the grain protein content (GPC) of wheat. For instance, Kumar et al. (2011) successfully incorporated the *Gpc-B1* gene, known for its role in enhancing GPC, into ten elite wheat genotypes. The MAS-derived progenies exhibited significantly higher GPC levels (ranging from 14.83% to 17.85%) without compromising yield, indicating the potential of MAS in developing high-GPC wheat varieties suitable for future hybridization programs. Similarly, Vishwakarma et al. (2014) utilized marker-assisted backcrossing (MABC) to introduce the *Gpc-B1* gene into the wheat cultivar HUW468, resulting in improved lines with increased GPC. Additionally, MAS has been employed in the development of near-isogenic lines

(NILs) to enhance grain weight (GW) in wheat. Kumari et al. (2019) transferred three QTLs for grain weight into NILs, resulting in lines with significant differences in thousand-grain weight (TGW), validating the QTL effects. The successful application of MAS in these and other studies underscores its critical role in improving wheat varieties, addressing both current and future challenges in wheat breeding. Some of the studies related to MAS are given in Table 1.

**Table 1** Marker-assisted breeding studies in wheat.

| *Genotyping markers* | *Traits/Characters* | *References* |
|---|---|---|
| SSR | TGW, SW, SHI, GWPS, NGPS, and HI | Golabadi et al., 2011 |
| DArT and SSR | Root angle and root number (seminal) | Christopher et al., 2013 |
| RAPD, SSR, SRAP, ISSR, and STS | CL, SH, root (FW, DW, number, length) seedling fresh and dry weight, stem and leaves fresh & dry weight | Zhang et al., 2013 |
| TRAP, SRAP, and SSR | ABA content | Barakat et al., 2015 |
| SSR | WC, CMS, and net photosynthesis | Malik and Malik, 2015 |
| SNP | Yield and biomass | Merchuk-Ovnat et al., 2016 |
| DArTs, SSRs, and ESTSSRs | SNPP, KNPS, and TKW | Xu et al., 2017 |
| SNP (GBS) | KNPS and TKW | Ballesta et al., 2019 |
| GWAS | for Normalized Difference Vegetation Index, LCC, LR, and DB | Condorelli et al., 2018 |
| SSR | Spike length and GPS | Fatima et al., 2018 |
| DArTs eq SNPs | GY and its components | Sukumaran et al., 2018 |
| 90K SNP chip | Reproductive growth | Dolferus et al., 2019 |
| 90K SNP chip | CMS & flag leaf (area, length, & width) | Khanna-Chopra et al., 2020 |
| SNPs and SSRs | Yield-related traits | Li et al., 2019 |
| SNP, SSR, GBS markers, DArT, and ISBP | Days to heading, GFD, GY, and TGW | Tura et al., 2020 |
| AS-PCR, AFLP | HMWG | de Bustos et al., 2001 |
| SSR, AFLP | GPC, Kernel Harvest Index | Sun et al., 2010; Bordes et al., 2011 |
| DArT, SSR | GPC | Li et al., 2012 |
| SNP | GPC | Liu et al., 2018 |
| SNP | GPC | Fatiukha et al., 2020 |
| SNP | GPC, flour protein, baking quality traits | Zhang-Biehn et al., 2021 |
| SNP | GPC | Jiang et al., 2021 |
| SNP | GPC, Zeleny Sedimentation Test | Suliman et al., 2021 |
| SNP | Gluten aggregation properties | Wu et al., 2022 |
| SNP | GPC | White et al., 2022 |

Although MAS has achieved notable success, its effectiveness in enhancing complex traits, for example, stress adaptation and yield potential, has been limited. This is largely due to the intricate genetic architecture of these traits and their strong dependence on environmental interactions (Araus et al., 2008). The polygenic nature of characteristics such as drought tolerance, coupled with the variability of environmental conditions, poses significant challenges in breeding efforts. Nonetheless, MAS supports identifying genomic regions and uncovering the mechanisms linked to cereal stress tolerance for developing crop varieties resilient to extreme environmental factors (Wani et al., 2020). To overcome the limitations of MAS in addressing complex traits, a multidisciplinary approach that integrates physiological understanding, advanced phenotyping techniques, and molecular tools is essential (Araus et al., 2008). In light of the problems presented by climate change and the pressing necessity to enhance agricultural output to satisfy the needs of an increasing global population, a comprehensive strategy is essential for augmenting breeding efficiency (Kamboj et al., 2020; Araus et al., 2008). Furthermore, advancements in high-throughput techniques are expected to increase the precision of molecular breeding in cereals. These innovations accelerate the process of breeding by implementing rapid screening of large populations for beneficial genes, thereby developing improved grain varieties (Randhawa et al., 2013). The combination of these advanced knowledges with conventional breeding approaches offers a suitable pathway for improving cereal crops.

## 3. Genome Editing Using CRISPR/Cas9 Techniques

CRISPR/Cas9 technology has brought a transformative shift in genome editing for cereal crops, offering unmatched precision, simplicity, and cost-effectiveness. Originating from a bacterial adaptive immune system, this RNA-guided method enables precise and efficient gene modifications by inducing targeted double-stranded DNA breaks. The mechanism involves the application of Cas9 nuclease to a genomic location that is specific using a guide RNA (gRNA), where the Cas9 enzyme creates a break in the DNA. The cell's intrinsic repair system then restores the break, leading to specific genetic alterations. In comparison to earlier genome-editing techniques such as Zinc Finger Nucleases (ZFNs) and Transcription Activator-Like Effector Nucleases (TALENs), CRISPR/Cas9 is more user-friendly and versatile, as it utilizes RNA–DNA base pairing, making both its design and implementation (Cermak et al., 2011; Basu et al., 2023).

The utilization of CRISPR/Cas9 in cereal crops has showcased its potential to enhance vital agronomic traits, such as quality, yield, and stress resilience (Erdoğan et al., 2023). CRISPR/Cas9 has been applied to modify genes in rice associated with water stress responses, thereby improving drought tolerance and enhancing adaptability to arid conditions (Hillary and Ceasar, 2019). Similarly, in wheat, this genome-editing approach alters susceptibility genes, resulting in disease-resistant varieties with improved defense against pathogens (Ansari et al., 2020). Furthermore, CRISPR/Cas9 enhances the nutritional composition (increased essential amino acids (EAA) and vitamin content in maize) of cereals, contributing to efforts aimed at reducing malnutrition in developing regions (Ansari et al., 2020).

The adaptability of CRISPR/Cas9 across different cereal crops highlights its potential to tackle food security challenges while supporting sustainable agriculture. Through precise genetic modifications, this technology facilitates the development of improved crop varieties that can endure the negative impacts of climate change and help meet the increasing global food demand (Basu et al., 2023). The complex hexaploid genome of wheat has traditionally posed significant challenges for conventional breeding methods, but CRISPR/Cas9 offers a powerful solution, allowing targeted and precise genetic modifications that were previously difficult to achieve. Wheat is improved using CRISPR/Cas9 in the enhancement of grain size, a critical factor in determining yield. Targeted mutagenesis of genes such as TaSBEIIa increases amylose content in wheat varieties (higher resistant starch content) which is beneficial for human health due to its positive effects on digestion and glycemic control. The mutant lines also display dosage-dependent effects on the properties, composition, and structure of starch, indicating a profound impact on grain quality (Li et al., 2021).

Recently, CRISPR/Cas9 technology is used to alter genes concerned with metabolic pathways, thereby enhancing the nutritional profile of wheat grains. Enhancing the amylose content in wheat offers significant health benefits and addresses specific dietary needs, particularly in regions where wheat is a staple crop (Li et al., 2021). CRISPR/Cas9 has also played a crucial role in improving tolerance for stress in wheat, encompassing both abiotic and biotic factors. By modifying stress-responsive genes, researchers have developed high-environmental-resilience wheat varieties for salinity and drought. For instance, targeted editing of genes such as *TaERF3* and *TaDREB2* has resulted in wheat lines with improved tolerance to these stresses, contributing to stable yields under extreme conditions (Nazir et al., 2022; Jogam et al., 2021). This advancement is particularly relevant to climate change, which is expected to exacerbate environmental stresses on crops. CRISPR/Cas9 is also instrumental in developing wheat cultivars with enhanced resistance to diseases, further strengthening crop resilience. By knocking out susceptibility genes, researchers have produced wheat lines with improved resistance to pathogens such as powdery mildew. For example, the TaMLO gene, associated with mildew susceptibility, was edited to develop mildew-tolerant wheat varieties, achieving a mutation frequency of up to 28.50% in the mildew-resistance locus. This modification substantially diminishes the dependence on chemical pesticides and encourages the adoption of sustainable agricultural practices (Tripathi et al., 2020; Shan et al., 2014). Similarly, the *TaEDR1* gene has been altered using CRISPR/Cas9 to improve wheat's tolerance to powdery mildew (Char and Yang, 2020). Furthermore, CRISPR/Cas9 has been utilized to induce targeted mutations in other important genes, such as *Tapx1*, which exhibited a 9% mutation rate, and *TaLox2*, with a 45% mutation rate. These genetic modifications have significantly contributed to improving disease resistance in wheat (Shan et al., 2014).

CRISPR/Cas9 is also utilized to enhance yield-related traits by modifying genes responsible for grain characteristics. Targeted editing of genes such as *TaDEP1*, *TaNAC2*, *TaGW2*, and *TaGASR7* has resulted in improvements in grain weight, area, length, and width, leading to wheat varieties with significantly higher yield potential compared to their wild counterparts (Wang et al., 2018). Applications of the CRISPR/Cas9 technique in wheat are given in Table 2. These advancements are critical for food security globally. CRISPR/Cas9 has significantly enhanced the toolkit for wheat improvement by enabling precise genetic modifications that improve grain size, nutritional content, stress tolerance, and disease resistance. These advancements are essential for breeding wheat varieties that can tackle food security concerns and support sustainable agriculture, with climate resilience. Consequently, CRISPR/Cas9 is a pivotal technology in contemporary wheat improvement programs. The comprehensive application of CRISPR/Cas9 in targeting multiple genes and traits underscores its potential to transform wheat breeding practices and participate in worldwide endeavors to ensure food security.

**Table 2** Role of CRISPR/Cas9 system in wheat.

| *Genetic approach* | *Targeted gene* | *References* |
|---|---|---|
| CRISPR/Cas9 | *Tainox* and *Tapds* | Upadhyay et al. (2013) |
| | *TaGW2*, *TaGLW7*, *TaCKX2-1*, and *TaGW8* | Zhang et al. (2019a) |
| | *TaGASR7 (A1*, *B1*, and *D1)* | Hamada et al. (2018) |
| | *TdGASR7* | Liang et al. (2018) |
| | *TaGW2 (B1*, *D1*, and *A1)* | Liang et al. (2017) |
| | *TaABCC6*, *TansLTP*, and *TaNFXL1* | Cui et al. (2019) |
| | *GW2-B*, *ASN2-A*, and *PinB-D* | Brandt et al. (2020) |
| | *TaGASR7 (A1*, *B1*, and *D1)* | Zhang et al. (2016) |
| | *TaMLO (D1*, *A1*, and *B1)* | Wang et al. (2014) |
| | *LOX2* | Shan et al. (2014) |
| | *TaPDS* | Howells et al. (2018) |

The deployment of CRISPR/Cas9 technology in agriculture presents both significant opportunities and complex ethical and regulatory challenges that must be managed to ensure its responsible use and public acceptance. Off-target effects are a primary ethical concern, as they may result in unanticipated characteristics or susceptibilities by causing unintended genetic alterations outside the intended target site (Rodríguez, 2016). This raises questions about the precision and safety of genome editing, necessitating rigorous measures to minimize such risks. Additionally, the ecological impact of introducing crops that have undergone genetic editing in the environment is a critical consideration.

Moreover, the socio-economic implications of CRISPR/Cas9 in agriculture cannot be overlooked. The proprietary nature and cost of genome-editing tools may create barriers for small-scale farmers, exacerbating disparities between large agribusinesses and smaller operations (Bartkowski et al., 2018). Ensuring equitable access to this technology is essential to avoid widening the gap in agricultural development. The public's perception and acceptance of CRISPR-edited crops are key factors influencing their adoption, with ethical concerns and perceived risks shaping public trust. Clear and transparent communication regarding both the advantages and potential risks of genome editing is essential to fostering and sustaining public confidence (Shinwari et al., 2018).

On the regulatory front, the lack of a unified global framework poses challenges for the adoption of CRISPR-edited crops. Regulatory approaches vary significantly across regions, with some countries imposing stringent regulations, while others adopt more lenient policies. Harmonizing international guidelines could facilitate broader acceptance and commercialization of CRISPR-edited crops (Eş et al., 2019). Regulatory processes must also incorporate ethical and moral considerations, particularly when addressing the implications of genetic modifications that could affect future generations. A diverse range of stakeholders (ethicists, policymakers, scientists, and the public) are engaging in these discussions, which is important to developing balanced and informed regulations (Brokowski and Adli, 2019).

Intellectual property rights associated with CRISPR/Cas9 technology further complicate access, potentially restricting its use and benefits to a limited number of entities. Developing policies that promote fair access and inclusive agricultural development is critical to ensure that the advantages of CRISPR/Cas9 technology are widely shared (Rodríguez, 2016). While CRISPR/Cas9 offers transformative potential for agricultural advancement and food security, it is imperative to navigate the associated ethical and regulatory challenges with care. Harmonized international guidelines, transparent public engagement, and policies that address socio-economic disparities are essential to realizing the full benefits of this technology while mitigating its potential risks.

## 4. Genomic Selection

When combined with speed breeding, GS expedites wheat breeding by facilitating the prediction of breeding values through genome-wide markers. This, in turn, reduces the necessity for phenotypic data and shortens breeding cycles (Watson et al., 2019; Larkin et al., 2019). Effective GS implementation requires optimizing prediction models, marker density, and training populations. A dual approach that emphasizes both population enhancement and product development can facilitate the seamless integration of GS into breeding programs (Merrick et al., 2022). Recurrent selection approaches have also been proposed to maintain genetic variation and enhance cross-prediction, making GS a promising tool for increasing annual genetic gain (Bassi et al., 2016).

The genomic prediction precision in breeding programs is influenced by several key factors. These factors are heritability, marker density, and trait architecture population size. Various GS models, such as Best Linear Unbiased Prediction (BLUP) and Least Absolute Shrinkage and Selection Operator (LASSO), generally achieve higher accuracy in prediction with an increase in marker density. Conversely, methods that depend on Bayesian methodologies may encounter a sluggish convergence rate as a result of the Markov Chain Monte Carlo (MCMC) iterations necessary for precise estimation (Arruda et al., 2015; Norman et al., 2018). Despite the benefits of high marker density, there are economic constraints associated with incorporating extremely dense marker data, which necessitates a balance between cost and prediction accuracy. Low-density markers can still achieve high prediction accuracy if there is strong linkage disequilibrium (LD) among them. However, this is dependent on the specific trait and may vary based on factors such as heritability and trait architecture (Lorenz et al., 2011; Werner et al., 2018). It is typically recommended to employ a modest density of markers, such as a minimum of 2000 SNPs, to avoid significant reductions in prediction accuracy (Abed et al., 2018). Additionally, increasing multiplexing levels in genotyping can significantly reduce costs without compromising accuracy, such as using Genotyping-by-Sequencing (GBS) with high multiplexing compared to lower-density arrays (Abed et al., 2018).

The accuracy in prediction is also influenced by the positioning of SNPs, as SNPs located in intergenic regions often better capture haplotype diversity due to the presence of essential regulatory sequences (enhancers and promoters) (Barrett et al., 2012; Abed et al., 2018). Additionally, the accuracy of predictions is contingent upon the quality of SNP data, particularly those with minor allele frequency (MAF) values exceeding 0.1. The size of the population, especially the training population, influences the accuracy of genomic prediction. Smaller training populations often lead to decreased accuracy due to inadequate estimation of marker effects. It has been suggested that a training population size of approximately 2*Ne*L (where *Ne* = effective population size and *L* = genome size in Morgans) and a marker number of 10*Ne*L can achieve high prediction accuracy (0.9). Reducing the training population to 1*Ne*L may result in lower prediction accuracy (0.7), assuming unrelated or distantly related training and breeding populations (Meuwissen, 2009). However, when training and breeding populations are related, a smaller training population can still achieve high accuracy (Meuwissen, 2009).

The heritability of a trait has a substantial impact on accuracy of prediction, particularly for low heritability ($h^2 < 0.4$) traits. High heritability typically results in better prediction accuracies, while low heritability poses challenges in genomic selection, particularly in plants where most traits exhibit low to moderate heritability (Hayes et al., 2009). For traits with low heritability, a larger training population is often required to achieve similar prediction accuracy as traits with higher heritability, though cost constraints may limit this in some regions. Despite these challenges, models such as BLUP and its variations (GBLUP, RR-BLUP), along with Bayesian approaches and Reproducing Kernel Hilbert Space (RKHS) methods, have demonstrated robustness across different levels of heritability (Crossa et al., 2010; Michel et al., 2019). In scenarios with missing observations, combining data from multiple traits with varying heritability can improve prediction accuracy. For instance, Multi-Trait Genomic Selection (MTGS) models can leverage correlations between traits to enhance predictions for low heritability traits (Spindel et al., 2015).

Traditional selection methods have contributed to develop wheat varieties; however, the rate of genetic gain over time remains limited. GS offers a promising alternative by accelerating the breeding cycle and enabling the early identification of individuals with high breeding potential without requiring extensive phenotypic evaluations (Bernardo and Yu, 2007; Lorenz et al., 2011). With changing weather patterns and the appearance of novel insect biotypes and pathogens, developing biotic stress-tolerant wheat varieties are crucial (Juarez et al., 2013; Fones et al., 2020).

While MAS is effective for breeding qualitative resistance, it has shown limitations in addressing quantitative resistance, which involves multiple genes with smaller effects. GS has demonstrated its potential in strengthening cereal crops' resistance to biotic stresses, particularly against diseases such as powdery mildew, *Septoria tritici* blotch, FHB, and rust. Its effectiveness in enhancing disease resistance in wheat is reflected in the genetic prediction accuracy, which ranges from 0.14 to 0.85 for these diseases (Rutkoski et al., 2012; Sarinelli et al., 2019). *Fusarium graminearum*-caused FHB is a threat to wheat production, which lowers both the yield and grain quality while also leading to mycotoxins accumulation such as deoxynivalenol (DON) (Milus and Parsons, 1994). Symptoms include premature bleaching of spikelets and pinkish, shriveled grains, leading to economic losses (Goswami and Kistler, 2004; Wegulo et al., 2015). Resistance to FHB involves major QTLs (Fhb1, Fhb2, Fhb4, Fhb5, and Qfhs.nau-2DL) (Guo et al., 2015). Type I resistance in wheat prevents the initial infection of FHB, while Type II resistance limits the spread of the fungus within the plant (Steiner et al., 2017). GS has shown significant potential in enhancing FHB resistance. A study involving 322 wheat genotypes evaluated various predictive models. The results indicated that Random Forest and RKHS models were the most accurate, with the highest accuracy in prediction for DON when both targeted and whole-genome markers were used (Rutkoski et al., 2012). Hoffstetter et al. (2016) observed improved prediction accuracy when selecting appropriate marker and training subsets. In another study, RR-BLUP, elastic net, and LASSO models were assessed using marker densities ranging from 500 to 4,500 SNPs. It identified RR-BLUP as the most effective model, with optimal marker sets ranging between 1,500 and 3,000 SNPs and training populations of fewer than 192 genotypes (Arruda et al., 2015). In the Pacific Northwest, research on spring wheat demonstrated high prediction accuracy for FHB incidence, with genotypes closely related to the training population achieving an accuracy of 0.60, further confirming the effectiveness of GS in improving resistance to FHB (Dong et al., 2018).

GS studies have primarily focused on grain yield, influenced by factors such as tiller count, the number of grains per panicle (NGPP), and TGW. The precision of genomic prediction for these traits is contingent upon the heritability, training population, and models employed. For example, genomic predictions of maize cereal yield have demonstrated accuracies ranging from 0.28 to 0.78 (Yabe et al., 2018). The genetic association between the training and breeding population is influenced by the efficacy of GS. Additionally, research has demonstrated that GS, when combined with phenotypic selection, can significantly increase genetic gain for wheat grain yield (Longin et al., 2014; Lozada and Carter, 2020). Newer schemes such as GS Rapid have been used to improve selection gains, particularly in hybrid breeding programs (Marulanda et al., 2016). GS has been used to boost genetic gains, with studies showing varied outcomes. In a comparison of 2,325 European winter wheat genotypes, RKHS, an epistatic model, outperformed additive models such as RR-BLUP, especially in larger populations (He et al., 2016). For 659 inbred lines, RR-BLUP showed average prediction accuracy of $r = 0.38$ over five breeding cycles, improving to $r = 0.41$ after removing outliers (Michel et al., 2016). GS combined with heritability indices achieved a prediction accuracy of 0.48 for grain yield, surpassing traditional phenotypic selection and GAS (Michel et al., 2017). Training populations of 50 genotypes yielded high prediction accuracy, with a training size of 350 achieving an average accuracy of 0.64, but with no significant gains beyond this size (Sarinelli et al., 2019).

Abiotic stresses, for example, extreme temperatures, drought, and flooding, present major challenges to wheat production, with climate change further intensifying these threats (Qin et al., 2011). Complex inheritance patterns, low heritability, and substantial environmental influences frequently restrict conventional breeding methods for abiotic stress tolerance (Bernardo, 2008). While yield QTLs have been identified and incorporated into wheat breeding programs using molecular markers under stressful conditions, their effectiveness is often limited by interactions with the genetic background and environmental influences (Ribaut and Ragot, 2007; Almeida et al., 2013). Research on wheat, maize, and rice has highlighted the advantages of GS over MAS, particularly in multi-environment models, where GS has demonstrated greater prediction efficiency for abiotic stress tolerance (Cerrudo et al., 2018; Sukumaran et al., 2018; Bhandari et al., 2019). Additionally, Sun et al. (2017) reported that the incorporation of secondary high-throughput traits has improved genomic prediction models accuracy under stress conditions by approximately 70%.

Quality wheat traits, for example, grain color, size, and protein content, have diverse genetic architectures, ranging from oligogenic to polygenic control. GS has been extensively used to improve quality traits of wheat, including milling and quality of flour, with studies indicating better prediction accuracies in multi-family populations compared to biparental ones (Heffner et al., 2011). Due to the inverse correlation between yield and protein content, multi-trait genomic selection has been employed to achieve an optimal balance between these traits, yielding promising results (Michel et al., 2019). The accuracy of quality trait predictions is influenced by the genetic diversity of the germplasm and the relationship between the training and prediction populations (Zhao et al., 2015). New research has demonstrated the potential for rapid enhancements in mineral content and grain yield through the use of Bayesian regression and Bayesian Ridge Regression (BRR), with mineral content prediction accuracies as high as 0.55 (Joukhadar et al., 2021). Achieving optimal genomic prediction accuracy necessitates striking a balance between trait architecture, heritability, population size, and marker density. By considering these factors and using appropriate models and techniques, researchers can improve prediction outcomes and enhance the efficiency of genomic selection programs.

GS has greatly advanced genetic gains in breeding, aided by affordable Next-Generation Sequencing (NGS) technologies and high-density SNP genotyping. However, improvements are still needed, including better imputation methods, incorporation of G x E interactions, and multi-trait data. Regular updates to training sets and precise trait phenotyping are crucial. A focused effort on methodology, data management, and skill development is essential for optimizing GS.

## 5. Omics Technologies for Wheat Improvement

Omics technology is a contemporary molecular instrument that is indispensable for comprehending the functional genomic systems of organisms. It includes the profiling of expressed transcripts and translated proteins, in addition to DNA sequencing (Zenda et al., 2021). The term "omics" is derived from the Greek affix "-ome", which means "whole", and indicates scientific disciplines that investigate a variety of biological molecules within comprehensive biological systems (Zenda et al., 2021). The advancement of bioinformatics and NGS has significantly accelerated the application of omics technologies (Faryad et al., 2021; Kaur et al., 2021). Omics technologies have revolutionized plant breeding, particularly in cereal crops such as wheat, by offering comprehensive insights into their genetic, epigenetic, proteomic, and metabolomic landscapes. These advancements enable the gene identification and mapping that are associated with key agronomic traits, stress resilience, and yield in major cereal crops, as well as other food and fiber plants (Naqvi et al., 2024; Yang et al., 2021). Systems biology integration with multi-omics data enhances understanding of molecular regulatory networks that support climate-resilient crop development (Yang et al., 2021).

The sequencing of wheat's complex genome has played a crucial role in its improvement, enabling Genome-Wide Association Studies (GWAS) and the identification of QTLs linked to key traits (Kaur et al., 2021). Epigenomics contributes to the discovery of epigenetic markers associated with traits such as drought tolerance and disease resistance by examining heritable changes in gene function influenced by environmental factors. This knowledge supports the development of more resilient wheat varieties (Yang et al., 2021). Proteomics, by analyzing proteins under various stress conditions, has discovered important proteins involved in metabolic processes, photosynthesis, and stress responses, all of which are critical for improving wheat's production and stress tolerance (Sehgal et al., 2023). Metabolomics, reflecting the biochemical activities within wheat, has been used to study metabolite profiles under environmental stresses and during different developmental stages, identifying crucial metabolic pathways for growth, stress tolerance, and grain quality, thus providing targets for genetic improvement and breeding programs (Khakimov et al., 2014).

Advancements in genome sequencing and omics technologies have significantly contributed to the generation of comprehensive genomic and transcriptomic datasets for various agricultural species. These developments have greatly improved discovery of gene, analysis of gene expression, MAS, and the incorporation of important traits into high-value crops (Cortés and López-Hernández, 2021). These advancements enable the routine identification of the molecular and genetic underpinnings of agriculturally significant traits (Scossa et al., 2021). When combined with advanced approaches, omics technologies increased the rate of development of elite, climate-resilient cultivars. These improved varieties exhibit enhanced nutritional quality, greater resistance to both abiotic and biotic stresses, and higher productivity (Gao, 2021; Singh et al., 2021). Precision breeding, driven by omics data, enables tailored strategies by comprehensively understanding gene, protein, and metabolite interactions (Pazhamala et al., 2021). The challenges persist in integrating complex omics datasets and effectively applying this knowledge to crop improvement (Faryad et al., 2021; Yang et al., 2021; Zhang et al., 2022). However, omics-driven strategies hold great potential for addressing global food security and malnutrition challenges (Faryad et al., 2021).

# 6. Biofortification for Improved Nutritional Quality

A sustainable approach called biofortification aims to compensate for micronutrient deficiencies by raising the levels of vitamins and minerals in staple foods that people in underprivileged communities often consume (Kumar et al., 2024). The primary components of these techniques are genetic biofortification strategies and agronomic methods. Numerous techniques are included in genetic biofortification, such as genome editing, transgenesis, conventional breeding, mutagenesis, molecular breeding, and metabolic engineering (Kumar et al., 2024). Biofortification of wheat with higher Fe and Zn levels presents a promising solution to widespread micronutrient deficiencies (Borrill et al., 2014; Roy et al., 2022). This sustainable strategy focuses on developing wheat varieties with enhanced Fe and Zn concentrations in the grain. Research by Velu et al. (2019) and Roy et al. (2022) is currently exploring genetic approaches, including genomic selection and MAS, to improve mineral content in wheat. Wild relatives and progenitors of wheat, such as *Triticum monococcum*, *Triticum dicoccoides*, and *Aegilops tauschii*, have been identified as valuable genetic resources due to their naturally high Fe and Zn concentrations (Cakmak et al., 2004; Velu et al., 2019). Velu et al. (2019) have demonstrated that it is feasible to cultivate wheat varieties that produce high yields while concurrently increasing their zinc and iron contents. Genetic modification has also proven effective, with transgenic methods such as overexpressing TaVIT2 and OsNAS2 genes doubling zinc and tripling iron bioavailability in wheat (Harrington et al., 2022). Agronomic biofortification, particularly the foliar application of micronutrient fertilizers, further enhances grain nutrient levels, boosting Zn concentration by up to 21% (Aziz et al., 2019). Together, these strategies aim to develop nutrient-rich wheat varieties to combat micronutrient deficiencies in populations reliant on wheat as a staple food. Narendra Modi, the Prime Minister of India, recently introduced 17 biofortified crop varieties to the nation during the recent commemoration of World Food Day. "HI 1633," "MACS 4058," and "DBW 303" were among the wheat varieties that were designated for this purpose. These varieties are enriched with high levels of iron, zinc, and protein. Additionally, "HD 3298" was highlighted for its high iron content, and "DDW 48" for its elevated protein levels (Kumar et al., 2024).

Challenges remain in enhancing mineral uptake, translocation, and bioavailability in wheat grains (Roy et al., 2022). Integrating genomic resources and knowledge from model crops is crucial for advancing wheat biofortification efforts (Borrill et al., 2014). Nanotechnology is a new agricultural instrument that has the potential to overcome the constraints of traditional agronomic biofortification and genetic biofortification. Nanofertilizers have the potential to enhance crop yield by improving nitrogen metabolism, seedling development, protein synthesis, photosynthesis, and the efficient translocation of nutrients from the roots to the foliage. Nanomaterials are particularly well-suited for this application due to their exceptional characteristics, including their high surface-to-volume ratio, delayed and controlled nutrient release, and high sorption capacity. Nanofertilizers guarantee efficient assimilation and prolonged nutrient delivery by encapsulating nutrients, resulting in superior crop growth in comparison to conventional fertilizers (Khan et al., 2021).

Biofortification is a sustainable and cost-effective solution to malnutrition by increasing the micronutrient content of staple crops such as wheat, addressing micronutrient deficiencies in regions with limited dietary diversity. By enhancing Fe and Zn levels, biofortified wheat can reduce iron deficiency anemia, improve immune function, and support cognitive development, particularly among susceptible groups (pregnant women and children) (Wakeel et al., 2018). This approach aligns with global health initiatives and provides significant economic benefits by reducing the need for costly supplementation programs and lowering the health burden associated with micronutrient deficiencies. Biofortification also effectively addresses hidden hunger, delivering essential nutrients naturally and reaching remote populations often inaccessible to conventional fortification efforts (Khush et al., 2012; Mushtaq and Nazir, 2021). Successful biofortification projects, such as those involving maize, rice, wheat, and other crops, have already shown measurable improvements in nutritional status in developing regions (Kiran et al., 2022). Ultimately, the biofortification of wheat is a crucial strategy for improving public health, ensuring nutritional security and supporting long-term economic and social development.

## Conclusion

The growing worldwide population and the rising demand for food highlight the urgent need to increase cereal production, particularly wheat, which plays a fundamental role in global diets and economies. As we face challenges such as climate change, disease pressure, and nutritional deficiencies, the role of innovative genetic approaches becomes paramount. This chapter has highlighted the transformative potential of molecular technologies, including molecular marker-assisted breeding, CRISPR/Cas9 genome editing, and genomic selection. These tools are not only accelerating improved wheat variety development but also fostering resilience against environmental stresses and enhancing nutritional profiles through biofortification. The incorporation of omics technologies expands the potential for crop improvement by providing a deeper understanding of the genetic and biochemical foundations of essential traits. By leveraging these advanced methodologies, researchers are better equipped to tackle the multifaceted challenges facing modern agriculture. As we move forward, the application of these genetic innovations will be crucial in ensuring that cereal production keeps pace with the growing demand, thereby contributing significantly to global food security and the agricultural systems sustainability. The future of cereal agriculture, particularly wheat, depends on our continued commitment to harnessing the full potential of these cutting-edge technologies, ultimately leading to a more resilient and nourished global population.

## References


Abed, A., Pérez-Rodríguez, P., Crossa, J. and Belzile, F. (2018). When less can be better: how can we make genomic selection more cost-effective and accurate in barley? *Theoretical and Applied Genetics*, *131*, 1873–1890. doi:10.1007/ S00122-018-3120-8

Almeida, G.D., Makumbi, D., Magorokosho, C., Nair, S., Borém, A., Ribaut, J.-M., et al. (2013). QTL mapping in three tropical maize populations reveals a set of constitutive and adaptive genomic regions for drought tolerance. *Theoretical and Applied Genetics 126*, 583–600.

Ansari, W.A., Chandanshive, S.U., Bhatt, V., Nadaf, A.B., Vats, S., Katara, J.L. and Deshmukh, R. (2020). Genome editing in cereals: approaches, applications and challenges. *International Journal of Molecular Sciences*, *21*(11), 4040.

Araus, J.L., Slafer, G.A., Royo, C. and Serret, M.D. (2008). Breeding for yield potential and stress adaptation in cereals. *Critical Reviews in Plant Science*, *27*(6), 377–412.

Arruda, M.P., Brown, P.J., Lipka, A.E., Krill, A.M., Thurber, C. and Kolb, F.L. (2015). Genomic selection for predicting Fusarium head blight resistance in a wheat breeding program. *The Plant Genome*, *8*(3).

Aziz, M.Z., Yaseen, M., Abbas, T., Naveed, M., Mustafa, A., Hamid, Y. and Saeed, Q. (2019). Foliar application of micronutrients enhances crop stand, yield and the biofortification essential for human health of different wheat cultivars. *Journal of Integrative Agriculture*, *18*(6), 1369–1378.

Ballesta, P., Mora, F. and Del Pozo, A. (2019). Association mapping of drought tolerance indices in wheat: QTL-rich regions on chromosome 4A. *Scientia Agricola*, *77*(2), e20180153.

Baloch, F.S., Altaf, M.T., Liaqat, W., Bedir, M., Nadeem, M.A., Cömertpay, G. and Sun, H.J. (2023). Recent advancements in the breeding of sorghum crop: current status and future strategies for marker-assisted breeding. *Frontiers in Genetics*, *14*, 1150616.

Barakat, M.N., Saleh, M.S., Al-Doss, A.A., Moustafa, K.A., Elshafei, A.A., Zakri, A.M. and Al-Qurainy, F.H. (2015). Mapping of QTLs associated with abscisic acid and water stress in wheat. *Biologia Plantarum*, *59*, 291–297.

Barrett, L.W., Fletcher, S., and Wilton, S.D. (2012). Regulation of eukaryotic gene expression by the untranslated gene regions and other non-coding elements. *Cellular and Molecular Life Sciences*, *69*, 3613–3634. doi:10.1007/S00018-012-0990-9

Bartkowski, B., Theesfeld, I., Pirscher, F. and Timaeus, J. (2018). Snipping around for food: economic, ethical and policy implications of CRISPR/Cas genome editing. Geoforum, 96, 172–180.

Bassi, F.M., Bentley, A.R., Charmet, G., Ortiz, R. and Crossa, J. (2016). Breeding schemes for the implementation of genomic selection in wheat (*Triticum* spp.). *Plant Science*, *242*, 23–36.

Basu, U., Riaz Ahmed, S., Bhat, B.A., Anwar, Z., Ali, A., Ijaz, A. and Mushtaq, M. (2023). A CRISPR way for accelerating cereal crop improvement: progress and challenges. *Frontiers in Genetics*, *13*, 866976.

Bernardo, R. and Yu, J. (2007). Prospects for genomewide selection for quantitative traits in maize. *Crop Science*, *47*(3), 1082-1090.

Bhandari, A., Bartholomé, J., Cao-Hamadoun, T.-V., Kumari, N., Frouin, J., Kumar, A., et al. (2019). Selection of trait-specific markers and multi-environment models improves genomic predictive ability in rice. *PLoS One*, *14*, e0208871.

Bhat, A.A., Gupta, S.K., Singh, S.K., Meraj, G., Kumar, P., Kanga, S. and Sajan, B. (2024). Simulating climatic patterns and their impacts on the food security stability system in Jammu, Kashmir and adjoining regions, India. *Climate*, *12*(7), 99.

Bordes, J., Ravel, C., Le Gouis, J., Lapierre, A., Charmet, G. and Balfourier, F. (2011). Use of a global wheat core collection for association analysis of flour and dough quality traits. *Journal of Cereal Science*, *54*(1), 137–147.

Borrill, P. (2020). Blurring the boundaries between cereal crops and model plants. *New Phytol. 228*(6), 1721–1727.

Borrill, P., Connorton, J.M., Balk, J., Miller, A.J., Sanders, D. and Uauy, C. (2014). Biofortification of wheat grain with iron and zinc: integrating novel genomic resources and knowledge from model crops. *Frontiers in plant science*, 5, 53.

Boruah, T., Devi, B., Chetia, T., Choubey, K., Talukdar, K., Ansari, M.J. and Nayik, G.A. (2023). Future Aspects of Grain Quality and Role of Technologists in Its Management. In *Cereal Grains* (pp. 309–327). CRC Press.

Brandt, K.M., Gunn, H., Moretti, N. and Zemetra, R.S. (2020). A streamlined protocol for wheat (*Triticum aestivum*) protoplast isolation and transformation with CRISPR-Cas ribonucleoprotein complexes. *Frontiers in Plant Science*, *11*, 769.

Brokowski, C. and Adli, M. (2019). CRISPR ethics: moral considerations for applications of a powerful tool. *Journal of Molecular Biology*, *431*(1), 88–101.

Cermak, T., Doyle, E.L., Christian, M., Wang, L., Zhang, Y., Schmidt, C., et al. (2011). Efficient design and assembly of custom TALEN and other TAL effector based constructs for DNA targeting. *Nucleic Acids Reseacrh*, *39*(12), e82. doi:10.1093/nar/gkr218.

Cerrudo, D., Cao, S., Yuan, Y., Martinez, C., Suarez, E.A., Babu, R., et al. (2018). Genomic selection outperforms marker-assisted selection for grain yield and physiological traits in a maize doubled haploid population across water treatments. *Frontiers in Plant Science*, *9*, 366. doi:10. 3389/FPLS.2018.00366/BIBTEX

Char, S.N. and Yang, B. (2020). Genome editing in grass plants. *aBIOTECH*, *1*(1), 41–57. doi:10.1007/s42994-019-00005-x

Christopher, J., Christopher, M., Jennings, R., Jones, S., Fletcher, S., Borrell, A. and Hammer, G. (2013). QTL for root angle and number in a population developed from bread wheats (*Triticum aestivum*) with contrasting adaptation to water-limited environments. *Theoretical and Applied Genetics*, *126*, 1563–1574.

Condorelli, G.E., Maccaferri, M., Newcomb, M., Andrade-Sanchez, P., White, J. W., French, A.N. and Tuberosa, R. (2018). Comparative aerial and ground based high throughput phenotyping for the genetic dissection of NDVI as a proxy for drought adaptive traits in durum wheat. *Frontiers in Plant Science*, *9*, 349736.

Cortés, A.J. and López-Hernández, F. (2021). Harnessing crop wild diversity for climate change adaptation. *Genes*, *12*:783.

Crossa, J., Campos, G.D.L., Pérez, P., Gianola, D., Burgueno, J., Araus, J.L. and Braun, H.J. (2010). Prediction of genetic values of quantitative traits in plant breeding using pedigree and molecular markers. *Genetics*, *186*(2), 713–724.

Cui, X., Balcerzak, M., Schernthaner, J., Babic, V., Datla, R., Brauer, E.K., et al. (2019). An optimised CRISPR/Cas9 protocol to create targeted mutations in homoeologous genes and an efficient genotyping protocol to identify edited events in wheat. *Plant Methods*, *15*(1), 119–212.

de Bustos, A., Rubio, P., Soler, C., García, P. and Jouve, N. (2001). Marker assisted selection to improve HMW-glutenins in wheat. In *Wheat in a Global Environment: Proceedings of the 6th International Wheat Conference, 5–9 June 2000, Budapest, Hungary* (pp. 171–176). Springer Netherlands.

Divte, P.R., Sharma, N., Parveen, S., Devika, S. and Anand, A. (2022). Cereal Grain Composition under Changing Climate. In *Climate Change and Crop Stress* (pp. 329–360). Academic Press.

Dixit, A.K., Sen, B., Bijla, S., Maiti, S. and Kathayat, B. (2023). Contribution of Wheat in Global Food Security in Changing Climatic Conditions: Challenges Ahead and Coping Strategies. In *Wheat Science* (pp. 107–124). CRC Press.

Dolferus, R., Thavamanikumar, S., Sangma, H., Kleven, S., Wallace, X., Forrest, K. and Cullis, B. (2019). Determining the genetic architecture of reproductive stage drought tolerance in wheat using a correlated trait and correlated marker effect model. *G3: Genes, Genomes, Genetics*, *9*(2), 473–489.

Dong, H., Wang, R., Yuan, Y., Anderson, J., Pumphrey, M., Zhang, Z. and Chen, J. (2018). Evaluation of the potential for genomic selection to improve spring wheat resistance to Fusarium head blight in the Pacific Northwest. *Frontiers in Plant Science*, *9*, 911.

Dreisigacker, S., Sukumaran, S., Guzmán, C., He, X., Lan, C., Bonnett, D. and Crossa, J. (2016). Molecular marker-based selection tools in spring bread wheat improvement: CIMMYT experience and prospects. *Molecular Breeding for Sustainable Crop Improvement. 2*, 421–474.

Erdoğan, İ., Cevher-Keskin, B., Bilir, Ö., Hong, Y. and Tör, M. (2023). Recent developments in CRISPR/Cas9 genome-editing technology related to plant disease resistance and abiotic stress tolerance. *Biology*, *12*(7), 1037.

Eş, I., Gavahian, M., Marti-Quijal, F. J., Lorenzo, J. M., Khaneghah, A.M., Tsatsanis, C. and Barba, F.J. (2019). The application of the CRISPR-Cas9 genome editing machinery in food and agricultural science: current status, future perspectives, and associated challenges. *Biotechnology Advances*, *37*(3), 410–421.

Faryad, A., Aziz, F., Tahir, J., Kousar, M., Qasim, M. and Shamim, A. (2021). Integration of OMICS technologies for crop improvement. *Protein and Peptide Letters*, *28*(8), 896–908.

Fatima, S., Chaudhari, S.K., Akhtar, S., Amjad, M.S., Akbar, M., Iqbal, M.S. and Shehzad, T. (2018). Mapping QTLs for yield and yield components under drought stress in bread wheat (*Triticum aestivum* L.). *Applied Ecology & Environmental Research*, *16*(4), 4431–4453.

Fatiukha, A., Filler, N., Lupo, I., Lidzbarsky, G., Klymiuk, V., Korol, A.B. and Krugman, T. (2020). Grain protein content and thousand kernel weight QTLs identified in a durum × wild emmer wheat mapping population tested in five environments. *Theoretical and Applied Genetics*, *133*, 119–131.

Fones, H.N., Bebber, D.P., Chaloner, T.M., Kay, W.T., Steinberg, G. and Gurr, S.J. (2020). Threats to global food security from emerging fungal and Oomycete crop pathogens. *Nat. Food*, *1*, 332–342.

Gao, C. (2021). Genome engineering for crop improvement and future agriculture. *Cell 184*, 1621–1635.

Gautam, T., Amardeep, Saripalli, G., Rakhi, Kumar, A., Gahlaut, V. and Gupta, P.K. (2021). Introgression of a drought insensitive grain yield QTL for improvement of four Indian bread wheat cultivars using marker assisted breeding without background selection. *Journal of Plant Biochemistry and Biotechnology*, *30*, 172–183.

Gautam, T., Dhillon, G.S., Saripalli, G., Singh, R., V.P., Prasad, P. and Gupta, P.K. (2020). Marker-assisted pyramiding of genes/QTL for grain quality and rust resistance in wheat (*Triticum aestivum* L.). *Molecular Breeding*, *40*, 1–14.

Golabadi, M., Arzani, A., Mirmohammadi Maibody, S.A.M., Sayed Tabatabaei, B.E. and Mohammadi, S.A. (2011). Identification of microsatellite markers linked with yield components under drought stress at terminal growth stages in durum wheat. *Euphytica*, *177*, 207–221.

Goswami, R.S. and Kistler, H.C. (2004). Heading for disaster: Fusarium graminearum on cereal crops. *Molecular Plant Pathology*, *5*(6), 515–525.

Guo, J., Zhang, X., Hou, Y., Cai, J., Shen, X., Zhou, T. and Kong, L. (2015). High-density mapping of the major FHB resistance gene Fhb7 derived from *Thinopyrum ponticum* and its pyramiding with Fhb1 by marker-assisted selection. *Theoretical and Applied Genetics*, *128*, 2301–2316.

Gupta, S.K., Charpe, A., Koul, S., Prabhu, K.V. and Haq, Q.M.R. (2005). Development and validation of molecular markers linked to an Aegilops umbellulata-derived leaf-rust-resistance gene, Lr9, for marker-assisted selection in bread wheat. *Genome*, *48*(5), 823–830.

Hamada, H., Liu, Y., Nagira, Y., Miki, R., Taoka, N. and Imai, R. (2018). Biolistic-delivery-based transient CRISPR/Cas9 expression enables in planta genome editing in wheat. *Scientific Reports 8*(1), 14422–14427.

Harrington, S., Connorton, J., Nyangoma, N., McNelly, R., Morgan, Y., Aslam, M., Sharp, P., Johnson, A., Uauy, C. and Balk, J. (2022). A two-gene strategy increases iron and zinc concentrations in wheat flour, improving mineral bioaccessibility. *Plant Physiology*, 191, 528–541.

Hayes, B.J., Bowman, P.J., Chamberlain, A.J. and Goddard, M.E. (2009). Invited review: genomic selection in dairy cattle: progress and challenges. *Journal of Dairy Science*, *92*, 433–443. doi:10.3168/JDS.2008-1646

He, S., Schulthess, A.W., Mirdita, V., Zhao, Y., Korzun, V., Bothe, R. and Jiang, Y. (2016). Genomic selection in a commercial winter wheat population. *Theoretical and Applied Genetics*, *129*, 641–651.

Heffner, E.L., Jannink, J.L. and Sorrells, M.E. (2011). Genomic selection accuracy using multifamily prediction models in a wheat breeding program. *The Plant Genome*, *4*(1), 65–75.

Hillary, V.E. and Ceasar, S.A. (2019). Application of CRISPR/Cas9 genome editing system in cereal crops. *The Open Biotechnology Journal*, *13*(1), 173–179.

Hoffstetter, A., Cabrera, A., Huang, M. and Sneller, C. (2016). Optimizing training population data and validation of genomic selection for economic traits in soft winter wheat. *G3: Genes, Genomes, Genetics*, *6*(9), 2919–2928.

Howells, R.M., Craze, M., Bowden, S. and Wallington, E.J. (2018). Efficient generation of stable, heritable gene edits in wheat using CRISPR/Cas9. *BMC Plant Biology*, *18*(1), 215.

Ibrahim, A.A., Abd-Ellatif, S., Abdel Razik, E.S.S. and Salem, K.F. (2024). Biodiversity of Cereal Crops and Utilization in Food and Nutritional Security. In *Sustainable Utilization and Conservation of Plant Genetic Diversity* (pp. 31—61). Singapore: Springer Nature Singapore.

Idrees, M. and Irshad, M. (2014). Molecular markers in plants for analysis of genetic diversity: a review. *European Academic Research*, *2*(1), 1513–1540.

Jiang, P., Zhang, P., Wu, L., He, Y., Li, C., Ma, H. and Zhang, X. (2021). Linkage and association mapping and Kompetitive allele-specific PCR marker development for improving grain protein content in wheat. *Theoretical and Applied Genetics*, *134*(11), 3563–3575.

Jogam, P., Sandhya, D., Kumar, P., Allini, V.R., Abbagani, S. and Alok, A. (2021). Genetic transformation methods and advancement of CRISPR/Cas9 technology in wheat. In CRISPR and RNAi systems (pp. 253–275).

Joukhadar, R., Thistlethwaite, R., Trethowan, R.M., Hayden, M.J., Stangoulis, J., Cu, S., et al. (2021). Genomic selection can accelerate the biofortification of spring wheat. *Theoretical and Applied Genetics*, *134*, 3339–3350.

Juarez, M., Legua, P., Mengual, C.M., Kassem, M.A., Sempere, R.N., Gómez, P., et al. (2013). Relative incidence, spatial distribution and genetic diversity of cucurbit viruses in eastern Spain. *Annals of Applied Biology*, *162*, 362–370.

Kamboj, D., Kumar, S., Mishra, C.N., Srivastava, P., Singh, G. and Singh, G.P. (2020). Marker assisted breeding in cereals: progress made and challenges in India. *Journal of Cereal Research*, *12*(2), 85–102.

Kaur, B., Sandhu, K.S., Kamal, R., Kaur, K., Singh, J., Röder, M.S. and Muqaddasi, Q.H. (2021). Omics for the improvement of abiotic, biotic, and agronomic traits in major cereal crops: applications, challenges, and prospects. *Plants*, *10*(10), 1989.

Khakimov, B., Bak, S. and Engelsen, S. (2014). High-throughput cereal metabolomics: current analytical technologies, challenges and perspectives. *Journal of Cereal Science*, 59, 393–418.

Khan, M.K., Pandey, A., Hamurcu, M., Gezgin, S., Athar, T., Rajput, V.D., ... and Minkina, T. (2021). Insight into the prospects for nanotechnology in wheat biofortification. Biology, *10*(11), 1123.

Khanna-Chopra, R., Singh, K., Shukla, S., Kadam, S. and Singh, N.K. (2020). QTLs for cell membrane stability and flag leaf area under drought stress in a wheat RIL population. *Journal of Plant Biochemistry and Biotechnology*, *29*, 276–286.

Khush, G., Lee, S., Cho, J. and Jeon, J. (2012). Biofortification of crops for reducing malnutrition. *Plant Biotechnology Reports*, 6, 195–202.

Kiran, A., Wakeel, A., Mahmood, K., Mubaraka, R., Hafsa and Haefele, S.M. (2022). Biofortification of staple crops to alleviate human malnutrition: contributions and potential in developing countries. *Agronomy*, *12*(2), 452.

Kumar, J., Jaiswal, V., Kumar, A., Kumar, N., Mir, R.R., Kumar, S. and Gupta, P.K. (2011). Introgression of a major gene for high grain protein content in some Indian bread wheat cultivars. *Field Crops Research*, *123*(3), 226–233.

Kumar, J., Mir, R.R., Kumar, N., Kumar, A., Mohan, A., Prabhu, K.V. and Gupta, P.K. (2010). Marker-assisted selection for pre-harvest sprouting tolerance and leaf rust resistance in bread wheat. *Plant Breeding*, *129*(6), 617–621.

Kumar, J., Saini, D.K., Kumar, A., Kumari, S., Gahlaut, V., Rahim, M.S. and Roy, J. (2024). Biofortification of *Triticum* species: a stepping stone to combat malnutrition. *BMC Plant Biology*, *24*(1), 668.

Kumar, R., Sharma, V., Suresh, S., Ramrao, D.P., Veershetty, A., Kumar, S., et al. (2021). Understanding omics driven plant improvement and de novo crop domestication: some examples. *Front. Genet.* 12:415.

Kumar, S., Singh, A., Bist, C.M.S. and Sharma, M. (2024). Advancements in genetic techniques and functional genomics for enhancing crop traits and agricultural sustainability. *Briefings in Functional Genomics*, elae017.

Kumari, S., Mir, R.R., Tyagi, S., Balyan, H.S. and Gupta, P.K. (2019). Validation of QTL for grain weight using MAS-derived pairs of NILs in bread wheat (Triticum aestivum L.). *Journal of Plant Biochemistry and Biotechnology*, *28*, 336–344.

Larkin, D.L., Lozada, D.N. and Mason, R.E. (2019). Genomic selection—considerations for successful implementation in wheat breeding programs. *Agronomy*, *9*(9), 479.

Lateef, D.D. (2015). DNA marker technologies in plants and applications for crop improvements. *Journal of Biosciences and Medicines*, *3*(5), 7–18.

Li, J., Cui, F., Ding, A.M., Zhao, C.H., Wang, X.Q., Wang, L. and Wang, H.G. (2012). QTL detection of seven quality traits in wheat using two related recombinant inbred line populations. *Euphytica*, *183*, 207–226.

Li, J., Jiao, G., Sun, Y., Chen, J., Zhong, Y., Yan, L. and Xia, L. (2021). Modification of starch composition, structure and properties through editing of TaSBEIIa in both winter and spring wheat varieties by CRISPR/Cas9. *Plant Biotechnology Journal*, *19*(5), 937–951.

Li, L., Mao, X., Wang, J., Chang, X., Reynolds, M. and Jing, R. (2019). Genetic dissection of drought and heat-responsive agronomic traits in wheat. *Plant, Cell & Environment*, *42*(9), 2540–2553.

Liang, Z., Chen, K., Li, T., Zhang, Y., Wang, Y., Zhao, Q., et al. (2017). Efficient DNA-free genome editing of bread wheat using CRISPR/Cas9 ribonucleoprotein complexes. *Nature Communications*, *8*(1), 14261–14265.

Liang, Z., Chen, K., Zhang, Y., Liu, J., Yin, K., Qiu, J.L., et al. (2018). Genome editing of bread wheat using biolistic delivery of CRISPR/Cas9 in vitro transcripts or ribonucleoproteins. *Nature Protocols*, *13*(3), 413–430.

Liu, J., Feng, B., Xu, Z., Fan, X., Jiang, F., Jin, X. and Wang, T. (2018). A genome-wide association study of wheat yield and quality-related traits in southwest China. *Molecular Breeding*, *38*, 1–11.

Liu, S., Banik, M., Yu, K., Park, S.J., Poysa, V. and Guan, Y. (2007). Marker-assisted selection (MAS) in major cereal and legume crop breeding: current progress and future directions. *International Journal of Plant Breeding and Genetics*, *1*(2), 75–78.

Longin, C.F.H., Reif, J.C. and Würschum, T. (2014). Long-term perspective of hybrid versus line breeding in wheat based on quantitative genetic theory. *Theoretical and Applied Genetics*, *127*, 1635–1641. doi:10.1007/S00122-014-2325-8

Lorenz, A.J., Chao, S., Asoro, F.G., Heffner, E.L., Hayashi, T., Iwata, H. and Jannink, J.L. (2011). Genomic selection in plant breeding: knowledge and prospects. *Advances in Agronomy*, *110*, 77–123.

Lozada, D.N. and Carter, A.H. (2020). Genomic selection in winter wheat breeding using a recommender approach. *Genes*, *11*, 1–14. di:10.3390/ GENES11070779

Malik, S. and Malik, T.A. (2015). Genetic mapping of potential QTLs associated with drought tolerance in wheat. *JAPS: Journal of Animal & Plant Sciences*, *25*(4), 1032.

Marulanda, J.J., Mi, X., Melchinger, A.E., Xu, J.-L., Würschum, T. and Longin, C.F.H. (2016). Optimum breeding strategies using genomic selection for hybrid breeding in wheat, maize, rye, barley, rice, and Triticale. *Theoretical and Applied Genetics*, *129*, 1901–1913. doi:10.1007/s00122-016-2748-5

Merchuk-Ovnat, L., Barak, V., Fahima, T., Ordon, F., Lidzbarsky, G.A., Krugman, T. and Saranga, Y. (2016). Ancestral QTL alleles from wild emmer wheat improve drought resistance and productivity in modern wheat cultivars. *Frontiers in Plant Science*, *7*, 452.

Merrick, L.F., Herr, A.W., Sandhu, K.S., Lozada, D.N. and Carter, A.H. (2022). Utilizing genomic selection for wheat population development and improvement. *Agronomy*, *12*(2), 522.

Meuwissen, T.H. (2009). Accuracy of breeding values of 'unrelated' individuals predicted by dense SNP genotyping. *Genetics Selection Evolution*, *41*, 1–9.

Michel, S., Ametz, C., Gungor, H., Epure, D., Grausgruber, H., Löschenberger, F. and Buerstmayr, H. (2016). Genomic selection across multiple breeding cycles in applied bread wheat breeding. *Theoretical and Applied Genetics*, *129*, 1179–1189.

Michel, S., Ametz, C., Gungor, H., Akgöl, B., Epure, D., Grausgruber, H., ... and Buerstmayr, H. (2017). Genomic assisted selection for enhancing line breeding: merging genomic and phenotypic selection in winter wheat breeding programs with preliminary yield trials. *Theoretical and Applied Genetics*, *130*(2), 363–376.

Michel, S., Löschenberger, F., Ametz, C., Pachler, B., Sparry, E. and Bürstmayr, H. (2019). Combining grain yield, protein content and protein quality by multi-trait genomic selection in bread wheat. *Theoretical and Applied Genetics*, *132*, 2767–2780.

Miedaner, T. and Korzun, V. (2012). Marker-assisted selection for disease resistance in wheat and barley breeding. *Phytopathology*, *102*(6), 560–566.

Milus, E.A. and Parsons, C.E. (1994). Evaluation of foliar fungicides for controlling Fusarium head blight of wheat Plant Disease, *78*(7), 697–699.

Muitire, C., Kamutando, C. and Moyo, M. (2021). *Building Stress Resilience of Cereals under Future Climatic Scenarios: 'The Case of Maize, Wheat, Rice and Sorghum'*. London, UK: IntechOpen.

Mujtaba, S.M., Faisal, S., Khan, M.A., Shirazi, M.U. and Khan, M.A. (2018). Evaluation of drought tolerant wheat genotypes using morpho-physiological indices as screening tools. *Pakistan Journal of Botany*, *50*(1), 51–58.

Mushtaq, Z. and Nazir, A. (2021). Biofortification: way forward toward micronutrient deficiency. *EQA – International Journal of Environmental Quality*, *42*, 36–41.

Naqvi, R.Z., Mahmood, M.A., Mansoor, S., Amin, I. and Asif, M. (2024). Omics-driven exploration and mining of key functional genes for the improvement of food and fiber crops. *Frontiers in Plant Science*, *14*, 1273859.

Nazir, R., Mandal, S., Mitra, S., Ghorai, M., Das, N., Jha, N., Majumder, M., Pandey, D. and Dey, A. (2022). CRISPR/Cas genome-editing toolkit to enhance salt stress tolerance in rice and wheat. *Physiologia Plantarum*, e13642.

Norman, A., Taylor, J., Edwards, J. and Kuchel, H. (2018). Optimising genomic selection in wheat: effect of marker density, population size and population structure on prediction accuracy. *G3: Genes, Genomes, Genetics*, *8*(9), 2889–2899.

Pandurangan, S., Workman, C., Nilsen, K. and Kumar, S. (2021). Introduction to Marker-Assisted Selection in Wheat Breeding. In *Accelerated Breeding of Cereal Crops* (pp. 77–117). New York, NY: Springer US.

Pazhamala, L.T., Kudapa, H., Weckwerth, W., Millar, A.H. and Varshney, R.K. (2021). Systems biology for crop improvement. *The plant genome*, *14*(2), e20098.

Qin, F., Shinozaki, K. and Yamaguchi-Shinozaki, K. (2011). Achievements and challenges in understanding plant abiotic stress responses and tolerance. *Plant and Cell Physiology 52*, 1569–1582.

Raheem, D., Dayoub, M., Birech, R. and Nakiyemba, A. (2021). The contribution of cereal grains to food security and sustainability in Africa: potential application of UAV in Ghana, Nigeria, Uganda, and Namibia. *Urban Science*, *5*(1), 8.

Rai, A., Mahendru-Singh, A., Raghunandan, K., Kumar, T.P.J., Sharma, P., Ahlawat, A.K. and Sivasamy, M. (2019). Marker-assisted transfer of PinaD1a gene to develop soft grain wheat cultivars. *3 Biotech*, *9*, 1–10.

Rai, N., Bellundagi, A., Kumar, P.K., Kalasapura Thimmappa, R., Rani, S., Sinha, N. and Prabhu, K.V. (2018). Marker-assisted backcross breeding for improvement of drought tolerance in bread wheat (*Triticum aestivum* L. em Thell). *Plant Breeding*, *137*(4), 514–526.

Randhawa, H.S., Asif, M., Pozniak, C., Clarke, J.M., Graf, R.J., Fox, S.L. and Spaner, D. (2013). Application of molecular markers to wheat breeding in Canada. *Plant Breeding*, *132*(5), 458–471.

Ribaut, J.-M. and Ragot, M. (2007). Marker-assisted selection to improve drought adaptation in maize: the backcross approach, perspectives, limitations, and alternatives. *Journal of Experimental Botany*, *58*, 351–360.

Rodriguez, E. (2016). Ethical issues in genome editing using Crispr/Cas9 system. *Journal of Clinical Research and Bioethics*, *7*(2).

Roy, C., Kumar, S., Ranjan, R.D., Kumhar, S.R. and Govindan, V. (2022). Genomic approaches for improving grain zinc and iron content in wheat. *Frontiers in Genetics*, *13*, 1045955.

Rutkoski, J., Benson, J., Jia, Y., Brown-Guedira, G., Jannink, J.L. and Sorrells, M. (2012). Evaluation of genomic prediction methods for Fusarium head blight resistance in wheat. *The Plant Genome*, 5(2), 51–61.

Sarinelli, J.M., Murphy, J.P., Tyagi, P., Holland, J.B., Johnson, J.W., Mergoum, M. and Brown-Guedira, G. (2019). Training population selection and use of fixed effects to optimize genomic predictions in a historical USA winter wheat panel. *Theoretical and Applied Genetics*, *132*, 1247–1261.

Scossa, F., Alseekh, S. and Fernie, A.R. (2021). Integrating multi-omics data for crop improvement. *Journal of Plant Physiology*, *257*, 153352.

Sehgal, D., Dhakate, P., Ambreen, H., Shaik, K.H.B., Rathan, N.D., Anusha, N.M., ... and Vikram, P. (2023). Wheat omics: advancements and opportunities. Plants, *12*(3), 426.

Shahane, A.A. and Shivay, Y.S. (2024). Cereal crops: A valid option for bio-fortification and development of nutrient-dense food in developing and developed countries in Asia and Africa. Advances in Agronomy, 186, 205–276.

Shan, Q., Wang, Y., Li, J. and Gao, C. (2014). Genome editing in rice and wheat using the CRISPR/Cas system. *Nature Protocols*, *9*(10), 2395–2410. doi:10.1038/nprot. 2014.157

Sharma, A., Srivastava, P., Mavi, G.S., Kaur, S., Kaur, J., Bala, R. and Singh, G.P. (2021). Resurrection of wheat cultivar PBW343 using marker-assisted gene pyramiding for rust resistance. *Frontiers in Plant Science*, *12*, 570408.

Shinwari, Z., Tanveer, F. and Khalil, A. (2018). Ethical issues regarding CRISPR-mediated genome editing. *Current Issues in Molecular Biology*, *26*, 103–110.

Singh, R.K., Muthamilarasan, M. and Prasad, M. (2021). Biotechnological approaches to dissect climate-resilient traits in millets and their application in crop improvement. J. Biotechnol. 327, 64–73.

Singh, R.P., Hodson, D.P., Jin, Y., Lagudah, E.S., Ayliffe, M.A., Bhavani, S. and Hovmøller, M.S. (2015). Emergence and spread of new races of wheat stem rust fungus: continued threat to food security and prospects of genetic control. *Phytopathology*, *105*(7), 872–884.

Singh, R., Datta, D., Priyamvada, Singh, S. and Tiwari, R. (2009). A diagnostic PCR based assay for stripe rust resistance gene Yr10 in wheat. *Acta Phytopathologica et Entomologica Hungarica*, *44*(1), 11–18.

Song, L., Wang, R., Yang, X., Zhang, A. and Liu, D. (2023). Molecular markers and their applications in marker-assisted selection (MAS) in bread wheat (*Triticum aestivum* L.). *Agriculture*, *13*(3), 642.

Spindel, J., Begum, H., Akdemir, D., Virk, P., Collard, B., Redoña, E., et al. (2015). Correction: genomic selection and association mapping in rice (*Oryza sativa*): effect of trait genetic architecture, training population composition, marker number and statistical model on accuracy of rice genomic selection in elite, tropical rice breeding lines. *PLOS Genetics 11*, e1005350.

Srivastava, N. (2019). Molecular and biotechnological tools in developing abiotic stress tolerance in wheat. *Wheat Production in Changing Environments: Responses, Adaptation and Tolerance*, 283–341.

Steiner, B., Buerstmayr, M., Michel, S., Schweiger, W., Lemmens, M. and Buerstmayr, H. (2017). Breeding strategies and advances in line selection for Fusarium head blight resistance in wheat. *Tropical Plant Pathology*, *42*, 165–174.

Sukumaran, S., Jarquin, D., Crossa, J. and Reynolds, M. (2018). Genomic-enabled prediction accuracies increased by modeling genotype × environment interaction in durum wheat. *Plant Genome, 11*, 170112.

Sukumaran, S., Reynolds, M. P. and Sansaloni, C. (2018). Genome-wide association analyses identify QTL hotspots for yield and component traits in durum wheat grown under yield potential, drought, and heat stress environments. *Frontiers in Plant Science*, *9*, 81.

Suliman, S., Alemu, A., Abdelmula, A.A., Badawi, G.H., Al-Abdallat, A. and Tadesse, W. (2021). Genome-wide association analysis uncovers stable QTLs for yield and quality traits of spring bread wheat (*Triticum aestivum*) across contrasting environments. *Plant Gene*, *25*, 100269.

Sun, J., Rutkoski, J.E., Poland, J.A., Crossa, J., Jannink, J.L. and Sorrells, M.E. (2017). Multitrait, random regression, or simple repeatability model in high-throughput phenotyping data improve genomic prediction for wheat grain yield. *Plant Genome*, *10*, 1–12.

Sun, X., Marza, F., Ma, H., Carver, B.F. and Bai, G. (2010). Mapping quantitative trait loci for quality factors in an inter-class cross of US and Chinese wheat. *Theoretical and Applied Genetics*, *120*, 1041–1051.

Tripathi, L., Ntui, V.O. and Tripathi, J.N. (2020). CRISPR/Cas9-based genome editing of banana for disease resistance. *Current Opinion in Plant Biology*, *56*, 118–126. doi:10. 1016/j.pbi.2020.05.003

Tura, H., Edwards, J., Gahlaut, V., Garcia, M., Sznajder, B., Baumann, U. and Fleury, D. (2020). QTL analysis and fine mapping of a QTL for yield-related traits in wheat grown in dry and hot environments. *Theoretical and Applied Genetics*, *133*(1), 239–257.

Upadhyay, S.K., Kumar, J., Alok, A. and Tuli, R. (2013). RNA-guided genome editing for target gene mutations in wheat. *G3: Genes, Genomes, Genetics*, *3*(12), 2233–2238.

Vagndorf, N., Kristensen, P.S., Andersen, J.R., Jahoor, A. and Orabi, J. (2018). Marker-assisted breeding in wheat. *Next Generation Plant Breeding*, *1*, 3–22.

Velu, G., Crespo Herrera, L., Guzman, C., Huerta, J., Payne, T. and Singh, R.P. (2019). Assessing genetic diversity to breed competitive biofortified wheat with enhanced grain Zn and Fe concentrations. *Frontiers in Plant Science*, *9*, 1971.

Verma, S., Gupta, A.R.S.S.H., Yalla, S., Shreya, Patel, P.J., Sharma, R. and Donga, A. (2024). Integrating Marker-Assisted (MAS) and Genomic Selection (GS) for Plant Functional Trait Improvement. In *Plant Functional Traits for Improving Productivity* (pp. 203–215). Singapore: Springer Nature Singapore.

Vinci, G., Ruggieri, R., Ruggeri, M. and Zaki, M.G. (2022, September). Application of Life Cycle Assessment (LCA) to Cereal Production: An Overview. In *IOP Conference Series: Earth and Environmental Science* (Vol. 1077, No. 1, p. 012004). IOP Publishing.

Vishwakarma, M.K., Mishra, V.K., Gupta, P.K., Yadav, P.S., Kumar, H. and Joshi, A.K. (2014). Introgression of the high grain protein gene Gpc-B1 in an elite wheat variety of Indo-Gangetic plains through marker assisted backcross breeding. *Current Plant Biology*, *1*, 60–67.

Wakeel, A., Farooq, M., Bashir, K. and Ozturk, L. (2018). "Micronutrient Malnutrition and Biofortification: Recent Advances and Future Perspectives," *Plant Micronutrient Efficiency*, pp. 225–243.

Wang, W., Pan, Q., He, F., Akhunova, A., Chao, S., Trick, H., et al. (2018). Transgenerational CRISPR-Cas9 activity facilitates multiplex gene editing in allopolyploid wheat. *The CRISPR Journal*, *1*(1), 65–74. doi:10.1089/crispr.2017.0010

Wang, X., Xu, Y., Hu, Z. and Xu, C. (2018). Genomic selection methods for crop improvement: current status and prospects. *The Crop Journal*, *6*, 330–340. doi:10.1016/j. cj.2018.03.001

Wang, Y., Cheng, X., Shan, Q., Zhang, Y., Liu, J., Gao, C., et al. (2014). Simultaneous editing of three homoeoalleles in hexaploid bread wheat confers heritable resistance to powdery mildew. *Nature Biotechnology*, *32*(9), 947–951.

Wani, S.H., Choudhary, J.R., Choudhary, M., Rana, M. and Gosal, S.S. (2020). Recent advances in genomics assisted breeding for drought stress tolerance in major cereals. *Journal of Cereal Research*, *12*(1), 1–12.

Watson, A., Hickey, L.T., Christopher, J., Rutkoski, J., Poland, J. and Hayes, B.J. (2019). Multivariate genomic selection and potential of rapid indirect selection with speed breeding in spring wheat. *Crop Science*, *59*(5), 1945–1959.

Wegulo, S.N., Baenziger, P.S., Nopsa, J.H., Bockus, W.W. and Hallen-Adams, H. (2015). Management of Fusarium head blight of wheat and barley. *Crop Protection*, *73*, 100–107.

Werner, C.R., Voss-Fels, K.P., Miller, C.N., Qian, W., Hua, W., Guan, C.Y. and Qian, L. (2018). Effective genomic selection in a narrow-genepool crop with low-density markers: Asian rapeseed as an example. *The Plant Genome*, *11*(2), 170084.

White, J., Sharma, R., Balding, D., Cockram, J. and Mackay, I.J. (2022). Genome-wide association mapping of Hagberg falling number, protein content, test weight, and grain yield in UK wheat. *Crop Science*, *62*(3), 965–981.

Wu, Z., Qiu, H., Tian, Z., Liu, C., Qin, M., Li, W. and Hou, J. (2022). Uncovering the genetic basis of gluten aggregation parameters by genome-wide association analysis in wheat (*Triticum aestivum* L.) using GlutoPeak. *BMC Plant Biology*, *22*(1), 493.

Xu, Y.F., Li, S.S., Li, L.H., Ma, F.F., Fu, X.Y., Shi, Z.L. and An, D.G. (2017). QTL mapping for yield and photosynthetic related traits under different water regimes in wheat. *Molecular Breeding*, *37*, 1–18.

Yabe, S., Yoshida, H., Kajiya-Kanegae, H., Yamasaki, M., Iwata, H., Ebana, K., et al. (2018). Description of grain weight distribution leading to genomic selection for grain-filling characteristics in rice. *PLoS One 13*, e0207627. doi:10.1371/ JOURNAL.PONE.0207627

Yang, Y., Saand, M.A., Huang, L., Abdelaal, W.B., Zhang, J., Wu, Y. and Wang, F. (2021). Applications of multi-omics technologies for crop improvement. *Frontiers in Plant Science*, *12*, 563953.

Zenda, T., Liu, S., Dong, A., Li, J., Wang, Y., Liu, X., ... and Duan, H. (2021). Omics-facilitated crop improvement for climate resilience and superior nutritive value. *Frontiers in Plant Science*, 12, 774994.

Zhang, A., Liu, Y., Wang, F., Li, T., Chen, Z., Kong, D., et al. (2019a). Enhanced rice salinity tolerance via CRISPR/Cas9-targeted mutagenesis of the OsRR22 gene. *Molecular Breeding 39*(3), 47.

Zhang, H., Gou, F., Zhang, J., Liu, W., Li, Q., Mao, Y., et al. (2016). TALEN-mediated targeted mutagenesis produces a large variety of heritable mutations in rice. *Plant Biotechnology J. 14*(1), 186–194.

Zhang, R., Zhang, C., Yu, C., Dong, J. and Hu, J. (2022). Integration of multi-omics technologies for crop improvement: status and prospects. *Frontiers in Bioinformatics*, 2, 1027457.

Zhang, X., Tang, B., Yu, F., Li, L., Wang, M., Xue, Y. and Qiu, F. (2013). Identification of major QTL for waterlogging tolerance using genome-wide association and linkage mapping of maize seedlings. *Plant Molecular Biology Reporter*, *31*, 594–606.

Zhang-Biehn, S., Fritz, A.K., Zhang, G., Evers, B., Regan, R. and Poland, J. (2021). Accelerating wheat breeding for end-use quality through association mapping and multivariate genomic prediction. *The Plant Genome*, *14*(3), e20164.

Zhao, Y., Mette, M.F. and Reif, J.C. (2015). Genomic selection in hybrid breeding. *Plant Breeding*, *134*, 1–10.

CHAPTER 6

# Improvement Strategies Addressing Malnutrition: Augmenting Bioavailability and Biofortification in Food Crops

Mohana Kumara P[1*], Pavan PR[2], Chaitra Kulkarni[1], Sankalpa KB[3], Mahantesha BN Naika[4] and Belaghihalli N Gnanesh[5]

## 1. Introduction

Nutrition is crucial for maintaining a robust immune system, as its proper functioning depends on the intake of essential nutrients, including minerals such as zinc and selenium, as well as vitamins C, D, and E, along with various bioactive compounds. Malnutrition, whether in the form of undernutrition or overnutrition, is often linked to inflammation, which can increase susceptibility to infections and weaken immune defences (Morales et al., 2024). Essential minerals such as magnesium, zinc, copper, iron, and selenium are critical for maintaining immune competence and regulating inflammation. Deficiencies in these minerals can compromise immune responses and are particularly relevant in autoimmune thyroid disorders due to their roles in modulating immune tolerance and oxidative stress. Contemporary dietary patterns and agricultural practices have contributed to reduced mineral intake, potentially increasing inflammatory risk and altering immune cell metabolism. While adequate intake is generally achieved through a balanced diet, certain populations may require medically supervised supplementation. However, excessive intake should be avoided, as it may adversely affect immune function (Stefanache et al., 2023). Globally, malnutrition remains a critical issue, with approximately 149.2 million children under the age of five experiencing stunting, 45.4 million suffering from wasting, and 38.9 million classified as obese. Additionally, more than 40% of the global adult population (2.2 billion individuals) is either overweight or obese (UNICEF, 2022). The National Family Health Survey (2019–2021) reveals a

1 Department of Biotechnology and Crop Improvement, KRCCH, University of Horticultural Sciences, Arabhavi, Bagalkot, India.
2 Department of Fruit Science, KRCCH, University of Horticultural Sciences, Arabhavi, Bagalkot, India.
3 Food Process Engineering, DSLD CHEFT, Devihosur, Haveri, India.
4 Centre for Biotechnology Research, Department of BCI, College of Horticulture, University of Horticultural Sciences, Bagalkot, Karnataka, India.
5 Genetics & Plant Breeding, ICAR JSS – Krishi Vigyan Kendra, Suttur, Mysore, Karnataka, India.
* Corresponding author: monapatelpgatti@gmail.com

DOI: 10.1201/9781003584100-6

decline in malnutrition rates among children under the age of five. That is, stunting has fallen from 38.4% to 35.5%, wasting from 21.0% to 19.3%, and underweight prevalence has fallen from 35.8% to 32.1%. Furthermore, the malnutrition rates among women between the ages of 15 and 49 have also dropped, from 22.9% to 18.7% (Sethi et al., 2020).

Historically, human diets consisted of a diverse range of food sources. However, in recent times, there has been a significant shift toward diets dominated by animal-based products, refined fats, oils, sugar, and alcohol, particularly in urbanized and economically developed societies. This dietary transition is rapidly spreading to urbanizing nations, contributing to an increased risk of noncommunicable diseases such as obesity, diabetes, hypertension, cardiovascular diseases, and stroke. Additionally, these dietary patterns raise concerns about higher resource consumption, disparities in access to nutritious food, and challenges in food security (Pandey et al., 2020). The rising consumption of processed foods has also been linked to a decline in gut microbiota diversity, which plays a crucial role in digestion, nutrient bioavailability, and overall assimilation (Requena, Martínez-Cuesta, and Peláez 2018). An intake rich in key biomolecules—carbohydrates, amino acids, vitamins, proteins, lipids, and minerals—plays a vital role in maintaining the level of energy and boosting overall well-being. Additionally, nutraceutical foods also have other health benefits, especially in the prevention and management of different diseases. Epidemiologic research shows that the consumption of nine servings a day of fruits and vegetables, or 400 to 600 grams, is associated with a reduced occurrence of common cancers and degenerative illnesses (Heber and Bowerman 2001; Liu 2013). In spite of the association between body mass index, caloric intake, and patterns of eating, food availability and accessibility improvements may not always come with improved nutritional status (Samaddar et al., 2020). To counter these challenges, this chapter focuses on policies that seek to improve the nutritional value of crops, increase bioavailability and bioaccumulation of nutrients, and reduce malnutrition. It also examines current literature on food crop quality improvement and considers the power of conventional and novel breeding methods to guarantee nutritional quality in crops. Finally, the chapter ventures into various methods for enhancing the nutritional content of food crops, as illustrated in Figure 1.

## 2. Diverse Food and Landraces

Landraces are critical genetic resources that support the sustainability of traditional agro-ecosystems, making a valuable contribution to food security, nutrition, and livelihoods in the long term (Puneeth et al., 2024). In order to have a stable, sustainable food supply, it is essential to evolve dietary practices, reduce food loss and waste, and assist in the diversification of food systems (Walter et al., 2019; Díaz et al., 2019). Recent studies highlight the potential of underexploited species and wild relatives of domesticated crops as ecologically sustainable and climate-resilient food resources. Plant genetic resources (PGR) are key to agricultural development and world food security (Brown 1987). The continued landrace cultivation by farming communities has played a key role in preserving the genetic makeup of crops, with these landraces commonly being described as folk varieties, farmer's varieties, heirloom cultivars, heritage crops, and primitive cultivars (Pironon et al., 2019; Maxted, Hunter, and Ríos 2020).

Indigenous food systems, deeply rooted in cultural traditions, spiritual values, and ecological knowledge, are critical to promoting sustainable diets and strengthening global food security. These systems are embedded in diverse environments and are characterized by the management of biodiverse landscapes that yield a wide range of wild and cultivated foods (Kuhnlein and Chotiboriboon, 2022). Such foods are often nutrient-rich, locally adapted, and resilient to climatic stressors, contributing significantly to dietary diversity and environmental sustainability. Despite their central role in conserving agrobiodiversity and sustaining ecosystems, indigenous communities frequently face threats including climate variability, socio-economic marginalization, and health disparities. Recognizing, valuing, and safeguarding indigenous food knowledge is essential not only for enhancing the resilience and sovereignty of these communities but also for informing

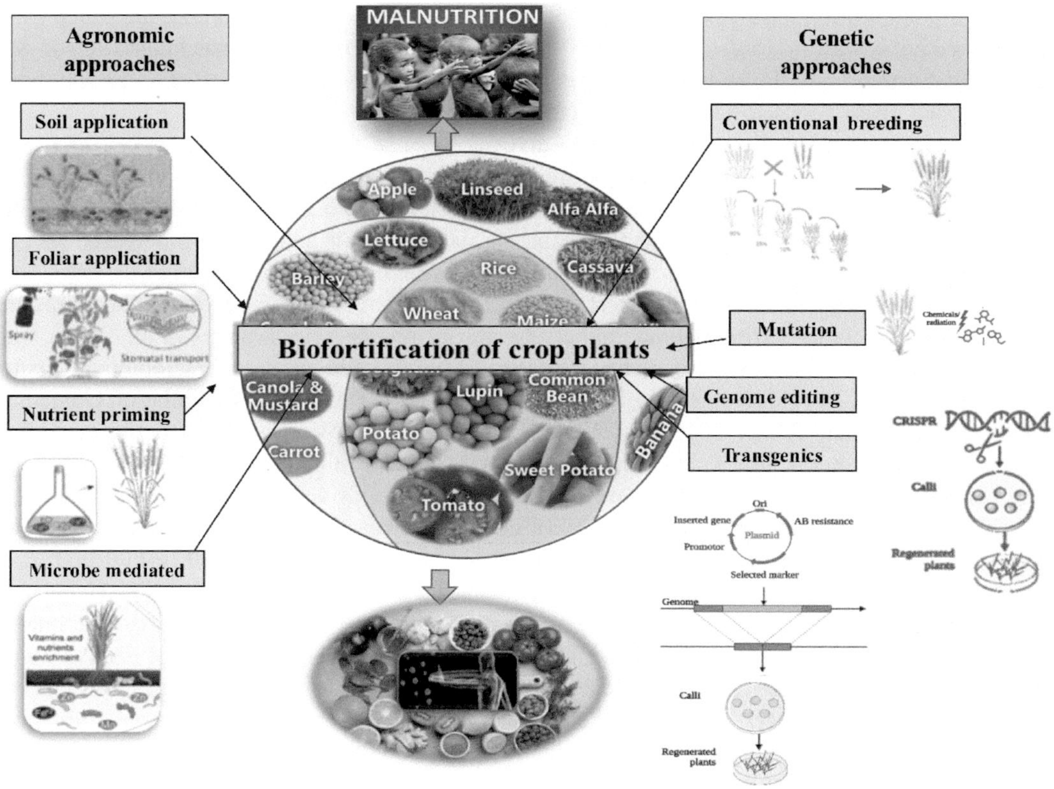


**Fig. 1** Agronomic and genetic approaches in crop biofortification.

broader efforts to create sustainable, equitable, and culturally relevant food systems at a global scale (Holdridge et al., 2020). Indigenous communities residing in biodiverse regions often possess extensive traditional ecological knowledge related to indigenous foods, which are frequently rich in essential micronutrients such as calcium, iron, zinc, and various vitamins. Despite this, many communities experience low dietary diversity and inadequate nutrient intake (Ghosh-Jerath et al., 2020). Factors influencing indigenous foods consumption include taste preferences, seasonal availability, and cultural significance, while barriers such as climate variability and increased access to commercial foods limit their use.

Wild edible plants (WEPs), though underutilized, contribute to both household nutrition and income generation. Strengthening the use of indigenous foods presents a valuable opportunity to enhance nutritional status and address malnutrition in these populations (Mishra et al., 2021). Extensive documentation highlights the use of diverse WFPs, such as leafy greens, fruits, tubers and mushrooms, which contribute significantly to dietary diversity and household income, especially in remote and resource-constrained areas (Abdullah et al., 2021). Although WFPs may not replace staple crops, they are important sources of essential micronutrients and serve to complement macronutrient intake in local diets. The traditional knowledge systems that govern the identification, harvesting, and use of WFPs are deeply embedded in indigenous cultures but are increasingly at risk due to modernization and shifting food systems (Campbell et al., 2021). Preserving this knowledge and integrating WFPs into national and global food security and biodiversity conservation strategies is essential for fostering sustainable, resilient, and inclusive food systems (Scott, Carolan, and Long, 2024).

Comprehensive knowledge of macronutrients, water- and fat-soluble vitamins, carotenoids, minerals, and other essential dietary components is critical to ensuring nutritional adequacy. While balanced diets should include a variety of nutrient-dense foods, certain wild plant species contain anti-nutritional compounds that can hinder nutrient absorption and bioavailability. These compounds—such as oxalates, phytates, cyanogenic glycosides, nitrates, and phenolic substances—vary widely in concentration across plant species and can pose health risks if consumed without proper processing. For example, cassava contains high levels of cyanogenic glycosides, which release cyanide, a potent respiratory toxin (Ogbadoyi et al., 2011). Food processing methods including soaking, germination, boiling, autoclaving, fermentation, and biotechnological interventions have been shown to reduce or eliminate these compounds effectively (Soni et al., 2022). Continued research is essential to develop and optimize techniques that target heat-stable anti-nutrients while preserving the nutritional integrity of food sources.

## 3. Crop Nutrient Enhancement through Agronomic and Genetic Approaches

**Agronomic approaches**

- **Soil application:** Application of mineral fertilizers directly to the soil to improve nutrient availability and uptake efficiency in crops.
- **Foliar application:** Spraying nutrient-enriched solutions on plant foliage to enhance nutrient absorption via stomatal and cuticular pathways.
- **Nutrient priming:** Pre-treatment of seeds by soaking in nutrient-enriched solutions prior to sowing to stimulate early growth and improve nutrient density.
- **Microbe-mediated biofortification:** Harnessing beneficial soil microorganisms (e.g., mycorrhizae, rhizobacteria) to enhance nutrient solubilization, bioavailability, and uptake in plants.

**Genetic approaches**

- **Conventional breeding:** Crossbreeding of high-nutrient crop varieties to develop cultivars with enhanced micronutrient profiles.
- **Mutation breeding:** Inducing genetic mutations through chemical agents or radiation to generate nutritionally superior crop genotypes.
- **Genome editing:** Precise modification of genes involved in nutrient biosynthesis and accumulation using tools such as CRISPR/Cas systems.
- **Transgenic approaches:** Introduction of foreign genes from other species to improve nutritional traits; a key example is Golden Rice, engineered to synthesize β-carotene in rice endosperm.

## 4. Agronomic and Physiological Practices

The nutritional and nutraceutical value of crops is determined most importantly by the occurrence of key vitamins, trace elements, free amino acids, proteins, carbohydrates, oils, antioxidants, and bioactive substances, which include peptides, phenolic compounds, terpenoids, and pigments. Incorporation of a number of different farming practices, such as precision mineral nutrition, biofortification, environmental changes, and use of biochemical and microbiological elicitors, can improve these quality characteristics significantly. Optimization of cultivation conditions, accurate fertilizer control, and use of natural or synthetic elicitors, with the consideration of climatic and soil parameters, are some of the strategies that largely contribute to enhancing crop quality (Cabrera-De la Fuente et al., 2018).

Agronomic biofortification, especially by means of using micronutrient-fortified fertilizers, is a key method in enhancing crop nutrient contents and, in the process, human health outcomes (Cakmak and Kutman 2018). This method focuses on maximizing the use of fertilizer application and mineral mobilization to increase the concentration of primary micronutrients in edible plant parts. Whereas macronutrients such as nitrogen, phosphorus, and potassium were the pillars of the Green Revolution, the challenge in the modern era is reducing micronutrient deficiencies. The traditional soil application tends to cause micronutrient fixation, hence making foliar application a more effective option. Some of the successes in this field include zinc fertilization programs in Turkey and selenium biofortification programs in Finland. Agronomic biofortification stands out as an economical, low-cost approach that is applicable to a wide variety of crops and provides a real-world alternative to genetic modification technologies, which are usually crop-specific and resource-exhaustive (Bhardwaj et al., 2022).

This approach has been useful for iron and zinc (Fe and Zn, respectively) amelioration in the staple crops of rice and wheat. Foliar applications of Fe resulted in a dramatic boost in Fe concentrations of rice grains, while combining Fe with urea increases Fe levels in wheat (Yuan et al., 2013; Aciksoz et al., 2011). Likewise, Zn foliar sprays, especially complementary to soils, have increased rice Zn content more effectively while improving some anti-nutritional factors such as phytic acid (Sangeetha et al., 2022). The implementation of Zn-fortified Nitrogen: Phosphorous: Pottassium (NPK) fertilizers in Turkey and selenium biofortification initiatives in Finland has shown promising results. Also, there is evidence that biofertilizers, mycorrhizal fungi, and plant-growth-promoting rhizobacteria improve nutrient uptake and crop productivity (Ramesh, Choimes, and Schachtman, 2004; de Andrade et al., 2023).

Biofortification using plant-growth-stimulating *rhizobacteria*, that is, *Azospirillum brasilense* and *Azotobacter vinelandii*, improved the nutritional composition of canola oil by enriching oleic acid and proteins (Nosheen, Bano, and Ullah, 2011). Absorption of selenium (Se) in mustard and potato plants was increased using foliar treatments as well as applying rhizospheric bacteria. Particularly, ZnO, $ZnSO_4$, and selenite/selenate formulations have effectively enhanced the levels of zinc (Zn) and selenium in potatoes (El-Ramady et al., 2020). In addition, biofortification interventions have also shown promising results in other crops such as tomatoes, quinoa, broccoli, and carrots, where foliar sprays of iron (Fe), zinc (Zn), and selenium (Se) have significantly enhanced their nutritional value (Lata-Tenesaca et al., 2023; Rahim et al., 2020).

## 4.1 Application Techniques in Agronomic Biofortification

Applying micronutrients to soil is crucial for replenishing essential nutrients; however, it is often less efficient and cost-effective, potentially leading to nutrient accumulation over time. Among the various application techniques, banding is more efficient than broadcasting. Foliar application, regarded as the most effective method, minimizes nutrient losses and enhances nutrient absorption, thereby increasing the concentrations of Zn, Fe, and Mn in grains. This approach is particularly advantageous in arid regions and alkaline soils (Dhaliwal et al., 2022).

Nutri-priming or seed priming consists of soaking seeds in nutrient-rich solutions to enhance germination, root development, and yield. Particularly, zinc nutri-priming using $ZnSO_4$ is found to significantly increase the concentration of zinc in the grains of chickpeas and wheat. Not only is this method economical, but it is also environmentally friendly since farmers are not economically burdened while trying to improve the nutrient content. Besides, nano-priming has shown to improve seed performance through enhanced nutrient and water absorption (Veena and Puthur, 2022). The effectiveness of priming techniques is dictated by the type of crop and environmental factors. However, these techniques are possibly not well known by the farmers and might shorten the shelf life of the seeds (Sharma and Ram, 2024).

Biofertilizers which consist of microbial inoculants are largely self-sufficient and inexpensive. These organisms, including cyanobacteria and *Bacillus aryabhattai*, assist in the biofortification of corn, wheat, and soybean crops by enhancing their zinc bioavailability. In addition, *achlorophyllous* microorganisms further increase the supplies of bioavailable iron, zinc, copper as well as other vital micronutrients such as feeders for plants (Dhiman et al., 2023).

Microgreens include the juvenile developmental stage of a variety of plants, including grains, vegetables, beans, and herbs, and they are usually cut between a week and four weeks after sprouting. These plants at such a tender age are rich in bioactive constituents, thus bestowing an array of health benefits such as being antioxidative, anti-inflammatory, and anticancer. Their health benefits can be improved even more by the use of different growing techniques such as hydroponics, nanoparticle application, and other solutions that are designed to be rich in nutrients because they improve the absorption of essential elements such as iron, zinc, and selenium. Agronomic practices based on the application of fertilizers such as NPK and microbial inoculants comprising Bacillus and *Rhizobium* greatly increase the intake of nutrients and enhance the growth of the plants. Also, priming seeds with cold plasma and incorporating nanofortification with nanoparticles, ZnO as well as FeO, augmented the concentration of nutrients, along with enhanced germination rates and better-quality plants (Gupta et al., 2023).

# 5. Genetic Approaches

## *5.1 Conventional Breeding Approaches*

Plant breeding, which is the scientific field for domesticating wild plant species, is still making strides by enhancing plant genotypes to increase productivity and climate change resilience. The aim of biofortification through plant breeding is to use genetic variation between close crop species to increase the concentration and bioavailability of vital minerals in crops (Sheoran et al., 2022; Dhaliwal et al., 2022). Priority crops for biofortification are major cereals, pulses, oilseeds, fruits, vegetables, and legumes, where efforts are concentrated on enhancing contents of essential nutrients such as zinc (Zn), iron (Fe), magnesium (Mg), selenium (Se), folic acid, carotenoids, and vitamin A (Poletti and Sautter, 2005).

Plant breeders can effectively utilize germplasm from primary, secondary, and tertiary gene pools to detect key genes required for the development of biofortified crop varieties (Jha and Warkentin 2020). A lot of research has been done to evaluate genetic diversity for micronutrient content (Boy et al., 2017; Dutta et al., 2020; Govindaraj et al., 2020). To this point, traditional plant breeding methods have been able to generate 290 different biofortified crop varieties, which are grown in more than 30 nations, whereas over 60 countries are undertaking the assessment of many varieties for formal release (Ofori et al., 2022; Dhaliwal et al., 2022). A familiar illustration of successful traditional breeding is quality protein maize (QPM) which is universally embraced by farmers (Sheoran et al., 2022; Singh et al., 2016).

A plant species genetic diversity among cultivars is an important factor that governs its ability to take up and use nutrients. A better understanding of the genetic diversity underlying micronutrient concentrations would facilitate the breeding of plant genotypes with higher yields and increased micronutrient levels. The genetic basis for micronutrient enrichment has been studied in a range of crops including cassava (*Manihot esculenta* Crantz), wheat (*Triticum aestivum* L. and *Triticum durum* Desf.), rice (*Oryza sativa* L.), maize (*Zea mays* L.), beans (*Phaseolus vulgaris* L.), and sweet potatoes (*Ipomoea batatas* L.) (Graham, Welch, and Bouis 2001). Conventional biofortification programs have substantially boosted the value of β-carotene in orange-fleshed sweet potatoes by ranging from $<2$ mg $L^{-1}$ in landraces to 30–100 mg $L^{-1}$ (Dhaliwal et al., 2022). Additionally, researchers have used genome-wide association studies (GWAS) to pinpoint specific genomic regions in maize that are linked to higher levels of zinc and iron content (Hindu et al., 2018).

Minerals-deficient soil for the cultivation of food crops is also a significant factor for micronutrient deficiencies, which is causing nutritional stress in human populations (Gaikwad et al., 2020). Plant breeding for biofortification remains one of the most cost-effective and sustainable ways of reducing malnutrition. A variety of staple crops with improved micronutrient content have been developed through conventional breeding techniques. Genetic biofortification is the choice and crossing of crop plants that produces the desired combinations of genes enriched with nutrients that are readily bioavailable and have high absorption efficiency. Plant genetic biofortification strategies aim to increase adoptive elements in edible parts so they can be part of the daily diet, and hence can contribute to improving nutrition in the general population and combat malnutrition, but in recent years, new directions have been combined with conventional breeding approaches, such as speed breeding, genomic editing, transgenic approaches, and integrated omics, all of which further effort to improve crop nutritional profiles.

## 5.2 Mutation

Mutation breeding focuses on enhancing traits that have not been naturally selected or improved through traditional plant breeding methods. While most induced mutations tend to be harmful, applying precise selection techniques enables the identification and recovery of beneficial mutants. From 170 plant species in more than 60 nations, more than 3,000 mutant varieties have been formally released. Mutation induction is particularly advantageous for vegetatively propagated crops, where genetic diversity is limited, as it facilitates the development of valuable genetic variants. Several mutant genes have contributed to improving the nutritional profile of commercially cultivated crops. For example, varieties of QPM are extremely advantageous for human and animal nutrition since they have about double the amounts of lysine and tryptophan as their parent lines. Recent advancements in crop biofortification have led to the identification of mutant rice and cassava varieties with improved nutritional profiles. In rice, five giant embryo mutants demonstrated significantly elevated levels of crude lipids, proteins, essential amino acids, vitamins B1, B2, and E, as well as key minerals compared to conventional varieties (Zhang et al., 2007). Additional studies have identified giant embryo rice lines, such as TNG71-GE, with enhanced concentrations of γ-oryzanol and tocols, compounds known for their antioxidant properties (Jeng et al., 2012). In cassava, a mutant with small starch granules and high amylose content was developed, offering distinct starch characteristics suitable for diverse industrial and nutritional applications (Ceballos et al., 2008).

The Targeting Induced Local Lesions in Genomes (TILLING) approach is a breakthrough in mutant breeding that has transformed the detection and use of gene variations for important agronomic characteristics. Induced mutations in crops such as melons have increased shelf life and encouraged unisexual flower formation, which has increased productivity in India (Bhattacharya et al., 2023).

The bioavailability, absorption, and overall nutritional composition of nutrients are essential aspects of crop-quality assessment. Numerous crops include antinutritional substances that impede the digestion and absorption of essential minerals and proteins. Phytic acid, gossypol, goitrogens, glucosinolates, lectins, oxalic acid, saponins, raffinose, tannins, enzyme inhibitors, alkaloids, β-N-oxalyl amino alanine (BOAA), and hydrogen cyanide (HCN) are integral to plant metabolic pathways and are often associated with crucial growth regulation mechanisms (Purohit et al., 2023).

By identifying both spontaneous and induced mutations in staple crops, Pramitha et al., (2021) have shown that significant progress has been achieved in reducing phytic acid levels in donor crops. Each of the three low phytic acid (lpa) mutants in maize (Ragi et al., 2022) had unique traits associated with phytic acid reduction. Because of a mutation that affected the early biosynthesis route of phytic acid involving myoinositol, the lpa1 mutant showed reduced levels of phytic acid with little buildup of myoinositol phosphates. On the other hand, the lpa2 and lpa3 mutants accumulated myoinositol phosphate and myoinositol intermediates, respectively, and showed decreased phytic acid concentration (Ramakrishna et al., 2024).

## 5.3 Transgenics/Molecular Breeding

Compared to traditional breeding procedures, which usually take more than ten years to produce economically viable cultivars with certain desirable features, transgenic and molecular breeding approaches have made substantial progress. Conventional breeding takes a long time because it requires several generations of crossings and backcrosses to include quantitative trait loci (QTLs) linked to the desired traits. Advanced genetic technologies have made it possible for some crop species to collect important genetic resources that are essential for creating superior variations. However, the time-intensive nature of traditional breeding remains a major challenge, restricting the ability to respond swiftly to the dynamic demands of global agriculture (Rojas-Vásquez and Gatica-Arias, 2020).

To improve food security, scientists can modify crop genomes through genetic modification to address nutritional inadequacies. The primary objective of transgenic breeding is to boost crops micronutrient content. To increase iron levels, for example, wheat has been genetically modified to overexpress genes such OsNAS2 and TaFer1-A, which greatly increases iron accumulation (Borg et al., 2012). Likewise, rice has been biofortified by introducing genes such as CrtI and PSY to increase vitamin A levels (Shahbaz et al., 2020). The nutritional profile of staple crops is improved by these changes, which also aid in reducing deficits in susceptible groups.

Transgenic breeding seeks to maximize the protein and amino acid content of crops in addition to micronutrient enrichment. Tyrosine and lysine, two important amino acids, have been efficiently boosted since the Ama1 gene was introduced. Furthermore, to increase their protein content, legumes such as soybeans and chickpeas have undergone genetic modification (Kaur et al., 2020). This tactic is especially pertinent to communities whose diets don't contain enough sources of protein.

The goal of biofortification is also to lessen anti-nutrients, which obstruct the absorption of vital minerals. Phytic acid levels have been successfully reduced by the use of gene silencing methods, increasing the bioavailability of iron and zinc. For instance, it has been shown that inhibiting the phyA gene can improve the nutritional value of a variety of cereal grains (Aggarwal et al., 2018).

Transgenic crops have made significant strides by incorporating several genes to increase their nutritious content. The addition of genes such as PSY and ferritin variations to rice has raised its iron and vitamin A, $B_1$, and $B_6$ levels (Roychoudhury and Bhowmik, 2020). Similar genetic changes have been made to maize by introducing bacterial genes to increase the grain's vitamin and amino acid content. Increased vitamin and mineral concentrations have also been achieved in sorghum and barley by targeted gene overexpression (Garg et al., 2018).

Genetic changes have been used to improve the nutritional value of soybeans and chickpeas. The addition of NAS2 and Glycine max ferritin genes to chickpeas has increased their iron bioavailability. While the addition of maize zein protein has increased the quantities of both cysteine and methionine, the cystathionine γ-synthase gene has been crucial in raising methionine levels in soybeans (Kumar et al., 2018). Moreover, the elevation of the dihydrodipicolinic acid synthase and aspartokinase genes has led to a larger lysine content, while the O-acetylserine sulfhydrylase gene has raised cysteine levels (Shahzad et al., 2021). To lower α-linolenic acid levels and encourage its conversion into the more stable γ-linolenic acid, recent studies additionally inhibit the ω-3 FAD3 gene and target pro-vitamin A enhancement through the carotene desaturase gene (Prabavathy et al., 2022). Additionally, the synthesis of isoflavones in soybeans has been greatly enhanced by the activation of the maize C1 and R transcription factors (Sohn et al., 2021). While lupins have benefited from better amino acid profiles via genes derived from sunflowers, common beans have demonstrated improved methionine content with the insertion of proteins derived from Brazil nuts (Jha et al., 2022).

Targeting certain genes and metabolic pathways, transgenic breeding has been widely used to increase the nutritional value of a variety of crops. While the ferritin (FER1) gene has been targeted to improve iron storage in lettuce (Mir et al., 2020), the inositol pentakisphosphate 2-kinase (IPK1) gene has been modified to raise iron levels in common beans (Silva et al., 2021). The quantity of folate in potatoes has increased by the use of genes such as FPGS and HPPK/DHPS (Roychoudhury 2021). Provitamin A levels in sweet potatoes have also been demonstrated to rise with the insertion of the CrtB, CrtI, and CrtY genes (Singh and Singh, 2020). The Orange (Or) gene affects the amount of β-carotene that accumulates in cauliflower, whereas the PSY gene has been utilized to raise the amount of carotenoid.

Numerous genes have been used to boost crops with vitamins and minerals. Genes such as TaFer1-A, GmFER, OsNAS21, and NAM-B1 have been shown to increase iron levels in wheat (Malik and Maqbool, 2020). Genes such as CrtI, bacterial PSY, and CrtB have been used to enrich vitamin A (Wang et al., 2014), while OsNAS2, PhyA, and NAM-B11 have been used to increase the zinc content of rice and other crops (Malik and Maqbool, 2020). Genes such as AtNAS1, OsFer2, AtIRT1, Pvferritin, Afphytase, and OSIRT1 have been shown to increase iron bioavailability in rice, whereas IDS31, OsNAS3, OsNAS1, OsNAS2, and HVNAAT-A have been shown to increase zinc levels (Kawakami and Bhullar, 2018). *Aspergillus phytase*, *Aspergillus phy2*, and *GmFER* have been used to increase iron accumulation in maize (Garg et al., 2018), while *Zmpsy1* has been used to target β-carotene levels (Decourcelle et al., 2015) (Table 1).

**Table 1** Nutrients and specific genes associated with the biofortification of crops.

| *Crop* | *Targeted Genes* | *Nutrient* | *References* |
|---|---|---|---|
| Wheat | TaFer1-A, GmFER, OsNAS21, NAM-B1 | Iron | Malik and Maqbool (2020) |
| Wheat | Crtl, Bacterial PSY, CrtB | Vitamin A | Wang et al. (2014) |
| Wheat | OsNAS2, PhyA, NAM-B11 | Zinc | Malik and Maqbool (2020) |
| Rice | AtNAS1, OsFer2, AtIRT1, Pvferritin, Afphytase, OSIRT1 | Iron | Kawakami and Bhullar (2018) |
| Rice | IDS31, OsNAS3, OsNAS1, OsNAS2, HVNAAT-A | Zinc | Kawakami and Bhullar (2018) |
| Rice | Carotene desaturase, daffodil PSY | Vitamin A | Trijatmiko et al. (2016) |
| Maize | GmFER, Aspergillus phytase, Aspergillus phy2 | Iron | Garg et al. (2018) |
| Maize | crtI, crtB | Vitamin A | Wang et al. (2014) |
| Maize | Zmpsy 1 | β-carotene | Decourcelle et al. (2015) |
| Soybean | Phytase | Iron & Zinc | Malik and Maqbool (2020) |
| Soybean | Carotene desaturase, bacterial PSY, crtB, crtW, bkt1 | Vitamin A | Kim et al. (2012) |
| Sorghum | PSY-1, PMI, CRT-I, LPA-1 | Vitamin A | Lipkie et al. (2013) |
| Chickpea | GmFER, NAS2 | Iron | Tan (2018) |
| Barley | Phytase, AtZIP1 | Iron | Malik and Maqbool (2020) |
| Barley | Zn transporter gene, DHPS | Zinc | Garg et al. (2018) |
| Barley | HvCs1F | β-glucans | Burton et al. (2011) |
| Common bean | Inositol pentakisphosphate 2-kinase (IPK1) | Iron | Silva et al. (2021) |
| Lettuce | Ferritin (FER1) | Iron | Mir et al. (2020) |
| Potato | HPPK/DHPS and/or FPGS | Folate | Roychoudhury (2021) |
| Sweet potato | CrtB, CrtI, and CrtY (lycopene beta-cyclase) | Provitamin A | Singh and Singh (2020) |
| Cauliflower | Orange (Or) gene | β-carotene | Singh and Singh (2020) |
| Apple | PSY gene | Carotenoid | Ampomah-Dwamena et al. (2015) |

Enhancing nutrient bioavailability has also been accomplished through the use of transgenic techniques. In comparison to non-transgenic types, β-carotene bioaccessibility has been improved in cassava, sweet potato, melon, sorghum, and potato. Processing techniques have had a big impact on these advancements. In potatoes, for example, bacterial biosynthetic pathways have increased the synthesis of β-carotene by 3,600 times (Lipkie et al., 2013). The amounts of zeaxanthin have significantly increased due to the inhibition of certain genes, such as ZEP, in potatoes. Tomato nutritional levels, especially those of β-carotene and lycopene, have also been improved via metabolic engineering (Diretto et al., 2007).

Transgenic techniques have been developed to maximize the absorption and storage of vital minerals including iron (Fe) and zinc (Zn), in addition to improving β-carotene and folate (Das et al., 2020). The iron content of crops has considerably risen due to the overexpression of genes involved in Fe absorption. Furthermore, *Chlamydomonas reinhardtii* genes have been successfully incorporated into sweet potatoes and cassava to increase iron buildup in their edible parts (Sheoran et al., 2022). These developments show how genetic engineering may be used to biofortify foods lacking in certain micronutrients, providing a viable way to improve nutrition worldwide.

## 5.4 Genome Editing/Silencing

Genome editing tools such as zinc finger nucleases (ZFNs), transcription activator-like effector nucleases (TALENs), meganucleases (MNs), and CRISPR/Cas9 (clustered regularly interspaced short palindromic repeats) have transformed agricultural breeding by making it possible to make precise gene edits (Bhagtaney and Sundarrajan, 2023). These tools cause specific double-stranded breaks (DSBs) in DNA, which are then repaired by either nonhomologous end-joining (NHEJ) or homology-directed repair (HDR). While NHEJ tends to produce random insertions or deletions, causing gene knockouts, HDR permits more accurate insertions or replacement of genes (Svitashev et al., 2020). The use of genome editing has promoted advances in crop characteristics, for example, the elimination of anti-nutrients and improvement in nutritional content, which play a role in biofortification and food security. These technologies are currently being employed to develop crops with enhanced disease resistance, greater resilience to environmental stressors, and improved nutritional profiles, such as soybeans with higher oleic acid content and tomatoes enriched with gamma-aminobutyric acid (GABA) (Siddique, 2022).

By modulating gene expression and modifying metabolic pathways, genome editing allows for targeted trait enhancement while maintaining other desirable agronomic characteristics. Of these technologies, the CRISPR/Cas system has been the most popular because of its high accuracy, affordability, and ease of use relative to ZFNs and TALENs. The system uses a guide RNA and Cas9 nuclease to target and edit DNA sequences specifically. There are also mutants like dead Cas9 (dCas9) that enable gene control and base editing without creating DSBs. On the other hand, base editors, such as cytosine base editors (CBEs) and adenine base editors (ABEs), facilitate programmed nucleotide substitutions. CRISPR-driven genome editing has shown great success in various crops, including wheat, which has enhanced iron content (TaVIT2), soybeans that have increased carotenoid levels (GmPDS11, GmPDS18), maize that reduces phytic acid (ZmIPK1A, ZmIPK, ZmMRP4), and cassava, which boosts carotenoid content (MePDS) (Kadam et al., 2023).

Genome editing has made huge strides in the breeding of fruit crops, especially with the fine-tuning of CRISPR/Cas9, which allows for more precise edits across various fruit species (Montecillo et al., 2020). This progress has enabled the creation of transgene-free modifications in apples and has also made it possible to introduce genetic changes in cucumbers, wild strawberries, and watermelons through backcrossing (Eckerstorfer et al., 2020). Ensuring the sustained transfer of desirable changes through the germline is still a major difficulty, though, especially for woody species that are propagated vegetatively, such as fruit trees. In these species, self-incompatibility and the need for outcrossing make it difficult to achieve homozygosity at the edited locus (Thapliyal et al., 2023).

Researchers have developed a number of CRISPR/Cas9 system derivatives to address these issues. This includes CRISPR-associated nucleases such as Cas12a/Cpf1, CRISPR base editors (both CBEs and ABEs), and CRISPR/Cas9 ribonucleoprotein (RNP) technology (Mushtaq et al., 2021). The unique feature of the CRISPR RNP technique is that it introduces purified RNPs straight into plant protoplasts, eliminating the need for transgene integration. This method has shown a lot of promise in increasing grapevine and apple disease resistance. Additionally, watermelon has successfully been made herbicide resistant through the use of CRISPR base editing (Hussain et al., 2021).

## 5.5 QTL Mapping

QTLs play a key role in enhancing crop biofortification. They help us pinpoint and map the specific genomic regions linked to various agronomic traits. This is especially important in the realm of genetic biofortification, as it addresses malnutrition by increasing the micronutrient levels in our staple crops. For instance, in wheat, researchers have discovered 21 new QTLs that are linked to grain iron, zinc, protein content, and thousand kernel weight, spread across 13 chromosomes. Notably, stable QTLs such as QGFe.IARI-7D.1 and QGPC.IARI-7D.2, along with some pleiotropic regions on chromosome 7D, are looking like great candidates for marker-assisted selection (MAS). Plus, eleven recombinant inbred lines (RILs) with promising QTL combinations have been spotted as potential donor lines for breeding wheat varieties that pack a nutritional punch (Krishnappa et al., 2021). In addition, seven QTLs with significant linkage to grain protein content (PC) have been located, three of which are new QTLs-qPC3.1, qPC5.1, and qPC9.1. The confirmation of four reported QTLs (qPC3 and qPC8) has established their importance, especially qPC3.1, qPC5.1, qPC9.1, and qPC12.1, as important genetic tools for enhancing protein content and yield in rice breeding programs (Pradhan et al., 2019).

In the context of anthocyanin biosynthesis, which contributes to red, blue, and purple pigmentation in vegetables such as cauliflower, broccoli, and cabbage, various QTLs have been linked to these traits. In cauliflower, the BoMYB2 gene plays a key role in anthocyanin biosynthesis through the MYB-bHLH-WD40 complex, with cyanidin-3-(6-p-coumaryl)-sophoroside-5-glucoside identified as essential for purple pigmentation (Yu et al., 2019). While several QTLs involved in color regulation have been mapped, challenges remain in ensuring stability during processing. Similarly, in broccoli, purple pigmentation is controlled by loci characterized by Yu et al., (2019) and is influenced by cold treatment. Genes such as BoPAL, BoDFR, and BoMYB114 are identified as key regulators of this trait. In cabbage, pigmentation is governed by structural genes, including CHS, F3H, and DFR, as well as hormone-regulating genes such as BoNCED2 and BoACS11, which contribute to anthocyanin biosynthesis. When discussing carrot pigmentation and sugar content, researchers have identified three primary alleles (Y, $Y_1$, and $Y_2$) that are essential for determining root color. For instance, orange carrots are known to have the genotype yyy2y2. Additionally, the P1 gene is responsible for the buildup of anthocyanins in purple carrots. On top of that, single nucleotide polymorphism (SNP) markers are closely associated with the Y2 gene, while other candidate genes such as ZDS2 and ZEP are linked to how β-carotene content is regulated (Singh et al., 2020).

## 5.6 Multi Omics Approach

Traditional breeding methods have done a great job of creating biofortified crop varieties (Satyavathi et al., 2021). However, they might not be enough to meet the growing nutritional needs of our ever-expanding global population. To enhance biofortification efforts, the integration of omics technologies is essential (Varshney et al., 2021). These approaches aim to evaluate germplasm for its micronutrient content and leverage GWAS to pinpoint molecular markers associated with nutritional traits. By using these markers, we can enhance specific cultivars through genomics-assisted breeding techniques or transgenic methods to boost their nutrient profiles (Chandra et al., 2020). It's crucial to integrate various omics fields—such as genomics, transcriptomics, proteomics, metabolomics, and ionomics—to develop crops that can tackle future nutritional challenges effectively.

Advancements in genomics have facilitated the establishment of various marker systems, with linkage mapping and GWAS being key tools in identifying genetic factors linked to nutrient content. Linkage mapping focuses on biparental populations to detect major genes associated with nutritional traits, whereas GWAS examines genetic variation across diverse panels to uncover novel genomic regions related to these traits (Roorkiwal et al., 2022). Together, these strategies contribute significantly to enhancing nutrient profiles through QTL mapping, GWAS, and MAS, reshaping crop improvement programs.

Breakthroughs in advanced sequencing tools have revolutionized transcriptome analysis, enabling the identification of genes involved in nutrient biosynthesis across different crop species (Mishra et al., 2019). For example, RNA sequencing (RNA-seq) has been utilized to analyze transcript abundance at various growth stages. A study on three sorghum cultivars identified more than 3,000 differentially expressed genes linked to nutritional attributes, revealing variations in carbohydrate, amino acid, and flavonoid metabolism (Zhou et al., 2022). Notably, the alpha-amylase 3 gene was found to be associated with differences in protein and fat content in sorghum grains, highlighting its potential for improving nutritional traits (Rhodes et al., 2017).

The integration of proteomics and metabolomics is crucial for characterizing biomolecules that influence crop nutrition. These methodologies facilitate the refinement of nutritional profiles in crops such as millets by identifying key biological pathways. For example, a non-targeted metabolomics analysis of 61 sorghum accessions successfully distinguished temperate and tropical varieties based on variations in phenolic acid and fatty acid composition (Ramalingam et al., 2021). Such insights are instrumental in developing biofortified sorghum varieties with enhanced health benefits.

Additionally, ionomics, which focuses on elemental profiling, is essential for quantifying mineral concentrations in crops (Huang and Salt, 2016). Analytical techniques such as atomic absorption spectroscopy (AAS) and inductively coupled plasma optical emission spectroscopy (ICP-OES) have been widely used to accurately measure grain micronutrient levels. Platforms such as the ionomic HUB provide valuable datasets for researchers working to improve mineral transport mechanisms in crops such as millets (Baxter et al., 2007). By leveraging these omics-based approaches, researchers can effectively develop biofortified crops, ultimately improving nutrient availability within food systems.

## 5.7 Pathway Engineering

Biofortification is a targeted strategy aimed at enhancing the micronutrients in crops through selective breeding approaches or genetic engineering, providing a sustainable promise to improving human nutrition, particularly in regions with limited dietary diversity. Unlike conventional interference such as food fortification and supplementation, biofortified crops offer a long-term and cost-effective approach, as they can be grown without the need for continuous external inputs. Advances in genetic engineering have significantly improved iron uptake and accumulation in various crops. For example, the introduction of the Naat-A (nicotianamine synthase) gene with 35S promoter in rice has been revealed to enhance iron efficiency by stimulating siderophore synthesis, thereby increasing iron bioavailability (Rao and Annadana, 2017). Similarly, the overexpression of zinc transporter proteins from Arabidopsis has been successfully employed to improve both zinc and iron absorption in transgenic barley (Ramesh, Choimes, and Schachtman, 2004). The incorporation of ferritin genes from soybean and *Phaseolus vulgaris*, using endosperm-specific promoters like glutelin, has resulted in a remarkable boost in iron levels in rice seeds up to twofold (Swamy et al., 2023).

Efforts to reduce anti-nutritional factors have also been explored, such as the expression of phytase genes in maize and wheat, which degrade phytates and thereby enhance iron bioavailability

(Sahu et al., 2024). Furthermore, the overexpression of metallothioneins and phytoferritin genes has shown great potential in enhancing absorption of iron and zinc, reflecting a multi-dimensional approach to increasing both nutrient content and bioavailability in staple crops.

Vitamin A is a crucial nutrient that supports vision and growth, and it comes from carotenoids such as β-carotene, which are mainly found in fruits and vegetables. When we eat these carotenoids, our bodies convert them into retinal acid, essential for seeing in low light, and retinoic acid, which is vital for cellular growth. Unfortunately, vitamin A deficiency is a significant issue, especially among children and pregnant women, leading to serious health problems like night blindness (Bhende, 2021). Genetic engineering has played a key role in tackling vitamin A deficiency by boosting the carotenoid biosynthesis pathway in staple crops. A notable example is Golden Rice, created by adding genes from daffodils and a bacterium to enhance the production of phytoene synthase (PSY) and carotene desaturase (crtI). This innovation allows the rice to accumulate β-carotene at levels reaching up to 14.5 μg/g in the endosperm (Nisar et al., 2015). Similarly, maize modified with the ZmPsy and crtI genes showed impressive β-carotene levels, ranging from 15 to 59 μg/g in optimized varieties (Bakshi et al., 2023). Other crops such as cassava and tomato have also seen promising results from genetic modifications. For example, introducing the crtB and AtDxs genes in cassava led to a 20- to 30-fold increase in β-carotene levels, while tomato plants engineered with crtB and crtL genes showed a threefold increase, which was further boosted by adding AtOr and CaFib genes (Beyene et al., 2018). However, some of these modifications have unintentionally reduced other important nutrients, such as chlorophyll and tocopherol, highlighting the need for ongoing refinement to ensure a well-rounded nutritional profile.

Folate (Vitamin $B_9$) biofortification focuses on increasing its accumulation in crops to address dietary deficiencies, particularly among vulnerable populations. Folate plays a crucial role in plant processes such as photorespiration and chlorophyll biosynthesis, and it is vital for human health in preventing neural tube defects and cardiovascular diseases (DellaPenna, 2007). Key enzymes involved in folate biosynthesis, including GTP cyclohydrolase I (GCHI) and aminodeoxychorismate synthase (ADCS), have been overexpressed in plants such as Arabidopsis, Mexican bean, and tomato, resulting in substantial increases in folate precursors and moderate improvements in folate levels. For example, the overexpression of GCHI in tomato plants led to a twofold increase in folate content, whereas the combined expression of GCHI and ADCS enhanced folate levels by up to 25-fold in ripe fruit (Díaz De La Garza et al., 2004).

Cobalamin (Vitamin $B_{12}$) biofortification poses a significant challenge, as this vitamin is synthesized exclusively by certain bacteria and archaea, making it virtually absent in plant-based foods. This deficiency presents dietary challenges for vegans, necessitating supplementation. Vitamin $B_{12}$ is crucial for DNA synthesis, neurotransmitter production, and fatty acid metabolism, and its deficiency can lead to serious health conditions such as pernicious anemia and cognitive decline. Due to the complexity of its biosynthesis—requiring approximately 30 enzymatic steps—direct genetic engineering of $B_{12}$ production in crops has largely been unsuccessful (Watanabe, 2007). However, some promising strategies have emerged, such as pathway engineering in plants such as Lepidium sativum (garden cress), where cultivation in $B_{12}$-enriched media enabled the accumulation of $B_{12}$ in vacuoles. Despite its effectiveness, this method relies on $B_{12}$ sourced from bacterial fermentation, posing challenges for large-scale application due to cost and resource constraints (Lawrence et al., 2018).

Additionally, L-ascorbic acid (Vitamin C) is a major objective for biofortification due to its essential role in collagen synthesis, immune function, and antioxidant activity. These biofortification strategies emphasize the importance of leveraging advanced agricultural and biotechnological approaches to enhance micronutrient content in crops, ultimately contributing to improved human health, and some of the important biofortified crop varieties are listed in Table 2.

**Table 2** Crop varieties improved for their nutritional value through a range of breeding methods.

| *Sl. No.* | *Crop (Variety)* | *Enriched Nutritional Traits* | *Region(s) Released* | *Breeding Method* |
|---|---|---|---|---|
| 1 | Rice (IR68144) | High iron content (up to 15 ppm) | Asia, Africa | Conventional breeding |
| 2 | Rice (Golden Rice—GR2E) | Provitamin A (β-carotene) | Philippines, Bangladesh | Transgenic |
| 3 | Rice (Zinco Rice MS) | Zinc content (27.4 ppm) | India | Conventional breeding |
| 4 | Wheat (Zinc Shakti, WB02) | Elevated zinc levels (up to 42 ppm) | India, Nepal | Conventional breeding |
| 5 | Maize (Obatanpa Plus) | High provitamin A content | Sub-Saharan Africa, Latin America | Conventional breeding |
| 6 | Pearl Millet (ICTP-8203 Fe—Dhanashakti) | Iron-rich (up to 84 ppm) and zinc-enriched (up to 46 ppm) | India | Conventional breeding |
| 7 | Cassava (UMUCASS 36, 37, 38) | High provitamin A (β-carotene) content | Nigeria | Conventional breeding |
| 8 | Sweet Potato (Kabode, Bhu Sona) | Elevated provitamin A levels | Africa, India | Conventional breeding |
| 9 | Beans (Manteca, BIO101, BIO107) | High iron and zinc content | Latin America, Africa | Conventional breeding |
| 10 | Lentil (Pusa Ageti Masoor) | Higher iron (65 ppm) | India | Conventional breeding |
| 11 | Banana (high provitamin A varieties) | Increased provitamin A content | Uganda | Transgenic |

## 6. Conclusions and Future Perspectives

The global challenge of ensuring that our food is nutritionally adequate, especially when it comes to essential micronutrients that support immune function, is a pressing issue we can't ignore. Biofortification and enhancing nutrient bioavailability in food crops have gained a lot of attention as promising solutions. A multidisciplinary approach is required to address micronutrient malnutrition, which includes actions such as the use of mineral fertilizers, the release of transgenic crops, and better plant breeding. Techniques such as gene overexpression, transgenes from wild relatives, and CRISPR/Cas9 gene editing open up new possibilities for the creation of crops with higher nutritional value. Still, fighting hidden hunger through biofortification involves not only the process of creating novel cultivars, it also calls for the penetration of a wide sociopolitical and economic reality in order for the products to be accepted and consumed. Rolling out successful programs of biofortification faces great challenges: biofortified variety shortages, large costs of production, lack of public knowledge and perception, deficient channels of distribution, and manifold political and administrative obstacles. However, through strategic planning, strong implementation, and adequate regulatory frameworks, biofortified crops can contribute remarkably to better health outcomes for hundreds of millions of underprivileged communities in India at a minimal research and development cost.

## References

Abdullah, Abdullah, Shujaul Mulk Khan, Andrea Pieroni, Aminul Haq, Zahoor Ul Haq, Zeeshan Ahmad, Shazia Sakhi, et al., 2021. "A Comprehensive Appraisal of the Wild Food Plants and Food System of Tribal Cultures in the Hindu Kush Mountain Range: A Way Forward for Balancing Human Nutrition and Food Security." *Sustainability (Switzerland)* 13 (9). https://doi.org/10.3390/su13095258.

Aciksoz, Seher Bahar, Atilla Yazici, Levent Ozturk, and Ismail Cakmak. 2011. "Biofortification of Wheat with Iron through Soil and Foliar Application of Nitrogen and Iron Fertilizers." *Plant and Soil* 349 (1–2): 215–225. https://doi.org/10.1007/s11104-011-0863-2.

Aggarwal, Sipla, Anil Kumar, Kaushal K. Bhati, Gazaldeep Kaur, Vishnu Shukla, Siddharth Tiwari, and Ajay K. Pandey. 2018. "RNAi-Mediated Downregulation of Inositol Pentakisphosphate Kinase (IPK1) in Wheat Grains Decreases Phytic Acid Levels and Increases Fe and Zn Accumulation." *Frontiers in Plant Science* 9. https://doi.org/10.3389/fpls.2018.00259.

Ampomah-Dwamena, Charles, Nicky Driedonks, David Lewis, Maria Shumskaya, Xiuyin Chen, Eleanore T. Wurtzel, Richard V. Espley, and Andrew C. Allan. 2015. "The Phytoene Synthase Gene Family of Apple (Malus x Domestica) and Its Role in Controlling Fruit Carotenoid Content." *BMC Plant Biology* 15 (1). https://doi.org/10.1186/s12870-015-0573-7.

Andrade, Luana Alves de, Carlos Henrique Barbosa Santos, Edvan Teciano Frezarin, Luziane Ramos Sales, and Everlon Cid Rigobelo. 2023. "Plant Growth-Promoting Rhizobacteria for Sustainable Agricultural Production." *Microorganisms* 11 (4). https://doi.org/10.3390/microorganisms11041088.

Bakshi, Suman, Suchita Kamble, Sanjay J. Jambhulkar, Pradeep Bhati, and Uttam Kumar. 2023. "Transgenic Approaches for Nutritional Quality Improvement of Cereals and Pseudocereals." *Developing Sustainable and Health-Promoting Cereals and Pseudocereals: Conventional and Molecular Breeding*, 285–313. https://doi.org/10.1016/B978-0-323-90566-4.00004-7.

Baxter, Ivan, Mourad Ouzzani, Seza Orcun, Brad Kennedy, Shrinivas S. Jandhyala, and David E. Salt. 2007. "Purdue Ionomics Information Management System. An Integrated Functional Genomics Platform." *Plant Physiology* 143 (2): 600–611. https://doi.org/10.1104/pp.106.092528.

Beyene, Getu, Felix R. Solomon, Raj D. Chauhan, Eliana Gaitán-Solis, Narayanan Narayanan, Jackson Gehan, Dimuth Siritunga, et al., 2018. "Provitamin A Biofortification of Cassava Enhances Shelf Life but Reduces Dry Matter Content of Storage Roots Due to Altered Carbon Partitioning into Starch." *Plant Biotechnology Journal* 16 (6): 1186–1200. https://doi.org/10.1111/pbi.12862.

Bhagtaney, Lekha, and Priya Sundarrajan. 2023. "An Overview of Tools for Genome Editing: ZFNs, Mega Nucleases, and Talens." *CRISPR/Cas-Mediated Genome Editing in Plants*, 37–64. https://doi.org/10.1201/9781003331759-3.

Bhardwaj, Ajay Kumar, Sukirtee Chejara, Kapil Malik, Raj Kumar, Ashwani Kumar, and Rajender Kumar Yadav. 2022. "Agronomic Biofortification of Food Crops: An Emerging Opportunity for Global Food and Nutritional Security." *Frontiers in Plant Science* 13. https://doi.org/10.3389/fpls.2022.1055278.

Bhattacharya, Anjanabha, Vilas Parkhi, Bhavesh Palan, and Bharat Char. 2023. "Mutagenesis and TILLING in the Era of Precise Genome Editing." *TILLING and Eco-TILLING for Crop Improvement*, 1–34. https://doi.org/10.1007/978-981-99-2722-7_1.

Bhende, B.B. 2021. "Vitamin A: A Review Article." *Asian Journal of Nursing Education and Research* 11 (4): 582–584. https://doi.org/10.52711/2349-2996.2021.00135.

Borg, Søren, Henrik Brinch-Pedersen, Birgitte Tauris, Lene Heegaard Madsen, Behrooz Darbani, Shahin Noeparvar, and Preben Bach Holm. 2012. "Wheat Ferritins: Improving the Iron Content of the Wheat Grain." *Journal of Cereal Science* 56 (2): 204–213. https://doi.org/10.1016/j.jcs.2012.03.005.

Boy, E., J.D. Haas, N. Petry, C.I. Cercamondi, J.B. Gahutu, S. Mehta, J.L. Finkelstein and R.F. Hurrell. 2017. "Efficacy of Iron-Biofortified Crops." *African Journal of Food, Agriculture, Nutrition and Development* 17 (2): 11879–11892. https://doi.org/10.18697/ajfand.HarvestPlus03.

Brown, A.H.D. 1987. "Plant Genetic Resources: An Introduction to Their Conservation and Use." *Trends in Genetics* 3:82. https://doi.org/10.1016/0168-9525(87)90181-8.

Burton, R.A., Collins, H.M., Kibble, N.A., Smith, J.A., Shirley, N.J., Jobling, S.A. and Fincher, G.B. 2011. "Over-Expression of Specific HvCslF Cellulose Synthase-like Genes in Transgenic Barley Increases the Levels of Cell Wall (1,3)." *Plant Biotechnol Journal* 9 (2): 117–135.

Cabrera-De la Fuente, Marcelino, Susana González-Morales, Antonio Juárez-Maldonado, Paola Leija-Martínez, and Adalberto Benavides-Mendoza, 2018. "Plant Nutrition and Agronomic Management to Obtain Crops With Better Nutritional and Nutraceutical Quality." *Therapeutic Foods*, 99–140. https://doi.org/10.1016/b978-0-12-811517-6.00004-0.

Cakmak, I. and U.B. Kutman. 2018. "Agronomic Biofortification of Cereals with Zinc: A Review." *European Journal of Soil Science* 69 (1): 172–180. https://doi.org/10.1111/ejss.12437.

Campbell, Donovan, Alex A. Moulton, David Barker, Tashana Malcolm, Lance Scott, Adrian Spence, Jhannel Tomlinson and Tiffany Wallace, 2021. "Wild Food Harvest, Food Security, and Biodiversity Conservation in Jamaica: A Case Study of the Millbank Farming Region." *Frontiers in Sustainable Food Systems* 5. https://doi.org/10.3389/fsufs.2021.663863.

Ceballos, Hernán, Teresa Sánchez, Kay Denyer, Adriana P. Tofiño, Elvia A. Rosero, Dominique Dufour, Alison Smith, Nelson Morante, Juan C. Pérez and Brendan Fahy, 2008. "Induction and Identification of a Small-Granule, High-Amylose Mutant in Cassava (Manihot Esculenta Crantz)." *Journal of Agricultural and Food Chemistry* 56 (16): 7215–7222. https://doi.org/10.1021/jf800603p.

Chandra, Ajay Kumar, Amarjeet Kumar, Alka Bharati, Rini Joshi, Aparna Agrawal and Sumit Kumar. 2020. "Microbial-Assisted and Genomic-Assisted Breeding: A Two Way Approach for the Improvement of Nutritional Quality Traits in Agricultural Crops." *3 Biotech* 10 (1). https://doi.org/10.1007/s13205-019-1994-z.

Das, Priyanka, Sanghamitra Adak and Arun Lahiri Majumder. 2020. "Genetic Manipulation for Improved Nutritional Quality in Rice." *Frontiers in Genetics* 11. https://doi.org/10.3389/fgene.2020.00776.

Decourcelle, Mathilde, Laura Perez-Fons, Sylvain Baulande, Sabine Steiger, Linhdavanh Couvelard, Sonia Hem, Changfu Zhu, et al., 2015. "Combined Transcript, Proteome, and Metabolite Analysis of Transgenic Maize Seeds Engineered for Enhanced Carotenoid Synthesis Reveals Pleotropic Effects in Core Metabolism." *Journal of Experimental Botany* 66 (11): 3141–3150. https://doi.org/10.1093/jxb/erv120.

DellaPenna, Dean. 2007. "Biofortification of Plant-Based Food: Enhancing Folate Levels by Metabolic Engineering." *Proceedings of the National Academy of Sciences of the United States of America* 104 (10): 3675–3676. https://doi.org/10.1073/pnas.0700640104.

Dhaliwal, Salwinder Singh, Vivek Sharma, Arvind Kumar Shukla, Vibha Verma, Manmeet Kaur, Yashbir Singh Shivay, Shahida Nisar, et al., 2022. "Biofortification—A Frontier Novel Approach to Enrich Micronutrients in Field Crops to Encounter the Nutritional Security." *Molecules* 27 (4). https://doi.org/10.3390/molecules27041340.

Dhiman, Karuna, Deepka Sharma, Reena Kumari and Parul Tomar. 2023. "Biofortification of Crops Using Microbes – A Promising Sustainable Agriculture Strategy." *Journal of Plant Nutrition* 46 (12): 2912–2935. https://doi.org/10.1080/01904167.2022.2160755.

Díaz De La Garza, Rocío, Eoin P. Quinlivan, Sebastian M.J. Klaus, Gilles J.C. Basset, Jesse F. Gregory and Andrew D. Hanson. 2004. "Folate Biofortification in Tomatoes by Engineering the Pteridine Branch of Folate Synthesis." *Proceedings of the National Academy of Sciences of the United States of America* 101 (38): 13720–13725. https://doi.org/10.1073/pnas.0404208101.

Díaz, Sandra, Josef Settele, Eduardo S. Brondízio, Hien T. Ngo, John Agard, Almut Arneth, Patricia Balvanera, et al., 2019. "Pervasive Human-Driven Decline of Life on Earth Points to the Need for Transformative Change." *Science* 366 (6471). https://doi.org/10.1126/science.aax3100.

Diretto, Gianfranco, Salim Al-Babili, Raffaela Tavazza, Velia Papacchioli, Peter Beyer and Giovanni Giuliano. 2007. "Metabolic Engineering of Potato Carotenoid Content through Tuber-Specific Overexpression of a Bacterial Mini-Pathway." *PLoS ONE* 2 (4). https://doi.org/10.1371/journal.pone.0000350.

Dutta, Suman, Vignesh Muthusamy, Firoz Hossain, Aanchal Baveja, Rashmi Chhabra, Sunil K. Jha, Devendra K. Yadava and Rajkumar U. Zunjare. 2020. "Analysis of Genetic Variability for Retention of Kernel Carotenoids in Sub-Tropically Adapted Biofortified Maize under Different Storage Conditions." *Journal of Cereal Science* 93. https://doi.org/10.1016/j.jcs.2020.102987.

Eckerstorfer, Michael, Marion Dolezel, Anita Greiter, Marianne Miklau, Andreas Heissenberger and Ricarda A. Steinbrecher. 2020. "Risk Assessment of Plants Developed by New Genetic Modification Techniques (NGMs): Biosafety Considerations for Plants Developed by Genome Editing and Other New Genetic Modification Techniques (NGMs) and Considerations for Their Regulation: Final Report O." *BfN-Skripten* 592:204.

El-Ramady, H, S.E.D. Faizy, N. Abdalla, H. Taha, D.S. Eva, et al. 2020. "Selenium and Nano-Selenium Biofortification for Human Health." https://edoc.hu-berlin.de/handle/18452/25606%0Ahttps://edoc.hu-berlin.de/bitstream/handle/18452/25606/soilsystems-04-00057-v2.pdf?sequence=4.

Gaikwad, Kiran B., Sushma Rani, Manjeet Kumar, Vikas Gupta, Prashanth H. Babu, Naresh Kumar Bainsla and Rajbir Yadav. 2020. "Enhancing the Nutritional Quality of Major Food Crops Through Conventional and Genomics-Assisted Breeding." *Frontiers in Nutrition* 7. https://doi.org/10.3389/fnut.2020.533453.

Garg, Monika, Natasha Sharma, Saloni Sharma, Payal Kapoor, Aman Kumar, Venkatesh Chunduri and Priya Arora. 2018. "Biofortified Crops Generated by Breeding, Agronomy, and Transgenic Approaches Are Improving Lives of Millions of People around the World." *Frontiers in Nutrition* 5. https://doi.org/10.3389/fnut.2018.00012.

Ghosh-Jerath, Suparna, Ridhima Kapoor, Archna Singh, Shauna Downs, Satabdi Barman and Jessica Fanzo. 2020. "Leveraging Traditional Ecological Knowledge and Access to Nutrient-Rich Indigenous Foods to Help Achieve SDG 2: An Analysis of the Indigenous Foods of Sauria Paharias, a Vulnerable Tribal Community in Jharkhand, India." *Frontiers in Nutrition* 7. https://doi.org/10.3389/fnut.2020.00061.

Govindaraj, Mahalingam, Kedar N. Rai, Anand Kanatti, Hari D. Upadhyaya, Harshad Shivade and Aluri S. Rao. 2020. "Exploring the Genetic Variability and Diversity of Pearl Millet Core Collection Germplasm for Grain Nutritional Traits Improvement." *Scientific Reports* 10 (1). https://doi.org/10.1038/s41598-020-77818-0.

Graham, Robin D., Ross M. Welch and Howarth E. Bouis. 2001. "Addressing Micronutrient Malnutrition through Enhancing the Nutritional Quality of Staple Foods: Principles, Perspectives and Knowledge Gaps." *Advances in Agronomy* 70:77–142. https://doi.org/10.1016/s0065-2113(01)70004-1.

Gupta, Astha, Tripti Sharma, Surendra Pratap Singh, Archana Bhardwaj, Deepti Srivastava and Rajendra Kumar. 2023. "Prospects of Microgreens as Budding Living Functional Food: Breeding and Biofortification through OMICS and Other Approaches for Nutritional Security." *Frontiers in Genetics* 14. https://doi.org/10.3389/fgene.2023.1053810.

Heber and Bowerman. 2001. "Applying Science to Changing Dietary Patterns." *Journal of Nutrition* 131 (11 SUPPL.): 3078S–3081S. http://0-ovidsp.ovid.com.wam.city.ac.uk/ovidweb.cgi?T=JS&PAGE=reference&D=emed5&NEWS=N&AN=2001408771.

Hindu, Vemuri, Natalia Palacios-Rojas, Raman Babu, Willy B. Suwarno, Zerka Rashid, Rayalcheruvu Usha, Gajanan R. Saykhedkar and Sudha K. Nair. 2018. "Identification and Validation of Genomic Regions Influencing Kernel Zinc and Iron in Maize." *Theoretical and Applied Genetics* 131 (7): 1443–1457. https://doi.org/10.1007/s00122-018-3089-3.

Holdridge, Genevieve A., Fausto O. Sarmiento, Suzanne E. Pilaar Birch, Bynum Boley, James K. Reap, Eric A. Macdonald, María Navarro, Sarah L. Hitchner and John W. Schelhas. 2020. "Feeding Futures Framed: Rediscovering Biocultural Diversity in Sustainable Foodscapes." *The Elgar Companion to Geography, Transdisciplinarity and Sustainability*, 235–251. https://doi.org/10.4337/9781786430106.00023.

Huang, Xin Yuan and David E.E. Salt. 2016. "Plant Ionomics: From Elemental Profiling to Environmental Adaptation." *Molecular Plant* 9 (6): 787–797. https://doi.org/10.1016/j.molp.2016.05.003.

Hussain, Amjad, Xiao Ding, Muna Alariqi, Hakim Manghwar, Fengjiao Hui, Yapei Li, Junqi Cheng, Chenglin Wu, Jinlin Cao, and Shuangxia Jin. 2021. "Herbicide Resistance: Another Hot Agronomic Trait for Plant Genome Editing." *Plants* 10 (4). https://doi.org/10.3390/plants10040621.

Jeng, Toong Long, Yi Ju Shih, Pei Tzu Ho, Chia Chi Lai, Yu Wen Lin, Chang Sheng Wang and Jih Min Sung. 2012. "γ-Oryzanol, Tocol and Mineral Compositions in Different Grain Fractions of Giant Embryo Rice Mutants." *Journal of the Science of Food and Agriculture* 92 (7): 1468–1474. https://doi.org/10.1002/jsfa.4728.

Jha, Ambuj B. and Thomas D. Warkentin. 2020. "Biofortification of Pulse Crops: Status and Future Perspectives." *Plants* 9 (1). https://doi.org/10.3390/plants9010073.

Jha, Uday C., Harsh Nayyar, Swarup K. Parida, Rupesh Deshmukh, Eric J.B. von Wettberg and Kadambot H.M. Siddique. 2022. "Ensuring Global Food Security by Improving Protein Content in Major Grain Legumes Using Breeding and 'Omics' Tools." *International Journal of Molecular Sciences* 23 (14). https://doi.org/10.3390/ijms23147710.

Kadam, Shaila, Dongho Lee and Pallavi Dhiman. 2023. "Genome-Editing Approaches for Biofortification of Cereal Crops." *Biofortification in Cereals: Progress and Prospects*, 93–126. https://doi.org/10.1007/978-981-19-4308-9_4.

Kaur S., Kumari A., Singh P., Kaur L., Sharma N. and Garg M. 2020. "Biofortification in Pulses. In: Tilak Raj S, Rupesh D, Humira S, Editors," Advances in Agri-Food Biotechnology." *Advances in Agri-Food Biotechnology.* New York, NY: Springer.

Kawakami, Yuta and Navreet K. Bhullar. 2018. "Molecular Processes in Iron and Zinc Homeostasis and Their Modulation for Biofortification in Rice." *Journal of Integrative Plant Biology* 60 (12): 1181–1198. https://doi.org/10.1111/jipb.12751.

Kim, Mi Jin, Jae Kwang Kim, Hye Jeong Kim, Jung Hun Pak, Jai Heon Lee, Doh Hoon Kim, Hong Kyu Choi, et al. 2012. "Genetic Modification of the Soybean to Enhance the β-Carotene Content through Seed-Specific Expression." *PLoS ONE* 7 (10): e48287. https://doi.org/10.1371/journal.pone.0048287.

Krishnappa, Gopalareddy, Nagenahalli Dharmegowda Rathan, Deepmala Sehgal, Arvind Kumar Ahlawat, Santosh Kumar Singh, Sumit Kumar Singh, Ram Bihari Shukla, et al. 2021. "Identification of Novel Genomic Regions for Biofortification Traits Using an SNP Marker-Enriched Linkage Map in Wheat (Triticum Aestivum L.)." *Frontiers in Nutrition* 8. https://doi.org/10.3389/fnut.2021.669444.

Kuhnlein, Harriet V. and Sinee Chotiboriboon. 2022. "Why and How to Strengthen Indigenous Peoples' Food Systems With Examples From Two Unique Indigenous Communities." *Frontiers in Sustainable Food Systems* 6. https://doi.org/10.3389/fsufs.2022.808670.

Kumar, Manoj, Mohd Aslam Yusuf, Manisha Nigam, and Manoj Kumar. 2018. "An Update on Genetic Modification of Chickpea for Increased Yield and Stress Tolerance." *Molecular Biotechnology* 60 (8): 651–663. https://doi.org/10.1007/s12033-018-0096-1.

Lata-Tenesaca, Luis Felipe, Renato de Mello Prado, Gabriela Eugenia Ajila-Celi, Dalila Lopes da Silva, José Sidnaldo Pinzetta Junior and Ben Hur Mattiuz. 2023. "Iron Biofortification in Quinoa: Effect of Iron Application Methods on Nutritional Quality, Anti-Nutrient Composition, and Grain Productivity." *Food Chemistry* 404. https://doi.org/10.1016/j.foodchem.2022.134573.

Lawrence, Andrew D., Emi Nemoto-Smith, Evelyne Deery, Joseph A. Baker, Susanne Schroeder, David G. Brown, Jennifer M.A. Tullet, et al. 2018. "Construction of Fluorescent Analogs to Follow the Uptake and Distribution of Cobalamin (Vitamin B 12 ) in Bacteria, Worms, and Plants." *Cell Chemical Biology* 25 (8): 941-951.e6. https://doi.org/10.1016/j.chembiol.2018.04.012.

Lipkie, Tristan E, Fabiana F De Moura, Zuo-Yu Zhao, Marc C Albertsen, Ping Che, Kimberly Glassman and Mario G Ferruzzi. 2013. "Bioaccessibility of Carotenoids from Transgenic Provitamin A Biofortified Sorghum." *The FASEB Journal* 27 (S1). https://doi.org/10.1096/fasebj.27.1_supplement.638.20.

Liu, Rui Hai. 2013. "Health-Promoting Components of Fruits and Vegetables in the Diet." *Advances in Nutrition* 4 (3). https://doi.org/10.3945/an.112.003517.

Malik, K.A. and Maqbool, A. 2020. "Transgenic Crops for Biofortification." *Front Sustain Food Syst.*

# CHAPTER 7

# Millets: An Emerging and Promising Climate-Resilient Nutri-Cereals of the Globe

Gami RA[1*] Kugashiya KG[2], Patil JV[3], Ganapathy KN[3] and Belaghihalli N Gnanesh[4]

## 1. Introduction

The globalization and industrialization of agricultural practices have significantly contributed to the increase in food supply. Nevertheless, these developments have also led to negative consequences, particularly the reduction of crop biodiversity, which poses a threat to the survival of minor crops. The continuous decline in the diversity of food crops is a potential threat to food security on a global scale. Currently, three cereals, namely, rice, wheat, and maize are responsible for major percent of the world's food supply. These crops are insufficient in essential amino acids and minerals that are critical for balanced diet, resulting in a lack of vital nutrients in human nutrition, notably vitamins and minerals including zinc (Zn), magnesium (Mg), and iron (Fe). The global issue of hidden hunger is intensifying, particularly among the populations of developing nations. What is the answer to this challenge? The adoption of alternative crops in agriculture is critical for safeguarding the global food supply, addressing hidden hunger, and increasing farmers' earnings. Millet crops are particularly important in this regard (Kumar et al., 2021 and Gami et al., 2023).

Millets are often referred to as "miracle crops" due to their wide array of benefits. These include their role as a source of food and value-added products, their use as forage, their contribution to agricultural diversity, their low nutrient requirements, enhanced carbon sequestration as $C_4$ plants, their capacity to mitigate erosion in arid areas, and their ability to ensure sufficient food and nutrition for smallholder farmers residing in challenging environmental conditions (Kumar and Singh 2015). Millets utilize the $C_4$ photosynthetic pathway, which benefits plants by enhancing their efficiency in using water and nutrients. This pathway also allows for more variable appropriation of biomass, decreases hydraulic conductivity per unit leaf area, supports improved growth, and boosts tolerance to high temperature (Sage and Zhu, 2011). They are well-suited for cultivation in harsh climates typical of arid and semi-arid areas, requiring minimal water compared to other cereal crops. Furthermore, they often manage to survive in poor soil conditions. Millets are exceptionally


1 ICAR-AICRP on Sorghum, Centre for Millets Research, S.D. Agricultural University, Deesa, Gujarat, India.
2 Centre for Millets Research, S.D. Agricultural University, Deesa, Gujarat, India.
3 ICAR-IIMR, Hyderabad, Telangana, India.
4 Genetics and Plant Breeding, ICAR JSS – Krishi Vigyan Kendra, Suttur, Mysore, Karnataka, India.
Email: kiran@sdau.edu.in[2]; drpatiljv@gmail.com; ganapathy@millets.res.in[3]; gnaneshbn@gmail.com[4]
* Corresponding author: ramangami@sdau.edu.in


DOI: 10.1201/9781003584100-7

nutritious and serve as a rich source of essential nutrients, with a wide range of applications in human consumption, animal feed, pharmaceuticals, and commercial starch manufacturing. Due to these significant attributes, they are frequently referred to as "climate-resilient crops," "miracle grains," or "crops of the future" (Paschapur et al., 2021 and Tripathi et al., 2021).

As hardy and climate-smart crops, millets are particularly well-suited for environments characterized by drought and high temperatures. They demonstrate adaptability to a variety of temperature and moisture conditions, serving as a vital source of food and animal feed for numerous dryland farmers, especially in developing nations. The primary crops that can grow in arid regions include sorghum and millets. A key feature of these crops is their remarkable resilience to both persistent and sporadic droughts, which are often caused by insufficient or unpredictable rainfall. Notably, millets stand out as the sole cereal crop capable of flourishing in dry environments, needing merely 350–400 mm of rainfall each year (Reddy, 2018).

Historically, millets have been a crop of choice during periods of famine risk. In low-rainfall regions, they yield a modest but more consistent harvest compared to other crops. Small millets are classified as coarse grains and serve as a food source in circumstances where other grains are either difficult to cultivate or expensive to purchase. Consequently, these millets have predominantly been consumed by impoverished and marginalized communities. The tough outer seed coat and distinct flavor of these millets contribute to their limited appeal among consumers of wheat and rice (Malleshi, 1989).

## *1.1 Groups of Millets*

Millets are primarily categorized into two distinct groups: (A) "major millets," namely, pearl millet [*Pennisetum glaucum* (L.)] and sorghum [*Sorghum bicolor* (L.)]; (B) "minor or small millets," namely, foxtail millet [*Setaria italica* (L.) Beauv.], finger millet [*Eleusine coracana* (L.) Gaertn.], barnyard millet (*Echinochloa* spp.), proso millet [*Panicum miliaceum* (L.)], little millet [*Panicum sumatrense* Roth ex. Roem. and Schult.], and kodo millet [*Paspalum scrobiculatum* (L.)], which are cultivated globally. The list of major and minor millets with their botanical names, chromosome numbers, ploidy level, and exclusive features is provided in Table 1.

**Table 1** List of millets cultivated in the world with their basic information.

| *English Name* | *Botanical Name* | *Ch. No. (Ploidy)* | *Origin* | *Important Features* |
|---|---|---|---|---|
| **(A) Major millets** | | | | |
| Pearl millet | *Pennisetum glaucum* (L.) R. Br. | 14 (2x) | Ethiopia | Grown in arid and semi-arid areas, this plant serves as a significant source of folic acid and exhibits tolerance to elevated temperatures |
| Sorghum | *Sorghum bicolor* (L.) | 20 (2x) | Africa | Tolerates high-temperature and moisture stress better than any other crops |
| **(B) Minor millets** | | | | |
| Finger millet | *Eleusine coracana* (L.) | 36 (4x) | East Africa | Wider adaptability, potential source of calcium |
| Foxtail/Italian millet | *Setaria italica* (L.) | 18 (2x) | Central Asia/ China | Short duration, adaptable to dry environments and soils with low fertility |
| Proso millet | *Panicum miliaceum* (L.) | 36 (4x) | Central Asia | Short duration, capable of withstanding heat and drought |

*Contd.*

**Table 1** *Contd.*

| *English Name* | *Botanical Name* | *Ch. No. (Ploidy)* | *Origin* | *Important Features* |
|---|---|---|---|---|
| Barnyard millet | *Echinochloa frumentacea* (L.) | 54 (6x) | Japan | Fastest growing habit, bulky fodder |
| Shama millet | *Echinochloa colona* (L.) | 54 (6x) | India | |
| Kodo millet | *Paspalum scrobiculatum* (L.) | 40 (4x) | India | Long duration, grown well in both shallow and deep soil, an abundant source of folic acid |
| Little millet | *Panicum sumatrense* (L.) | 36 (4x) | India (Eastern Ghat of India) | Short duration, withstands both waterlogging and drought |

Refrence: Rawat et al. (2021) and Paschapur et al. (2021)

## 2. Role of Climate on Global Food Security

As climate-resilient crops, millets demonstrate extensive adaptability, thriving in drought and heat stress conditions. These hardy cereals are predominantly found in the marginal environments of Africa and the arid and semi-arid tropical zones of Asia. They possess the ability to resist certain degrees of soil acidity and alkalinity, cope with moisture and temperature fluctuations, and adapt to varying soil fertility conditions. These crops are cultivated at elevations ranging from sea level up to 3000 meters, resulting in a variation of photoperiods from short to long days. Valued for their significance in human diets and as animal feed, these grains are primarily cultivated in areas with scarce rainfall, generally between 300 and 500 mm. In these conditions, crops such as rice and maize are prone to failure in the majority of years. Consequently, millet plays a crucial role in ensuring food security for impoverished rural populations residing in arid regions (Patil and Davada, 2023). Millets are often referred to as a poor man's crop due to their significant contributions to the diet of resource-limited farmers and consumers. These have often been called the coarse grains; their significant nutritional benefits have led to their reclassification as "nutria-millets" or "nutria-cereals." Millets are typically characterized as thermophilic, meaning they survive in warmer temperatures, and xerophilic, indicating their ability to reproduce with minimal water availability (Saxena et al., 2018). Millets demonstrate remarkable resilience to various abiotic stresses such as drought and heat. Their capacity to endure high temperatures and grow in drought-prone areas, especially in parts of hot summer, highlights their adaptability to climate change and their ability to counteract its detrimental impacts (Gupta et al., 2015). Millets demonstrate physiological sustainability in challenging environmental conditions, maintaining production levels due to their efficient use of water and nitrogen.

Climate change and the swift population rise pose significant threats to global food security (Lesk et al., 2016). In light of climate change, drought and extreme heat are prominent abiotic stressors that adversely affect crop yields and compromise global food security. In reaction to these challenges, plants have cultivated a range of adaptive strategies across morphological, physiological, and biochemical dimensions, facilitating their ability to either escape or acclimatize to harsh environmental conditions. It is important to note that even minimal exposure to heat and drought stress can detrimentally influence crop yields. Additionally, numerous studies have demonstrated that elevated temperatures and drought can decrease crop yields by as much as 50% (Lamaoui and Bekkaoui, 2016). The effects of climate change are expected to intensify the decline in agricultural productivity. According to the Intergovernmental Panel on Climate Change (IPCC) report, both air and ocean temperatures have risen, alongside an increase in greenhouse gas concentrations (IPCC

2014). These factors contribute to increased drought and heat stress and subsequently affect plant growth and crop yields (Bita and Gerats, 2013).

Cereal crops are extensively grown and serve as essential nutritional sources for the world's population, while also contributing notably to global warming. The carbon equivalent emissions associated with these crops are substantial, with wheat, rice, and maize producing 1000, 956, and 935 kg C/ha, respectively (Jain et al., 2016). On the other hand, the carbon footprints of minor cereal crops such as millets and sorghum are notably lower. This aspect is a fundamental reason that positions millets as a viable crop for contributing to the reduction of carbon emissions worldwide (Prasad and Staggenborg, 2009).

The distinctive morphological, physiological, and biochemical properties of millets make them an excellent and promising alternative crop to meet the challenges presented by climate change. Certain traits are evident in $C_4$ plants, which exhibit significantly enhanced photosynthetic efficiency. Additionally, the duration of their life cycle is notably short; for instance, millets complete their growth in approximately 12 to 14 weeks, whereas major cereals such as rice, wheat, and maize typically require an average of 20 to 24 weeks (Bandyopadhyay et al., 2017). The plant exhibits a short height, a limited leaf area with a high leaf area index, an upright leaf type characterized by efficient radiation use, thickened cell walls, and a dense root system (Li and Brutnell, 2011). There is a pronounced ability to generate higher dry matter yields alongside a robust tolerance to both drought and elevated temperatures. These crops are capable of surviving with limited nutrient inputs and reduced irrigation, demonstrating enhanced growth and productivity. Additionally, it can easily adjust to degraded soils that are saline, acidic, or affected by aluminium toxicity (Yadav and Rai, 2013). The root length of little millet tends to extend in response to stress conditions, particularly during periods of drought (Ajithkumar and Panneerselvam, 2014). At a minimum, they face vulnerabilities to biotic and abiotic stresses and have a lower dependence on synthetic fertilizers (Kole et al., 2015). The ability to adapt to a wide array of environmental factors leads to improved growth and productivity, especially in low-nutrient conditions, while minimizing the need for irrigation. Higher levels of biodiversity in millet cultivation can lead to a reduction in insect pests and climate risks, which in turn enhances the resilience of farmers (Talwar et al., 2019).

## 3. Response of Millets Against Various Stresses

### *3.1 Drought Stress*

In many countries, particularly those with arid and semi-arid climates, water scarcity is a prominent issue that leads to severe drought stress. Notably, the intensity and frequency of drought stress have risen in recent years (Tadele, 2016). Millets possess a deep root structure and an efficient water-usage mechanism, allowing them to tolerate prolonged dry conditions (Shrestha et al., 2023). Pearl millet and finger millet possess root systems that can penetrate the soil to depths of up to 3.6 meters and 2.7 meters, respectively, which are greater than the root depths found in many other crops. This adaptation enables them to access water that is found deep beneath the surface during periods of drought when surface water is scarce. A dense network of fine roots in the topsoil efficiently gathers and absorbs any moisture present. Additionally, some millet species have evolved specific adaptations, including hairy leaves and thicker cuticles, which help minimize water loss through transpiration (Paudel et al., 2024).

### *3.2 Heat Stress*

In general, when temperatures rise above the ideal level, the growth period of crops is shortened, resulting in less time for biomass accumulation, which ultimately leads to decreased yields of both grain and fodder. Research findings on stress suggest that alterations in climate variability, which are anticipated in a warmer climate, could have a more pronounced impact on yield than the changes in climate conditions. Additionally, the rising frequency of extreme climatic events during the more vulnerable stages of growth has been identified as a significant factor determining yield in hotter regions in the future (Yadav et al., 2016). Temperature serves as a critical environmental variable that influences reproductive processes. When high-temperature stress occurs during panicle development and flowering stages, it reduces floret fertility and seed-set percentage (Djanaguiraman et al., 2014; Djanaguiraman et al., 2018). Millets possess a deep-rooted system that enables them to access nutrient-rich soil layers that are inaccessible to many other crop species. This ability allows them to obtain vital minerals and nutrients from deeper soil strata, which are essential for their growth and development, particularly in the nutrient-deficient soils prevalent in hot and arid regions. The management of heat stress involves various mechanisms, including the stabilization of membranes, the scavenging of reactive oxygen species (ROS), the production of antioxidants, the regulation of osmotic solutes, and the synthesis of heat shock proteins (HSPs). Genetic improvements in heat tolerance are facilitated through both traditional and molecular breeding strategies (Yadav et al., 2016). The study of heat shock proteins in pearl millet examined their mechanisms, focusing on genotypic and developmental variations during seed germination (Howarth, 1994). It was found that millet seedlings were capable of producing the entire range of heat shock proteins after a two-hour exposure to a heat shock of 45°C. Additionally, the research revealed that the thermosensitivity of a genotype and the synthesis of heat shock proteins represent a complex interplay involving both transcriptional and translational regulation of RNA within the seed embryo.

### 3.3 *Salinity Stress*

The primary abiotic factor hindering the growth and productivity of millets is salinity. Furthermore, the rising global population necessitates researchers to create new and efficient approaches to increase crop production, ensuring food security in challenging environments. Several studies concerning millets have examined the critical role of mitigants and the threshold tolerance of these crops under salt stress conditions. Understanding the mechanisms of stress tolerance and implementing genetic manipulation in millets will further facilitate sustainable development efforts focused on enhancing crop performance in marginal and less productive areas. High levels of salinity create osmotic stress, thereby impeding plant growth. This phenomenon commonly results in the buildup of compatible osmolytes such as proline, sugars, and proteins in plants (Chapman et al., 2019). Considering the crucial role of mitigants in scavenging free radicals at biochemical and physiological levels during salinity stress, further exploration of their protective mechanisms may yield significant advancements in addressing the detrimental impacts on millet cultivation (Naveed et al., 2020).

### 3.4 *Waterlogging Stress*

The phenomenon of waterlogging, prevalent in regions with excessive rainfall, significantly stresses plants by affecting stomatal closure, disrupting photosynthesis, and modifying root architecture through the inhibition of gas diffusion. Approximately 16% of the yield in proso millet is lost as a consequence of waterlogging (Linkemer et al., 1998). Likewise, studies have indicated that waterlogging decreases finger millet yield, as it promotes an anaerobic metabolic process that is inefficient for adenosine triphosphate (ATP) production (Hossain and Uddin, 2011). Matsuura and colleagues conducted research in 2016 to explore how waterlogging affects proso millet, little millet, and foxtail millet. The results revealed that little millet demonstrates a higher level of tolerance to waterlogged conditions than the other types of millet (Matsuura et al., 2016, Choudhary et al., 2023).

### 3.5 *Biotic Stress*

Millets are less affected by major insect pests such as armyworms, shoot flies, and stem borers compared to other major cereals rice, maize, and wheat. This less vulnerability is largely a result of their waxy leaf structure, thick foliage, and various chemical defense mechanisms. The crop demonstrates an important resistance to various fungal diseases that commonly impact other cereals, including blast, smut, and mildew. For instance, pearl millet shows inherent resistance to leaf blast, a significant disease affecting rice. The elevated phenolic levels in finger millet also provide a natural defense against fungal infections. Their natural resilience makes them an ideal option for sustainable agricultural practices, particularly in regions characterized by limited resources or challenging climatic conditions (Paudel et al., 2024).

# 4. Nutritional Composition and Potential Health Benefits

## *4.1 Nutritional Profile*

- **Carbohydrates:** The carbohydrate composition of sorghum is made up of starch, soluble sugars, and various types of fiber, including pentosans, cellulose, and hemicellulose. Millet, on the other hand, categorizes its carbohydrates into nonstructural types, such as sugars, starch, and fructosans, and structural types, which consist of cellulose, hemicelluloses, and pectin. Starch serves as the main nonstructural carbohydrate (NSC) in sorghum, predominantly featuring amylopectin. Notably, certain mutants of sorghum produce either waxy starch, which is solely amylopectin, or starch that is rich in amylose (Boyer and Liu, 1983, Rao et al., 2017). The color of sorghum starches is influenced by the levels of pigments present in the pericarp and the leaves of the sorghum plant (Freeman and Watson, 1971, Subramanian et al., 1994, Rao et al., 2017).
- **Fiber (dietary fiber):** When compared to rice, millet crops are significantly higher in fiber content. Specifically, pearl millet contains 11.4 g of fiber per 100 g, while finger millet has 11.1 g per 100 g, making them the top contenders among millets. Fiber is critical for the effective operation of the gastrointestinal tract and helps to relieve constipation. The advantages of dietary fiber include promoting regular bowel movements and lowering blood cholesterol and sugar levels. A fiber-rich diet is crucial for preventing and managing conditions such as obesity, type 2 diabetes, heart disease, and cancer (Eshak et al., 2010). Additionally, its importance is highlighted by its effects on lowering blood sugar, lipids, and cholesterol levels. As a result, it contributes to the prevention of atherosclerosis (NIN, 2007).
- **Protein:** Millets are important sources of protein. The amount of protein present and its consumption can vary significantly due to agronomic conditions and environmental factors during grain development, along with the influence of genotype. The proteins found in millet are distributed as follows: 80% in the endosperm, 16% in the germ, and 3% in the pericarp (Taylor and Schussler, 1986). In terms of amino acid composition, millet exhibits a lower protein quality relative to other cereals. Lysine, an essential amino acid, serves as a limiting factor in millets. The germ is notably rich in albumins and globulins, while the endosperm contains kafirins and glutelins. Albumin, globulin, and glutelin fractions are particularly abundant in lysine and other essential amino acids. The storage proteins in maize, millets, and sorghum are categorized as prolamins, as they dissolve in alcohol–water mixtures. In terms of amino acid composition, the bran–germ material was notably high in lysine, arginine, and glycine, but it was deficient in other essential amino acids such as isoleucine, leucine, and phenylalanine relative to whole grain. Furthermore, the endosperm fractions exhibit a diverse range of total protein levels (Rao et al., 2017).
- **Fat:** Millets are characterized by their low-fat content, which includes both mono- and poly-unsaturated fatty acids. The fat levels in millet vary, with proso millet containing approximately 1.10% and pearl millet reaching up to 5.43%. Notably, pearl millet has a significant oil content, with about 4.2% of its composition being unsaturated fatty acids. This elevated fat content in pearl millet leads to a quicker rancidity of the flour, limiting its storage duration. Research indicates that the triglyceride profile in finger millet may lower the risk of developing duodenal ulcers (Gopalan et al., 2003, Kunyanga et al., 2013).

- **Mineral nutrients:** Millets serve as a valuable source of essential minerals, vitamins, and both micronutrients and macronutrients. The majority of millet varieties are particularly abundant in calcium (14–364 mg/100 g), iron (0.80–6.42 mg/100 g), zinc (1.40–2.76 mg/100 g), and phosphorus (101–289 mg/100 g). Notably, finger millet stands out as an exceptional source of calcium (>340 mg/100 g), magnesium (146 mg/100 g), and phosphorus (210 mg/100 g). Pearl millet is recognized for its high iron content (6.42 mg/100 g), along with significant levels of phosphorus (289 mg/100 g), zinc (2.76 mg/100 g), and magnesium (124 mg/100 g). Little millet also provides a substantial amount of iron (1.26 mg/100 g) and zinc (1.82 mg/100 g). Foxtail millet is known for its contributions of phosphorus and zinc, while barnyard millet is rich in iron (5.0 mg/100 g), zinc (3.0 mg/100 g), and phosphorus (280 mg/100 g). Furthermore, millets are excellent sources of B-complex vitamins, particularly thiamine ($B_1$) at 0.25–0.59 mg/100 g, riboflavin ($B_2$) at 0.05–0.28 mg/100 g, niacin ($B_3$) at 0.86–4.50 mg/100 g, and folic acid at 15.0–39.4 mg/100 g. Compared to other staple cereals such as rice, wheat, barley, and maize, pearl millet and barnyard millet exhibit higher iron content (5.0–6.42 mg/100 g) (NIN, 2007, Gopalan et al., 2016, Kumar et al., 2016, Kumar et al., 2021).
- **Anti-nutritional factors:** In addition to their essential nutrient profiles, millets contain various anti-nutritional compounds, such as phytates, tannins, polyphenols, and trypsin inhibitors. These anti-nutritional factors can diminish the bioavailability of minerals and may pose health risks. Since these compounds are predominantly found in the outer layers of the grains, their negative effects can be mitigated through various processing methods, both industrial and domestic, including milling, roasting, soaking, popping, puffing, malting, fermentation, and sprouting. Although milling reduces the overall nutrient content of millets, it can significantly enhance the bioavailability of certain nutrients. Research indicates that the germination of millet enhances the levels of methionine and cysteine. Additionally, the fermentation of finger millet has been shown to elevate net protein utilization (NPU), biological value (BV), and the concentrations of thiamine ($B_1$), riboflavin ($B_2$), and niacin ($B_3$). The bioavailability of iron is improved when weaning foods are made from roasted grains. Furthermore, soaking millet flour before cooking typically increases the availability of zinc. It has also been observed that popping finger millet can improve the digestibility of both protein and carbohydrates (Kumari and Sumathi 2002, Nkhata et al., 2018).

### 4.2 Health Benefits

Millet is a source of vital amino acids, including tryptophan, methionine, threonine, valine, and isoleucine. Isoleucine helps treat muscle pain, joint pain, blood formation, bone strength, and skin diseases. Valine helps to improve metabolism, build muscle, and repair body cells. In addition, amino acids maintain nitrogen content in the body. Most cereals are deficient in the functional amino acid called methionine, but it is present in millets, is beneficial in various functions of the body, removes excess fat from the body, and provides sulphur to the body, which is helpful for the production of glutathione that acts as a natural antioxidant for the body. Arginine and lysine actively release growth hormone (GH) (Gami et al., 2023). Due to their gluten-free nature, millets are a safe dietary option for those with gluten allergies and celiac disease. Furthermore, they are non-acidic, facilitating easier digestion. Studies indicate that communities that incorporate millets into their diets experience a reduced incidence of diabetes. (Saleh et al., 2013). The high fiber content of millet is a key factor in its diabetes-preventing properties. Moreover, specific antioxidant phenols in millets are believed to have beneficial effects against diabetes. Sorghum is characterized by a high concentration of phenolic compounds and antioxidants (Awika et al., 2004).

## 5. Genetic and Genomic Resources

Over the past few decades, traditional breeding methods have effectively produced biofortified varieties of millet (Rai et al. 2014; Satyavathi et al. 2021). Nevertheless, relying solely on these methods is inadequate to meet the rapidly growing population's future food and nutritional requirements. The development of biofortified varieties of sorghum and millet suitable for future climatic conditions can be greatly enhanced through the application of recent advancements in omics technologies (Varshney et al. 2021a). Additionally, significant genetic diversity exists regarding micronutrient concentrations within sorghum and millet germplasm; however, a comprehensive characterization of this variation remains to be conducted. Assessing the existing germplasm accessions for their micronutrient levels and identifying robust molecular markers through genome-wide association studies (GWAS) will aid in pinpointing quantitative trait loci (QTLs), candidate genes, and alleles that influence the nutritional traits of interest. The QTLs and genes identified can subsequently be integrated into superior cultivars using genomics-assisted breeding techniques or transgenic approaches to improve the micronutrient content of the grains (Varshney et al. 2021b). The introduction of high-throughput markers and multiplexed genotyping technologies has greatly facilitated marker-assisted breeding aimed at improving millets. In the future, gene editing techniques, including base editing and prime editing, are anticipated to offer promising avenues for enhancing millets' resilience against biotic and abiotic challenges. Such advancements may contribute to higher millet yields and reinforce food security in developing countries (Kumar et al., 2016; Krishna et al., 2022).

## 6. Processing and Value Addition

The processing can improve millet product quality and shelf life. The products have become more affordable, tasty, and attractive. Employment opportunities are increased through establishing large-scale industries, and foreign exchange can be earned through exports. With the help of various modern technologies such as agro-processing, biotechnology, and nanotechnology, millet produce can be transformed into innovative, more acceptable forms for consumers. Overall, advancements in technology present an opportunity to enrich millets with crucial nutrients and enhance the bioavailability of specific essential nutrients. This improvement not only elevates the nutritional quality of food but also enables farmers to command higher prices for their products, thereby increasing their income (Hamad 2012, Sood et al., 2015, Rao et al., 2011).

Conversely, consumers demonstrate a readiness to invest more in value-added products, as there is a growing demand for high-quality, nutritious millet items. This trend reflects an increasing awareness of health, with individuals aiming to prevent nutritional deficiencies and lifestyle-related illnesses. Due to increasing awareness, their preference toward millet consumption is rising daily. Thus, farmers are interested in reviving the cultivation of millet crops as they are getting enough affordable prices through increasing the value of millet. Millet processing is the traditional or nontraditional process of converting millet grain into an edible form with quality enhancement through various techniques. Various conventional processes, namely, dehulling, soaking, sprouting, roasting, popping, etc., have been traditionally performed in millets from ancient times to eliminate anti-nutritional factors and improve the taste in different millets. Among the various traditional techniques, the primary methods are grain soaking and decoration/dehulling (Gami et al., 2023, Kamaraddi and Shanthakumar, 2003).

Value addition refers to any process undertaken to enhance the worth of a product through both primary and secondary processing methods. In the context of millets, value addition encompasses activities such as destoning, grading, cleaning, and removing impurities from freshly harvested grains. The machinery employed in millet processing significantly alleviates the labor intensity and time required for these tasks. Dehulling is performed on minor millet varieties, while polishing is applied to major millets to create value-added products. Additionally, puffed and flaked millets are produced to generate revenue from underutilized marketable varieties, maximizing the potential of these grains (Hemasankari et al., 2023). Millet value addition converts millet grains into various food products using various techniques. Additional income can be generated by enhancing the value of millets through various products such as ready-to-eat foods, porridge, sweet items, papads, chapatis, bakery products, fermented dishes, and geographical location-specific dishes. Various processed and value-added millet products have been developed across a range of categories, including popped, flaked, puffed, extruded, roller-dried, fermented, and malted forms, as well as composite flours and weaning foods (Gami et al., 2023).

## 7. Major Challenges and Potential Benefits Associated with Millet Farming

Grain millets exhibit resilience to temperature variations and are rich in nutrients, positioning them as a viable solution to combat climate change and tackle food security challenges. Nevertheless, certain shortcomings warrant consideration. Specifically, millets tend to have lower yields, necessitating the introduction of high-yielding cultivars (Grovermann et al., 2018, Paudel et al., 2024). In world, a large number of individuals lack awareness regarding the health benefits of millets, which contributes to their low market demand. Additionally, the availability of millets in both traditional and modern retail environments, including online marketplaces, is limited, hindering consumer access. The processing of millets is fraught with challenges due to their grain size, shape, surface texture, hardness, and the adhesion between the husk and the grain.

Additionally, the same small millet crop can exhibit variations due to differences in varieties, cultivation methods, and microclimatic conditions in various production regions. Although pearl millet flour is noted for its superior nutritional content, its consumption is largely limited to specific regions globally, primarily because of its short shelf life and the likelihood of developing rancidity or unpleasant odors during storage (Rani et al., 2018, Paudel et al., 2024). A prevalent anti-nutritional element found in millet grains significantly hinders their complete utilization. Removing these anti-nutritional compounds from specific millet varieties before consumption is crucial, as their presence poses considerable health risks. Additionally, various undesirable quality characteristics, such as flour with a dark and dull appearance, elevated fiber content, extended cooking durations, a coarse and gritty texture, an astringent flavor, and a limited shelf life, further impede their use.

### *7.1 Opportunities*

1. Millets possess drought-resistant characteristics, hence are well-suited for cultivation in drought-affected regions (Hu et al., 2022, Jiang et al., 2023). Millets, which can thrive in diverse soils and climates, are an excellent crop choice for agricultural practitioners. They have a potential for higher dry matter yield and a strong resilience to drought and heat.
2. Millets have a rich nutrient profile, containing ample fiber, protein, vitamins, and minerals (Amadou et al., 2011, Chandrasekara and Shahidi, 2012).
3. Due to their natural gluten-free properties, they are appropriate for individuals with celiac disease or gluten sensitivity.
4. The processing of millet contributes to improved product quality and extended shelf life, leading to options that are more affordable, palatable, and visually appealing.

## 8. Conclusion

Millets represent a significant natural resource for humanity. $C_4$ crops demonstrate remarkable efficiency in utilizing water and nutrients for their growth. These crops exhibit a high degree of tolerance to elevated temperatures and, to a certain extent, to flooding conditions. Their ability to withstand salinity contributes to successful germination and robust seedling development. Due to their physiological adaptations that enable quick recovery from abiotic stresses such as drought and heat, millets are considered highly promising food sources in the context of climate change. During periods of climate variability, they often remain the only viable crop, making them an effective risk management strategy for farmers with limited resources. Millets can endure climate change's adverse effects, except for potential increases in ozone levels in urban areas.

Millet is an excellent source of nutrients, characterized by its sweet, nutty flavor, and is deemed one of the most digestible and nonallergenic grains. It stands out as one of the few grains that can alkalize the body. Unfortunately, many individuals are unaware of millet and its nutritional benefits. Nonetheless, millet is one of the best-kept secrets passed down from our ancestors. This small "grain" is gluten-free and loaded with vitamins and minerals. Although it is commonly referred to as a grain due to its appearance, it is technically classified as a seed. The advent of modern chemical analysis techniques has led to the incorporation of herbal medicine in the treatment of various diseases. Presently, millet crops are acknowledged for their substantial content of micronutrients, minerals, and phytochemicals, which are positively correlated with numerous health benefits such as antidiabetic, antioxidant, anticancer, antiaging, antihypertensive, and cardioprotective effects. The hypoglycemic properties and other beneficial characteristics of millet starch render it a valuable constituent of the food sector. As a result, the demand for millet in everyday diets has surged, reflecting an increase in public awareness regarding health and nutrition.

Value addition in millet entails converting millet grains into various food products utilizing different techniques. This enhancement can lead to increased income by producing items such as ready-to-eat foods, porridge, confections, papads, chapatis, bakery products, fermented dishes, and dishes that are specific to certain geographical areas.

The application of GWAS will enable the discovery of QTLs, candidate genes, and alleles that govern important nutritional traits. Once identified, these QTLs and genes can be introduced into high-performing cultivars using genomics-assisted breeding or transgenic methods to enhance the micronutrient content of grains. The genetic affinity of millets with other cereal grains, such as rice, wheat, and barley, enhances the possibility of introducing nutritional traits into these key cereal crops. Integrating diverse omics methodologies, including genomics, transcriptomics, proteomics, metabolomics, and ionomics, is critical for the future design of biofortified crops.

## References

Ajithkumar, I.P. and Panneerselvam, R. 2014. ROS scavenging system, osmotic maintenance, pigment and growth status of *Panicum sumatrense* Roth. under drought stress. Cell Biochem. Biophys. 68:587–595. doi: 10.1007/s12013-013-9746-x.

Amadou, I., Gbadamosi, O.S. and Guo-Wei, L. 2011. Millet-based traditional processed foods and beverages – a review. *Cereal Foods World* 56, 115–121.

Awika, J.M., Rooney, L.W. and Waniska, R.D. 2004. Anthocyanins from black sorghum and their antioxidant properties. *Food Chemistry* 90:293–301B.

Bandyopadhyay, T., Muthamilarasan M. and Prasad, M. 2017. Millets for next generation climate-smart agriculture. *Front Plant Sci* 8:1266.

Bita, C.E. and Gerats, T. 2013. Plant tolerance to high temperature in a changing environment: scientific fundamentals and production of heat stress-tolerant crops. *Frontiers in Plant Science* 4:273.

Boyer, C.D. and Liu, K.C. 1983. Starch and water-soluble polysaccharides from sugary endosperm of sorghum. *Phytochemistry* 22:2513–2515.

Chandrasekara, A. and Shahidi, F. 2012. Bioaccessibility and antioxidant potential of millet grain phenolics as affected by simulated in vitro digestion and microbial fermentation. *J. Funct. Foods* 4(1):226–237.

Chapman, J.M., Muhlemann, J.K., Gayomba, S.R. and Muday, G.K. 2019. RBOH-dependent ROS synthesis and ROS scavenging by plant specialized metabolites to modulate plant development and stress responses. *Chem. Res. Toxicol.* 32, 370–396.

Choudhary, P., Shukla, P. and Muthamilarasan, M. 2023. Genetic enhancement of climate-resilient traits in small millets: a review. *Comprehensive Reviews in Food Science and Food Safety* 12 (3):281–295.

Djanaguiraman, M., Perumal, R., Ciampitti, I.A., Gupta, S.K. and Prasad, P.V.V. 2018. Quantifying pearl millet response to high temperature stress: thresholds, sensitive stages, genetic variability and relative sensitivity of pollen and pistil. *Plant, Cell and Environment* 41, 993–1007.

Djanaguiraman, M., Prasad P.V.V., Murugan M., Perumal M. and Reddy U.K. 2014. Physiological differences among sorghum (*Sorghum bicolor* L. Moench) genotypes under high temperature stress. *Environ. Exp. Bot.* 100 43–54.

Eshak, E.S., Iso, H., Date, C., Kikuchi, S., Watanabe, Y., Wada, Y., Wakai, K. and Tamakoshi, A. 2010. Dietary fiber intake is associated with reduced risk of mortality from cardiovascular disease among Japanese men and women. *J Nutr.* 140:1445–1453. doi: 10.3945/jn.110.122358.

Freeman, J.E. and Watson, S.A. 1971. Influence of sorghum endosperm pigments on starch quality. *Cereal Sci. Today* 16:378.

Gami, R.A., Kugashiya, K.G., Vaghela, S.J., Patel, R.N., Patel, P.T., Solanki, S.D. and Chaudhary, J.N. 2023. Essentials of Millets (Shree Anna) – The Nutricereal (Production, Processing and Value addition). ISBN: 978-81-966161-0-6

Gopalan, C., Rama Sastri, B.V. and Balasubramanian, S.C. 2003. *Nutritive value of Indian foods*. Hyderabad, National Institute of Nutrition.

Gopalan, C., Ramashastri, B.V. and Balasubramanium, S.C. 2016. *Nutritive value of Indian foods*. National Institute of Nutrition, ICMR, Hyderabad.

Grovermann, C., Umesh, K.B., Quiedeville, S., Kumar, B.G. and Moakes, S. 2018. The economic reality of underutilized crops for climate resilience, food security and nutrition: assessing finger millet productivity in India. *Agriculture* 8(9), 131.

Gupta, S.K., Rai, K.N., Singh, P., Ameta, V.L., Gupta, S.K., Jayalekha, A.K. and Verma, Y.S. 2015. Seed set variability under high temperatures during flowering period in pearl millet (*Pennisetum glaucum* L. (R.) Br.). *Field Crops Research* 171, 41–53.

Hamad, S.H. 2012. The microbial quality of processed date fruits collected from a factory in Al-Hofuf City, Kingdom of Saudi Arabia. Emir. *Journal of Food Agriculture*. 24(2):105–112.

Hemasankari, P., Rao, B.D., Malathi, V.M., Kiranmai, E. and Rathnavathi, C.V. 2023. Value addition in millets through agricultural processing techniques. *Indian Farming* 73 (01):107–109.

Hossain, M.A. and Uddin, S.N. 2011. Mechanisms of waterlogging tolerance in wheat: morphological and metabolic adaptations under hypoxia or anoxia, Aust. *J. Crop. Sci.* 5 1094.

Howarth, C.J., Cavan, G.P., Skot, K.P., Layton, R.W.H., Hash, C.T. and Witcombe, J.R. 1994. Use of molecular markers in sorghum and pearl millet breeding for developing countries. *Proceedings of an ODA Plant Sciences Research Programme Conference*, 29thMarch–1stApril 1993, Norwich, UK, pp. 80–85.

Hu, Q., Zhao, Y., Hu, X., Qi, J., Suo, L., Pan, Y., et al., 2022. Effect of saline land reclamation by constructing the "raised field-shallow trench" pattern on agroecosystems in Yellow River Delta. Agric. *Water Manag.* 261:107345. doi: 10.1016/j.agwat.2021.107345

IPCC. 2014. Intergovernmental panel on climate change. In: Proceeding of the 5th Assessment Report, WGII, Climate Change 014: impacts, adaptation, and vulnerability. Cambridge University Press, Cambridge.

Jain, N., Arora, P., Tomer, R., Mishra, S.V., Bhatia, A., Pathak, H., Chakraborty, D., Kumar, V., Dubey, D. and Harit, R. 2016. Greenhouse gases emission from soils under major crops in northwest India. *Science of the Total Environment* 542:551–561.

Jiang, C., Wang, Y., Yang, Z. and Zhao, Y. 2023. Do adaptive policy adjustments deliver ecosystem-agriculture-economy co-benefits in land degradation neutrality efforts? Evidence from southeast coast of China. *Environ. Monit. Assess.* 195:1215. doi: 10.1007/s10661-023-11821-6

Kamaraddi, V. and Shanthakumar, G. 2003. Effect of incorporation of small millet flour to wheat flour on chemical, rheological and bread characteristics. In: *Recent trends in millet processing and utilization.* CCS Hisar Agricultural University, Hisar, India, pp. 74–81.

Kole, C., Muthamilarasan, M., Henry, R., et al. 2015. Application of genomics-assisted breeding for generation of climate resilient crops: progress and prospects. *Front. Plant Sci.* 6:563.

Krishna, T.P.A., Maharajan, T. and Ceasar, S.A. 2022. The role of membrane transporters in the biofortification of zinc and iron in plants. *Biol. Trace Elem. Res.* 3:1–15.

Kumar, A. and Singh, P. 2015. Millets: nutritional composition, some health benefits and processing – a review, *J. Food Sci. Technol.* 52(5):2481–2495.

Kumar, A., Metwal, M., Kaur, S., Gupta, A.K., Puranik, S., Singh, S., Singh, M., Gupta, S., Babu, K., Sood, S. and Yadav, R. 2016. Nutraceutical value of finger millet (*Eleusine corcana* L Gaertn) and their improvement using omics approaches. *Front. Plant Sci.* https://doi.org/10.3389/fpls.2016.00934

Kumar, A., Tripathi, M.K., Joshi, D. and Kumar, V. 2021. *Millets and Millet Technology.* ISBN:978-981-16-0675-5

Kumari, P.L. and Sumathi, S. 2002. Effect of consumption of finger millet on hyperglycemia in non-insulin dependent diabetes mellitus (NIDDM) subjects. *Plant Foods Hum. Nutr.* 57 (3–4):205–213.

Kunyanga, C.N., Imungi, J.K. and Velingiri, V. 2013. Nutritional evaluation of indigenous foods with potential food-based solution to alleviate hunger and malnutrition in Kenya. *J. Appl. Biosci.* 67:5277–5288. doi: 10.4314/jab.v67i0.95049

Lamaoui, J. and Bekkaoui. 2016. Heat and drought stresses in crops and approaches for their mitigation. *Frontiers in Chemistry* 6(26):1–14.

Lesk, C., Rowhani, P. and Ramankutty, N. 2016. Influence of extreme weather disasters on global crop production. *Nature* 529:84–87.

Li, P. and Brutnell, T.P., 2011. *Setaria viridis* and *Setaria italica*, model genetic systems for the Panicoid grasses. *J. Exp. Bot.* 62(9):3031–3037.

Linkemer, G., Board, J.E. and Musgrave, M.E., 1998. Waterlogging effects on growth and yield components in late-planted soybean, *Crop. Sci.* 38:1576–1584.

Malleshi, N.G. 1989. Processing of small millets for food and industrial uses, in *Small Millets in Global Agriculture* (eds. A. Seetharam, K.W. Riley, and G. Harinarayana), Oxford and IBH Publishing Co. Pvt. Ltd. New Delhi.

Matsuura, A., Murata, K. et al. 2016. Effect of pre- and post-heading waterlogging on growth and grain yield of four millets. *Plant Prod. Sci.* 19(3):348–359, https://doi.org/10.1080/1343943X.2016.1146907.

Naveed, U.M., Seerat, S., Aadil, R., Wasifa, H.S., Khalid, R.H. and Reiaz, U.R. 2020. Salt stress threshold in millets: perspective on cultivation on marginal lands for biomass. *Phyton.* 90(1):51–64.

NIN. 2007. Indian food composition tables. NIN and Nutritive value of Indian Foods. NIN.

Nkhata, S.G., Ayua, E., Kamau, E.H. and Shingiro, J.B. 2018. Fermentation and germination improve nutritional value of cereals and legumes through activation of endogenous enzymes. *Food Sci. Nutr.* 6:2446–2458.

Paschapur, A.U., Joshi, D., Mishra, K.K., Kant, L., Kumar, V. and Kumar, A. 2021. Millets for Life: A Brief Introduction. In: Kumar A., Tripathi M. K., Joshi D., Kumar V. (eds.) *Millets and Millet Technology*. Springer, Singapore. https://doi.org/10.1007/978-981-16-0676-2_1

Patil, J.V. and Davada, B.K. 2023. Millets the Future Crops for Climate Resilience and Nutritional Security. Lead paper published in National conclave on Promotion of Millets (Shree Anna) for Sustainable Agriculture and Nutritional Security Towards Global Prosperity: Key Challenges and Future Prospects. *Conclave Compendium* (Lead Paper and Abstract), ISBN: 978-81-966161-7-5, pp:141–160.

Paudel, M., Subedi, M. and Pandey, M.K. 2024. Millets: *The Miracle Grains of 21st Century*. Chapter 16, pp. 165–175. ISBN: 978-81-969534-8-5

Prasad, P.V. and Staggenborg, S.A. 2009. Growth and Production of Sorghum and Millets. In: *Soils, Plant Growth and Crop Production*. EOLSS Publishers Co., Ltd. Oxford, UK. Volume 2.

Rai, K.N., Patil, H.T., Yadav, O.P., Govindaraj, M., Khairwal, I.S., Cherian, B., et al. 2014. Dhanashakti: a high-iron pearl millet variety. *Ind. Farm.* 64:32–34.

Rani, S., Singh, R., Sehrawat, R., Kaur, B.P. and Upadhyay, A. 2018. Pearl millet processing: a review. *Nutrition & Food Science*, 48(1), 30–44.

Rao, B.R., Nagasampige, M.H and Ravikiran, M. 2011. Evaluation of nutraceutical properties of selected small millets. *J. Pharm. Bioallied Sci.* 3:277–279.

Rao, D.B., Bhaskarachary, K., Christina, G.D.A., Devi, G.S. and Tonapi, V.A. 2017. *Nutritional and Health Benefits of Millets*. ICAR – Indian Institute of Millets Research (IIMR) Rajendranagar, Hyderabad, pp. 112. ISBN: 81-89335-68-5

Rawat, L., Karnatak, A.K., Bisht, T.S. and Kukreti, A. 2021. Minor Millets: Profile and Ethnobotanical Scenario. In: Kumar A., Tripathi M.K., Joshi D. and Kumar V. (eds.). *Millets and Millet Technology*. Springer, Singapore. https://doi.org/10.1007/978-981-16-0676-2_3

Reddy, A. 2018. Millets – The Miracle Grains. *Millets and sorghum biology and genetic improvement*. J.V. Patil (Editor), Wiley Blackwell.

Sage, R.F. and Zhu, X.G. 2011. Exploiting the engine of $C_4$ photosynthesis. *J. Exp. Bot.* 62:2989–3000.

Saleh, A.S.M., Zhang, Q., Chen, J. and Shen. Q. 2013. Millet grains: nutritional quality, processing, and potential health benefits. *Comprehensive Reviews in Food Science and Food Safety.* 12:281–295.

Satyavathi, C.T., Ambawat, S., Khandelwal, V. and Srivastava, R.K. 2021. Pearl millet: a climate-resilient nutricereal for mitigating hidden hunger and provide nutritional security. *Front. Plant Sci.* 12:659938. doi: 10.3389/fpls.2021.659938

Saxena, R., Vanga, S.K., Wang, J., Orsat, V. and Raghavan, V. 2018. Millets for food security in the context of climate change: a review. *Sustainability* 10(7): 2228.

Shrestha, N., Hu, H., Shrestha, K. and Doust, A.N. 2023. Pearl millet response to drought: a review. *Frontiers in Plant Science,* 14, 1059574.

Sood, S., Khulbe, R.K., Gupta, A.K., Agrawal, P.K., Upadhyaya, H.D. and Bhatt, J.C. 2015. Barnyard millet – a potential food and feed crop of future. *Plant Breed.* 134:135–147.

Subramanian, V., Hoseney, R.C. and Bramel Cox, P. 1994. Factors affecting the color and appearance of sorghum starch. *Cereal Chem.* 71:275–278.

Tadele, Z. 2016. Drought Adaptation in Millets, InTech, London, UK.

Talwar, H.S., Kannababu, N., Venkatesh Bhat, B. and Tonapi, V.A. 2019. *In climate resilience in millets*. ICAR-IIMR, Hyderabad.

Taylor, J.R.N. and Schussler, L. 1986. The protein composition of the different anatomical parts of sorghum grain. *J. Cereal. Sci.* 4:361–369.

Tripathi, M.K. et al., 2021. Nutritional Composition of Millets. In: Kumar, A., Tripathi, M.K., Joshi, D., Kumar, V. (eds.), *Millets and Millet Technology*. Springer, Singapore. https://doi.org/10.1007/978-981-16-0676-2-5

Varshney, R.K., Bohra, A., Roorkiwal, M., Barmukh, R., Cowling, W.A., Chitikineni, A., et al. 2021a. Fast-forward breeding for a food-secure world. *Trends Genet*. 37:1124–1136. doi: 10.1016/j.tig.2021.08.002

Varshney, R.K., Bohra, A., Yu, J., Graner, A., Zhang, Q. and Sorrells, M.E. 2021b. Designing future crops: genomics-assisted breeding comes of age. *Trends Plant Sci*. 26:631–649. doi: 10.1016/j.tplants.2021.03.010

Yadav, A.K., Arya, R.K., Singh M.K., Kumar, D. and Panchta, R. 2016. Heat tolerance in pearl millet: a review. *Forage Res.* 42(2):65–81.

Yadav, O.P. and Rai, K.N. 2013. Genetic improvement of pearl millet in India. *Agric. Res.* 2(4):275–292.

# Chapter 8

# Underutilized Vegetable Crops for Food and Nutritional Security

Virupakshi Hiremata[1*], Ratnakar M Shet[2*] Raghavendra Gunnaiah[2], Belaghihalli N Gnanesh[3] and Sheshshayee MS[1]

## 1. Introduction

India has achieved impressive progress toward food production self-sufficiency with a record harvest of 305 million metric tons of food grains in 2022–2023. Despite this achievement, the country faces persistent challenges in addressing malnutrition, with 195 million undernourished children, according to the Food and Agriculture Organization (FAO, 2023). In response, there is a growing emphasis on increasing the production of fruits and vegetables, particularly those rich in essential nutrients. Although India ranks as the world's second-largest producer of vegetables, with an annual output of 66 million tons (Pradeepkumar et al., 2022), issues such as post-harvest losses, low purchasing power, and other logistical challenges prevent this production from fully benefiting the population.

To tackle these challenges, it is imperative to boost the per capita availability of vegetables as a vital micronutrient source. This can be achieved by increasing overall vegetable production or tapping into alternative sources of plant-based foods. One promising yet underexplored avenue is the utilization of locally available wild flora, which could ease the pressure on conventional food production systems. Many underutilized plants boast superior nutrient content, medicinal properties, and high yield potential. Despite these benefits, their use has been largely confined to traditional remedies or Ayurvedic preparations. Underutilized foods are typically less accessible, less frequently consumed, or regionally specific.

Among the cucurbit vegetables belonging to the family Cucurbitaceae, there are 96 genera and 1,000 species (Chomicki et al., 2020), with many species used as food, particularly in tropical and subtropical regions (Aberoumand and Deokule 2010). In India, cucurbits are grown on about 0.118 million hectares, producing 1,665 million tons annually (Indiastat, 2022). Cucurbits, placed first in rank in diversity, are associated with a lot of unrevealed nutrients and health benefits. Among such underutilized cucurbit vegetables are the Mangalore melon and spine gourd, which are found in Western Ghats. These crops, native to the Western Ghats (a biodiversity hotspot), offer valuable nutritional benefits and are crucial for ensuring food and nutrition security for vulnerable populations.

[1] University of Agricultural Sciences, Bengaluru, Karnataka, India.
[2] University of Horticultural Sciences, Bagalkote, Karnataka, India.
[3] ICAR JSS – Krishi Vigyan Kendra, Suttur 571129, Mysore, Karnataka, India.
* Corresponding author: virugpb2019@gmail.com, ratnakar.shet@uhsbagalkot.edu.in

DOI: 10.1201/9781003584100-8

These crops, despite their significant potential, remain largely neglected in mainstream agriculture. Mangalore melon and spine gourd are mainly focused in this chapter among the cucurbits vegetables.

## 2. Significance of Underutilized Vegetables

The growing recognition of underutilized vegetables in tackling poverty, enhancing financial stability, and addressing malnutrition is becoming increasingly effective. These vegetables can improve the quality of life for countless tribal communities and are essential to the ecosystems relied upon by impoverished rural populations. Rich in vitamins, minerals, and other health-boosting properties, such as strong antioxidant levels, underutilized vegetables significantly contribute to food diversity, leading to a more balanced source of micronutrients. Furthermore, these neglected crops are resilient against various biotic and abiotic stressors.

Vegetable crops that are underutilized can give impoverished communities crucial nutrients, which can be a lifesaver. Many of these often overlooked vegetables serve as abundant, organic sources of essential vitamins, minerals, and antioxidants. Rural inhabitants play a crucial role in diets and nutrition, having been valued for their health benefits, therapeutic properties, and nutritional value for generations. Traditionally, they are enjoyed both raw and cooked, and selling any surplus helps numerous rural families sustain their livelihoods Aberoumand and Deokule (2009).

Malnutrition and food scarcity are prevalent among poorer rural populations, but these vegetables can help alleviate poverty by creating jobs and income opportunities. They can enhance productivity and profitability for farm families in both urban and rural environments. Utilizing underutilized vegetable crops is one way to reduce the risk tied to relying heavily on a limited range of major crops, thus providing more options for edible foods. They contribute to sustainable living through improved household food security and are often convenient sources of nourishment for urban dwellers facing economic challenges.

Adapted to fragile ecosystems, these crops can help stabilize agroecosystems, particularly in steppes, mountainous areas, tropical forests, and arid or semi-arid regions. They offer diverse crop options that meet emerging market demands, ultimately boosting productivity and global food security (Jena et al., 2018).

### *2.1 Characteristics of Underutilized Vegetable Crops*

- **Embracing Cultural Heritage:** Strengthening the bond with the valued traditions of various nations by bringing back indigenous crops.
- **Celebrating Overlooked Gems:** Shedding light on the underexplored distribution, biology, cultivation, and applications of local and traditional crops, uncovering their hidden possibilities.
- **Harnessing Strong Adaptability:** Acknowledging the extraordinary capability of these crops to flourish in particular agroecological zones and less fertile lands, reflecting their natural resilience.
- **Promoting Seed Sovereignty:** Enhancing fragile or absent formal seed supply chains, ensuring the protection of crucial genetic resources of these important plants.
- **Valuing Enduring Traditions:** Safeguarding and encouraging the time-honored uses of these crops within local communities, highlighting their cultural and practical relevance.
- **Fostering Sustainable Growth:** Supporting traditional farming practices that reduce or eliminate reliance on external inputs, promoting eco-friendly agriculture.
- **Raising Consumer Awareness:** Engaging consumers by showcasing the distinct features of these crops, emphasizing their high nutritional value and health benefits.
- **Advocating for Policy Support:** Seeking greater recognition and backing from donors, legislators, extension services, researchers, and technology developers.
- **Providing Affordable Nutrition:** Offering low-cost and easily available food options, empowering communities with accessible nutritional resources.

- **Supporting Holistic Health:** Acknowledging the diverse roles of these crops, from their essential contribution to nutrition to their powerful healing capabilities.

## 3. Mangalore Melon (*Cucumis melo* ssp. *agrestis* var. *acidulus*): An Underappreciated Nutritional Powerhouse

Mangalore melon, also referred to as sour melon, acid melon, culinary melon, or vegetable melon, is a significant member of the Cucurbitaceae family. This diverse taxon encompasses a range of botanical and horticultural varieties (Swamy, 2017). *C. melo* is categorized into two subspecies based on the presence of pubescence on the hypanthium: *C. melo* ssp. *agrestis* (lacking pubescence) and *C. melo* ssp. *melo* (featuring pubescence) (Jeffrey, 1980).

Mangalore melon [*Cucumis melo* var. *acidulus* Naudin ($2n = 24$)] is widely cultivated and consumed in Southern India. In Karnataka, it is called Mangalore southekayi or Mangalore cucumber; however, it has been more appropriately renamed Mangalore melon to better reflect its closer ties to melons than to cucumbers (Shet et al., 2022). This melon serves as an essential component in sambar, a popular lentil soup from South India. It is called various names in different areas, including vellari in Kerala, (Van Rheede 2003). Madras cucumber in Tamil Nadu, Vidya (2012), and mage-kaayi or dosakaya in Andhra Pradesh and Telangana (Swamy, 2017) (Table 1). Mangalore melon flourishes in both rainfed and irrigated conditions, as well as in rice fallow areas during the Kharif and summer seasons. The fruits, which are botanically categorized as pepo (a fleshy, multi-seeded fruit that develops from a single flower with one ovary divided into several carpels), are sizeable and have an impressive storage life at room temperature. These fruits can last between six to twelve months when stored properly (Shet et al., 2022). In traditional farming methods, Mangalore melons are often suspended from the roof using banana pseudostem fibers, enabling them to be stored for up to one year for regular use.

Mangalore melon is morphologically diverse, particularly in terms of fruit traits. However, the genetic resources of this crop are under threat as commercial cultivars with higher yield and better quality are increasingly being cultivated, leading to the potential extinction of local landraces. Despite this, Mangalore melon strains hold valuable genes that could be utilized in breeding programs (Kacar et al., 2012). The crop is gaining importance due to its early maturation, high production potential, high nutritional value, taste, delicacy, long-keeping quality, and suitability for cultivation under both rainfed and irrigated conditions (Shet et al., 2022).

### *3.1 Botanical Descriptions of Mangalore Melon*

The genus *Cucumis* is a highly diverse group within the Cucurbitaceae family, comprising more than 30 species, among which cucumber (*C. sativus*, $2n = 2x = 14$) and melon (*C. melo*, $2n = 2x = 24$) are the most commercially exploited. The exact center of origin for melons (*C. melo*) remains a subject of debate. Wild forms of *C. melo* are found across Africa, Asia, and Australia, contributing to the uncertainty (Endl et al., 2018, Sebastian et al., 2010; Kerje and Grum, 2000). Some taxonomists have suggested Africa as the primary center of origin due to archeological evidence of melon remains dating back to 3700–3500 BC and the high number of wild *Cucumis* species in Africa, along with their diversity in chromosome numbers.

Contrarily, the significant diversity of melon landraces in India and East Asia, including both *C. melo* subsp. *melo* and *C. melo* subsp. *agrestis*, and their ability to cross with modern *C. melo* genotypes suggest that melons may have been domesticated in Asia (Naudin, 1856; Nhi et al., 2010; Kirkbride, 1993). Supporting this hypothesis is the near identity of plastid genome sequences of *C. melo* with the wild Himalayan species *Cucumis trigonus* and *Cucumis callosus*, indicating that the wild progenitor of melon could be found in India (Sebastian et al., 2010).

Recent research, however, complicates the picture. Studies clustering global melon accessions according to geographical regions, rather than the traditional grouping into *C.*

*melo* subsp. *melo* and *C. melo* subsp. *agrestis*, imply that there may be multiple centers of origin and domestication for melons (Esteras et al., 2013; Nimmakayala et al., 2016; Stepansky et al., 1999). A study constructing the lineage of *C. melo* using seven different genomic regions revealed at least two independent domestications of wild melons, one in Africa and another in Asia. The Asian domestication gave rise to *C. melo* subsp. *melo agrestis*, while the African domestication resulted in *C. melo* subsp. *meloides* (Endl et al., 2018).

The major melon cultivar groups, including *inodorus*, *cantaloupensis*, *reticulatus*, and *adzhur* melons of the conventional subsp. *melo* and *chinensis*, *conomon*, and *makuwa* melons of subsp. *agrestis*, are primarily found in India, Southeast Asia, Central Asia, and East Asia, all nested within the "Asian agrestis" clade. In contrast, African cultivars were grouped under the "African agrestis" clade. This suggests that the wild populations of melon in India and Africa were domesticated independently. The Asian *C. melo* subsp. *melo* f. *agrestis* is the ancestor of most modern market melon cultivars, while *C. melo* subsp. *meloides* is the wild progenitor of African cultivars, including *tibish*, *fadasi*, and *seinat* (Endl et al., 2018).

Interestingly, culinary melons from South India, such as the *acidulus* and *momordica* types, are distinctly grouped from other Indian melons and Southeast Asian groups. This distinction suggests that South Indian melons either originated separately or diverged from *C. melo* subsp. *melo* f. *agrestis* many years ago (Fergany et al., 2011; Nhi et al., 2010; Tanaka et al., 2007).

Within the cucurbit family, several classifications have been put forth for melons, both interspecific (Pitrat et al., 2000) and intraspecific (Stepansky et al., 1999). *Cucumis melo* is identified as a polymorphic taxon, comprising a diverse array of botanical and horticultural varieties (Swamy, 2017). *C. melo* is categorized into two subspecies based on the presence of hairs on the hypanthium: *C. melo* ssp. *agrestis* (hairless) and *C. melo* ssp. *melo* (hairy) (Jeffrey, 1980). The former subspecies is further broken down into five groups: conomon, makuwa, chinensis, and acidulous, and acidulus (Munger and Robinson, 1991). The latter subspecies contains ten groups, including cantaloupe, reticulatus, adana, chandalak, ameri, inodorus, flexuosus, chate, tibish, dudaim, and morren (Pitrat et al., 2000). In the past, South Indian melons that are not typically eaten as dessert were classified under the variety acidulus along with Southeast Asian melons, often referred to as oriental pickling melons or Chinese melons (Naudin, 1856). On the other hand, designated non-dessert melons from Eastern Asia into a distinct group are known as conomon.

Mangalore melons are frequently known as "conomon" melons (*C. melo* ssp. agrestis var. conomon) (Gondi et al., 2016; Kamagoud, 2018; Lakshmi et al., 2017a; Ramesh Babu and Hariprasad Rao, 2018; Silpa et al., 2019; Swamy, 2017; Vinutha et al., 2017). However, Mangalore melons are distinct from other "*conomon*" melons found in Southeast Asia. They feature larger fruits with a variety of shapes and sizes, possess a long shelf life of over six months, and have smaller seeds (Fergany et al., 2011; Koli and Murthy 2013; Manohar and Murthy, 2012; Roy et al., 2012) (Fig. 1).

## 4. Spine Gourd: A Potentially Valuable Crop

Spine gourd belongs to family Cucurbitaceae, chromosome no. $2n = 28$. It is a perennial dioecious plant, having great potential for income generation in Western Ghats and northeastern region of India. The spine gourd is generally found in the forest areas of Western Ghats in Karnataka and Kerala. It is also cultivated by the farmers of Assam, Tripura, Maharashtra, West Bengal, Orissa, Chhattisgarh, Jharkhand, and Bihar. Spine gourd, also known as teasel gourd, an underexploited and underutilized vegetable crop, is a perennial climbing plant that produces small, spiny fruits rich in antioxidants and known for their medicinal properties. *Momordica dioica* is known to be originated in the northeastern region, whereas *Momordica sahyadrica* is native to the Western Ghats. *M. sahyadrica*, a recently described species (Joseph et al., 2007), is endemic to the Western Ghats and cultivated to a limited extent in Karnataka's Malnad region. *M. sahyadrica*, with a wider range of pollinators, has a higher fertilization success rate and may have evolved from spine gourd in the Western Ghats, potentially making it a neoendemic species (Behera et al., 2010). It may also be of

**Fig. 1** Morphological features of Mangalore melon (*Cucumis melo* ssp. *agrestis* var. *acidulus*).

**Table 1** Botanical classification of Mangalore melon.

| | |
|---|---|
| **Kingdom** | Plantae |
| ***Clade*** | Tracheophytes |
| ***Clade*** | Angiosperms |
| ***Clade*** | Eudicots |
| ***Clade*** | Rosids |
| **Order** | Cucurbitales |
| **Family** | Cucurbitaceae |
| **Genus** | Cucumis |
| **Species** | *C. melo* |
| **Sub-species** | Agrestis |
| **Variety** | Acidulus |
| **Ploidy** | Diploid |
| **Chromosome number** | $2n = 24$ |
| **Binomial name** | *Cucumis melo* ssp. *agrestis* var. *acidulus* |
| **Vernacular names in Indian province** | |
| **Province** | **Vernacular names** |
| Andhra Pradesh, Telangana | Dosakaya, budamekaya, bollari |
| Karnataka | Sambar southe, mogghekayi, or Mangalore southe |
| Kerala | Vellari, kanivellari, vellarikka, Malabar cucumber |
| Tamil Nadu | Madras cucumber |
| Goa | Mogem |

(*Source:* Hiremata et al., 2023)

hybrid origin (Schaefer and Renner, 2011). The fruit and seed morphology of *M. sahyadrica* shares many similarities with the other species of spine gourd.

## 4.1 Botanical Descriptions of Spine Gourd

The genus *Momordica* is a significant source of both vegetables and traditional medicine, belonging to the tribe Joliffieae (Schrad.) of the family Cucurbitaceae. Originating from the Palaeotropics (Robinson and Decker-Walters, 1997), *Momordica* species are primarily found in the warm tropics of both hemispheres. The genus comprises around 60 species (Schaefer and Renner, 2011), with the majority being distributed in Africa, and about 10 species found in Southeast Asia. India is home to six species of *Momordica*, including the recently identified *M. sahyadrica* K.J. John & V.T. Antony (Joseph, 2005; Joseph et al., 2007; John and Antony, 2010; Bharathi and John, 2013) (Table 2).

The latest revision of Indian *Momordica* species (Chakravarthy, 1982; John and Antony, 2010; Joseph, 2005) categorizes four of these species as dioecious and two as monoecious. The monoecious taxa include *Momordica charantia* L. ($2n = 22$), commonly known as bitter gourd, and *Momordica balsamina* L. ($2n = 22$), also known as balsam apple. The dioecious species comprise *Momordica dioica* Roxb. ($2n = 28$), commonly referred to as spine gourd; *Momordica sahyadrica* ($2n = 28$); *Momordica cochinchinensis* Spreng. ($2n = 28$), known as sweet gourd or gac; and *Momordica subangulata* Blume ssp. *renigera* ($2n = 56$). Notably, Joseph (2005) did not mention *Momordica cymbalaria*, which some authors classify within the genus *Momordica*.

Among the *Momordica* species, *Momordica dioica* (spine gourd) and *Momordica cochinchinensis* (sweet gourd) are of particular interest, though they are primarily found in the wild and are underutilized as cultivated crops. These species are not only rich in nutritional value but also hold significant potential as vegetables. However, despite their benefits, they remain largely gathered from the wild, especially by tribal communities in the regions where they naturally occur.

The genus *Momordica* plays a crucial role in traditional medicine across its native range. Various species are used to treat ailments due to their bioactive compounds, which include cucurbitacins, saponins, and flavonoids. These compounds contribute to the medicinal properties of *Momordica* species, which have been used to manage conditions such as diabetes, inflammation, and gastrointestinal issues. For example, *Momordica charantia* (bitter gourd) is widely recognized for its hypoglycemic effects, making it a popular natural remedy for managing blood sugar levels in diabetes.

Despite the documented medicinal properties of *Momordica* species, their potential as vegetables remains underexplored. The fruits of *Momordica dioica* and *Momordica cochinchinensis* are particularly nutritious, offering a rich source of vitamins, minerals, and antioxidants. *Momordica cochinchinensis*, for instance, is notable for its high content of beta-carotene and lycopene, antioxidants that are beneficial for eye health and reducing the risk of chronic diseases. The seeds of *M. cochinchinensis* are also rich in essential fatty acids (Table 2), further enhancing the nutritional profile of the species.

However, the underutilization of *Momordica* species in agriculture poses a challenge to their conservation and broader application. Their cultivation is limited, and they are more commonly gathered from the wild, which can lead to overharvesting and potential depletion of natural populations. This scenario underscores the need for strategies to domesticate and cultivate these species on a larger scale, thereby ensuring their availability and sustainability as both food sources and medicinal plants.

The domestication and cultivation of *Momordica* species could significantly contribute to food security, particularly in regions where these plants are native. Their ability to thrive in tropical climates makes them well-suited for cultivation in areas with similar environmental conditions. Moreover, their resistance to pests and diseases, which are common challenges in agriculture, adds to their appeal as potential crops for sustainable farming practices.

**Table 2** Botanical classification of spine gourd.

| | |
|---|---|
| Kingdom | **Plantae** |
| Phylum | **Tracheophyta** |
| *Clade* | **Eudicots** |
| Class | **Magnoliopsida** |
| Order | **Cucurbitales** |
| Family | **Cucurbitaceae** |
| Genus | **Momordica** |
| Species | ***Sahyadrica, Dioica*** |
| Ploidy | **Diploid** |
| Chromosome number | **$2n = 28$** |

**Vernacular names of spine gourd in India**

| *Language* | *Names of spine gourd* |
|---|---|
| English | Teasel gourd, spine gourd, small bitter gourd |
| Bengali | Kartoli, kakrol |
| Assam | Batkarila |
| Hindi | Kantola, parora, kakora |
| Tamil | Paluppakkay |
| Telugu | Agakara, karkotaki |
| Malayalam | Venpaval |
| Kannada | Madahagala-Kaya |
| Sanskrit | Vahisi |
| Gujarati | Katwal |

(*Source:* Anjana et al., 2019)

# 5. Nutritional Aspects

## *5.1 Nutritional Profiling of Mangalore Melon*

Mangalore melon (*Cucumis melo* var. *acidulus*), commonly known as Mangalore southekayi, is an ideal summer vegetable crop valued for its use as a fresh vegetable and for pickling. The fruit, known for its cooling properties, is often used as a skin moisturizer and digestive agent. Its acidic nature makes it particularly preferred for pickling, with salted pickles being known to stimulate appetite. According to Jompitak (2002), every 100 g of the edible portion of oriental pickling melon fruit contains 96.5 g of water, 0.4 g of protein, 0.1 g of fat, 2.2 g of carbohydrates, 0.5 g of crude fiber, 9 mg of phosphorus, 23.3 mg of β-carotene, 0.04 mg of thiamine, 0.10 mg of riboflavin, 0.4 mg of niacin, and 18 mg of vitamin C.

Farmers have long recognized the medicinal value of Mangalore melon. Traditional knowledge suggests that consuming the flesh of Mangalore melon during summer helps reduce dehydration and body heat (Swamy and Nath, 2020). The juice extracted from the seeds is used to treat dyspepsia (indigestion) and to reduce body heat. In traditional practices, the fruits are also fed to cows and buffaloes to assist in the digestion of the newborn calf's umbilical cord (Swamy and Nath, 2020).

Both mature and immature fruits of Mangalore melon are cooked as vegetables. The off-white, turgid, and crunchy flesh of the melon retains its crunchiness even after cooking, making it a delicious and low-calorie vegetable. According to Swamy (2017), every 100 g of the edible portion of oriental pickling melon fruit contains 96.5 g of water, 0.65 g of protein, 0.11 g of fat, 3.63 g of carbohydrates, 1.67 g of sugars, 0.5 g of crude fiber, 13 mg of phosphorus, 23.3 mg of β-carotene, 0.027 mg of thiamine, 0.033 mg of riboflavin, 0.098 mg of niacin, 18 mg of vitamin C, 0.04 mg of vitamin $B_6$, 16 mg of calcium, 0.28 mg of iron, 13 mg of magnesium, 24 mg of potassium, and 0.20 mg of zinc (Table 3).

Consuming Mangalore melon has been linked to various health benefits, including a reduced risk of macular degeneration in the eyes. The seeds, which are often discarded due to their slightly bitter taste, are highly sought after in the nutrition and beauty industries for their oils and fatty acids. The seeds contain 34.68% oil on a dry weight basis, primarily composed of nine fatty acids with a high concentration of polyunsaturated fatty acids (80.54%). This composition makes them a potential source of edible oil (Manohar and Murthy, 2014).

Despite its high nutritional and medicinal values, the nutritional profiling of Mangalore melon accessions MS4 (golden yellow), MS21 (green), and MS78 (stripe skin) has been limited. These accessions were used for metabolite profiling, with different tissues such as skin, flesh, and seeds being analyzed. Untargeted metabolic profiling of the three Mangalore melon accessions (MS4, MS21, and MS78) and one variety of muskmelon revealed the differential accumulation of various classes of metabolites, including amino acids, vitamins, phenolic compounds, organic acids, flavonoids, terpenoids, and alkaloids. A total of 10,023 features (a probable metabolite or its derivative) were identified across the three tissues (skin, flesh, and seeds) and four genotypes. The highest number of features were identified in the skin (6,611), followed by the flesh (2,017) and seeds. Among the different Mangalore melon accessions, the highest number of features were identified in MS21, followed by MS4 and MS78.

## 5.2 Amino Acids and Vitamins

In the selected Mangalore melon accessions, ten amino acids and seven vitamins or their precursors were identified. Most of the amino acids and vitamins were found to be significantly higher in muskmelon compared to Mangalore melon accessions. However, Mangalore melon accessions were found to be higher in threonine, glutamine, and arginine. Among the vitamins, Mangalore melon accessions were rich in vitamin $B_3$ (niacin) and riboflavin, with higher levels than in muskmelon. Additionally, the seeds of Mangalore melon were found to be rich in vitamin $B_5$ (pantothenic acid).

The skin of Mangalore melon fruits is a rich source of flavonoids, amino acids, and vitamins. The highest number of flavonoids were found in the green fruits of MS21 (34), followed by the yellow fruits of MS4 (25). The skin of Mangalore melon also contained 10 (MS04), 09 (MS21), and 03 (MS78) amino acids or their precursors and derivatives. Notably, the skin of MS04 was found to have significantly higher amounts of a unique aromatic compound, toluquinol, which is uniquely found in cucurbits and has exhibited anti-cancerous activity. The skin of Mangalore melon is also a good source of vitamin $B_2$, with significantly higher accumulation, and vitamin $B_5$ (pantothenic acid) compared to the skin of muskmelon.

## 5.3 Fatty Acids and Peptides

The flesh of Mangalore melon fruits is a rich source of fatty acids, sugars, and amino acids. The flesh of MS4 is particularly rich in amino acids such as L-threonine, L-glutamine, and L-arginine. MS21 and MS78 are also good sources of peptides, with six peptides identified. However, the nutritional value of these peptides needs to be further investigated. The seeds of Mangalore melon were found to accumulate many peptides compared to the seeds of muskmelon. Of the identified features, more than 50% were peptide molecules, indicating higher amounts of proteins in the seeds. The nutritional value of these peptides requires further exploration.

**Table 3** The nutrients present in the Mangalore melon.

| *Nutrients* | *Total amount* |
|---|---|
| Water | 96.5 g |
| Protein | 0.65 g |
| Fat | 0.11 g |
| Carbohydrate | 3.63 g |
| Sugars | 1.67 g |
| Crude fiber | 0.5 g |
| Phosphorus | 13 mg |
| Beta-carotene | 23.3 mg |
| Thiamine | 0.027 mg |
| Riboflavin | 0.033 mg |
| Niacin | 0.098 mg |
| Vitamin C | 18 mg |
| Vitamin $B_6$ | 0.04 mg |
| Calcium | 16 mg |
| Iron | 0.28 mg |
| Magnesium | 13 mg |
| Potassium | 24 mg |
| Zinc | 0.20 mg |

(*Source:* Swamy, 2017)

## 5.4 *Traditional and Medicinal Uses of Spine Gourd*

Spine gourd has been valued for centuries in traditional Indian, Chinese, and African medicine, where it has been used as a food and for its medicinal properties. Ancient pharmacopoeias describe its use as an anthelmintic, laxative, digestive stimulant, and appetite enhancer. Various species of *Momordica* in Southeast Asia have been recognized for their medicinal properties, including their use as anthelmintics, vermifuges, cathartics, hypoglycemics, aphrodisiacs, antipyretics, antiulcerogenic agents, and hepatoprotectives. They have also been utilized in treating burns, bilious disorders, diabetes, cataracts, hypertension, leprosy, jaundice, snake bites, hemorrhoids, and piles (Table 4).

Despite its extensive traditional use, much of the indigenous knowledge about spine gourd remains undocumented. However, some ethnobotanical claims have recently been validated by phytochemical and animal studies. For example, tender fruits of spine gourd are consumed as health food for asthmatic patients and those with intestinal ulcers. The tender leaves and shoots of male plants are cooked as a leafy vegetable, particularly recommended for pregnant women and anemic patients among the "Paniya" community in Wayanad. Additionally, the medicinal uses of *M. sahyadrica* are well-known among the Malayarayar, Gowli, and Jain Kurba tribes, all of whom reside in forested areas. These tribes use tuber paste as an anti-inflammatory remedy for mastitis in milking cows and to treat painful eruptions, swellings, and breast inflammations in humans. In early pregnancy, tuber juice mixed with *Calotropis* leaves is used as an abortifacient.

## 5.5 *Nutritional Value of Spine Gourd*

Nutritionally, spine gourd has moderate levels of protein (19.38 mg/g), fat (4.7 mg/g), phenolic compounds (3.69 mg/g), and calorie content (4125/83 kcal/kg) (Table 5). It can be a good source of nutrients such as fiber, potassium, zinc, lipids, protein, and carbohydrates. Plant foods that derive more than 12% of their calorific value from protein are considered good protein sources (Asibey-

Berko and Tayie, 1999) and *M. sahyadrica* fits this criterion. Additionally, it is suitable for high-temperature food processing because of its low free-sugar concentration, which minimizes the risk of Maillard reactions and subsequent acrylamide formation.

## 5.6 Nutritional and Medicinal Properties of Spine Gourd

Several studies have confirmed that spine gourd is not only a nutritious vegetable but also widely used in folk medicine (Salvi, 2015). The kernels of its seeds are used in the varnish industry for producing drying oil, and the hexane-extracted compound from spine gourd has shown anti-feeding effects on *Spodoptera litura*. Additionally, spine gourd seed oil exhibits insecticidal properties, achieving 100% mortality with a 4% concentration (Bharathi et al., 2011).

Spine gourd contains numerous phytoconstituents and bioactive compounds, which are important in Western medicinal treatments and also have insecticidal effects. Phytochemical studies have isolated three triterpenes and two steroid compounds from spine gourd fruit (Jha et al., 2017; Talukdar and Hossain, 2014). The plant is rich in nutritional value, containing proteins, triterpenes, vitamin C, iodine, alkaloids, flavonoids, glycosides, amino acids, and trace amounts of manganese (Talukdar and Hossain, 2014). It is also a good source of carotene, thiamine, riboflavin, niacin (Salvi, 2015), chromium, and zinc (Talukdar and Hossain, 2014) (Table 6 and 7).

Nutritionally, 50 g of edible spine gourd fruit contains 42.1% moisture, 3.35 g carbohydrate, 1.9 g protein, 1.56 g fat, 1.5 g fiber, and 0.5 g minerals, including 16.5 mg calcium, 2.3 mg iron, and 4.21 mg phosphorus (Jha et al., 2017) (Table 5). Spine gourd has high market demand due to its nutritional value, long shelf life, ability to withstand long-distance transport, and good export potential (Rasul et al., 2004). Its fruit has numerous therapeutic properties, including diuretic, laxative, hepatoprotective, antivenomous, antihypertensive, anti-inflammatory, antiasthmatic, antileprosy, antidiabetic, and antidepressant effects. The leaves possess anthelmintic, aphrodisiac, antipyretic, anti-bronchitic, and analgesic properties, while the roots are used to treat stack, urinary, and bowel infections (Talukdar and Hossain, 2014) (Table 4).

In summary, spine gourds are not only important for their traditional and medicinal uses but also for their nutritional value. Despite the cultural taboos surrounding the cultivation of *M. sahyadrica*, it holds promise for both the food industry and the cosmetic and health care sectors. Spine gourd, with its rich nutrient profile and wide range of medicinal properties, continues to be a valuable crop with significant market potential.

**Table 4** Phytoconstituents of different plant parts of spine gourd and their effect.

| *S.No.* | *Plant parts* | *Extracted compound* | *Effect of each compound* |
|---|---|---|---|
| 1. | Fruits | Hexane extract | Have anti-inflammatory property and neuroprotective ability. |
| | | Methanol extract | Protect the liver cells from damages on hepatocytes; also have the anticancer effect. |
| | | n-Butanol extract | Can reduce pancreatic lipase activities. |
| | | Hexane and ethyl acetate extract | Provide protection against the anti-feeding activity of cotton leaf worm and also have anti-diabetic and antidepressant properties. |
| 2. | Roots | Alcoholic extraction | Inhibit the formation of free oxygen radical. |
| | | Ethanol extraction | Inhibit the growth and have anti-fertility effect. |
| | | Methanol | Anticancer effect. |
| 3. | Leaves | Methanol extract | Hepatoprotective and also have anti-hepatotoxicity effect. |
| | | Aqueous extract | Allelopathic activity on seedling growth and seed germination. |
| 4. | Seeds | Seed oil | Provide protection against Callosobruchus chinensis, and have antiallergic effect. |

(*Source:* Weerasinghe and Dahanayake, 2021)

**Table 5.** Nutritional composition in fruits of spine gourd (per 100 g edible fruit on dry weight basis).

| *Parameter* | *Nutrient composition* |
|---|---|
| Crude protein (g) | 19.38 |
| Crude lipid (g) | 4.70 |
| Carbohydrates (g) | 47.92 |
| Crude fiber (g) | 21.30 |
| Ash (g) | 6.70 |
| Calcium (mg) | 33.00 |
| Phosphorus (mg) | 42.00 |
| Iron (mg) | 4.60 |
| Calorific value (kcal) | 311.50 |

(*Source:* Tiwari et al., 2022)

**Table 6** Vitamin composition in fruits of spine gourd.

| *Vitamins* | *Vitamin composition (g/100g)* |
|---|---|
| Vitamin A | 2.5 |
| Vitamin $B_1$ (Thiamine) | 1.8 |
| Vitamin $B_2$ (Riboflavin) | 3.5 |
| Vitamin $B_3$ (Niacin) | 1.9 |
| Vitamin $B_5$ (Pantothenic acid) | 18.0 |
| Vitamin $B_6$ (Pyridoxine) | 4.3 |
| Vitamin $B_9$ (Folic acid) | 3.6 |
| Vitamin $B_{12}$ (Cyanocobalamin) | 4.0 |
| Vitamins $D_2$ & $D_3$ (Cholecalciferol) | 3.0 |
| Vitamin H (Biotin) | 6.5 |
| Vitamin K (Phytonadione) | 15.0 |

(*Source:* Jha et al., 2017)

**Table 7** Fatty acid composition in fruits of spine gourd.

| *Fatty acids* | *Value (%)* |
|---|---|
| Myristic acid | 3.58 |
| Palmitic acid | 12.15 |
| Stearic acid | 3.54 |
| Oleic acid | 56.25 |
| Linoleic acid | 22.51 |
| Alpha-linolenic acid | 1.94 |

(*Source:* Jha et al., 2017)

In summary, both Mangalore melon and spine gourd are rich in essential nutrients. Mangalore melon is an excellent source of vitamins A and C, which are crucial for maintaining healthy skin, vision, and immune function. It also contains dietary fiber, which aids in digestion. Spine gourd, on the other hand, is packed with antioxidants, vitamins, and minerals, including iron, calcium, and magnesium, making it a powerful food for combating anemia and bone-related disorders.

# 6. Role of Underutilized Vegetables in Food Security

## *6.1 The Role of Underutilized Vegetables in Enhancing Food Security: Focus on Mangalore Melon and Spine Gourd*

In regions like the Western Ghats, underutilized vegetables such as Mangalore melon (*Cucumis melo* var. *acidulus*) and spine gourd (*Momordica dioica*) play a crucial role in enhancing food security. These crops, though less commonly cultivated, offer significant potential for diversifying the food base, reducing reliance on major staples and providing year-round nutrition. By incorporating these vegetables into mainstream agriculture, India can take significant strides toward achieving food security and reducing malnutrition rates.

## *6.2 Diversifying the Food Base*

The overreliance on a few staple crops, such as rice and wheat, has made food systems vulnerable to climate change, pests, and diseases. Mangalore melon and spine gourd, adapted to the unique agroclimatic conditions of the Western Ghats, are resilient crops that can thrive in less favorable conditions where other crops may fail. By promoting the cultivation of these underutilized vegetables, farmers can diversify their crop portfolios, reducing the risk of crop failure and ensuring a more stable food supply.

Mangalore melon, traditionally grown in the Western Ghats, is known for its drought tolerance and ability to thrive in poor soils. It is typically cultivated during the monsoon season when water availability is limited. This makes it a valuable crop for regions prone to erratic rainfall patterns. Spine gourd, on the other hand, is a perennial climber that can be harvested for several years once established. Its ability to grow in marginal lands and under low-input conditions makes it an ideal crop for smallholder farmers.

## *6.3 Providing Year-Round Nutrition*

Underutilized vegetables such as Mangalore melon and spine gourd are rich in essential nutrients, making them important contributors to a balanced diet. Mangalore melon is a good source of vitamins A and C, which are crucial for immune function and skin health. It also contains significant amount of dietary fiber, which aids digestion and helps prevent chronic diseases such as diabetes and heart disease. Spine gourd is particularly high in vitamin C, iron, and other micronutrients that are often lacking in the diets of vulnerable populations.

Incorporating these vegetables into the daily diet can help address micronutrient deficiencies, which are prevalent in many parts of India. For example, iron deficiency anemia is a major public health concern, particularly among women and children. The high iron content in spine gourd can help combat this issue when included as part of a regular diet. Moreover, the diverse range of phytonutrients found in these vegetables offers additional health benefits, including antioxidant properties that protect against cellular damage.

## *6.4 Enhancing Food Security*

Integrating Mangalore melon and spine gourd into mainstream agriculture offers a sustainable approach to enhancing food security in India. These crops not only contribute to dietary diversity but also provide an additional source of income for farmers, thereby improving their livelihoods. By encouraging the cultivation and consumption of these underutilized vegetables, policymakers can support smallholder farmers, promote biodiversity, and strengthen the resilience of food systems.

Moreover, these crops have cultural significance in the Western Ghats, where they are often used in traditional ceremonies and local cuisine. Preserving and promoting these cultural practices can also play a role in maintaining the agricultural biodiversity of the region. Underutilized vegetables such as Mangalore melon and spine gourd have the potential to play a pivotal role in enhancing food security and reducing malnutrition in India. By recognizing their value and integrating them into agricultural policies and practices, India can build a more resilient and nutritionally secure future for its population.

## 7. Crop Improvement Activities for Underutilized Vegetables: Focus on Mangalore Melon and Spine Gourd

Underutilized vegetables such as Mangalore melon (*Cucumis melo* var. *acidulus*) and spine gourd (*Momordica dioica*) have traditionally received limited attention in crop improvement programs. However, recent interest in these crops, driven by their potential to contribute to food security and nutrition, has spurred research and development activities aimed at enhancing their agronomic performance, nutritional content, and resistance to pests and diseases. This document discusses ongoing efforts in crop improvement for these vegetables, highlighting breeding programs, germplasm evaluation, and the potential of these crops to contribute to sustainable agriculture.

### *7.1 Mangalore Melon*

Mangalore melon, a traditional crop of the Western Ghats, is known for its ability to thrive in harsh agroclimatic conditions, such as drought and poor soils. Despite its resilience, the crop has not been extensively improved through conventional breeding techniques until recently. The recent surge in interest has led to efforts to enhance its yield, shelf life, and resistance to pests and diseases, while maintaining or improving its nutritional value.

Mangalore melon is one of the less known and classified as a "orphan breeding crop" despite its nutritional value and breeding capabilities. There are various cultivars of the Mangalore melon that are grown along the coastline of Southern India. These have been characterized to allow for data collection of their growth and yield traits (Fergany et al., 2011; Manohar and Murthy, 2012). Despite the huge genetic variability South India offers, there has been negligence toward the genetic improvement of this crop. Mangalore melon has great prospects for future cultivation due to its many health benefits and resistance to biotic stress. To better serve the farmer and consumer along with the nations' interest, it is prudent to promote the cultivation of Mangalore melon by developing improved cultivars while focusing on the nutritional value and devising strategies for processing and value addition.

### *7.2 Breeding Programs*

The characterization and improvement of Mangalore melon varieties have become a focus of recent breeding programs. One such effort involved the characterization of the Mangalore melon cultivar "sambar southe" (SS-17). This cultivar was found to have dark green foliage and a monoecious sex form, which is typical of many melon species. The plant produces elongated fruits with smooth, thin white or light green skin, white firm flesh, and small yellow seeds. Unlike the sweeter and more aromatic dessert melons, "sambar southe" fruits are not sweet, not aromatic, and not climacteric, meaning they do not ripen further after being harvested. These characteristics, combined with the long shelf life of up to a year, make Mangalore melon an attractive candidate for breeding programs aimed at improving the shelf life of other melon varieties, such as cantaloupes, which typically have a shelf life of less than 15 days.

Mangalore melons also exhibit desirable agronomic traits, such as early flowering and a high number of primary branches per plant. The first male flowers appear approximately 26.9 days after sowing, followed by the first female flowers around 31 days after sowing. The plants produce monoecious flowers with a sex ratio of 9.06 (male:female), and the fruits are typically oval-shaped, with an average length of 19.12 cm and a width of 10.49 cm. The average fruit weight is 961.3 g, and the fruits have a peduncle length of 2.77 cm and a flesh thickness of 3.05 cm. Despite these favorable traits, Mangalore melons have certain undesirable characteristics, such as low sweetness, low flesh thickness, and a lack of aroma. These traits present challenges for breeders who aim to improve the crop's commercial viability (Table 8).

## 7.3 Crossability and Hybridization

The crossability of Mangalore melons with other melon varieties offers promising opportunities for crop improvement. Mangalore melons share the same chromosomal numbers as cantaloupe melons (*Cucumis melo* var. *cantaloupensis*), making them easily crossable. The crossability index of Mangalore melons with other melons ranges from 50% to 70%, indicating a good potential for successful hybridization. This crossability is particularly significant for breeding programs aimed at transferring desirable traits, such as the long shelf life of Mangalore melons, to commercially popular melon varieties (Kalgudi et al., 2021).

One of the key challenges in using Mangalore melons for breeding purposes is the presence of undesirable traits, such as low sweetness and low total soluble sugars. However, the identification of molecular markers linked to these and other traits, such as shelf life, could facilitate marker-assisted backcrossing (MABC) to introduce desirable traits from Mangalore melons into other melon varieties. For instance, breeders could use markers to track the inheritance of long shelf life and integrate this trait into dessert melons, thereby extending their marketability and reducing post-harvest losses.

Intraspecific and interspecific crossability studies have shown that Mangalore melons can hybridize with a wide range of *Cucumis melo* varieties, including *C. melo* var *cantaloupensis* (cantaloupe), *C. melo* var. *reticulatus* (netted melon), *C. melo* var. *inodorus* (winter melon), and others. These studies suggest that there is minimal sterility among progenies from these crosses, which is encouraging for the development of hybrid varieties. For example, hybrids between Mangalore melons and winter melons (*C. melo* var. *inodorus*) could combine the long shelf life of Mangalore melons with the desirable fruit qualities of winter melons.

## 7.4 Germplasm Evaluation and Resistance Traits

Germplasm evaluation is an essential component of crop improvement programs, providing the genetic material necessary for breeding efforts. In the case of Mangalore melons, several local genotypes have been identified with unique characteristics that could be valuable for breeding. For instance, the genotype "Sirsi local" was found to have a low incidence of cucurbit fruit fly infestation, powdery mildew, and downy mildew, making it a promising candidate for breeding disease-resistant varieties (Gondi et al., 2016). 80 germplasm accessions of Mangalore melon were collected from six South Indian states, namely, Karnataka, Kerala, Andhra Pradesh, Tamil Nadu, Telangana, and Goa. The 80 accessions/landraces were evaluated during 2018 and 2019 at the College of Horticulture, Sirsi (Karnataka). The landraces of Mangalore melon (*Cucumis melo* var. *acidulus*) found in Kerala are locally called as kanivellari, fruits that are elongated, golden yellow on maturity, with or without green patches, having average weight of 1.5 to 2 kg with white flesh, and have sweat odor like momordica. Whereas Andhra Pradesh and Telangana landraces are round, small to medium size, having average weight of 250 g to 750 g golden yellow with green patches to no stripe on maturity, commonly known as dosakaya. The flesh is thinner than any other state landraces with sour taste. However, Karnataka landraces have a lot of variability in fruit color, shape, and size. They have variable colors from green to yellow with white to green patches, with oblong to pyriform shape, weigh 2 to 2.5 kg, and have thicker, white, turgid, and crunchy flesh with good taste. In coastal Karnataka (Udupi, Mangalore, Kumta, Honnavar, Bhatkal, Kundhapur, etc.), usually observed are golden yellow fruits with white to green patches with medium size in each and every vegetable shop. In malnad region of Karnataka, especially in Sirsi and Sidhapur taluks of Uttara Kannada, green to dark green bigger size fruits are grown during summer which are stored up to 8–10 months till next crop (Shet et al., 2019).

Research conducted by the University of Agricultural Sciences (UAS), Bengaluru, evaluated 25 F1 hybrids of gourmet melons for quality, yield, and resistance to pests and diseases. The F1 hybrid "CMC GKVK 1 X CMC GKVK 2" demonstrated better performance in terms of fruit length and diameter, while "CMC GKVK 2 X CMC GKVK 4" performed well for fruit set percentage, number of fruits, and total fruit yield per vine. Additionally, "CMC GKVK 1 X CMC GKVK 12" exhibited high fruit flesh thickness and total soluble solids, which are important traits for market acceptance (Thyagaraj et al., 2014).

The diverse genetic resources of Mangalore melons offer valuable traits that can be harnessed for breeding purposes. For example, Southern Indian melon germplasm, particularly the *acidulus* and *momordica* varieties, display resistance to various pathogens, including Fusarium wilt, cucumber mosaic virus, and powdery mildew. Molecular markers linked to these resistance traits can be used to assess and incorporate disease resistance into commercial melon varieties, enhancing their resilience and reducing the need for chemical inputs (Manohar and Murthy, 2012).

**Table 8** The evaluation and characterization of Mangalore melon for morphological traits.

| *Traits* | *Description* |
|---|---|
| Germination (epigeal; hypogeal) | Epigeal |
| Duration (annual; perennial) | Annual (65–70 days) |
| Reproduction (sexual; asexual) | Sexual |
| Anthesis (4–8 am; 11–12 am; 7–9 am) | 7–9 am |
| Nature of roots (fibrous; tuberous) | Fibrous |
| Stigma color (green; yellow) | Yellow |
| Stamen (green; yellow) | Green |
| Seed shape (round, oval, oblong) | Oval |
| Stem pubescence (absent; presence) | Presence |
| Stem shape (angular; round) | Round |
| Flower color (light yellow; yellow; deep yellow) | Yellow |
| Fruit shape (oval; round; elongated; oblong) | All |
| Fruit color (green; green stripe; yellow; orange with white patches) | Green |
| Anthers (symmetrical; asymmetrical) | Symmetrical |
| Flesh color (orange; creamy white) | Creamy white |
| Stripes on fruit surface (absent; present) | Absent |

## 7.5 Crop improvement—Spine Gourd

Spine gourd is a unique vine that produces flowers between June and July, with fruiting extending from November to September. The plant is dioecious, bearing male and female flowers separately, a characteristic that makes it monosexual. Under natural environmental conditions, the fruit set rate is about 22%, but it can reach 100% with hand pollination (Sandilya et al., 2019). Each node on the vine produces both male and female floral buds, with male buds appearing in the second week of August and continuing to the first week of October, while female flowers open from the first or second week of September until the third week of October (Sandilya et al., 2019).

Male flowers of spine gourd are light yellow, approximately 2.8 cm long, with oblong or lanceolate petals. The flowers have five calyxes, which are linear or lanceolate, five corolla segments, and three stamens. Female flowers, on the other hand, have small yellow bracts and three nectar glands (Bawara et al., 2010). According to Sandilya et al. (2019), male flowers open around 4:00 am, and female flowers around 6:00 am, making the period between 5:00 am and 6:00 am ideal for hand pollination. The fruits are small, measuring 2–3 cm in diameter and weighing between 2.9 and 5 grams. They are short-beaked with a soft exocarp covered in spines. Immature fruits are green and turn yellow or light green as they mature (Salvi, 2015). The seeds, which are arranged along the central column of the fruit, are encased in a hard endocarp, providing resistance to pests such as caterpillars, gall flies, and root-knot nematodes (Anant et al., 2019).

The leaves of spine gourd are simple, broadly ovate, and deeply lobed, ranging from 3.8 to 10 cm in length (Bawara et al., 2010). The stem is round, branched, and furrowed, with elongated tendrils (Sandilya et al., 2019). The plant itself is a robust tendrillar climber with dark green foliage and spiny growth, reaching heights of 5–6 meters. The taproot is perennial, tuberous, and fusiform when young, becoming subglobose or irregularly bulged as it matures, measuring 10–18 cm long and 5–10 cm across, with a brownish outer skin and whitish-yellow inner flesh. The stems are stout, cylindrical, and quadrangular as they mature, with internodes measuring 5–10 cm. The nodes are quadrangular, blackish-green, and covered in long hairs. Tendrils are medium thick, unbranched, 8–15 cm long, with 4–5 cm of the base uncoiled, and the rest coiled.

The leaves of *M. sahyadrica* are heterophyllous, with petioles measuring 3–8 cm long and 1–1.5 mm in diameter. They are medium thick and longitudinally grooved above. The leaf blades are medium thick, ovate, and broadly triangular in outline, either 3–5 lobed or entire, and measure 8–18 cm. The base of the leaves is deeply cordate with a subangulate juncture with the petiole, sometimes hastate, and the apex is acute or acuminate. The margins of the leaves are highly variable, ranging from entire to undulate or coarsely crenulate. The lateral veins are arranged in 5–7 pairs, with the lower pair running close to the margin at the subangulate petiole juncture. The leaves are covered in short, scattered hairs on the outside and snowy white hairs on the inside.

Male flowers are axillary, solitary, or found in loose fascicles of 5–7 (and sometimes up to 15) flowers. In these fascicles, the lowermost flower is produced separately and earlier than the others. The peduncles are 2–5 cm long and dark green, while the pedicels are short, measuring 0.8–1 cm long, whitish green, and covered by an inflated bract that is 1–1.5 cm long, reniform, with cucullate margins. The calyx base is funnel-shaped, up to 8 mm long and 1 cm across, purplish black, with lobes that are free, up to 1.6 cm, yellowish-white at the center, and blackish-purple at the base and margins. The calyx lobes are elliptic-oblong, recurved at the apex, with scarious margins and apex, densely hairy within, and sparsely hairy without. The petals are five in number, free, fleshy, obovate, up to 4 × 1.5 cm, bright yellow with a greenish-yellow narrow base, and have prominent veins. Each petal bears a small tongue-like ciliate appendage near the base. The stamens are three in number, with two bearing a pair of anthers and the third with a single anther. The filaments are up to 3 mm long, and the anthers measure 2–3 × 1–2 mm, are extrorse, with dull black thecae that are 'S' shaped, and produce abundant orange pollen. The pollen grains are tricolpate.

Female flowers are solitary in the leaf axils, with peduncles measuring 0.5–2 cm, often less than 1 cm, and pedicels short, up to 5 cm long, subtended by a small rudimentary bract measuring 1–3 × 0.5–5 mm. The sepals are five in number, green, persistent, 0.8–1.3 × 1–3 mm, equal, lanceolate, acuminate at the apex, and densely glandular hairy within and without. The petals are five, free, fleshy, up to 4 × 2 cm, narrow, greenish-yellow and ciliate at the base, widening toward the middle, bright yellow, with 5–7 subparallel veins. The nectary is five in number, white, short cylindrical, alternating with the petals, and protected by a spur at the base of the petals. The ovary is inferior, oblong-ovoid, 1–1.5 × 0.2–0.4 cm, and densely covered with soft papillae of 1 mm length. The style measures up to 6 mm long, is whitish-yellow, and the stigma measures up to 4 × 9 mm, is cushiony, trifid, with each lobe further sub-lobed dichotomously (Table 9).

**Fig. 2** Morphological features of spine gourd (*M. sahyadrica*).

The fruits are broadly ellipsoid, ovoid to fusiform, or top-shaped, with a rounded blossom end and a rostrate distal end. They measure 5–7.5 × 3–4.2 cm, with a circumference of 9–12 cm, and weigh 35–50 grams. The fruits are dark green when unripe, turning bright orange upon ripening, and are densely covered with soft short spines measuring 2–4 mm long. The arils are sweet to taste, and the ripe fruits are aromatic with a slightly bitter flavor. The seeds are black, shiny when fresh but lose luster upon drying, round to slightly cogwheel-shaped, with warty dentate margins sculptured on the faces with irregular furrows and ridges. The seeds measure 0.2–0.3 × 0.2–0.3 cm, with a brittle seed coat, hard shell-like, and a very thin membrane that is smooth and blackish-green with conspicuous veins (Fig. 2). The endosperm is oily and distinctly aromatic when crushed (Joseph et al., 2007).

# 8. Other Underutilized Vegetable Crops

## *8.1 Amaranthus (Amaranthus viridis)*

Amaranthus can be either a short-lived perennial or an annual. Amaranthus is an inexpensive, high-protein, high-fiber food (Fig. 3). Vitamins, particularly pro-vitamin A, are abundant in it.

**Table 9** The evaluation and characterization of spine gourd for morphological traits.

| *Characterization of traits* | *M. sahyadrica/dioica* |
|---|---|
| Habit (annual, perennial) | Perennial |
| Reproduction (sexual, asexual) | Sexual |
| Sex expression (monoecious, dioecious ) | Dioecious |
| Germination (epigeal, hypogeal ) | Hypogeal |
| Roots (nontuberous, tuberous) | Tuberous |
| Adventitious root tubers (absent, present) | Absent |
| Stem pubescence (absent, presence) | Absent |
| Stem shape (angular, round) | Round |
| Petiole umbilical glands (absent, present) | Absent |
| Foliaceous bract (flower) (absent, present) | Present |
| Staminate flowers (solitary, in short racemes) | Solitary |
| Nature of flower bud bract (protective, nonprotective) | Protective |
| Anthesis (morning, evening) | Morning |
| Leaf lobing (Deeply lobed, angled) | Deeply lobed |
| Flower color | Yellow |
| Flower size (small, large) | Small |
| Corolla color (yellow, milky white) | Yellow |
| Bull's eye nectar guide (absent, present ) | Absent |
| Stigma color (green, yellow) | Yellow |
| Anther symmetry (symmetrical, asymmetrical) | Symmetrical |
| Fruit type (fleshy, dry) | Fleshy |
| Fruit shape (ovoid-oblongoid, pyriform) | Ovoid-oblongoid |
| Fruit ridges (absent, present ) | Absent |
| Fruit spines (absent, present) | Present |
| Tubercles on fruit surface (absent, present) | Absent |
| Fruit color at ripening (orange red, green ) | Orange red |
| Placenta color on ripening (red, white ) | Red |
| Aril on ripening (wet slimy, dry granular ) | Wet slimy |
| Intensity of bitterness (extreme, light/less/nil) | Light/less/nil |
| Seed sculpturing (present, absent ) | Present |
| Seed margin (dented, smooth) | Dented |
| Stem | Stout |
| Leaves | Thick, lobes prominent, dark green |
| Male flower—peduncle length (cm) | Up to 5 |
| Male flower—pedicel length (mm) | 10 |
| Male flower—sepal | Linear, 6 mm broad, margin undulate |
| Female flower—sepal | 7 mm long, margin undulate |
| Corolla (male & female) | Up to 8 cm in diameter, base yellow |

*Contd.*

**Table 9** *Contd.*

| *Characterization of traits* | *M. sahyadrica/dioica* |
|---|---|
| Petals (male & female) | Obovate, acute or acuminate |
| Stigma | Bright yellow, broadly "V" |
| Ovary | Hypanthium 8 mm long, apex |
| Fruit | Ovate-ellipsoid, with 2 blunt ends |
| Single fruit weight (g) | 35–50 |
| Seed length | 6.2–7.7 |
| Seed width | 5.4–6.9 |
| Thickness (mm) | 3.5–4.0 |
| No. of seeds/fruits | 35–50 |

Amaranthus leaves are rich in minerals, carotene, and phytochemicals including phenolic compound and isothiocyanates, which have potent antioxidant qualities and have been shown to help prevent and inhibit diseases such as arteriosclerosis, cancer, and aging.

**Fig. 3** Morphological features of amaranthus (*Amaranthus viridis*).

## *8.2 Asparagus (Asparagus officinalis)*

Asparagus is a herbaceous perennial plant, and in Hindi it is known as shatawar. Its tastiest parts are its soft, succulent branches, called "pears," which are eaten as vegetables and made into soups. It produces for ten to fifteen years. Asparagine is a white, crystalline chemical found in asparagus juice, and it has diuretic effects. It is very important in diet due to its high cellulose content, vitamins, and essential salts. It can be canned and frozen in large quantities (Fig. 4).

## *8.3 Lettuce (Lactuca sativa)*

Lettuce is the salad vegetable grown in the largest area. The head and scrumptious edges are eaten as salad. It was originally cultivated for its oil, which is extracted from the seeds. It is high in iron, calcium, and vitamin A. There are four botanical varieties of lettuce. (1) type of head, (2) type of cos, (3) type of leaf, (4) type of stem. This is an annually herbaceous plant giving simple grains (Fig. 5). It has antibacterial properties as well as protects against cancer, heart disease, improper sleep, and anxiety.

**Fig. 4** Morphological features of asparagus (*Asparagus officinalis*).

**Fig. 5** Morphological features of lettuce (*Lactuca sativa*).

**Fig. 6** Morphological features of elephant foot yam (*Amorphophallus paeoniifolius*).

## 8.4 Elephant Foot Yam (Amorphophallus paeoniifolius)

Also known as stink lily or white spot giant arum. In Hindi, it is called zamikand, suran, and sooran. It is a perennial herbaceous plant (Fig. 6). It's not just a vegetable, but one used to create pickles, chips, and other foods. It is also grown for its medicinal properties. It contains triacontane, stigmasterol, betulinic acid, and lupeol. It has blood-purifying qualities and is used to treat rheumatism, piles, tumors, and asthma. In elephant foot yams, calcium oxalate contributes acridity and irritates the mouth and throat.

## 8.5 Oriental Pickling Melon (C. melo var. conomon Thunb.)

The eastern pickled melon, or *C. melo* var. *conomon* Thunb., is the oldest type of cultivated melon. Kelong *C. melo* var. *conomon* includes the type conomon (pickling type), makuwa (dessert type), and chinensis (Munger and Robinson, 1991). Other sources report that in Korea and southern regions of China, the *conomon* and *makuwa* types have been cultivated (Kitamura et al., 1975). Unripe fruits in these varieties are eaten fresh (Fig. 7), cooked, or pickled, whereas smooth-skinned fruits are produced in the latter ripening stage with sweet, fragrant flesh. Juice prepared from the seeds serves as a remedy for dyspepsia. *Makuwa* and *conomon* types were believed to have evolved in India and entered China from the west through the "Silk Road" (Kitamura et al., 1976; Kato et al., 2001). They also suggested that the wild melon (var. *agrestis*) in China was the source of this melon.

**Fig. 7** Morphological features of oriental pickling melon (*Cucumis melo* var. *conomon* Thunb.).

## 8.6 Pointed Gourd (Trichosanthes dioica Roxb.)

Pointed gourd goes by the name "king of gourds." In Hindi, it is spelled parwal or pravar, and it is also commonly referred to as "green potato" in India. Its highly nutritious leaves are used as vegetables. It is a rich source of minerals, vitamins A and C, and carbohydrates. According to Nadkarnier (1996), its juice is used for the treatment of alopecia, liver enlargement, and subacute cases. This plant is a dioecious climber with a perennial root system (Fig. 8).

## 8.7 Snap Melon (Cucumis melo var. momordica)

Snap melon (*Cucumis melo* var. *momordica*), also known as "phut" or "phoont," is an indigenous crop to India, with a name that reflects its characteristic of splitting upon maturity. This unique trait has made it a recognizable and culturally significant fruit in various parts of the country. The snap melon is particularly popular in Kerala, coastal Karnataka, and Goa. Snap melon is highly valued not only for its culinary uses but also for its nutritional and medicinal properties (Fig. 9). The seeds of snap melon are rich in oil, containing between 12.5% to 39.1% oil content. This oil is edible and has

**Fig. 8** Morphological features of pointed gourd (*Trichosanthes dioica* Roxb.).

**Fig. 9** Morphological features of snap melon (*Cucumis melo* var. *momordica*).

a nutty flavor, though it is challenging to extract due to the small size and fibrous coating of the seeds. The fruit itself is known for its stomachic properties, aiding in digestion, while the flowers have expectorant and emetic qualities, making them useful in traditional medicine. In recent years, snap melon juice has gained popularity as a refreshing beverage, especially during the summer months. Its cooling effects make it an ideal drink for reducing body heat, and it is also used as a natural skin cleanser and a first-aid treatment for burns and abrasions.

## *8.8 Karchikai (Momordica cymbalaria)*

*Karchikai* is widely distributed in the tropical regions of Africa and India, particularly in the South Indian states of Andhra Pradesh, Karnataka, Maharashtra, and Tamil Nadu. Despite its adaptability to the Kharif and Rabi seasons, the crop is not regularly cultivated by farmers. It thrives in black soils, where it is often found as a weed in fields cultivated with sorghum, Bengal gram, and onion. Historically, *karchikai* was considered a weed, but its tubers have long been used for medicinal purposes (Fig. 10). Due to the increasing recognition of its nutritional value, the fruit of *M. cymbalaria* has recently gained attention as a vegetable.

The nutritional studies of *karchikai* fruits have revealed a rich composition of essential nutrients, making it a highly valuable food source. The fruits are particularly rich in calcium, potassium, and vitamin C, in addition to having a high crude fiber content. Notably, the calcium content of *karchikai* is reported to be three times higher than that of bitter gourd (*Momordica charantia*), and its ascorbic acid (vitamin C) and potassium levels are twice as high. These nutrient-rich fruits can help address nutritional deficiencies, especially in regions where calcium and vitamin C are lacking in the diet.

**Fig. 10** Morphological features of karchikai (*Momordica cymbalaria*).

*Karchikai* is a nutritional powerhouse and a medicinal plant with a wide range of therapeutic uses. Various parts of the plant, including the fruits, tubers, and leaves, have been found to contain beneficial compounds such as flavonoids, steroids, triterpenes, and saponins. The fruit extracts of *M. cymbalaria* have demonstrated antidiabetic, hypolipidemic, antiulcer, and antidiarrheal properties, making it a valuable natural remedy for these conditions (Grover et al., 2002; Parvati and Kumar, 2002; Rao et al., 1999; Koneri et al., 2007).

## Conclusion

Underutilized vegetables have an abundant amount of nutritional potential, therapeutic qualities, and resilience to harsh environmental circumstances. Due to incomplete information on their performance and input needs, an absence of planting materials, a lack of information on how they can fit into production processes, and the difficulty of producing vegetables locally like the major cultivated species, these underutilized vegetables are still neglected. To ensure future security of nutrition and food, it is vital to start an initiative on genetic resource exploration, management, utilization, and enhancement of these underutilized vegetable crops. Additionally, the production of underutilized veggies will address the lack of per capita consumption availability and the nutritional gap, while also creating jobs, raising rural residents' incomes, and possibly even boosting the country's economy.

By promoting the cultivation and consumption of these underutilized crops, India can diversify its food sources, improve nutritional outcomes, and support sustainable agricultural practices. Integrating these nutrient-rich plants into mainstream agriculture can help address malnutrition and enhance food security, particularly for vulnerable populations. Embracing the potential of these crops not only preserves biodiversity but also leverages traditional knowledge and practices, contributing to a more resilient and nutritious food system.

Mangalore melon and spine gourd represent underutilized vegetable crops with significant potential to enhance food and nutritional security in India. These two vegetable crops, widely grown in the Western Ghats and the northeast regions of India, offer a rich source of nutrients with high antioxidant content and medicinal properties. Their adaptability to local climatic conditions and potential for higher productivity make them a promising candidate for wider cultivation.

## References


Aberoumand, A. and Deokule, S.S. (2009). Studies on nutritional values of some wild edible plants from Iran and India. *Pakistan Journal of Nutrition*, 8(1), 26–31.

Aberoumand, A. and Deokule, S.S. (2010). Elements evaluation of some edible vegetables and fruits of Iran and India. *Asian Journal of Agricultural Sciences*, *2*(1), 35–37.

Anant, P., Painkra, K.L., Painkra, G.P., Tiwari, J.K. and Bhagat, P.K. (2019). Seasonal incidence and extent of damage by cucurbit fruit fly, Bactrocera Cucurbitae (Coq.) on spine gourd (*Momordica Dioica* Roxb.). *Journal of Plant Development Sciences*, *11*(9), 543–546.

Anjana, M., Swathi, V., Ramya Sai, A., Divya, N. and Sunisha, Y. (2019). A review on Momordica dioica fruits. *J. Adv. Plant Sci. 2*(2), 201.

Asibey-Berko, E. and Tayie, F.A.K. (1999). Proximate analysis of some under-utilized Ghanaian vegetables. *Ghana Journal of Science*, *39*, 91–96.

Bawara, B., Dixit, M., Chauhan, N.S., Dixit, V.K. and Saraf, D.K. (2010). Phyto-pharmacology of Momordica dioica Roxb. ex. Willd: a review. *International Journal of Phytomedicine*, *2*(1).

Behera, T.K., Behera, S., Bharathi, L.K., John, K.J., Simon, P.W. and Staub, J.E. (2010). Bitter gourd: botany, horticulture, breeding. *Horticultural Reviews*, *37*, 101–141.

Bharathi, L.K. and John, K.J. (2013). Taxonomy and Biosystematics. *Momordica* Genus in Asia – An Overview, Springer, New Delhi, pp. 45–60.

Bharathi, L.K., Munshi, A.D., Vinod, Chandrashekaran, S., Behera, T.K., Das, A.B., John, K.J. and Vishalnath. (2011). Cytotaxonomical analysis of *Momordica* L.(*Cucurbitaceae*) species of Indian occurrence. *Journal of Genetics*, 90, 21–30.

Chakravarthy, H.M. (1982). Fascicles of flora of India-11 Cucurbitacae. *Botanical Survey of India*, 136.

Chomicki, G., Schaefer, H. and Renner, S.S. (2020). Origin and domestication of Cucurbitaceae crops: insights from phylogenies, genomics and archaeology. *New Phytologist*, 226(5), 1240–1255.

Endl, J., Achigan-Dako, E.G., Pandey, A.K., Monforte, A.J., Pico, B. and Schaefer, H. (2018). Repeated domestication of melon (*Cucumis melo*) in Africa and Asia and a new close relative from India. *American Journal of Botany*, 105(10), 1662–1671.

Esteras, C., Formisano, G., Roig, C., Díaz, A., Blanca, J., Garcia-Mas, J., Gómez-Guillamón, M.L., López-Sesé, A.I., Lázaro, A., Monforte, A.J. and Picó, B. (2013). SNP genotyping in melons: genetic variation, population structure, and linkage disequilibrium. *Theoretical and Applied Genetics*, 126, 1285–1303.

FAO, (2023). FAOSTAT. Canada: Food and Agriculture Organization of the United Nations [cited 2024 Aug]. Available from: https://www.fao.org/faostat/en/

Fergany, M., Kaur, B., Monforte, A.J., Pitrat, M., Rys, C., Lecoq, H., Dhillon, N.P.S., Dhaliwal, S.S. (2011). Variation in melon (*Cucumis melo* L.) landraces adapted to the humid tropics of southern India. *Genetic Res. Crop Evolution*, 58, 225–243.

Gondi, S., Ganiger, V.M., Bhuvaneshwari, G., Madalageri, M.B., Kotikal, Y.K. and Manjunatha, G. (2016). Evaluation of oriental pickling melon (*Cucumis melo* L. var. *conomon*) genotypes for pest and disease reaction under northern dry zone of Karnataka. *Journal of Environmental Ecology*, *7*(1), 29–36.

Grover, J.K., Yadav, S., Vat, V. (2002). Medicinal plants of India with anti-diabetic potential. J Ethnopharmacol. 81: 81–100.

Hiremata, V., Shet, R.M., Gunnaiah, R., Anjappa, P. and Manjunathagowda, D.C. (2023). Mangalore melon (*Cucumis melo* ssp. *agrestis* var. *acidulus*): a neglected and underutilized vegetable of the Western Ghats of India. *Genetic Resources and Crop Evolution*, *70*(6), 1895–1902.

Indiastat. 2022. www.indiastat.com

Jeffrey, C. (1980). A review of the *Cucurbitaceae*. *Botanical Journal of the Linnean Society*, *81*(3), 233–247.

Jena, A.K., Deuri, R., Sharma, P. and Singh, S.P. (2018). Underutilized vegetable crops and their importance. *Journal of Pharmacognosy and Phytochemistry*, *7*(5), 402–407.

Jha, D.K., Koneri, R. and Samaddar, S. (2017). Potential bio-resources of *Momordica dioica* roxb: a review. *International Journal of Pharmaceutical Sciences Review and Research*, *45*(2), 203–209.

John, K.J. and Antony, V.T. (2010). A taxonomic revision of the genus Momordica L.(*Cucurbitaceae*) in India. *Indian Journal of Plant Genetic Resources*, *23*(02),172–184.

Jompitak, Y. (2002). Nature Food, Native Vegetable. Orient Store Press. Bangkok, 132.

Joseph J.K., Antony, V.T. and Roy, Y.C. (2007). On the occurrence, distribution and taxonomy of *Momordica subangulata* Blume subsp. *renigera* (G. Don) de Wilde in India. *Genetic Resources and Crop Evolution*, *54*, 1327–1332.

Joseph, J.K. (2005). Studies on ecogeography and genetic diversity of the genus Momordica L. India. Dissertation, Mahatma Gandhi University, Kottayam, Kerala.

Kato-Emori, S., Higashi, K., Hosoya, K., Kobayashi, T. and Ezura, H. (2001). in melon Cucumis melo L. reticulatus. Mol Genet Genomics, 265, 135–142.

Kattukunnel, J.J. and Antony, V.T. (2004). *Momordica sahyadrica* sp. nov. (Cucurbitaceae), an endemic species of Western Ghats of India. *Nordic Journal of Botany*, *24*(5), 539–542.

Kacar, Y., Simsek, O., Solmaz, I., Sari, N. and Mendi, Y. (2012). Genetic diversity among melon accessions (*Cucumis melo* L.) from Turkey based on SSR markers. *Genetic Molecular Research*, 11:4622–4631.

Kalgudi, A. (2021). Genetic improvement of muskmelon (*Cucumis melo* L.) through intraspecific hybridization with culinary melon (*Cucumis melo* var *acidulus*), Doctoral dissertation, University of Horticultural Sciences, Bagalkot.

Kamagoud, S., Ratnakar, M.S., Sandhyarani, N., Shivanand, H., Hanchinmani, C.N., and Prashanta, A. (2018), Assessment of genetic variability among oriental pickling melon (Cucumis melo var. conomon) genotypes. *Int. J. Chem. Studies, 6*(4), 2630–2633.

Kerje, T. and Grum, M.I.K.K.E.L. (2000). The origin of melon, Cucumis melo: a review of the literature. In *VII Eucarpia Meeting on Cucurbit Genetics and Breeding*, *510*, 37–44.

Kirkbride, J.H. (1993). *Biosystematic monograph of the genus Cucumis (Cucurbitaceae): botanical identification of cucumbers and melons*. Parkway Publishers, Inc.

Kitamura, T., T. Umemoto, Iwata and T. Akazawa. (1975). Studies on the storage of melon fruits. II. Changes of respiration and ethylene production during ripening with reference to cultivars. *J. Japan Soc Hort Sci.*, 44: 197–203.

Kitamura, T., Umemoto, T., Iwata, T. and Akazawa, T., (1976). Studies on the Storage of Melons II. Changes in Production of Volatiles during Ripening with Reference to Cultivars. Journal of the Japanese Society for Horticultural Science, *44*(4), 417–421.

Koli, S.P. and Murthy, H.N. (2013) Estimation of phenotypic divergence in a collection of Cucumis melo from Kerala State, Southern India. *Agric Sci* 4:21–31. https:// doi. org/ 10.4236/ as. 2013. 47A004

Koneri, R., Saraswati, C.D., Balaraman, R. and Ajeesha, E.A. (2007). Antiimplantation activity of the ethanolic root extract of Momordica cymbalaria Fenzl in rats. Indian Journal of Pharmacology, *39*(2), 90–96.

Lakshmi, LM., Lingaiah, H., Rao, A.M., Raju, T., Pitchaimuthu, M., and Gowda, D. (2017). Variability and genetic divergence studies for economic traits in indigenous oriental pickling melon (Cucumis melo var. conomon) genotypes. *Electronic J. Pl. Breeding*, 8, 365–370.

Manohar, S.H. and Murthy, H.N. (2012). Estimation of phenotypic divergence in a collection of *Cucumis melo*, including shelf-life of fruit. *Scientia Horticulturae*, *148*, 74–82.

Manohar, S.H. and Murthy, H.N. (2014). Fatty acid profile of *Cucumis melo* var. *acidulus* (culinary melon) seed oil. *Journal of the American Oil Chemists' Society*, *91*(5),815–816.

Munger, H.M. and Robinson, R.W. (1991) Nomenclature of *Cucumis melo* L. *Cucurbit Genet Coop Rep*, 14, 43–44.

Naudin, C. (1856). *Nouvelles recherches sur les caractères spécifiques et les variétés des plantes du genre Cucurbita*. V. Masson.

Nhi, P.T.P., Akashi, Y., Hang, T.T.M., Tanaka, K., Aierken, Y., Yamamoto, T., Nishida, H., Long, C. and Kato, K. (2010). Genetic diversity in Vietnamese melon landraces revealed by the analyses of morphological traits and nuclear and cytoplasmic molecular markers. *Breeding Science*, *60*(3), 255–266.

Nimmakayala, P., Tomason, Y.R., Abburi, V.L., Alvarado, A., Saminathan, T., Vajja, V.G., Salazar, G., Panicker, G.K., Levi, A., Wechter, W.P. and McCreight, J.D. (2016). Genome-wide differentiation of various melon horticultural groups for use in GWAS for fruit firmness and construction of a high resolution genetic map. *Frontiers in Plant Science*, *7*, 1437.

Parvathi, S. and Kumar, V.J.F. (2002). Studies on chemical composition and utilization of the wild edible vegetable athalakkai (Momordica tuberosa). *Plant Foods for Human Nutrition, 57*(3), 215–222.

Pitrat, M., Hanelt, P. and Hammer, K. (2000). Some comments on infraspecific classification of cultivars of melon. In *VII Eucarpia Meeting on Cucurbit Genetics and Breeding, 510*, 29–36.

Pradeepkumar, T., Peter, K.V. and Sangeetakutty, M. (2022). Seven decades of vegetables research in India. *International Journal of Innovative Horticulture, 11*(2), 135–157.

Ramesh Babu, R. and Hariprasad Rao, N. (2018). Studies on genetic variability, heritability and genetic advance in oriental pickling melon (*Cucumis melo*. var. *conomon*) genotypes. *International Journal of Pure Applied Biosciences, 6*, 1042–1047.

Rasul, M.G., Hiramatsu, M. and Okubo, H. (2004). Morphological and physiological variation in kakrol (*Momordica dioica* Roxb.).

Robinson, R.W. and Decker-Walters, D.S. (1997). Cucurbits. CABI, Wallingford, Oxford, 97–101.

Roy, A., Bal, S.S., Fergany, M., Kaur, S., Singh, H., Malik, A.A., Singh, J., Monforte, A.J. and Dhillon, N.P.S. (2012). Wild melon diversity in India (Punjab state). *Genetic Resources and Crop Evolution, 59*(5), 755–767.

Salvi, J. (2015). Nutritional composition of *Momordica dioica* fruits: as a wild vegetable. *Journal of Food and Pharmaceutical Sciences, 3*(2).

Sandilya, V.K., Kumar, R., Yadav, N.K., Ram, N. and Tiwari, J.K. (2019). Heterosis for yield attributing traits in spine gourd (*Momordica dioica* Roxb). *Journal of Pharmacognosy and Phytochemistry, 8*(6), 1847–1852.

Schaefer, H. and Renner, S.S. (2011). Phylogenetic relationships in the order Cucurbitales and a new classification of the gourd family (*Cucurbitaceae*). *Taxon, 60*(1), 122–138.

Sebastian, P., Schaefer, H., Telford, I.R. and Renner, S.S. (2010). Cucumber (*Cucumis sativus*) and melon (*C. melo*) have numerous wild relatives in Asia and Australia, and the sister species of melon is from Australia. *Proceedings of the National Academy of Sciences, 107*(32), 14269–14273.

Shet, R.M., Shantappa, T., Hongal. S., Veeresh, S.M., Sujatha, P., Gunnaiah R. and Jagadeesha R.C. (2019). Collecting and evaluation of culinary melon germplasm accessions in India. *South East Asian Vegetable Symposium*, Malacca, Malaysia, from 9–11 July 2019, 78–79.

Shet, R.M., Hongal, S., Shantappa, T., Veeresha, S.M. and Jagadeesha, R.C. (2022). Survey, collection and preliminary observation of culinary melon germplasm from Southern India. *Indian Journal of Plant Genetic Resources, 35*(1), 85–93.

Shet, R.M., Shantappa, T., Hongal, S., Gunnaiah, R. and Jagadeesha, R.C. (2022). Evaluation of postharvest shelf life of south Indian culinary melon (*Cucumis melo* var. *acidulus*) accessions. *Journal of Applied Horticulture, 24*(1).

Silpa, R., George, S.T., Anitha, P., Pradeepkumar, T., Sujatha, V.S., Dijee, B., Berin, P., and Laly John, C. (2020). Genetic divergence analysis in oriental pickling melon (Cucumis melo var. conomon Mak.). *Res. J. Agril. Sci, 11*(1), 115–119.

Stepansky, A., Kovalski, I. and Perl-Treves, R. (1999). Intraspecific classification of melons (*Cucumis melo* L.) in view of their phenotypic and molecular variation. *Plant Systematics and Evolution, 217*, 313–332.

Swamy, K.R.M. (2017). Origin, distribution and systematics of culinary cucumber (*Cucumis melo* subsp. *agrestis* var. *conomon*). *Journal of Horticultural Sciences, 12*(1),1–22.

Swamy, K.R.M. and Nath, P. (2020). Culinary melon of South India: a review. *Vegetable Science, 47*(2), 157–175.

Talukdar, S.N. and Hossain, M.N. (2014). Phytochemical, phytotherapeutical and pharmacological study of *Momordica dioica. Evidence-Based Complementary and Alternative Medicine, 2014*(1), 806082.

Tanaka, K., Nishitani, A., Akashi, Y., Sakata, Y., Nishida, H., Yoshino, H. and Kato, K. (2007). Molecular characterization of South and East Asian melon, *Cucumis melo* L. and the origin of Group *Conomon* var. *makuwa* and var. *conomon* revealed by RAPD analysis. *Euphytica, 153*, 233–247.

Thyagaraj, G.N., Puttaraju, T.B., Santhosh, N. and Padmaraja, S.R. (2014). Evaluation of F1 hybrids in Oriental pickling melon for yield and quality attributes. *Trends in Biosciences, 7*(17), 2518–2523.

Tiwari, J.K., Ameen, G., Sandilya, V.K., Kumar, S. and Kaushik, S.K. (2022). Cultivation practices of an underutilized, yet wholesome and remunerative vegetable: Spine gourd. *Indian Horticulture, 67*(2).

Van Rheede, H. (2003). Hortus Malabaricus. *English ed., KS Manilal.*

Vidya, B. (2012). Madras cucumber: Finding the origins of this South Indian vegetable. *The Economic Times*, December 9, 2012.

Vinutha, B., Kotikal, Y.K., Venkateshalu, V., Manjunath, G. and Patil, S. (2017). Insect-pests composition and natural enemies association on oriental pickling melon (Cucumis melo var. conomon). *J. Entomology and Zoology Studies*, 5, 1838–1840.

Weerasinghe, M.G.W.K. and Dahanayake, N. (2021) *Momordica dioica* Roxb. (Spine Gourd) – An underutilized vegetable and medicinal plant in Sri Lanka. *Int. J. Minor Fruits Med. Aromat. Plants*, 7, 100–104.

# CHAPTER 9

# Grafting Innovations in Vegetable Crops to Enhance Nutritional Quality for Food Security

Ruksana[1], Shoeb Ahmed[2], Jitendra Kumar[3*], Pradeep Singh[3] and Harish Pal Bhati[4]

## 1. Introduction

For many years, East Asians have used grafting as a traditional horticultural technique to overcome the challenges of extensive cultivation on limited arable land. The grafting method, introduced in the late 20th century to Europe and other regions, has evolved to include improved methods suitable for commercial production (Singh et al., 2017). This method includes the joining of parts of plants, typically the scion and rootstock, and produces a single plant that incorporates the advantageous characteristics of both components (Shivran et al., 2022; Sallaku et al., 2022; Loupit et al., 2023; Harris et al., 2023). Grafting enhances plant health, increases yield, and improves resistance to diseases and environmental stressors. Research highlights the effectiveness of grafting technology in increasing tolerance to heavy metal toxicity and abiotic factors, for example, extreme temperature, salt, and drought (Bahadur & Kumar, 2024; Faisal et al., 2024). For instance, grafted plants such as tomatoes show osmolytes, photosynthesis, and antioxidant higher levels, leading to improved fruit harvests and enhanced nutritional value, including elevated levels of β-carotene, lycopene, and ascorbic acid (Bahadur & Kumar, 2024; Faisal et al., 2024). Furthermore, grafting with suitable rootstocks positively impacts the nutritional quality of fruits, including protein content, contributing to overall improvements in vegetable crop production and quality (Kyriacou et al., 2020; Nordey et al., 2020; Majhi et al., 2023; Bahadur & Kumar, 2024; Faisal et al., 2024). Researchers have addressed global nutritional deficiencies in micronutrients, for example, minerals (such as Fe and Zn) and vitamins (A, C, and E), by grafting different vegetable crop varieties to create plants with higher yields and increased levels of vital nutrients such as minerals and vitamins (Ramakrishnan, 2002; Akhtar et al., 2013; Bailey et al., 2015; Kyriacou et al., 2017; Djidonou et al., 2016; Harika et al., 2017; Hollic & Kubota, 2022; Manjarrez et al., 2023). The important advantage of grafting is improvement in the root system. The rootstock, chosen for its robustness and efficiency in nutrient absorption, provides a stronger foundation for the scion, resulting in improved water and nutrient

[1] Eternal University, Baru Sahib, Sirmaur, Himachal Pradesh, India.
[2] Department of Genetics and Plant Breeding, Ch. Charan Singh University, Meerut, U.P., India.
[3] BRIC–National Agri-Food & Biomanufacturing Institute (NABI), Sector 81, Mohali, Punjab, India.
[4] Deparment of Botany, Shanti Devi Government College, Jewar, Gautam Buddha Nagar, U.P., India.
* Corresponding author: jitendra_biotech09@rediffmail.com

DOI: 10.1201/9781003584100-9

uptake and leading to healthier, more nutritious produce (Orosco-Alcalá et al., 2021; Chen et al., 2022; Remzeena & Anitha, 2023). After comparisons of grafted with non-grafted plants, grafted plants frequently have higher concentrations of antioxidants, vitamins, and minerals (Nordey et al., 2020). Grafting enhances disease resistance, particularly against pathogens (soilborne), for example, nematodes and *Fusarium* and *Verticillium* wilt, which are known to cause root knotting. The use of suitable rootstocks promotes the formation of a strong root system, improving plant resilience and overall health. As a result, dependence on chemical inputs such as fertilizers and pesticides is significantly reduced (Karthiga et al., 2023; Faisal et al., 2024; Naik et al., 2024). Grafted plants exhibit enhanced tolerance to both biotic and abiotic factors that result in improved growth, yield, and quality. This makes grafting a sustainable and efficient approach for reducing the adverse effects of pathogens and environmental challenges (Karthiga et al., 2023; Bahadur & Kumar, 2024; Faisal et al., 2024). Grafting growth in vegetables is supported by the demand and necessity for enhanced yields, resilience to abiotic stressors, and tolerance to soilborne diseases (Louws et al., 2010; Lee et al., 2010; Schwarz et al., 2010; Kumar et al., 2015; Rouphael et al., 2016). While grafting has traditionally not been employed to improve the quality of fruit, recent studies have demonstrated significant benefits. For example, grafting melons onto robust rootstocks resulted in a 57% to 90% increase in marketable yields, extended harvest periods, and reduced symptoms of sudden wilt (Ohletz & Loy, 2021). Similar to this, under high tunnel culture conditions, grafted pepper plants demonstrated increased rates of fruit production, vegetative growth, and survival (Soltan et al., 2017). Various grafting techniques are employed depending on the crop type and available facilities. These methods include (i) hole insertion, (ii) cleft grafting, (iii) pin grafting, (iv) tongue approach grafting, and (v) splice grafting (Lee et al., 2010). The success of grafting can be influenced by the diameter differences between scion and rootstock, and aligning these components can be challenging, potentially impacting grafting success (Lee et al., 2010). Innovations in grafting techniques and the development of new rootstock and scion combinations hold significant potential for producing more nutritious, resilient, and productive crops. These innovations are important to ensure food security globally as they promote a food supply that is stable and sustainable (Kyriacou et al., 2020; Shivran et al., 2022; Loupit et al., 2023; Caradonia et al., 2023). Grafting can enhance plant resilience to climate change and environmental stressors, potentially revolutionizing agricultural practices and addressing pressing global food system challenges. This chapter explores the techniques and innovations in vegetable grafting that enhance nutritional quality, providing an overview of how grafting improves nutrient uptake and utilization, discussing specific methods for different vegetable species, and highlighting recent research findings. By examining the benefits of grafting, this chapter underscores its potential to transform agricultural practices and support global food security. The methods of grafting and its benefits are depicted in Fig. 1.

## 2. Fundamentals of Grafting

Grafting is a refined horticultural technique that unites a rootstock and scion to create an integrated plant that combines the strengths of both components, enhancing plant resilience and productivity. Choosing the right rootstock is critical because it offers a strong root system that is necessary for healthy development and better uptake of nutrients and water, helping the plant endure environmental challenges such as salt, drought, and severe temperatures. Furthermore, to minimize the need for chemical treatments and advance sustainable agriculture, the selection of rootstocks is on the basis of resistance to diseases such as *Verticillium* wilt and *Fusarium* wilt (Tirupathamma et al., 2019). The selection of the scion is based on desirable characteristics, including fruit quality, yield potential, and resilience to environmental stresses, which significantly impact both the commercial and nutritional value of the final crop. For instance, in tomato grafting, scions are chosen for superior fruit characteristics and resistance to foliar diseases, ensuring the final product meets market demands (Nie & Wen, 2023). Various types of grafting methods are described in Table 1.

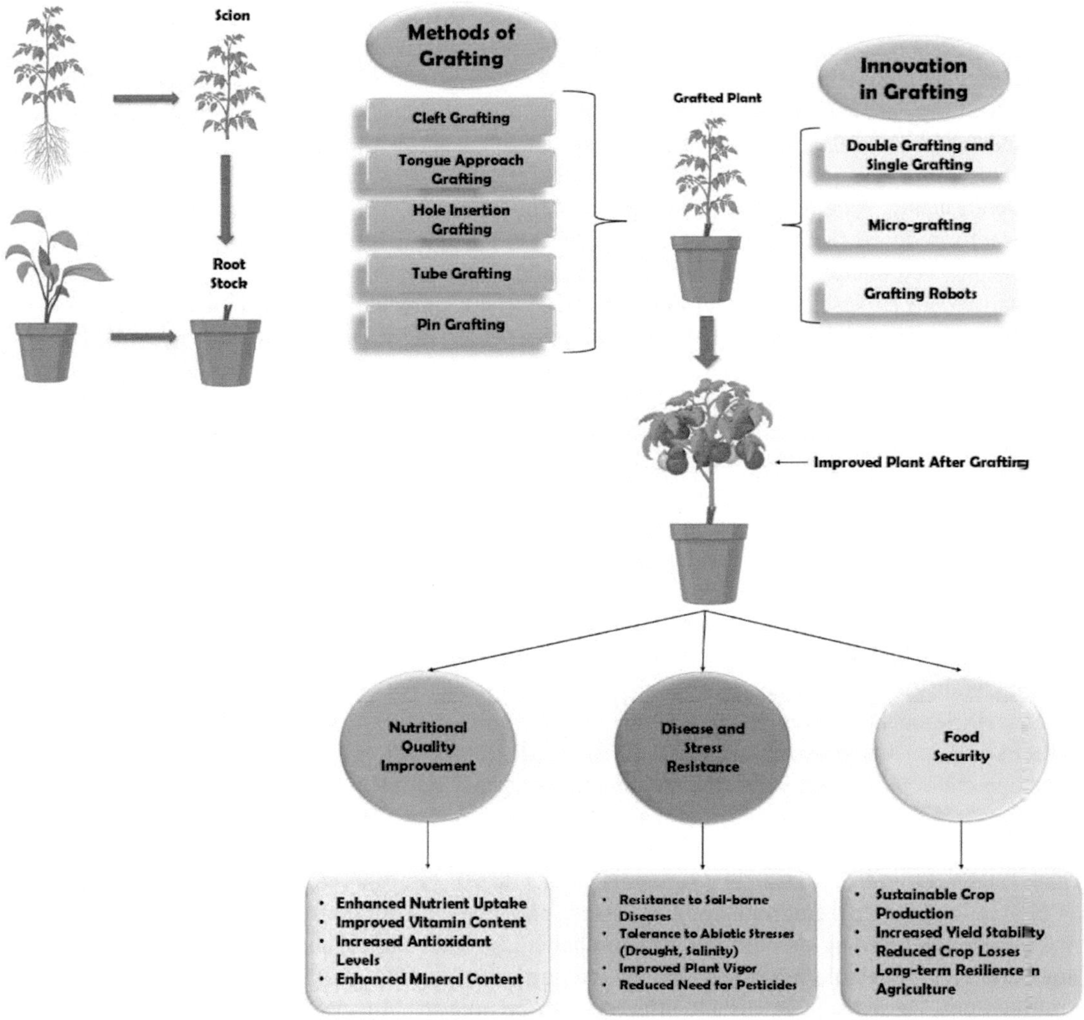


**Fig. 1** Different methods and importance of grafting in nutritional quality improvement and food security.

**Table 1** Different types of vegetable grafting.

| *Grafting Method* | *Description* | *Common Usage* | *General Features* |
|---|---|---|---|
| **Cleft Grafting** | Scion pruned with 2–4 true leaves; lower stem cut to a tapered wedge, placed into a split rootstock | Solanaceous crops | Widely used; also called apical or wedge grafting |
| **Tongue Approach Grafting** | Rootstock and scion of equal size; requires early sowing of scion seeds | Farmers and small nurseries | Labour-intensive; high survival rate; not suitable for hollow hypocotyls |
| **Hole Insertion/Top Insertion Grafting** | Scion inserted into a hollow hypocotyl of the rootstock | Cucurbits, especially watermelons | Requires optimal temperatures; popular in China |
| **Splice/Tube/One-Cotyledon Splice Grafting** | Scion cut and attached to the rootstock; can be done manually or with machines | Cucurbits and solanaceous vegetables | Most widely used by commercial transplant producers |
| **Pin Grafting** | Same as splice grafting, but uses pins to hold the graft | Various vegetables | Specialized pins are used to secure the graft |

In order to successfully fuse and generate a callus—which creates a vascular link between the two and permits the transfer of hormones, nutrients, and water—there must be the precise coherence of the layers of vascular cambium of rootstock and scion throughout the grafting process. The grafted plant's long-term survival and productivity depend on this relationship (Tsaballa et al., 2021; Rasool et al., 2020). Recent developments in molecular biology have demonstrated that intricate molecular communication involving the interchange of proteins, RNAs, and hormones that control growth, stress responses, and disease resistance is necessary for effective grafting between the rootstock and scion. These exchanges can induce epigenetic changes, enhancing the performance of the grafted plant (Jin et al., 2023). Understanding these interactions is key to optimizing grafting techniques and selecting compatible rootstock–scion combinations that maximize plant health and yield. Thus, grafting is a precise, technical process crucial for modern horticulture.

Grafting significantly influences the nutritional composition of plants, with profound implications for plant physiology and crop quality. The union of two genetically distinct plants in grafting creates a complex physiological network that enhances the concentration and distribution of both macronutrients and micronutrients across plant tissues. Nitrogen (N), Phosphorous (P), Potassium (K) contents rise due to better root systems of chosen rootstocks, which are more robust and effective at absorbing nutrients. This promotes rapid development and higher output in grafted plants. Grafted tomato plants, for instance, have demonstrated increased N and K concentrations, which have enhanced plant vigor and fruit productivity (Hollick & Kubota, 2022).

Grafting also significantly affects the micronutrient (such as Zn, Cu, and Fe) concentrations that are vital for physiological processes such as enzyme activation, chlorophyll synthesis, and stress response. The relationship of rootstock with scion can influence enzymatic activity in the roots, improving the plant's efficiency in absorbing essential micronutrients. This leads to higher concentrations of essential micronutrients in plant tissues, contributing to improved health and resilience (Sallaku et al., 2022). The graft union acts as a site of nutrient exchange, where the rootstock and scion communicate through vascular tissues, influencing nutrient redistribution due to the compatibility of the graft partners and the scion's specific physiological demands. In grafted melon plants, the rootstock can modulate nutrient uptake by altering the expression of nutrient transporter genes, and tailoring nutrient distribution to meet the scion's metabolic needs (Allevato et al., 2019).

The enhanced nutrient uptake and redistribution capabilities resulting from grafting directly improve plant health, stress tolerance, and yield potential. The higher accumulation of micronutrients (Fe and Zn) in the edible portions of grafted plants improves their nutritional quality, making grafted vegetables a more nutrient-rich option. This has significant implications for addressing micronutrient deficiencies in human diets and improving public health. As research continues to explore the interactions at the graft union, grafting emerges as a powerful tool for optimizing plant nutrition and improving the quality of horticultural crops, offering promising avenues for sustainable agriculture and food security.

Integrating grafting with complementary agricultural practices, such as biostimulants and mycorrhizal inoculation, is a pivotal strategy for optimizing nutrient absorption and enhancing plant health, productivity, and nutritional quality. Biostimulants—plant growth promoting rhizobacteria such as *Azospirillum brasilense*—enhance nutrient use efficiency (NUE) in grafted plants by modulating root architecture, increasing root surface area, and also improving the absorption of nutrients (N and K). This results in more efficient nutrient uptake, vigorous growth, and higher yields (Consentino et al., 2022). Additionally, biostimulants improve stress tolerance by inducing stress-related proteins and enzymes, further boosting the resilience and productivity of grafted crops. Inoculation with mycorrhizal fungi, especially arbuscular mycorrhizal (AM) fungi, enhances grafting by establishing symbiotic associations with plant roots. This relationship extends root access and improves the uptake of both macronutrients and micronutrients, even in less favorable soil conditions. Research shows that AM fungi significantly enhance nutrient uptake, particularly N,

P, Zn, and Cu, while also increasing concentrations of essential antioxidants such as ascorbic acid and chlorogenic acid, vital for plant health and nutritional quality (Sabatino et al., 2020).

The synergy between grafting, biostimulants, and mycorrhizal inoculation optimizes root function, nutrient uptake, and efficient nutrient use, leading to healthier, more productive plants. This integrated approach is especially beneficial in organic and sustainable farming, where minimizing chemical fertilizer use is crucial. By improving plants' ability to utilize naturally available nutrients, these practices reduce agriculture's environmental impact while maintaining high productivity and crop quality (Bayoumi et al., 2022). Additionally, supplying nutrient-dense, healthier food alternatives and improved nutrient absorption and utilization promotes global food security by treating micronutrient shortages and raising the concentrations of important nutrients and bioactive chemicals in produce. The integration of grafting with biostimulants and mycorrhizal inoculation enhances nutrient absorption and utilization in horticultural crops, contributing to sustainable agricultural practices by reducing chemical fertilizer dependence and increasing crop resilience under environmental stresses. This strategy presents a viable way to create resilient, high-yield crops that satisfy the needs of contemporary farming.

It is crucial to differentiate the function of the rootstock and the grafting process itself for determination of rootstock impact on fruit quality. Homografting, where plants are self-grafted, serves as a control to address this issue. Studies focused on melon have used homografts to compare with non-grafted controls and heterografts. After comparing non-grafted controls with heterografts, Aloni et al. (2011) determined that self-grafting in a hydroponic environment momentarily decreased oxidative damage that is induced by salinity and improvement in growth. Nevertheless, in Galia melon or pumpkin, no difference was shown in the root dry weight (RDW) and shoot dry weight (SDW) by Edelstein et al. (2011) between self-grafted (SG) and non-grafted (NG) treatments. Guan et al. (2014, 2015) and Schultheis et al. (2015) reported that SG yielded no significant differences in quality measurements or yield compared to NG controls. Despite these findings, some studies suggest that grafting can influence plant physiological responses, such as improved growth under stress and enhanced water transport (Orsini et al., 2013; Agele & Cohen, 2009). However, there is no conclusive evidence that homografting affects fruit quality at the reproductive stage. When comparing watermelons grafted onto different rootstocks, such as that of bottle gourd and squash to self-grafted plants, recent high-throughput sequencing revealed significant differential gene expression, highlighting the impact of heterografting on metabolism, hormone signalling, and stimuli responses (Liu et al., 2016).

## 3. Nutritional Enhancement through Grafting

Grafting has emerged as a significant technique in vegetable production for improving plant resilience, yield, and quality of nutrition of the produce. By carefully selecting compatible rootstock–scion combinations, growers can achieve significant improvements in the nutrient content of vegetables, which is essential for addressing dietary needs and improving public health outcomes. The following case studies show how grafting may improve the nutritional value of two popular vegetables: peppers and tomatoes.

Blossom-end rot (BER) in tomatoes, characterized by necrotic lesions due to localized calcium (Ca) deficiency, disrupts cell wall integrity and is exacerbated by environmental stresses such as irregular watering and nutrient imbalances. Grafting technology mitigates BER by enhancing Ca uptake and distribution. For example, grafting tomatoes onto the "FORTAMINO" rootstock significantly increases Ca content in the fruit, reducing BER incidence and improving overall mineral composition, as shown in studies by Manjarrez et al. (2023). This is attributed to the robust root system of "FORTAMINO," which efficiently absorbs and translocates calcium, along with other essential minerals such as K, Magnesium (Mg), and Fe, improving fruit quality, firmness, and shelf life. Further research demonstrates that combining grafting with Ca supplementation, especially under salinity stress, improves fruit quality by increasing calcium uptake and enhancing levels of vitamin C, total soluble solids (TSS), and β-carotene (Islam et al., 2023). This integration optimizes nutrient management, reduces physiological disorders such as BER, and enhances the nutritional value of tomato fruits.

Grafting is widely used in protected tomato cultivation despite concerns about costs and disease risks such as tobamoviruses (Luria et al., 2017). Rootstocks such as *Solanum lycopersicum* L. and interspecific hybrids are common, often paired according to breeder recommendations. Rootstocks affect fruit weight, size, yield, and shape, with vigorous rootstocks such as "Maxifort" increasing fruit size, while less vigorous ones such as "Brigeor" reducing it (Romano et al., 2000; Schwarz et al., 2013; Riga, 2015). Grafting also influences fruit color, especially the a* (redness) component linked to lycopene content, though results vary depending on the rootstock (Miskovic et al., 2016). Grafting lowers BER incidence, especially under stress; the amount of the decrease depends on (i) rootstock genotype and (ii) interactions of the rootstock with scion (Krumbein & Schwarz, 2013; Schwarz et al., 2013). However, some rootstocks, such as "Edkawi" or eggplant rootstocks, may increase BER by impairing Ca uptake (Fan et al., 2011). Rootstocks also affect textural attributes, with some, such as "Alligator," reducing firmness and others, such as "King Kong," enhancing it (Riga, 2015).

Grafting has varying effects on the acidity and sweetness of the fruit. While some researches (Matsuzoe et al., 1996; Khah et al., 2006; Savvas et al., 2011) show alterations in sugars and soluble solids content (SSC), others find no changes at all. Increased sweetness is associated with certain rootstocks under stress circumstances that concentrate sugars, including salinity or water scarcity (Oda et al., 1996; Fernández-García et al., 2004a,b). However, a lot of rootstocks lower sugar content and SSC, maybe as a result of dilution effects or greater plant vigor (Qaryouti et al., 2007). Grafted tomatoes tend to have higher levels of acidic content due to improved nutrient absorption (Turhan et al., 2011; Schwarz et al., 2013). Grafting's effect on tomato aroma and carotenoid content, particularly lycopene, varies with the scion–rootstock combination and environmental factors (Helyes et al., 2009; Nicoletto et al., 2013b). Grafting generally lowers ascorbic acid content, potentially due to redistribution within the plant or stress from the grafting process, though some rootstocks may increase it (Fernández-García et al., 2004a,b). Flavonoid content is variable, with some rootstocks enhancing phenolic compounds, often influenced by stress conditions (Slimestad et al., 2008; Sánchez-Rodríguez et al., 2012b). These findings underscore the need for further research into the effect of rootstocks on tomato fruit quality.

Grafting has been proven effective in enhancing sweet peppers' (*Capsicum annuum* L.) nutritional quality, particularly by increasing their micronutrient levels and antioxidant capacity. Peppers are rich in vitamins, minerals, and various bioactive compounds, making them an important component of a nutritious diet. Antioxidants, such as ascorbic acid, glutathione, and compounds of phenols, protect against oxidative stress, while micronutrients such as Fe, Zn, and Manganese (Mn) support essential physiological functions. Uresti-Porras et al. (2021) showed that grafting bell peppers onto specific rootstocks increased antioxidant levels, with higher ascorbic acid, total phenols, and glutathione compared to non-grafted controls. This process also enhanced Fe, Zn, and Mn content by 18.2%, 19.5%, and 34.7%, respectively. The integration of zinc oxide nanoparticles further amplified these benefits by promoting increased essential nutrients and bioactive compounds, leading to superior nutritional profiles. Grafting also improved levels of bioactive compounds such as β-carotene and phenolic acids, crucial for antioxidant properties. Gisbert-Mullor et al. (2020) found that grafting increased lycopene content and overall antioxidant capacity across different ripening stages in Spanish local pepper landraces, which are vital for protecting against chronic diseases such as cardiovascular disease and different cancers. The enhanced nutrient uptake and metabolic activity due to the rootstock's influence on the grafted plant facilitate efficient nutrient translocation and synthesis of secondary metabolites, leading to an increase in health-promoting compounds.

These enhancements in nutritional quality have significant implications for consumer health. Grafted peppers with higher concentrations of antioxidants, such as lycopene and ascorbic acid, offer protection against oxidative stress, which has been connected to chronic illnesses, including cancer, diabetes, and heart disease. Improved levels of Fe, Zn, and Mn support various physiological functions, making grafted peppers a valuable component of a healthy diet. With rising consumer demand for nutrient-rich and functional foods, grafting presents a promising approach for producing healthier, more resilient crops with enhanced health benefits. Grafting, combined with innovative technologies such as nanotechnology, may improve horticulture crops' nutritional value and contribute to the creation of more nourishing and sustainable systems of food. Sweet pepper is a major global vegetable crop valued for its diverse uses and phenotypic variety (FAO, 2020; Bosland et al., 2000; Rajesh Kumar et al., 2011). While modern breeding has produced high-performance hybrid cultivars with improved productivity and disease resistance, it has also led to genetic erosion, reducing the organoleptic diversity in fruits of pepper and increasing susceptibility to both biotic and abiotic stress factors (Votava et al., 2005; Rivera et al., 2016). Grafting is less prevalent in peppers, while being widely utilized in Cucurbitaceae and Solanaceae crops; this may be because commercial rootstocks offer only moderate economic benefits (Lee et al., 2010; Penella et al., 2015). Nonetheless, studies have demonstrated that grafting can affect the morphometric characteristics, yield, sweetness, acidity, and levels of functional compounds in pepper fruit (Leal-Fernández et al., 2013; Doñas-Uclés et al., 2014; Colla et al., 2008; Chávez-Mendoza et al., 2013; Sánchez-Torres et al., 2016). Pepper aroma, an important quality parameter for consumers (Whitfield & Last, 1991; Eggink et al., 2012; Lama et al., 2016), is also influenced by grafting. Quality, encompassing external appearance and nutritional attributes, is increasingly valued for its health benefits, such as delaying senescence and reducing cancer and cardiovascular disease risks (Alissa & Ferns, 2017; Grosso et al., 2017). Functional foods, rich in antioxidants, can counteract oxidative stress (Diplock et al., 1998). With consumer demand for enhanced sensory and nutraceutical qualities growing, particularly in landraces (Casañas et al., 2017), grafting provides a means to improve the phytochemical content of peppers, making them more appealing and beneficial for health (Howard et al., 2000; Zhuang et al., 2012). However, the effect of grafting on phytochemical content in peppers varies in the literature (Kyriacou et al., 2017; Sánchez-Torres et al., 2016; Rouphael et al., 2010; Chávez-Mendoza et al., 2015).

## 4. Quality Considerations in Grafted Vegetables

The quality and marketability of grafted vegetables are significantly shaped by their enhanced nutritional and nutraceutical properties, which align with rising consumer demand for foods that offer both basic nutrition and additional health benefits. Grafted vegetables often exhibit superior nutritional profiles, with increased concentrations of essential nutrients such as vitamin C, K, and Ca due to improved nutrient uptake facilitated by the rootstock (Dias, 2021). In addition, these vegetables have greater concentrations of bioactive substances, for example, flavonoids, phenolic acids, carotenoids, and antioxidants, which support their anti-inflammatory, anti-carcinogenic, and antioxidant properties. For instance, grafted tomatoes show higher lycopene levels, while grafted peppers have increased β-carotene and phenolic compounds, enhancing their health benefits (Gisbert-Mullor et al., 2020). The improved taste, texture, and color further enhance their market appeal, meeting consumer expectations for nutritious and flavorful produce (Ferrari et al., 2021).

Compounds such as lycopene and β-carotene in grafted tomatoes and peppers are linked to reduced risks of prostate cancer and improved immune function (Arya et al., 2019; Ahmad et al., 2020). The continuous consumption of grafted vegetables, which are rich in bioactive compounds, also offers anti-aging benefits by combating oxidative stress, thus promoting longevity and reducing age-related diseases (Dias, 2019). Moreover, flavonoids in these vegetables exhibit anti-inflammatory and anti-carcinogenic properties, lowering harmful impact of chronic diseases and emphasizing the importance of grafted vegetables in a preventive health strategy (Marino et al., 2023). Incorporating grafted vegetables into daily diets leverages their enhanced nutraceutical properties, making them essential for a health-conscious lifestyle that supports overall well-being and disease prevention (Sharma et al., 2021).

*Solanum melongena* L. (eggplant) and its related species play a crucial role in rootstock development, with *Solanum torvum* being the most commonly used rootstock. Additionally, various interspecific hybrids, such as *Solanum incanum* and *Solanum melongena* × *Solanum aethiopicum*, have also been explored as potential rootstocks (Gisbert et al., 2011a,b; Marsic et al., 2014; Sabatino et al., 2016). The effect of grafting on the quality of eggplant fruit has not always been reported; reports have been impacted by harvest maturity, rootstock–scion interactions, and experimental conditions (open-field vs. greenhouse) (Rouphael et al., 2010; Kyriacou et al., 2016).

Studies using morphometric analysis reveal that grafting frequently has little effect on the average fruit weight. For instance, whether grafted onto *S. torvum, S. incanum* × *S. melongena*, or *S. melongena* × *S. aethiopicum*, cultivar "Black Beauty" showed comparable mean weight of fruit (Gisbert et al., 2011b). Likewise, there was no discernible variation in the average fruit weight of other landraces grafted onto *S. torvum* (Sabatino et al., 2016). However, grafting "Black Bell" using *S. torvum* in an environment without soil was found to significantly improve fruit weight in several trials (Cassaniti et al., 2011).

A heritable trait such as fruit shape showed minimal changes due to grafting, with non-significant or minor effects on shape index observed (Cassaniti et al., 2011). Conflicting reports exist on fruit peel color, often attributed to variations in harvest maturity and sampling practices (Moncada et al., 2013). Textural properties, such as firmness, were negatively affected in some grafted eggplants, likely due to altered water uptake and pest pressures (Arvanitoyannis et al., 2005; Cassaniti et al., 2011). Regarding taste, grafting often does not improve sweetness or acidity, with some studies noting reduced sugar content and sensory acceptability in grafted plants (Lee et al., 2010; Khah, 2011; Arvanitoyannis et al., 2005). Functional compounds such as phenolics, associated with antioxidant activity, showed variable results. For instance, increased total polyphenol content was observed in some grafted eggplants (Sabatino et al., 2016), while other studies reported higher phenolic content in non-grafted plants (Moncada et al., 2013). Rootstock–scion combinations and environmental factors, such as sun radiation, might impact anthocyanin accumulation and, therefore, the concentration of flavonoids such as anthocyanins (Awad et al., 2001; Marsic et al., 2014).

## 5. Advancements in Molecular Breeding and Biotechnology and Their Applications in Grafting

Grafting joins the rootstock and scion of two plants, forming a unified vascular system. This process starts with tissue connection at graft points, leading to callus formation, cell wall integration, and vascular development. In *Arabidopsis*, this occurs within a week, with the reconnection of the phloem, then followed by the reconnection of the root and xylem (Kollmann & Glockmann, 1985; Yin et al., 2012; Melnyk et al., 2015; Melnyk et al., 2017; Melnyk et al., 2018). Root responses to the shoot are facilitated by plasmodesmata, even before full vascular connection. Successful grafting is typically within the same plant family, though distant species can also be grafted with cellulases aiding adhesion (Notaguchi et al., 2020; Kurotani et al., 2020; Wulf et al., 2020). Auxin is essential for wound healing and vascular reconnection because it activates auxin transporter and response genes at graft junctions (Melnyk et al., 2015, 2018; Yu et al., 2017; Aloni et al., 2008). As demonstrated by studies on tobacco (Hirata & Yagishita, 1986; Yagishita et al., 1990; Ohta, 1991; Liu et al., 2006; Stegemann and Bock, 2009; Stegemann et al., 2012; Tsaballa et al., 2013; Fuentes et al., 2014), grafting can cause phenotypic variations known as graft hybridization, where the genetic material movement takes place between scion and rootstock. Small RNAs that affect gene expression and epigenetic control, such as miR399, miR172, miR156, and miR395, are mobile between shoots and roots (Pant et al., 2008; Martin et al., 2009; Buhtz et al., 2010; Bhogale et al., 2014). DNA methylation is influenced by 24-nt siRNAs in recipient tissues in *Arabidopsis* by its migration toward root from shoot, by directing gene transcription silencing and RdDM (Molnar et al., 2010; Melnyk et al., 2011; Borges & Martienssen, 2015; Xie & Yu, 2015; Lewsey et al., 2016). Viral resistance can benefit from sRNA mobility, which is more effective from scion to rootstock (Kasai et al., 2013; Nakamura et al., 2015; Spanò et al., 2015).

Epigenetic modifications (e.g. DNA methylation) occur in grafted plants. Heritable alterations in Solanaceae and Cucurbitaceae have been seen by interspecies grafting (Chen et al., 2006; Wu et al., 2013; Avramidou et al., 2015; Cao et al., 2016). Reciprocal grafting in cucumber and pumpkin reveals alterations in stress-induced miRNA expression, whereas cucurbita grafting modifies DNA methylation and miRNA expression, impacting fruit quality (Xanthopoulou et al., 2019). In Solanaceae, sRNA movement leads to transcriptional silencing and stable trait inheritance (Kasai et al., 2016). Grafting influences gene expression and developmental traits, with significant variations depending on the combination, including changes related to immune response, oxidative stress, quality of fruit, and resistance to stress (Tsaballa et al., 2011; Cookson et al., 2014; Kumari et al., 2015; Ntatsi et al., 2017; Zhao et al., 2018; Melnyk et al., 2018; Zhang et al., 2019; Wang et al., 2019; Wei et al., 2019; Aslam et al., 2020; Spanò et al., 2020). By identifying important proteins and gene expression changes that affect graft union formation and adaptation, molecular breeding—which includes marker assisted selection (MAS) and genomic selection (GS)—along with advances in proteomics and transcriptomics, improves grafting efficiency (Saleh et al., 2019; Shivran et al., 2022; Jin et al., 2023). These integrated approaches drive advancements in grafting technology, promoting sustainable and resilient agricultural practices.

Genes such as cellulases, which support tissue adhesion and cell wall reconstruction, facilitate grafting. This may help overcome interfamily grafting barriers and create chimeric plants that benefit from partners who are distantly related (Kurotani et al., 2020; Notaguchi et al., 2020). Cellular division and differentiation are key to the grafting process, which require effective cell communication and hormone involvement. During grafting and wounding reactions, auxin is crucial and interacts with other hormones (gibberellin, cytokinin, ethylene, and jasmonic acid) (Nanda & Melnyk, 2018). The genes that are responsive to auxin are active in rootstocks below the junction of the graft in *Arabidopsis*, indicating that the vascular reconnection takes place after the signal of the auxin that is transmitted from the scion toward rootstock (Melnyk et al., 2015, 2018). In the genes that code for auxin efflux protein, induction takes place in scions just above the graft junction. The genes of auxin transporters are important for auxin transport, wound healing, and vascular development. Recent studies on watermelons grafted onto squash have identified changes in auxin transporter gene expression levels during the grafting process (Yu et al., 2017). Furthermore, studies in which melon is grafted on the rootstock of Cucurbita indicate that auxin influences root deterioration in auxin-sensitive rootstocks by inducing the synthesis of reactive oxygen species (ROS) and ethylene hormones (Aloni et al., 2008).

Chromatin transfer from rootstocks to scions and then to seed progenies was shown in early investigations (Ohta, 1991). Random amplification of polymorphic DNA (RAPD) markers specific to the rootstock were used to distinguish graft-induced variations from the scion (Taller et al., 1998). It is reported by Stegemann and Bock (2009) that tobacco grafts with distinct antibiotic markers showed reporter genes and acquired resistance to both antibiotics, suggesting horizontal gene transfer. Stegemann et al. (2012) found that chloroplast DNA from *Nicotiana tabacum* was successfully transferred to *Nicotiana glauca* or *Nicotiana benthamiana* through grafting, resulting in stably inherited plastid genomes. Fuentes et al. (2014) demonstrated that grafting *Nicotiana glauca* scions expressing kanamycin resistance (*npt*II gene) onto *Nicotiana tabacum* rootstocks expressing hygromycin resistance (*hpt* gene) created a plant resistant to both antibiotics, indicating genetic recombination. It has been well demonstrated that messenger RNA (mRNA) moves through phloem and that grafting may be used to examine RNA transport (Spiegelman et al., 2013; Turnbull & Lopez-Cobollo, 2013). Recent research in *Arabidopsis* has shown that the 24-nt siRNAs generated by Dicer-like protein (DCL3), which travels from the shoot to the root. In the root, these siRNAs guide transposable elements, and repetitive DNA through RdDM leads to the gene transcriptional silencing (Molnar et al., 2010; Borges & Martienssen, 2015; Xie & Yu, 2015). Specific genomic locations may be targeted for transcriptional suppression and DNA methylation using mobile 24-nt siRNAs (Melnyk et al., 2011; Lewsey et al., 2016). The mobility of sRNAs is more efficient from scion to rootstock, moving through the phloem and plasmodesmata, though siRNAs from rootstock can also travel to scion and reduce viroid infection (Kasai et al., 2013; Nakamura et al., 2015). Various molecular and biochemical approaches are shown in Table 2.

**Table 2** Molecular and biochemical approaches involved in grafting.

| *Aspect* | *Description* | *References* |
|---|---|---|
| **Root Responses and Plasmodesmata** | Plasmodesmata facilitate root responses to the shoot even before full vascular connection. | Yin et al., 2012; Kollmann & Glockmann, 1985 |
| **Auxin's Role in Grafting** | Auxin is crucial in vascular reconnection and wound healing, with auxin transporter and response genes activated at graft junctions. | Melnyk et al., 2015, 2018; Yu et al., 2017; Aloni et al., 2008 |
| **Graft Hybridization** | Grafting can induce phenotypic variations, with genetic material moving between rootstock and scion. | Liu et al., 2006; Hirata & Yagishita, 1986; Yagishita et al., 1990; Taller et al., 1998, 1999) |
| **Small RNAs (sRNAs) Mobility** | sRNAs, including miRNAs, move between shoots and roots, impacting gene expression and epigenetic regulation. | Borges & Martienssen, 2015; Xie & Yu, 2015 |
| **Epigenetic Changes** | Grafting induces epigenetic changes, including DNA methylation, with potential for heritable changes in interspecies grafting. | Wu et al., 2013; Avramidou et al., 2015 |
| **Molecular Breeding and Technology** | Molecular breeding, also incorporating MAS and GS, enhances grafting efficiency through proteomics and transcriptomics. | (Shivran et al., 2022; Jin et al., 2023). |
| **Cellulases in Grafting** | Cellulases aid in cell wall reconstruction and tissue adhesion, potentially overcoming interfamily grafting barriers. | (Kurotani et al., 2020; Notaguchi et al., 2020). |
| **Hormone Involvement in Grafting** | Auxin, along with other hormones (cytokinin, gibberellin, and ethylene), plays a key role in cell division, communication, and differentiation during grafting. | (Nanda & Melnyk, 2018). |
| **Chromatin Movement** | Early studies showed chromatin movement from rootstock to scion and seed, suggesting genetic transfer during grafting. | (Ohta, 1991; Taller et al., 1998). |
| **Messenger RNA (mRNA) Movement** | mRNA movement through the phloem has been documented, indicating RNA transport's role in grafting processes. | Spiegelman et al., 2013; Turnbull & Lopez-Cobollo, 2013 |
| **Grafting-Induced Gene Silencing** | siRNAs from rootstock can move to the scion, reducing viroid infection and contributing to silencing of transcription. | (Kasai et al., 2013; Nakamura et al., 2015). |

## 6. Innovations in Grafting Techniques

Recent innovations in grafting technology have significantly improved both the success rate and efficiency of the process, crucial for meeting the growing demand in commercial vegetable production. Traditional grafting methods, though effective, are labour-intensive and variable. To address these issues, researchers have developed automated grafting systems and advanced micrografting techniques. Automated grafting robots, utilizing machine vision and artificial intelligence (AI), perform precise cuts and alignments, enhancing efficiency and consistency, particularly in large-scale operations (Yan et al., 2022). Micrografting, involving precision tools for delicate plants, reduces graft stress and improves success rates, facilitating the examination of long-range signaling and systemic reaction (Tsutsui et al., 2020). Techniques such as single grafting, which joins one scion to a rootstock, and double grafting, which combines multiple scions or rootstocks, cater to specific horticultural needs by improving traits such as disease resistance and nutrient uptake (Shiwani, 2020). Micrografting is also valuable for studying plant physiology at a microscopic level, revealing nutrient and hormone movement between rootstock and scion (Li et al., 2019). Grafting robots automate the process, reducing labor costs and ensuring high success rates, crucial for large-scale agriculture (Zhao et al., 2020). These advancements enhance grafting efficiency and also improve the quality of nutrition in vegetables. Grafting increases the uptake of essential nutrients and antioxidants, addressing soil nutrient deficiencies and enhancing produce quality (Sabatino et al., 2020). Furthermore, improved stress tolerance in grafted plants ensures stable yields and high-quality produce under adverse conditions, contributing to global food security (Bayoumi et al., 2022). Overall, these innovations are revolutionizing vegetable production by enhancing grafting processes and nutritional quality, crucial for sustainable food security. Some of the innovations in grafting are summarized in Table 3.

**Table 3** Recent innovations in vegetable grafting.

| *Innovation* | *Description* | *Key Features* | *Examples/Notes* |
|---|---|---|---|
| **Double-Grafted and Single-Grafted Tomato** | Using the cleft grafting method, tomato scions are joined with potato rootstocks, creating a hybrid plant referred to as the "Pomato" | Yields more than 500 cherry tomatoes with a total soluble solids (TSS) content of 100 Brix.; double grafting with two scions | Examples: Indigo Rose, Brandywine, Sun Sugar; marketed in the USA in 2010 |
| **Micro-grafting** | *In vitro* grafting utilizes minute micro-explants obtained from meristematic tissues | Eliminates viruses; used to study grafting physiology and contacts of cell-to-cell | Enables the fast multiplication of virus-free plants, but is costly |
| **Grafting Robots** | Full automation models that can graft seedlings at a high rate, with automatic selection of matching rootstocks and scions | Grafts 1,000 seedlings per hour; increases success rate | First commercial model for cucurbits available in 1993; developed in the Netherlands and Japan |

# Conclusion

This chapter explored the critical aspects of grafting technology in modern horticulture, particularly its application in vegetable crops. It detailed the grafting process, including rootstock and scion selection and their molecular and biochemical interactions. Case studies highlighted how grafting enhances nutritional content, such as increasing calcium in tomatoes and boosting antioxidant capacity in peppers. Innovations such as grafting robots, double and single grafting methods, and micrografting were reviewed, emphasizing their role in improving success rates, efficiency, and nutritional quality. The chapter also examined the contributions of molecular breeding, biotechnology, and epigenetics to advancing grafting technology, showcasing their importance in crop enhancement and sustainable agriculture (Tsaballa et al., 2021; Jin et al., 2023). The implications of grafting innovations for food security and nutritional quality are significant. Grafting improves the tolerance of crops to various biotic and abiotic stress factors, ensuring consistent yields and improving food security. It also improves the nutritional content of vegetables, increasing concentrations of essential vitamins, minerals, and antioxidants, thereby better meeting global dietary needs. Combining grafting with biotechnological advancements, such as biostimulants and precision breeding, further elevates its potential to produce nutrient-dense vegetables vital for public health (Consentino et al., 2022; Gisbert-Mullor et al., 2020).

Future research in grafting technology should focus on understanding the molecular and epigenetic mechanisms of rootstock–scion interactions, integrating advanced "omics" technologies for deeper insights, and optimizing grafting robots to enhance scalability and cost-effectiveness. Additionally, studying the long-term effects of grafting under changing climate conditions will be crucial for ensuring its viability in sustainable agriculture (Kapazoglou et al., 2021). In conclusion, grafting technology holds significant potential for improving food security and nutrition-related changes, with ongoing research and innovation being key to its successful application.

## References

Agele, S. and Cohen, S. (2009). Effect of genotype and graft type on the hydraulic characteristics and water relations of grafted melon. *J. Plant Interact.* 4, 59–66. doi: 10.1080/17429140802311671

Ahmad, N., Mailk, A., Hussain, K., Malik, G., Wani, S., Rather, A. and Magaray, M. (2020). Nutraceutical and bioactive healthy compounds in vegetable crops. *International Journal of Chemical Studies*, 8, 3078–3086.

Akhtar, S., Ismail, T., Atukorala, S. and Arlappa, N. (2013). Micronutrient deficiencies in South Asia – current status and strategies. *Trends in Food Science and Technology*, 31, 55–62.

Alissa, E.M. and Ferns, G.A. (2017). Dietary fruits and vegetables and cardiovascular diseases risk. *Critical Reviews in Food Science and Nutrition*, *57*(9), 1950–1962.

Allevato, E., Mauro, R.P., Stazi, S.R., Marabottini, R., Leonardi, C., Ierna, A. and Giuffrida, F. (2019). Arsenic accumulation in grafted melon plants: Role of rootstock in modulating root-to-shoot translocation and physiological response. *Agronomy*, *9*(12), 828.

Aloni, B., Karni, L., Deventurero, G., Cohen, R., Katzir, N., Edelstein, M., et al. (2011). The use of plant grafting and plant growth regulators for enhancing abiotic stress tolerance in vegetable transplants. *Acta Hortic.* 898, 255–264. doi: 10.17660/actahortic.2011.898.31

Aloni, B., Karni, L., Deventurero, G., Levin, Z., Cohen, R., Katzir, N., et al. (2008). Possible mechanisms for graft incompatibility between melon scions and pumpkin rootstocks. *Acta Hortic.* 782, 313–324. doi: 10.17660/ ActaHortic.2008.782.39

Arvanitoyannis, I.S., Khah, E.M., Christakou, E.C. and Bletsos, F.A. (2005). Effect of grafting and modified atmosphere packaging on eggplant quality parameters during storage. *Intern. J. Food Sci. Techn.* 40, 311–322. doi: 10.1111/ j.1365-2621.2004.00919.x

Arya, M., Reshma, U., Thampi, S., Anaswara, S. and Sebastian, K. (2019). Nutraceuticals in vegetables: new breeding approaches for nutrition, food and health: a review. *Journal of Pharmacognosy and Phytochemistry*, 8, 677–682.

Aslam, A., Zhao, S., Azam, M., Lu, X., He, N., Li, B., et al. (2020). Comparative analysis of primary metabolites and transcriptome changes between ungrafted and pumpkin-grafted watermelon during fruit development. *PeerJ* 8:e8259. doi: 10.7717/peerj.8259

Avramidou, E., Kapazoglou, A., Aravanopoulos, F. A., Xanthopoulou, A., Ganopoulos, I., Tsaballa, A., et al. (2015). Global DNA methylation changes in Cucurbitaceae inter-species grafting. *Crop Breed. Appl. Biotechnol.* 15, 112–116. doi: 10.1590/1984-70332015v15n2n20

Awad, M.A., de Jager, A., van der Plas, L.H.W. and van der Krol, A.R. (2001). Flavonoid and chlorogenic acid changes in skin 'Elstar' and 'Jonagold' apples during development and ripening. *Sci. Hortic.* 90, 69–83. doi: 10.1016/s0304-4238(00)00255-7

Bahadur, A. and Kumar, R. (2024). Grafting in tomato for improving abiotic stress tolerance, yield and quality traits. *Vegetable Science*, *51*, 22–33.

Bailey, R., Jr, K. and Black, R. (2015). The epidemiology of global micronutrient deficiencies. *Annals of Nutrition and Metabolism*, 66, 22–33.

Bayoumi, Y., Shalaby, T., Abdalla, Z. F., Shedeed, S. H., Abdelbaset, N., El-Ramady, H., & Prokisch, J. (2022). Grafting of vegetable crops in the era of nanotechnology: A photographic mini review. *Environment, Biodiversity and Soil Security*, 6(2022), 133–148.

Bhogale, S., Mahajan, A.S., Natarajan, B., Rajabhoj, M., Thulasiram, H.V. and Banerjee, A.K. (2014). MicroRNA156: a potential graft-transmissible microRNA that modulates plant architecture and tuberization in Solanum tuberosum ssp. andigena. *Plant Physiol.* 164, 1011–1027. doi: 10.1104/pp.113.230714

Borges, F. and Martienssen, R.A. (2015). The expanding world of small RNAs in plants. *Nat. Rev. Mol. Cell Biol.* 16, 727–741. doi: 10.1038/nrm4085

Bosland, P.W. and Votava, E.J. (2000). Peppers: Vegetables and Spice Capsicums. Crop Production Science in Horticulture 12; CAB International Publishing: Wallingford, UK, 2000.

Buhtz, A., Pieritz, J., Springer, F. and Kehr, J. (2010). Phloem small RNAs, nutrient stress responses, and systemic mobility. *BMC Plant Biol.* 10:64. doi: 10.1186/ 1471-2229-10-64

Cao, L., Yu, N., Li, J., Qi, Z., Wang, D. and Chen, L. (2016). Heritability and reversibility of DNA methylation induced by in vitro grafting between Brassica juncea and B. oleracea. *Sci. Rep.* 6:27233. doi: 10.1038/srep27233

Caradonia, F., Francia, E., Alfano, V. and Ronga, D. (2023). Grafting and plant density influence tomato production in organic farming system. *Horticulturae*, *9*(6), 669.

Casañas, F., Simó, J., Casals, J. and Prohens, J. (2017). Toward an evolved concept of landrace. *Frontiers in Plant Science*, *8*, 145.

Cassaniti, C., Giuffrida, F., Scuderi, D. and Leonardi, C. (2011). Effect of rootstock and nutrient solution concentration on eggplant grown in a soilless system. *J. Food Agric. Environ.* 9, 252–256.

Chávez-Mendoza, C., Sánchez, E., Carvajal-Millán, E., Munõz-Márquez, E. and Guevara-Aguillar, A. (2013). Characterization of the nutraceutical quality and antioxidant activity in bell pepper in response to grafting. *Molecules* 18, 15689–15703. doi: 10.3390/molecules181215689

Chávez-Mendoza, C., Sanchez, E., Muñoz-Marquez, E., Sida-Arreola, J.P. and Flores-Cordova, M.A. (2015). Bioactive compounds and antioxidant activity in different grafted varieties of bell pepper. *Antioxidants*, *4*(2), 427–446.

Chen, L.-P., Ge, Y.-M. and Zhu, X.-Y. (2006). Artificial synthesis of interspecific chimeras between tuber mustard (Brassica juncea) and cabbage (Brassica oleracea) and cytological analysis. *Plant Cell Rep.* 25, 907–913. doi: 10.1007/ s00299-006-0150-5

Chen, X., Guo, P., Wang, Z., Liang, J., Li, G., He, W. and Zhen, A. (2022). Grafting improves growth and nitrogen-use efficiency by enhancing NO3− uptake, photosynthesis, and gene expression of nitrate transporters and nitrogen metabolizing enzymes in watermelon under reduced nitrogen application. *Plant and Soil*, *480*(1), 305–327.

Colla, G., Rouphael, Y., Cardarelli, M., Temperini, O., Rea, E., Salerno, A., et al. (2008). Influence of grafting on yield and fruit quality of pepper (Capsicum annuum L.) grown under greenhouse conditions. *Acta Hortic.* 782, 359–363. doi: 10.17660/actahortic.2008.782.45

Consentino, B.B., Sabatino, L., Vultaggio, L., Rotino, G.L., La Placa, G.G., D'Anna, F., ... and De Pasquale, C. (2022). Grafting eggplant onto underutilized solanum species and biostimulatory action of Azospirillum brasilense modulate growth, yield, nue and nutritional and functional traits. *Horticulturae*, *8*(8), 722.

Cookson, S.J., Clemente Moreno, M.J., Hevin, C., Nyamba Mendome, L.Z., Delrot, S., Magnin, N., et al. (2014). Heterografting with nonself rootstocks induces genes involved in stress responses at the graft interface when compared with autografted controls. *J. Exp. Bot.* 65, 2473–2481. doi: 10.1093/jxb/eru145

Dias, J. (2019). Nutritional Quality and Effect on Disease Prevention of Vegetables. *Food and Nutrition Sciences,* 83–112.

Dias, J.D.S. (2021). Minerals and nutraceuticals in vegetables and their human health benefits. *Highlights on Medicine and Medical Science*, 12, 27–71.

Diplock, A.T., Charuleux, J.L., Crozier-Willi, G., Kok, F.J., Rice-Evans, C., Roberfroid, M. and Vina-Ribes, J. (1998). Functional food science and defence against reactive oxidative species. *British Journal of Nutrition*, *80*(S1), S77–S112.

Djidonou, D., Simonne, A.H., Koch, K.E., Brecht, J.K. and Zhao, X. (2016). Nutritional quality of field-grown tomato fruit as affected by grafting with interspecific hybrid rootstocks. *HortScience*, *51*(12), 1618–1624.

Doñas-Uclés, F., Jiménez-Luna, M.M., Gongora-Corral, J.A., Pérez-Madrid, D., Verde-Fernandez, D. and Camacho-Ferre, F. (2014). Influence of three rootstocks on yield and commercial quality of 'Italian sweet' pepper. *Ciênc. Agrotecnol.* 38, 538–545. doi: 10.1590/s1413-70542014000600002

Edelstein, M., Plaut, Z. and Ben-Hur, M. (2011). Sodium and chloride exclusion and retention by non-grafted and grafted melon and Cucurbita plants. *J. Exper. Bot.* 62, 177–184. doi: 10.1093/jxb/erq255

Eggink, P.M., Maliepaard, C., Tikunov, Y., Haanstra, J.P.W., Bovy, A.G. and Visser, R.G.F. (2012). A taste of sweet pepper: volatile and non-volatile chemical composition of fresh sweet pepper (Capsicum annuum) in relation to sensory evaluation of taste. *Food Chemistry*, *132*(1), 301–310.

Faisal, M., Kamran, M., Arshad, A., Maqsood, M.J., Rehman, K., Usman, S. and Farooq, P. (2024). Mitigate the impact of various abiotic stress by using grafting in tomato (Solanum lycopersicum L.) and other vegetables: a comprehensive review. *Asian Journal of Research in Crop Science*, *9*(2), 35–43.

Fan, M., Bie, Z., Krumbein, A. and Schwarz, D. (2011). Salinity stress in tomatoes can be alleviated by grafting and potassium depending on the rootstock and K-concentration employed. *Sci. Hortic.* 130, 615–623. doi: 10.1016/j. scienta. 2011.08.018

Fernández-García, N., Martínez, V. and Carvajal, M. (2004a). Effect of salinity on growth, mineral composition, and water relations of grafted tomato plants. *J. Plant Nutr. Soil Sci.* 167, 616–622. doi: 10.1002/jpln.200420416

Fernández-García, N., Martínez, V., Cerda, A. and Carvajal, M. (2004b). Fruit quality of grafted tomato plants grown under saline conditions. *J. Hortic. Sci. Biotechnol.* 79, 995–1001. doi: 10.1080/14620316.2004.11511880

Ferrari, V., Rodríguez, G., González, M., Vicente, E., Giménez, G., Cabrera, D. and Ibañez, F. (2021). Nutritional Quality and Nutraceutical Potential of Fruits and Vegetables as a tool for Genetic Breeding Programs. *Agrociencia Uruguay*, 25(NSPE2)

Food and Agriculture Organization Faostat. Food and Agriculture Data; Food and Agriculture Organization: Rome, Italy, 2018; Available online: http://www.fao.org/faostat/en/#data/QC (accessed on 1 April 2020).

Fuentes, I., Stegemann, S., Golczyk, H., Karcher, D. and Bock, R. (2014). Horizontal genome transfer as an asexual path to the formation of new species. *Nature* 511, 232–235. doi: 10.1038/nature13291

Gisbert, C., Prohens, J. and Nuez, F. (2011a). Performance of eggplant grafted onto cultivated and wild materials and hybrids of eggplant and tomato. *Intern. J. Plant Prod.* 5, 367–380.

Gisbert, C., Prohens, J., Raigón, M.D., Stommel, J. and Nuez, F. (2011b). Cassaniti, C., Giuffrida, F., Scuderi, D. and Leonardi, C. (2011). Effect of rootstock and nutrient solution concentration on eggplant grown in a soilless system. *J. Food Agric. Environ.* 9, 252–256.

Gisbert-Mullor, R., Ceccanti, C., Gara Padilla, Y., López-Galarza, S., Calatayud, Á., Conte, G. and Guidi, L. (2020). Effect of grafting on the production, physico-chemical characteristics and nutritional quality of fruit from pepper landraces. *Antioxidants*, *9*(6), 501.

Grosso, G., Micek, A., Godos, J., Pajak, A., Sciacca, S., Galvano, F. and Boffetta, P. (2017) Health risk factors associated with meat, fruit and vegetable consumption in cohort studies: a comprehensive meta-analysis. *PLoS One* 2017, 12, e0183787.

Guan, W., Zhao, X., Dickson, D.W., Mendes, M.L. and Thies, J. (2014). Rootknot nematode resistance, yield, and fruit quality of specialty melons grafted onto Cucumis metulifer. *HortScience* 49, 1046–1051.

Guan, W., Zhao, X., Hubler, J.H., Huber, J.D. and Sims, A.C. (2015). Instrumental and sensory analyses of quality attributes of grafted specialty melons. *J. Sci. Food. Agric.* 95, 2989–2995. doi: 10.1002/jsfa.7050

Harika, R., Faber, M., Samuel, F., Kimiywe, J., Mulugeta, A. and Eilander, A. (2017). Micronutrient status and dietary intake of iron, vitamin A, iodine, folate and zinc in women of reproductive age and pregnant women in Ethiopia, Kenya, Nigeria and South Africa: a systematic review of data from 2005 to 2015. *Nutrients*, 9.

Harris, Z.N., Pratt, J.E., Kovacs, L.G., Klein, L.L., Kwasniewski, M.T., Londo, J.P. and Miller, A.J. (2023). Grapevine scion gene expression is driven by rootstock and environment interaction. *BMC Plant Biology*, *23*(1), 211.

Helyes, L., Lugasi, A., Pogonyi, A. and Pek, Z. (2009). Effect of variety and grafting on lycopene content of tomato (Lycopersicon lycopersicum L. Karsten) fruit. *Acta Aliment.* 38, 27–34. doi: 10.17660/actahortic.2006.712.62

Hirata, Y. and Yagishita, N. (1986). Graft-induced changes in soybean storage proteins. I. Appearance of the changes. *Euphytica* 35, 395–401. doi: 10.1007/ BF00021847 Huang, Y., Bie, Z

Hollick, J.R. and Kubota, C. (2022). Effect of self-and inter-cultivar grafting on growth and nutrient content in sweet basil (Ocimum basilicum L.). *Frontiers in Plant Science*, *13*, 921440.

Howard, L.R., Talcott, S.T., Brenes, C.H. and Villalon, B. (2000). Changes in phytochemical and antioxidant activity of selected pepper cultivars (Capsicum species) as influenced by maturity. *Journal of Agricultural and Food Chemistry*, *48*(5), 1713–1720.

Islam, M., Jahan, K., Sen, A., Urmi, T., Haque, M., Ali, H., Siddiqui, M. and Murata, Y. (2023). Exogenous application of calcium ameliorates salinity stress tolerance of tomato (Solanum lycopersicum L.) and enhances fruit quality. *Antioxidants*, 12.

Jin, Q., Chachar, M., Ahmed, N., Zhang, P., Chachar, Z., Geng, Y., ... and Chachar, S. (2023). Harnessing epigenetics through grafting: revolutionizing horticultural crop production. *Horticulturae*, *9*(6), 672.

Kapazoglou, A., Tani, E., Avramidou, E., Abraham, E., Gerakari, M., Megariti, S., Doupis, G. and Doulis, A. (2021). Epigenetic changes and transcriptional reprogramming upon woody plant grafting for crop sustainability in a changing environment. *Frontiers in Plant Science*, 11.

Karthiga, P., Senthilkumar, S., Srivignesh, S. and Manivannan, S. (2023). Inevitable role of grafting in improvement of cucurbitaceous crops: a review. *Agricultural Reviews*. DOI: 10.18805/ag.R-2602

Kasai, A., Bai, S., Hojo, H. and Harada, T. (2016). Epigenome editing of potato by grafting using transgenic tobacco as siRNA donor. *PLoS One* 11:e0161729. doi: 10.1371/journal.pone.0161729

Kasai, A., Sano, T. and Harada, T. (2013). Scion on a stock producing siRNAs of potato spindle tuber viroid (PSTVd) attenuates accumulation of the viroid. *PLoS One* 8:e57736. doi: 10.1371/journal.pone.0057736

Khah, E.M. (2011). Effect of grafting on growth, performance and yield of aubergine (Solanum melongena L.) in greenhouse and open-field. *Intern. J. Plant Prod.* 5, 359–366.

Khah, E.M., Kakava, E., Mavromatis, A., Chachalis, D. and Goulas, C. (2006). Effect of grafting on growth and yield of tomato (Lycopersicon esculentum Mill.) in greenhouse and open-field. *J. Appl. Hortic.* 8, 3–7

Kollmann, R. and Glockmann, C. (1985). Studies on graft unions. I. Plasmodesmata between cells of plants belonging to different unrelated taxa. *Protoplasma* 124, 224–235. doi: 10.1007/BF01290774

Krumbein, A. and Schwarz, D. (2013). Grafting: A possibility to enhance health promoting and flavour compounds in tomato fruits of shaded plants? *Sci. Hortic.* 149, 97–107. doi: 10.1016/j.scienta.2012.09.003

Kumar, P., Edelstein, M., Cardarelli, M., Ferri, E. and Colla, G. (2015). Grafting affects growth, yield, nutrient uptake, and partitioning under cadmium stress in tomato. *HortScience*, *50*(11), 1654–1661.

Kumari, A., Kumar, J., Kumar, A., Chaudhury, A. and Singh, S.P. (2015). Grafting triggers differential responses between scion and rootstock. *PLoS One* 10:e0124438. doi: 10.1371/journal.pone.0124438

Kurotani, K., Wakatake, T., Ichihashi, Y., Okayasu, K., Sawai, Y., Ogawa, S., et al. (2020). Host-parasite tissue adhesion by a secreted type of β-1, 4-glucanase in the parasitic plant Phtheirospermum japonicum. *Commun. Biol.* 3:407.

Kyriacou, M.C., Colla, G. and Rouphael, Y. (2020). Grafting as a sustainable means for securing yield stability and quality in vegetable crops. *Agronomy*, *10*(12), 1945.

Kyriacou, M.C., Rouphael, Y., Colla, G., Zrenner, R. and Schwarz, D. (2017). Vegetable grafting: The implications of a growing agronomic imperative for vegetable fruit quality and nutritive value. *Frontiers in Plant Science*, 8, 741.

Kyriacou, M.C., Soteriou, G.A., Rouphael, Y., Siomos, A.S. and Gerasopoulos, D. (2016). Configuration of watermelon fruit quality in response to rootstock-mediated harvest maturity and postharvest storage. *J. Sci. Food Agric.* 96, 2400–2409. doi: 10.1002/jsfa.7356

Lama, K., Alkalai-Tuvia, S., Perzelan, Y. and Fallik, E. (2016). Nutritional qualities and aroma volatiles of harvested red pepper fruits stored at suboptimal temperatures. *Scientia Horticulturae*, *213*, 42–48.

Leal-Fernández, C., Godoy-Hernández, H., Núñezcolín, C.A., Anaya-López, J.A., Villalobos-Reyes, S. and Castellanos, J.Z. (2013). Morphological response and fruit yield of sweet pepper (Capsicum annuum L.) grafted onto different commercial rootstocks. *Biol. Agric. Hortic.* 29, 1–11. doi: 10.1080/01448765. 2012.746063

Lee, J.M., Kubota, C., Tsao, S.J., Bie, Z., Echevarria, P.H., Morra, L. and Oda, M. (2010). Current status of vegetable grafting: diffusion, grafting techniques, automation. *Scientia Horticulturae*, *127*(2), 93–105.

Lewsey, M.G., Hardcastle, T.J., Melnyk, C.W., Molnar, A., Valli, A., Urich, M.A., et al. (2016). Mobile small RNAs regulate genome-wide DNA methylation. *Proc. Natl. Acad. Sci. U. S. A.* 113, E801–E810. doi: 10.1073/pnas.1515072113

Li, Y., Sun, W., Liu, F., Cheng, J., Zhang, X., Zhang, H. and Zhao, Y. (2019). Methods for grafting Arabidopsis thaliana and Eutrema salsugineum. *Plant Methods*, 15.

Liu, H.Y., Zhu, Z.J., Diao, M. and Guo, Z.P. (2006). Characteristic of the sugar metabolism in leaves and fruits of grafted watermelon during fruit development. *Plant Phys. Commun.* 42, 835–840.

Liu, N., Yang, J.H., Fu, X.X., Zhang, L., Tang, K., Guy, K.M., et al. (2016). Genome-wide identification and comparative analysis of grafting-responsive mRNA in watermelon grafted onto bottle gourd and squash rootstocks by high-throughput sequencing. *Mol. Genet. Genomics* 291, 621–633. doi: 10.1007/ s00438-015-1132-5

Loupit, G., Brocard, L., Ollat, N. and Cookson, S.J. (2023). Grafting in plants: recent discoveries and new applications. *Journal of Experimental Botany*, *74*(8), 2433–2447.

Louws, F.J., Rivard, C.L. and Kubota, C. (2010). Grafting fruiting vegetables to manage soilborne pathogens, foliar pathogens, arthropods and weeds. *Sci. Hortic.* 127, 127–146. doi: 10.1016/j.scienta.2010.09.023

Luria, N., Smith, E., Reingold, V., Bekelman, I., Lapidot, M., Levin, I., et al. (2017). A new Israeli Tobamovirus isolate infects tomato plants harboring Tm-22 resistance genes. *PLoS One* 12:e0170429. doi: 10.1371/journal.pone.0170429

Majhi, P.K., Bhoi, T.K., Sahoo, K.C., Mishra, N., Tudu, S., Das, S. and Saini, V. (2023). Understanding the Genetics and Genomics of Vegetable Grafting to Ensure Yield Stability. In *Smart Plant Breeding for Vegetable Crops in Post-genomics Era* (pp. 69–98). Singapore: Springer Nature Singapore.

Peralta Manjarrez, R.M., Delgado Martínez, R., Benavides Mendoza, A., Juárez Maldonado, A. and Cabrera De la Fuente, M. (2023). Calcium, potassium, and magnesium affect the nutritional value of tomato grafted fruits grown in a nutrient film technique system. *Agriculture*, *13*(12), 2189.

Marino, P., Pepe, G., Basilicata, M., Vestuto, V., Marzocco, S., Autore, G., Procino, A., Gomez-Monterrey, I., Manfra, M. and Campiglia, P. (2023). Potential role of natural antioxidant products in oncological diseases. *Antioxidants*, 12.

Marsic, N.K., Mikulik-Petkovsek, M. and Stampar, F. (2014). Grafting influences phenolic profile and carpometric traits of fruits of greenhouse-grown eggplant (Solanum melongena L.). *J. Agric. Food Chem.* 62, 10504–10514. doi: 10.1021/ jf503338m

Martin, A., Adam, H., Díaz-Mendoza, M., Żurczak, M., González-Schain, N.D. and Suárez-López, P. (2009). Graft-transmissible induction of potato tuberization by the microRNA miR172. *Development* 136, 2873–2881. doi: 10.1242/ dev.031658

Matsuzoe, N., Hiromi, A., Hanada, K., Ali, M., Okubo, H. and Fujideda, K. (1996). Fruit quality of tomato plants grafted on Solanum rootstocks. *J. Jpn. Soc. Hortic. Sci.* 65, 73–80. doi: 10.2503/jjshs.65.73

Melnyk, C.W. (2017). Plant grafting: insights into tissue regeneration. *Regeneration* 4, 3–14. doi: 10.1002/reg2.71

Melnyk, C.W., Gabel, A., Hardcastle, T.J., Robinson, S., Miyashima, S., Grosse, I., et al. (2018). Transcriptome dynamics at Arabidopsis graft junctions reveal an intertissue recognition mechanism that activates vascular regeneration. *Proc. Natl. Acad. Sci. U. S. A.* 115, E2447–E2456. doi: 10.1073/pnas.1718263115

Melnyk, C.W., Molnar, A., Bassett, A. and Baulcombe, D.C. (2011). Mobile 24 nt small RNAs direct transcriptional gene silencing in the root meristems of Arabidopsis thaliana. *Curr. Biol.* 21, 1678–1683. doi: 10.1016/j.cub.2011.08.065

Melnyk, C.W., Schuster, C., Leyser, O. and Meyerowitz, E.M. (2015). A developmental framework for graft formation and vascular reconnection in Arabidopsis thaliana. *Curr. Biol.* 25, 1306–1318. doi: 10.1016/j.cub.2015.03.032

Miskovic, A., Ilic, O., Bacanovic, J., Vujasinovic, V. and Kukiæ, B. (2016). Effect of eggplant rootstock on yield and quality parameters of grafted tomato. *Acta Sci. Polon. Hortic. Cul.* 15, 149–159.

Molnar, A., Melnyk, C.W., Bassett, A., Hardcastle, T.J., Dunn, R. and Baulcombe, D.C. (2010). Small silencing RNAs in plants are mobile and direct epigenetic modification in recipient cells. *Science* 328, 872–875. doi: 10.1126/science.1187959 .

Moncada, A., Miceli, A., Vetrano, F., Mineo, V., Planeta, D. and D'Anna, F. (2013). Effect of grafting on yield and quality of eggplant (Solanum melongena L.). *Sci. Hortic.* 149, 108–114. doi: 10.1016/j.scienta.2012.06.015

Naik, S.A., Hongal, S.V., Hanchinamani, C.N., Manjunath, G., Ponnam, N., Shanmukhappa, M.K. and Kumar, P. (2024). Grafting bell pepper onto local genotypes of Capsicum spp. as rootstocks to alleviate bacterial wilt and root-knot nematodes under protected cultivation. *Agronomy*, *14*(3), 470.

Nakamura, S., Hondo, K., Kawara, T., Okazaki, Y., Saito, K., Kobayashi, K., et al. (2015). Conferring high-temperature tolerance to nontransgenic tomato scions using graft transmission of RNA silencing of the fatty acid desaturase gene. *Plant Biotechnol. J.* 14, 783–790. doi: 10.1111/pbi.12429

Nanda, A.K. and Melnyk, C.W. (2018). The role of plant hormones during grafting. *J. Plant Res.* 131, 49–58.

Nicoletto, C., Tosini, F. and Sambo, P. (2013b). Effect of grafting on biochemical and nutritional traits of 'Cuore di bue' tomatoes harvested at different ripening stages. *Acta Agric. Scand. B* 63, 114–122. doi: 10.1080/09064710.2012.729606

Nie, W. and Wen, D. (2023). Study on the applications and regulatory mechanisms of grafting on vegetables. *Plants*, *12*(15), 2822.

Nordey, T., Schwarz, D., Kenyon, L., Manickam, R. and Huat, J. (2020). Tapping the potential of grafting to improve the performance of vegetable cropping systems in sub-Saharan Africa: a review. *Agronomy for Sustainable Development*, *40*(4), 23.

Notaguchi, M., Kurotani, K., Sato, Y., Tabata, R., Kawakatsu, Y., Okayasu, K., et al. (2020). Cell–cell adhesion in plant grafting is facilitated by β-1,4-glucanases. *Science* 369, 698–702. doi: 10.1126/science.abc3710

Ntatsi, G., Savvas, D., Papasotiropoulos, V., Katsileros, A., Zrenner, R.M., Hincha, D.K., et al. (2017). Rootstock sub-optimal temperature tolerance determines transcriptomic responses after long-term root cooling in rootstocks and scions of grafted tomato plants. *Front. Plant Sci.* 8:911. doi: 10.3389/ fpls.2017.00911

Oda, M., Nagata, M., Tsuji, K. and Sasaki, H.S.O. (1996). Effects of scarlet eggplant rootstock on growth, yield, and sugar content of grafted tomato fruits. *J. Jpn. Soc. Hortic. Sci.* 65, 531–536. doi: 10.2503/jjshs.65.531

Ohletz, J.L. and Loy, J.B. (2021). Grafting melons increases yield, extends the harvest season, and prevents sudden wilt in New England. *HortTechnology hortte*, *31*(1), 101–114. Retrieved Jul 25, 2024,

Ohta, Y. (1991). Graft-transformation, the mechanism for graft-induced genetic changes in higher plants. *Euphytica* 55, 91–99. doi: 10.1007/BF00022565

Orosco-Alcalá, B.E., Núñez-Palenius, H.G., Díaz-Serrano, F., Pérez-Moreno, L., Valencia-Posadas, M., Trejo-Tellez, L.I. and Valiente-Banuet, J.I. (2021). Grafting improves salinity tolerance of bell pepper plants during greenhouse production. *Horticulture, Environment, and Biotechnology*, *62*(6), 831–844.

Orsini, F., Sanoubar, R., Oztekin, G.B., Kappel, N., Tepecik, M., Quacquarelli, C., et al. (2013). Improved stomatal regulation and ion partitioning boosts salt tolerance in grafted melon. *Funct. Plant Biol.* 40, 628–636. doi: 10.1071/FP12350

Pant, B.D., Buhtz, A., Kehr, J. and Scheible, W.R. (2008). MicroRNA399 is a longdistance signal for the regulation of plant phosphate homeostasis. *Plant J.* 53, 731–738. doi: 10.1111/j.1365-313x.2007.03363.x

Penella, C., Nebauer, S.G., Quinones, A., San Bautista, A., López-Galarza, S. and Calatayud, A. (2015). Some rootstocks improve pepper tolerance to mild salinity through ionic regulation. *Plant Science*, *230*, 12–22.

Qaryouti, M.M., Qawasmi, W., Hamdan, H. and Edwan, M. (2007). Tomato fruit yield and quality as affected by grafting and growing system. *Acta Hortic.* 741, 199–206. doi: 10.17660/actahortic.2007.741.22

Rajesh Kumar, R.K., Neeraj Dwivedi, N.D., Singh, R.K., Sanjay Kumar, S.K., Rai, V.P. and Major Singh, M.S. (2011). A review on molecular characterization of pepper for capsaicin and oleoresin. *International Journal of Plant Breeding and Genetics*, Vol. 5, No. 2, 99–110

Ramakrishnan, U. (2002). Prevalence of micronutrient malnutrition worldwide. *Nutrition Reviews*, 60 5 Pt 2, S46–52 .

Rasool, A., Mansoor, S., Bhat, K., Hassan, G., Baba, T., Alyemeni, M., Alsahli, A., El-Serehy, H., Paray, B. and Ahmad, P. (2020). Mechanisms underlying graft union formation and rootstock–scion interaction in horticultural plants. *Frontiers in Plant Science*, 11.

Remzeena, A. and Anitha, P. (2023). Vegetable grafting: a green technique to combat biotic and abiotic stresses. *Vegetos*, 1–10.

Riga, P. (2015). Effect of rootstock on growth, fruit production and quality of tomato plants grown under low temperature and light conditions. *Hortic. Environ. Biotechnol.* 56, 626–638. doi: 10.1007/s13580-015-0042-0

Rivera, A., Monteagudo, A.B., Igartua, E., Taboada, A., García-Ulloa, A., Pomar, F. and Silvar, C. (2016). Assessing genetic and phenotypic diversity in pepper (Capsicum annuum L.) landraces from North-West Spain. *Scientia Horticulturae*, *203*, 1–11.

Romano, D., Paratore, A. and Vindigni, G. (2000). "Caratteristiche dei frutti di pomodoro in rapporto a diversi portinnesti," in Proceedings of the Workshop on 'Applicazione di Tecnologie Innovative per il Miglioramento dell' Orticoltura Meridionale', Rome: Consiglio Nazionale delle Ricerche, 113–114.

Rouphael, Y., Rea, E., Cardarelli, M., Bitterlich, M., Schwarz, D. and Colla, G. (2016). Can adverse effects of acidity and aluminum toxicity be alleviated by appropriate rootstock selection in cucumber? *Front. Plant Sci.* 7:1283. doi: 10.3389/fpls.2016.01283

Rouphael, Y., Schwarz, D., Krumbein, A. and Colla, G. (2010). Impact of grafting on product quality of fruit vegetables. *Scientia Horticulturae*, *127*(2), 172–179.

Sabatino, L., Iapichino, G., Consentino, B.B., D'Anna, F. and Rouphael, Y. (2020). Rootstock and arbuscular mycorrhiza combinatorial effects on eggplant crop performance and fruit quality under greenhouse conditions. *Agronomy*, *10*(5), 693.

Sabatino, L., Iapichino, G., Maggio, A., D'anna, E., Bruno, M. and D'Anna, F. (2016). Grafting affects yield and phenolic profile of Solanum melongena L. landraces. *J. Integr. Agric.* 15, 1017–1024. doi: 10.1016/s2095-3119(15) 61323-5

Saleh, S., Staes, A., Deborggraeve, S. and Gevaert, K. (2019). Targeted Proteomics for Studying Pathogenic Bacteria. *PROTEOMICS*, 19.

Sallaku, G., Rewald, B., Sandén, H. and Balliu, A. (2022). Scions impact biomass allocation and root enzymatic activity of rootstocks in grafted melon and watermelon plants. *Frontiers in Plant Science*, 13.

Sánchez-Rodríguez, E., Ruiz, J.M., Ferreres, F. and Moreno, D.A. (2012b). Phenolic profiles of cherry tomatoes as influenced by hydric stress and rootstock technique. *Food Chem.* 134, 775–782. doi: 10.1016/j.foodchem.2012.02.180

Sánchez-Torres, P., Raigón, M.D., Gammoudi, N. and Gisbert, C. (2016). Effects of grafting combinations on the nutritional composition of pepper fruit. *Fruits* 71, 249–256. doi: 10.1051/fruits/2016014

Savvas, D., Savva, A., Ntatsi, G., Ropokis, A., Karapanos, I., Krumbein, A., et al. (2011). Effects of three commercial rootstocks on mineral nutrition, fruit yield, and quality of salinized tomato. *J. Plant Nutr Soil Sci.* 174, 154–162. doi: 10.1002/jpln.201000099

Schultheis, J., Thompson, W. and Hassel, R. (2015). Specialty melon yield and quality response to grafting in trials conducted in the Southeastern United States. *Acta Hortic.* 1086, 269–278. doi: 10.17660/actahortic.2015.1086.34

Schwarz, D., Öztekin, G.B., Tüzel, Y., Brückner, B. and Krumbein, A. (2013). Rootstocks can enhance tomato growth and quality characteristics at low potassium supply. *Sci. Hortic.* 149, 70–79. doi: 10.1016/j.scienta.2012.06.013

Schwarz, D., Rouphael, Y., Colla, G. and Venema, J.H. (2010). Grafting as a tool to improve tolerance of vegetables to abiotic stresses: thermal stress, water stress and organic pollutants. *Sci. Hortic.* 127, 162–171. doi: 10.1016/j.scienta.2010.09.016

Sharma, S., Katoch, V., Kumar, S. and Chatterjee, S. (2021). Functional relationship of vegetable colors and bioactive compounds: implications in human health. *The Journal of Nutritional Biochemistry*, 108615.

Shivran, M., Sharma, N., Dubey, A.K., Singh, S.K., Sharma, N., Sharma, R.M. and Singh, R. (2022). Scion–rootstock relationship: molecular mechanism and quality fruit production. *Agriculture*, *12*(12), 2036.

Shiwani, K. (2020). Perspectives of grafting in Solanaceous vegetables: a review. *International Journal of Chemical Studies*, 8, 698–702.

Singh, H., Kumar, P., Chaudhari, S. and Edelstein, M. (2017). Tomato grafting: a global perspective. *HortScience*, *52*(10), 1328–1336.

Slimestad, R., Fossen, T. and Verhuel, M.J. (2008). The flavonoids of tomatoes. *J. Agric. Food Chem.* 56, 2436–2441. doi: 10.1021/jf073434n

Soltan, M.M., ElAidy, F.A., Scheerens, J.C. and Kleinhenz, M. (2017). Grafting, scion and rootstock effects on survival rate, vegetative growth and fruit yield of high tunnel-grown grafted pepper (Capsicum annuum L.) plants. *Adv. Crop Sci. Technol*, *5*, 312.

Spanò, R., Ferrara, M., Montemurro, C., Mulè, G., Gallitelli, D. and Mascia, T. (2020). Grafting alters tomato transcriptome and enhances tolerance to an airborne virus infection. *Sci. Rep.* 10:2538.

Spanò, R., Mascia, T., Kormelink, R. and Gallitelli, D. (2015). Grafting on a non-transgenic tolerant tomato variety confers resistance to the infection of a Sw5-breaking strain of tomato spotted wilt virus via RNA silencing. *PLoS One* 10:e0141319 doi: 10.1371/journal.pone.0141319.t001

Spiegelman, Z., Golan, G. and Wolf, S. (2013). Don't kill the messenger: long distance trafficking of mRNA molecules. *Plant Sci.* 213, 1–8.

Stegemann, S. and Bock, R. (2009). Exchange of genetic material between cells in plant tissue grafts. *Science* 324, 649–651. doi: 10.1126/science.1170397

Stegemann, S., Keuthe, M., Greiner, S. and Bock, R. (2012). Horizontal transfer of chloroplast genomes between plant species. *Proc. Nat. Acad. Sci. U.S.A* 109, 2434–2438. doi: 10.1073/pnas.1114076109

Taller, J., Hirata, Y., Yagishita, N., Kita, M. and Ogata, S. (1998). Graft-induced genetic changes and the inheritance of several characteristics in pepper (Capsicum annuum L.). *Theor. Appl. Genet.* 97, 705–713. doi: 10.1007/s001220050946

Taller, J., Yagishita, N. and Hirata, Y. (1999). Graft-induced variants as a source of novel characteristics in the breeding of pepper (Capsicum annuum L.). *Euphytica* 108, 73–78. doi: 10.1023/A:1003681913996

Tirupathamma, T.L., Ramana, C.V., Naidu, L.N. and Sasikala, K. (2019). Vegetable grafting: A multiple crop improvement methodology. *Curr. J. Appl. Sci. Technol*, 33, 1–10.

Tsaballa, A., Athanasiadis, C., Pasentsis, K., Ganopoulos, I., Nianiou-Obeidat, I. and Tsaftaris, A. (2013). Molecular studies of inheritable grafting induced changes in pepper (Capsicum annuum) fruit shape. *Sci. Hortic.* 149, 2–8. doi: 10.1016/j.scienta.2012.06.018

Tsaballa, A., Pasentsis, K., Darzentas, N. and Tsaftaris, A.S. (2011). Multiple evidence for the role of an ovate-like gene in determining fruit shape in pepper. *BMC Plant Biol.* 11:46. doi: 10.1186/1471-2229-11-46

Tsaballa, A., Xanthopoulou, A., Madesis, P., Tsaftaris, A. and Nianiou-Obeidat, I. (2021). Vegetable grafting from a molecular point of view: the involvement of epigenetics in rootstock–scion interactions. *Frontiers in Plant Science*, *11*, 621999.

Tsutsui, H., Yanagisawa, N., Kawakatsu, Y., Ikematsu, S., Sawai, Y., Tabata, R., ... and Notaguchi, M. (2020). Micrografting device for testing systemic signaling in Arabidopsis. *The Plant Journal*, 103(2), 918–929.

Turhan, A., Ozmen, N., Serbeci, M.S. and Seniz, V. (2011). Effects of grafting on different rootstocks on tomato fruit yield and quality. *Hortic. Sci.* 38, 142–149. doi: 10.1007/s00122-015-2462-8

Turnbull, C.G. and Lopez-Cobollo, R.M. (2013). Heavy traffic in the fast lane: long-distance signalling by macromolecules. *New Phytol.* 198, 33–51.

Uresti-Porras, J., Fuente, M., Benavides-Mendoza, A., Olivares-Sáenz, E., Cabrera, R. and Juárez-Maldonado, A. (2021). Effect of graft and nano ZnO on nutraceutical and mineral content in bell pepper. *Plants*, 10.

Votava, E.J., Baral, J.B. and Bosland, P.W. (2005). Genetic diversity of chile (Capsicum annuum var. annuum L.) landraces from northern New Mexico, Colorado, and Mexico. *Economic Botany*, *59*(1), 8–17.

Wang, H., Zhou, P., Zhu, W. and Wang, F. (2019). De novo comparative transcriptome analysis of genes differentially expressed in the scion of homografted and heterografted tomato seedlings. *Sci. Rep.* 9:20240.

Wei, C., Li, M., Qin, J., Xu, Y., Zhang, Y. and Wang, H. (2019). Transcriptome analysis reveals the effects of grafting on sweetpotato scions during the full blooming stages. *Genes. Genomics* 41, 895–907.

Whitfield, F.B. and Last, J.H. Vegetables. In Volatile Compounds in Foods and Beverages; Maarse, H., Ed.; Marcle Dekker: New York, NY, USA, 1991; pp. 203–281.

Wu, R., Wang, X.R., Lin, Y., Ma, Y.Q., Liu, G., Yu, X.M., et al. (2013). Interspecies grafting caused extensive and heritable alterations of DNA methylation in Solanaceae plants. *PLoS ONE* 8:e61995. doi: 10.1371/journal.pone. 0061995

Wulf, K.E., Reid, J.B. and Foo, E. (2020). What drives interspecies graft union success? Exploring the role of phylogenetic relatedness and stem anatomy. *Physiol. Plant.* 170, 132–147. doi: 10.1111/ppl.13118

Xanthopoulou, A., Tsaballa, A., Ganopoulos, I., Kapazoglou, A., Avramidou, E., Aravanopoulos, F.A., et al. (2019). Intra-species grafting induces epigenetic and metabolic changes accompanied by alterations in fruit size and shape of Cucurbita pepo L. *Plant Growth Regul.* 87, 93–108. doi: 10.1007/ s10725-018-0456-7

Xie, M. and Yu, B. (2015). siRNA-directed DNA methylation in plants. *Curr. Genomics* 16, 23–31. doi: 10.2174/13892029 15666141128002211

Yagishita, N., Hirata, Y., Mizukami, H., Ohashi, H. and Yamashita, K. (1990). Genetic nature of low capsaicin content in the variant strains induced by grafting in Capsicum annuum L. *Euphytica* 46, 249–252. doi: 10.1007/ BF000272247

Yan, G., Feng, M., Lin, W., Huang, Y., Tong, R. and Cheng, Y. (2022). Review and prospect for vegetable grafting robot and relevant key technologies. *Agriculture*, *12*(10), 1578.

Yin, H., Yan, B., Sun, J., Jia, P., Zhang, Z., Yan, X., et al. (2012). Graft-union development: a delicate process that involves cell–cell communication between scion and stock for local auxin accumulation. *J. Exp. Bot.* 63, 4219–4232. doi: 10.1093/jxb/ers109

Yu, C., Dong, W., Zhan, Y., Huang, Z.-A., Li, Z., Kim, I.S., et al. (2017). Genome-wide identification and expression analysis of ClLAX, ClPIN and ClABCB genes families in Citrullus lanatus under various abiotic stresses and grafting. *BMC Genet.* 18:33. doi: 10.1186/s12863-017-0500-z

Zhang, G., Mao, Z., Wang, Q., Song, J., Nie, X., Wang, T., et al. (2019). Comprehensive transcriptome profiling and phenotyping of rootstock and scion in a tomato/potato heterografting system. *Physiol. Plant.* 166, 833–847. doi: 10.1111/ppl.12858

Zhao, L., Liu, A., Song, T., Jin, Y., Xu, X., Gao, Y., et al. (2018). Transcriptome analysis reveals the effects of grafting on sugar and α-linolenic acid metabolisms in fruits of cucumber with two different rootstocks. *Plant Physiol. Biochem.* 130, 289–302. doi: 10.1016/j.plaphy.2018.07.008

Zhao, Y., Feng, D., Zhang, C., Sun, Q. and Zhao, D. (2016). Virtual Prototype Simulation and Analysis of a Grafting Robot for Saplings. International Journal of Simulation–Systems, Science & Technology, *17*(33).

Zhuang, Y., Chen, L., Sun, L. and Cao, J. (2012). Bioactive characteristics and antioxidant activities of nine peppers. *Journal of Functional Foods*, *4*(1), 331–338.

# Chapter 10

# Advancements in Plasma Technology for Enhancing Crop Resilience to Environment Stress

Mahantesha BN Naika[1*] Sowmya KS[2], Nethra N[2], Arvindkumar[3] and Belaghihalli N Gnanesh[4]

## 1. Introduction

Quality seeds and resilient plants are fundamental inputs for sustainable farming, especially in the context of shrinking global arable land to approximately 0.17 hectares per person (Lal, 2016). Also, coupled with deteriorating soil health, a rapidly growing population of nearly 8 billion (Washington and Kopnina, 2022; Anonymous, 2022) and changing environmental conditions create opportunities for new agricultural ventures to work (Hemathilake and Gunathilake, 2022).

Crop productivity is a key determinant of farmers' livelihood and global food security. It is influenced by multiple interacting factors, including biological, environmental, socioeconomic, and management considerations. The biological factors include the plant's genetic makeup, initial seed quality, and growing environment, which encompass beneficial microorganisms, pollinators, weeds, pests, and diseases. Important environmental factors include temperature, rainfall, wind speed, and relative humidity, which determine the type and abundance of biota present in a given region. In addition, crop management practices, including soil fertility management, irrigation strategies, and pest control, play a significant role in determining yield potential. However, the increasing unpredictability of climate patterns, the emergence of resistant pest and pathogen strains, and the degradation of natural resources due to unsustainable agricultural practices have intensified environmental stressors, posing significant challenges to crop productivity.

Environmental stressors significantly impact crop's physiology and productivity (Figure 1). Abiotic stressors, such as drought and high temperatures, reduce water availability to plant tissues. Although plants attempt to cope through stomatal regulation, osmoregulation, and hormonal adjustments, extreme conditions severely limit $CO_2$ uptake, thereby reducing photosynthesis. Elevated temperatures denature proteins, disrupt enzyme activity, and increase transpiration rates,

[1] Centre for Biotechnology Research, Department of Biotechnology and Crop Improvement, College of Horticulture, University of Horticulture Sciences, Bagalkot, Karnataka, India.
[2] AICRP on Seed (Crops), National Seed Project, University of Agricultural Sciences, GKVK, Bangalore, Karnataka, India.
[3] College of Horticulture, Bidar, UHS, Bagalkot, Karnataka, India.
[4] Genetics & Plant Breeding, ICAR JSS – Krishi Vigyan Kendra, Suttur, Mysore, Karnataka, India.
* Corresponding author: mahantesha.bnn@uhsbagalkot.edu.in

DOI: 10.1201/9781003584100-10

leading to excessive water loss. In contrast, cold stress causes ice crystal formation within plant tissues, damaging cell structures and disrupting metabolic functions. Salinity stress induces osmotic stress and ionic stress, decreasing water potential and making it difficult for roots to absorb water. Excessive sodium ($Na^+$) and chloride ($Cl^-$) ions further disrupt ion balance and nutrient uptake, impairing overall plant metabolism. Soil contamination, particularly from heavy metals such as lead (Pb) and cadmium (Cd) from industrial discharge and pesticide residues, disrupts essential cellular processes and causes toxicity (Mareri et al., 2022). Additionally, UV radiation and pollutants exacerbate oxidative stress by inflicting damage on DNA, proteins, and lipids. Collectively, abiotic stresses result in stunted growth, reduced photosynthetic efficiency, impaired nutrient uptake, and metabolic dysfunction. Prolonged exposure to these adverse conditions can cause irreversible damage, ultimately resulting in plant death (Zhang et al., 2023; Parvathi et al., 2019; Srii et al., 2023).

Biotic stressors, including pathogens (bacteria, fungi, viruses), pests (insects, rodents, snails), interspecies competition, and invasive weeds, pose significant stress to plants affecting their metabolism and productivity (Figure 1). When pathogens attack plants, they release toxins and virulence factors that enter the plant system, disrupting cell membranes and damaging plant cells and tissues leading to cell death or necrosis. These pathogens interfere with cell signaling, suppressing the plant's natural defense mechanisms. They alter hormone-regulating pathways, disrupting normal growth and leading to symptoms such as crown gall formation and abnormal tissue development. Fungi secrete toxic secondary metabolites (mycotoxins) that weaken cell walls, disrupt plant metabolism, and produce molecules that suppress plant immunity, making the plant more vulnerable to further infections. On the other hand, pests cause physical damage by chewing, sucking, or boring into plant tissues, disrupting the flow of water and nutrients. This damage creates entry points for pathogens, further weakening the plant. Pests also interfere with plant metabolism and hormonal regulation, often leading to abnormal growth, deformities, or gall formation. Meanwhile, weeds compete with crops for nutrients, water, sunlight, and space, leading to competition stress. Additionally, some weeds release allelopathic compounds that suppress crop growth and development. The combined impact of these biotic stressors significantly reduces plant metabolic efficiency, weakens defense mechanisms, and results in significant yield losses (Shameer et al., 2019; Nishad et al., 2020; Naika et al., 2013). Without effective intervention, yield loss could reach up to 50–70% due to abiotic factors and 40–60% due to biotic factors. Moreover, these stressors often act synergistically, exacerbating crop damage; thus, there is an urgent need for innovation to curb the environmental stressors (Umar et al., 2021).

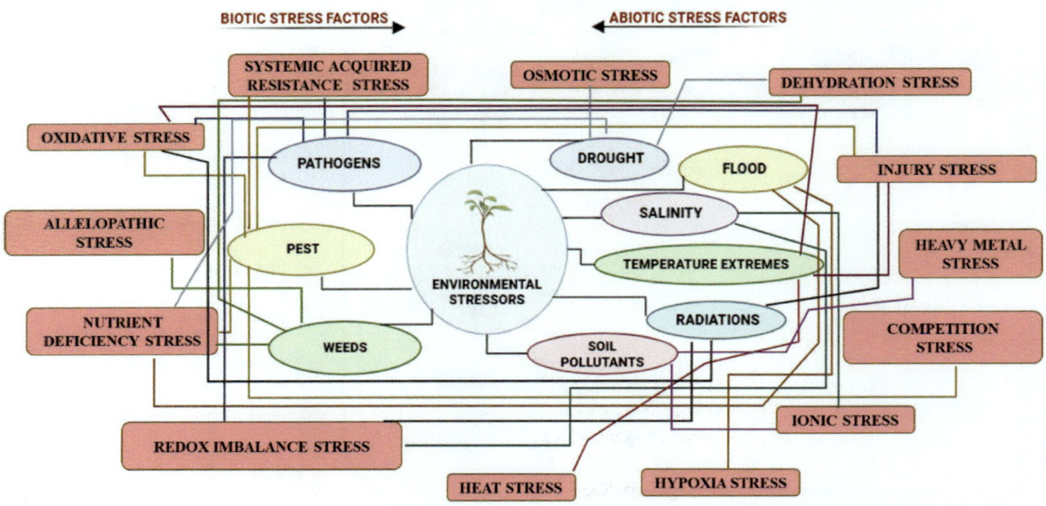


**Fig. 1** Illustration of biotic and abiotic stressors and their associated plant stress responses.

## 2. Cold Plasma Treatment Overview

Cold plasma also known as non-thermal plasma (NTP), room-temperature plasma, cold atmospheric plasma (CAP), or atmospheric pressure plasma (APP) is an ionized gas, the fourth state of matter. In nature, cold plasma occurs naturally during phenomena such as lightning, auroras, and even the upper atmospheric layers. It can also be artificially generated by applying high voltage to a gas at room temperature under atmospheric pressure, a process that produces the glowing effects commonly observed in neon lights and bulbs.

Plasma, an ionized gas, comprises a mixture of neutral particles, charged particles, electrons, reactive oxygen species (ROS) such as superoxide anion ($O_2^-$), hydrogen peroxide ($H_2O_2$), hydroxyl radical (•OH), singlet oxygen ($^1O_2$), and ozone ($O_3$), as well as reactive nitrogen species (RNS), that is, nitric oxide (NO), peroxynitrite ($ONOO^-$), nitrogen dioxide ($NO_2$), and nitrate radicals ($NO_3$). This highly reactive mixture operates near room temperature, with its composition influenced by parameters such as voltage, frequency, humidity, gas flow rate, and the choice of gases such as argon, oxygen, nitrogen, helium, or atmospheric air (Waskow et al., 2021; Nishanth et al., 2020; Krapivina et al., 1994).

Plasma is generated by applying a potential difference of a few hundred volts to a few kilovolts between electrodes in a reactor filled with inert gas (Bormashenko et al., 2012). The emitted electrons from the cathode collide with gas molecules, initiating excitation, ionization, and dissociation reactions. Excitation collisions produce excited species, contributing to the characteristic "glow" of plasma, while ionization collisions create ion–electron pairs that sustain the plasma. Dissociation reactions generate highly reactive radicals, enhancing the chemical reactivity of the plasma environment.

Cold plasma treatment is an exposure of ionized gas to biological entities at room temperature, triggering oxidative and nitrative stress responses and signal transduction (Figure 2). This process leads to surface modifications that enhance water absorption and nutrient uptake, and metabolic alterations that activate stress-responsive pathways and seed disinfection by eliminating pathogens, pests, and microbes. The electric field influences cellular processes, defense mechanisms, and gene expression stimulating seedling growth and development, improving germination rate and enhancing plant resilience across a wide range of environmental conditions (Holubová et al., 2020; Thakur and Vasudevan, 2021; Barjasteh *et al.*, 2024).

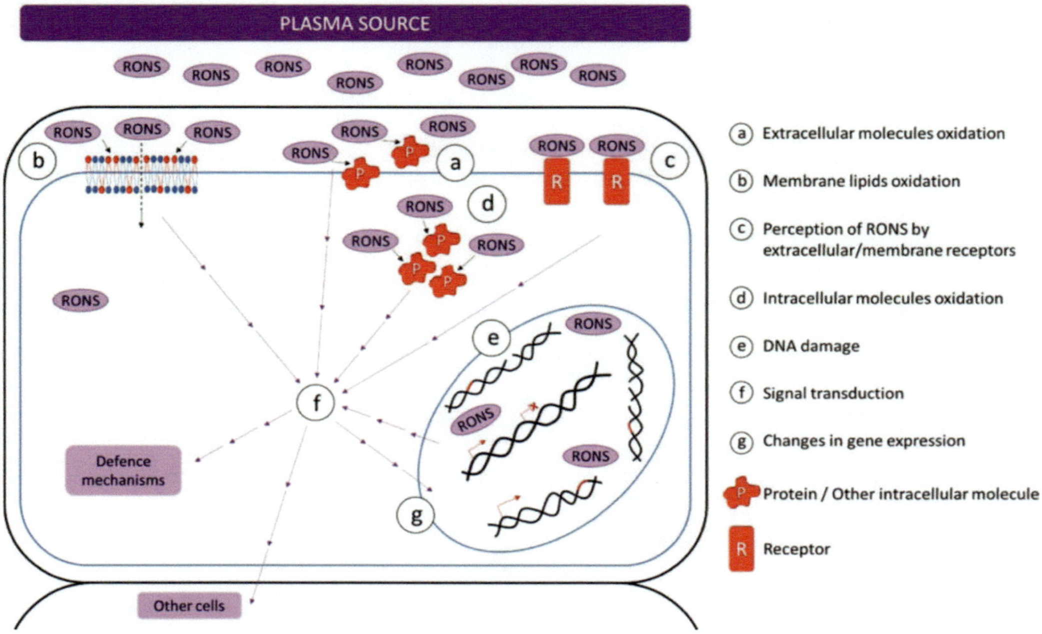


**Fig. 2** Illustration of possible interactions of cold plasma particles with cells (courtesy: Holubová et al., 2020).

## 2.1 Cold Plasma Treatment Types

Cold plasma treatments are versatile, offering applications based on exposure type, sample size, sample sensitivity, and intended outcomes. This technology is broadly categorized into two approaches: direct and indirect cold plasma treatments. While biological samples can encompass any part of the plant, seeds are the most commonly treated due to their robust seed coat, low metabolic activity, and efficient repair mechanisms (Radchuk and Borisjuk, 2014; El-Maarouf-Bouteau, H., 2022). The intricate physical and physiological characteristics of seeds make them an ideal model for plasma treatment compared to other plant parts.

### 2.1.1 Direct Treatment

In direct plasma treatment, the biological material is directly exposed to the plasma source, enabling immediate interaction with reactive species, the electric field, and UV radiation. This method provides localized and rapid effects, enhancing seed properties such as wettability, surface roughness or etching, and conductivity leading to faster imbibition and activation of metabolic pathways, healthier seedling establishment, and robust root development (Thirumdas et al., 2017). However, prolonged or high-dose exposure may pose a risk of physical or thermal damage.

Commonly used direct plasma techniques include dielectric barrier discharge plasma, atmospheric pressure plasma, radio frequency discharge plasma, and corona discharge plasma (Figure 3). Other methods, such as microwave plasma, gliding arc plasma, wave-heated plasma, arc discharge plasma, and piezoelectric direct discharge plasma, operate on similar principles and are employed based on specific application requirements (Shelar et al., 2022; Waskow et al., 2021).

1. Dielectric barrier discharge (DBD) plasma is generated by applying a high-voltage alternating current (kV) between two electrodes, with at least one electrode covered by a dielectric barrier. DBD can be categorized into volume dielectric barrier discharge (VDBD), where plasma forms by ionizing the entire gas volume between the electrodes, producing localized micro-discharges, and surface dielectric barrier discharge (SDBD), wherein the surface of dielectric material plasma formation occurs, resulting in more uniform and larger area plasma. Typically, one electrode operates at a lower frequency (approximately 100 kHz) or with alternating current (AC), while the other is grounded, enabling controlled plasma generation (Bian and Wu, 2018; Piferi et al., 2021).
2. Atmospheric pressure plasma (APP) is generated at atmospheric pressure without the need for a vacuum chamber. APP operates under nonequilibrium conditions, where electrons in the plasma are highly energetic while the gas remains at or near room temperature. This characteristic makes APP ideal for treating heat-sensitive materials, offering maximum plasma density and efficient treatment (Shelar et al., 2022).
3. Glow discharge (GD) or radio frequency discharge (RFD) plasma operates in low-pressure environments to ensure uniform treatment. GD utilizes direct current (DC) or low-frequency RF (<100 kHz) electric field between two electrodes to ionize the gas. RFD uses alternating current (AC) at radio frequencies (0.04 MHz to 13.56 MHz) to excite the gas molecules. RFD is further classified as capacitive-coupled plasma (CCP), where electrodes directly couple with plasma or inductively coupled plasma (ICP), where plasma forms in an electromagnetic field generated by an induction coil. These techniques generate minimal heat, making them suitable for temperature-sensitive applications (Bogaerts and Gijbels, 2000; Farouk et al., 2010)
4. Corona discharge (CD) plasma is generated at atmospheric pressure when a high-voltage or low-frequency electric field is applied to a sharply curved electrode. The surrounding gas molecules are ionized, creating a strong electric field and localized plasma discharge, creating a visible glow or corona around the electrode. CD results in nonuniform plasma distribution, leading to high plasma density in specific regions, which can risk damaging sensitive materials (Wang et al., 2020).

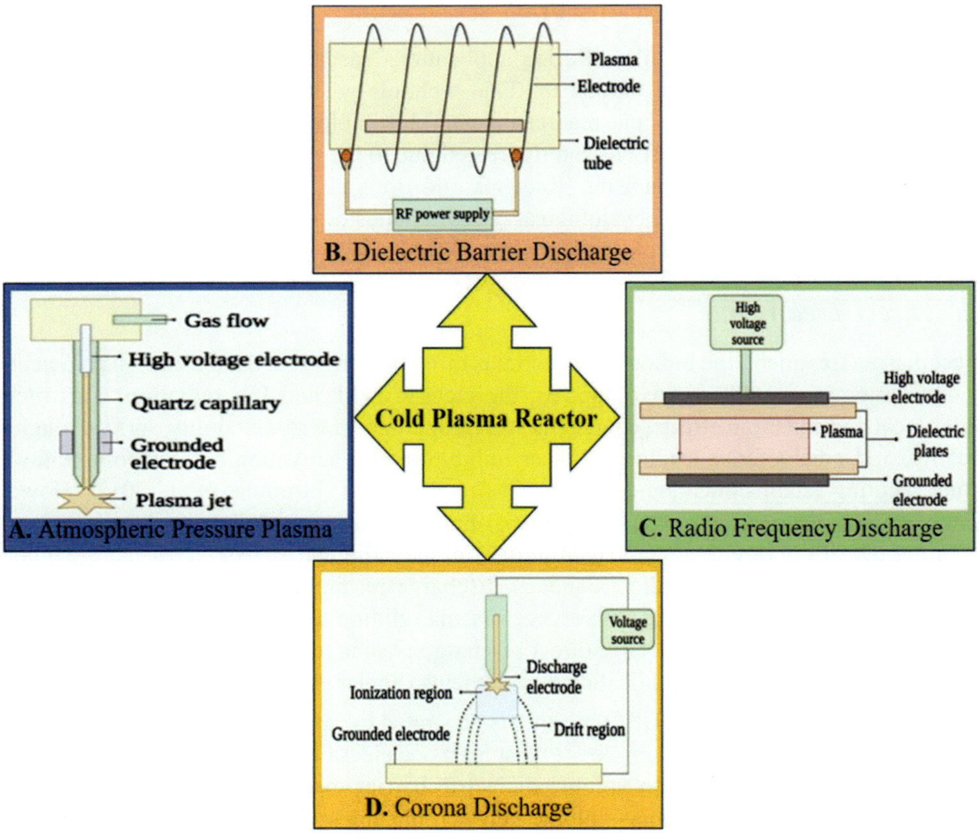


**Fig. 3** A schematic representation of various plasma reactors.

(A) APP is plasma generated at atmospheric pressure, implemented in open-air setups.
(B) DBD is plasma generated by applying high voltage across electrodes separated by a dielectric material, preventing continuous arc formation.
(C) RFD is plasma generated by applying high-frequency current (RF) to the gas, exciting the gas molecules and ionizing them.
(D) The CD is plasma generated by applying high voltage to an electrode with a sharp curvature, creating a strong electric field that ionizes the surrounding gas (courtesy: Shelar et al., 2022).

### *2.1.2 Indirect Treatment*

Indirect plasma treatment involves plasma generation in a separate chamber under atmospheric conditions, using gases such as air, nitrogen, oxygen, or argon as feeding precursors. The generated reactive species (reactive oxygen and nitrogen species (RONS)) interact with the surrounding medium such as water (plasma-activated water (PAW)), phosphate-buffered saline (PBS), or other activated solution media, to create a plasma-activated medium (PAM). This process induces modifications without exposing the material to intense UV radiation, charged particles, or electric fields (Judée et al., 2018; Thirumdas et al., 2017). PAM, enriched with RONS such as hydrogen peroxide, nitric oxide, and hydroxyl radicals, is then used to treat the target material without direct exposure to the plasma source.

Plasma jets, plasma pens, or torches are often preferred type of plasma reactors used for indirect plasma treatment over DBD, RF plasma, etc., due to their flexibility in targeting treatment areas outside the device (Zhou et al., 2020; Corbella et al., 2023; Guo et al., 2021; Chauvin et al., 2017). The specific effects of indirect plasma treatment depend on the type of plasma source, medium composition, and treatment parameters (Perrotti et al., 2022; Adhikari et al., 2020; Lukes et al., 2014).

In agriculture, PAW has significant applications. PAW undergoes chemical change due to RONS leading to reduced pH, increased oxidation–reduction potential, enhanced conductivity, and enhanced concentration of nitrates (Thirumdas et al., 2017; Chauvin et al., 2017; Mizoi et al., 2022). These alterations improve seed germination, seedling vigor, enzyme activity, and pesticidal removal while helping plants withstand environmental stressors. Studies on various crops have demonstrated that PAW can enhance growth metrics, biochemical composition, and stress resilience. Furthermore, PAW's ability to disinfect seeds and inactivate microbes without harmful chemicals makes it a sustainable alternative to conventional fertilizers and pesticides, aligning with environmental goals and reducing the agricultural sector's reliance on synthetic chemicals (Porto et al., 2018; Kučerová et al., 2019; Bafoil et al., 2018; Wang et al., 2023; Sajib et al., 2020; Thiyagarajan et al., 2013; Zhao et al., 2021; Soni et al., 2021; Čurila et al., 2023; Chou and Ting, 2023).

This versatile approach ensures the safe and efficient treatment of agricultural materials while supporting sustainable farming practices. By improving seed quality, germination rates, and resilience to environmental challenges, indirect plasma treatment and PAW are paving the way for eco-friendly innovations in agriculture. The adaptability of plasma reactors and the ability to tailor RONS composition for specific applications further enhance their relevance and effectiveness in diverse agricultural applications.

## 3. Mechanism of Cold Plasma Treatment for Enhancing Crop Resilience

Cold plasma treatment's efficacy stems from the intricate interplay of physical and biological processes. The composition and concentration of active agents generated are influenced by the plasma setup, shaping their interactions with biological samples. RONS initiate biochemical responses, including hormonal regulation and gene expression changes in treated samples. Additionally, factors such as ozone, heat, and electromagnetic fields contribute to physical and biochemical modifications, amplifying treatment effects. These interactions of active agents induce physical, physiological, and molecular alterations, influencing seed germination, plant growth, and stress tolerance (Bilea et al., 2024; Suschek, 2024; Yan et al., 2021).

Plasma generation within reactors is governed by fundamental physics principles, with treatment effectiveness critically dependent on operational parameters including voltage, pressure, capacitor configuration, electrode distance, charge particle density, and gas flow rate. Standardizing these parameters is essential, as these variables directly influence plasma characteristics and biological sample exposure time. These factors regulate the production, stability, and activity of reactive species, thereby determining their interactions with biological materials and the resulting treatment outcomes (Adhikari et al., 2020; Tian et al., 2024).

Voltage and frequency are primary operational parameters that control energy input, directly influencing electron energy, plasma density, and the generation of RONS. Higher voltage increases electron energy, producing a wider range of reactive species, whereas frequency modulates electron energy distribution and charged particle density, shaping the composition and concentration of reactive species formed. Power ($P = VI$), defined as the product of voltage and current, determines overall plasma intensity. Additionally, pressure and gas flow rate are essential for maintaining plasma stability by controlling ionization efficiency, electron collisions, and distribution of reactive species. The exposure duration further dictates the extent of surface modification and biological effects; however, overexposure can negatively impact seed quality or induce cellular and DNA damage. These parameters exhibit complex interdependencies, necessitating precise optimization to achieve maximum plasma treatment efficacy for targeted applications (Baek et al., 2016; Das et al., 2022).

DBD systems are widely employed plasma reactor systems due to their operational simplicity and versatility. The use of a dielectric material between the electrodes prevents electrical arc formation, stabilizing plasma generation at atmospheric pressure. In contrast, low-pressure plasma systems operate under conditions governed by Paschen's law, which defines the breakdown voltage required to initiate a discharge. This law dictates a nonlinear relationship between pressure and breakdown voltage, showing that the breakdown voltage in gas is a function of the product of the gas pressure and electrode gap distance. Sustaining discharges in low-pressure systems require higher voltage and reduced pressure (Kang et al., 2023; Gómez-Ramírez et al., 2017). These differences highlight the need for precise control of operational parameters in plasma applications to achieve optimal discharge conditions and maximize treatment effectiveness with minimal damage.

## 3.1 Factors Affecting Cold Plasma Treatment

The effectiveness of plasma treatment depends on the interplay of major active agents. The plasma setup and configuration determine the operational parameters that generate these active agents, which are critical for treatment standardization. The effectiveness of plasma treatment depends on the interplay of major active agents which are critical for treatment standardization (Figure 4). It is important to note that the response of biological materials to cold plasma treatment can vary significantly depending on their physiological and biochemical characteristics, that is, the response of cereal seeds may differ from that of pulses, oilseeds, vegetables, or medicinal plants (Waskow et al., 2021).

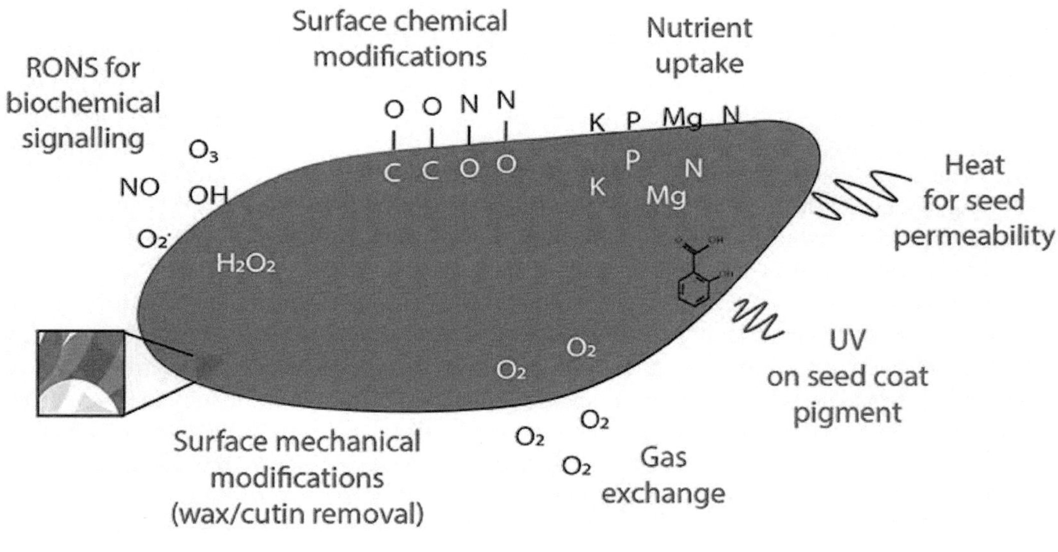


**Fig. 4** Summary of possible mechanical, chemical, and biochemical interactions of plasma components with seed surface (courtesy: Waskow et al., 2021).

### 3.1.1 Physical Factors

Physical factors are majorly involved in surface structure modification facilitating an increase in seed wettability, nutrient uptake, and gas exchange, contributing to better germination, field establishment, and increased responsiveness to conventional seed enhancement practices. Significant physical factors include heat, electrical field, UV radiation, and visible light.

#### 3.1.1.1 Heat

Heat generation is an inherent aspect of plasma treatment, with different plasma species exhibiting distinct temperature distributions. In a nonequilibrium cold plasma environment, the gas temperature remains near room temperature, while electron temperature can reach several electron volts (1–10 eV), significantly influencing interactions with treated seeds. Factors such as plasma design, power input, treatment duration, and gas composition determine variations in heat generation. Cold plasma treatments typically maintain temperatures between 30°C and 40°C, but excessive power and frequency can elevate these temperatures to above 90°C (De Felice et al., 2023; Butscher et al., 2016). The temperature of plasma setup is measured using infrared thermometers, thermal imaging cameras, fiber-optic sensors, or spectroscopic methods enabling optimization of treatment parameters.

Heat is generated in a plasma setup, although having limited reference to treatment significance. It is observed that a slight increase in temperature can improve surface permeability, facilitating better penetration of reactive species and promoting processes such as cell growth and tissue regeneration. This can accelerate seed germination and emergence. However, it is critical to manage heat generation carefully to prevent thermal damage, such as DNA and cellular damage, protein denaturation, and moisture loss. Strategies such as using shorter exposure times, employing cooling mechanisms with high gas flow rates (Pawlat et al., 2018; Fadhlalmawla et al., 2019), positioning seeds away from the discharge region to minimize localized heat exposure, and modulating voltage discharge using burst mode at low frequencies can help maintain a moderate and controlled temperature range in plasma setups, reducing the risk of thermal damage while enhancing treatment effectiveness (Nishime et al., 2020; Kitazaki et al., 2014).

#### 3.1.1.2 Electromagnetic Fields (EMFs)

The electromagnetic field is the driving force behind ionization and plasma stability. It arises from electrically charged objects, with stationary charges generating an electric field and moving charges (electric current) producing a magnetic field. The interaction between these fields accelerates electrons within the gas, leading to frequent collisions that trigger ionization. This process, known as electron impact ionization, generates additional free electrons and ions, sustaining plasma activity. Further collisions between ions and neutral gas molecules result in the formation of reactive species, which play a vital role in plasma-induced biochemical effects.

The strength of the electric field, regulated by voltage, frequency, and amplitude, influences gas ionization and affects plasma characteristics such as temperature, plasma density, and the production and composition of ROS. The waveform (pulsed or continuous) can also influence plasma properties and treatment outcomes. Typically, biological samples are treated with high-frequency plasma at voltages ranging from a few hundred volts to several kilovolts, but the effectiveness of the treatment largely depends on the duration of exposure, requiring precise control to optimize outcomes while minimizing potential damage (Murali et al., 2024; Waskow et al., 2021; Kletschkus et al., 2020).

EMFs influence various physiological and metabolic processes in seeds, enhancing seed coat permeability, promoting water uptake, improving gas exchange, and modulating metabolism. They also alter protein expression and increase antioxidant activity, strengthening seed vigor, accelerating seedling establishment, and improving resistance to oxidative stress. These effects contribute to higher yields, enhanced defense mechanisms, and improved crop performance (Shelar et al., 2022; Pauzaite et al., 2018; Chuesaard et al., 2023; Mildaziene et al., 2018; Ahmed et al., 2020). Additionally, ionized gas interactions in plasma treatment facilitate ozone production, which may further influence seed physiology.

#### 3.1.1.3 UV Radiation and Visible Light

Cold plasma treatment generates a broad spectrum of electromagnetic radiation, including ultraviolet (UV), infrared, and visible light. While the primary effects on seed physiology are attributed to RONS, high-energy photons and ozone act as catalytic agents, amplifying plasma-induced responses. The emission of high-energy photons contributes to the characteristic glow of plasma discharge.

Pre-sowing seed treatment with UV-B and UV-C radiation has been reported to accelerate germination; however, it may also reduce germination percentage, dry mass, grain filling, and yield (Stefanello et al., 2023; Foroughbakhch Pournavab et al., 2019). In case of plasma treatment, UV radiation, particularly in the 220–350 nm range, is generated by oxygen- and nitrogen-based plasma species and can influence seed germination through photochemical reactions, surface modification, and DNA damage (Kumaş et al., 2025; Tuteja et al., 2001).

The role of UV radiation in cold plasma seed treatment remains complex, with its effects largely dependent on exposure duration and interaction with other plasma-induced factors. While direct UV exposure alone may have minimal impact on seed wettability or germination, its synergy with ROS and ozone can influence seed physiology by modifying the seed coat, enhancing water uptake, surface modification, sterilization, and triggering stress responses. Additionally, UV radiation is linked to photomorphogenesis, potentially affecting cell differentiation and metabolic activity. However, prolonged or intense exposure may induce DNA damage and oxidative stress, negatively impacting seed viability and overall plant development (Ivankov et al., 2021; Jiang et al., 2014).

### *3.1.2 Chemical Factors*

#### 3.1.2.1 Surface Modification

Surface modification involves both mechanical and chemical modification. It is a physicochemical factor that has a significant impact on biological samples. Physical modifications include mechanical scarification, surface roughening, sputtering, and etching. Chemical modifications involve interactions with electromagnetic fields, RONS generation, and radiations, which contribute to the adsorption of reactive species, introduce new functional groups, and influence charge distribution on the seed surface.

Physical modification induces seed coat scarification and alters surface topography, leading to increased seed roughness. The bombardment of ionized gas molecules removes a thin layer of materials, such as waxes and oils, from the seed surface. This etching process enhances seed porosity, facilitating greater water uptake, which in turn helps in breaking seed dormancy, accelerating imbibition, improving biocompatibility with biopolymers or coatings, and enhancing seedling establishment (Alves Junior et al., 2016; Dufour et al., 2021).

Chemical modifications on the seed surface involve redox reactions that alter chemical composition based on gas composition, exposure duration, and plasma parameters. ROS play a crucial role in oxidizing organic molecules, leading to the formation of carbonyl, carboxyl, hydroxyl, and other functional groups. Similarly, RNS species can induce nitration reactions (Randeniya and de Groot, 2015; Waskow et al., 2021; Hasanuzzaman et al., 2020). UV radiation can synergistically interact with ROS generated within the plasma, amplifying their effects on seed surface chemistry. The combined impact of chemical and physical modifications significantly alters seed surface properties, improving overall seed performance.

#### 3.1.2.2 Gas Composition

A plasma treatment setup ensures uniform gas circulation, which is essential for maintaining consistent plasma discharge and achieving uniform treatment. The gas composition in the plasma system can include atmospheric air, oxygen ($O_2$), nitrogen ($N_2$), helium (He), argon (Ar), or a mixture of these gases. The choice of gas directly influences the type and concentration of reactive species generated, ultimately determining the treatment's effectiveness.

Atmospheric air, primarily composed of nitrogen and oxygen, generates both ROS and RNS, leading to diverse biological effects. Oxygen-rich plasma enhances oxidation and microbial sterilization, which can influence respiration, germination, and seedling establishment. In contrast, RNS species are more directly linked to cellular processes such as protein synthesis and photosynthesis due to nitrogen's role in these metabolic pathways. Treatments using inert gases such as argon or helium generate fewer RONS while minimizing seed surface damage. Selecting the appropriate gas composition in synergy with plasma treatment parameters contributes to key biochemical and molecular processes in plants (Shaban et al., 2024; Kumar et al., 2024).

#### 3.1.2.3 Nutrient Uptake

Plasma-treated biological samples exhibit enhanced nutrient uptake due to surface modifications that increase seed hydrophilicity. This improved surface property facilitates better mobilization and absorption of nutrients into the plant system. Studies using scanning electron microscopy (SEM) coupled with energy-dispersive X-ray spectroscopy (EDX), X-ray photoelectron spectroscopy (XPS), and X-ray fluorescence spectroscopy (XRF) reveal that plasma treatment leads to lipid oxidation, removal of surface waxes, and cutin degradation. These irreversible chemical modifications improve water and nutrient absorption efficiency. The increased nutrient uptake promotes better seedling establishment, higher biomass production, and enhanced crop yield while also improving plant resilience to environmental stress (Waskow et al., 2021; Pérez-Pizá et al., 2019).

### *3.1.3 Biochemical Factors*

#### 3.1.3.1 Reactive Oxygen and Nitrogen Species

ROS and RNS are byproducts of cellular metabolism. They have a dual nature; at lower concentration, they act as signaling molecules that are directly involved in seed and plant growth and development, but at higher dosage, they cause lipid peroxidation and oxidative damage resulting in cellular dysfunction and even cell death. Antioxidants including superoxide dismutase, catalase, peroxidases, glutathione, and ascorbate help in regulating RONS production and scavenging in plants.

Cold plasma treatment generates a complex mixture of reactive species (RONS) by utilizing various gas composition and ionization and excitation of gas molecules through application of electromagnetic field. RONS, when acts as a signaling molecule, can initiate and regulate complex biochemical process, thus regulating various aspects of plant growth including cell division, elongation, and differentiation (Su et al., 2024). It can activate enzymes involved in reserve mobilization and stimulate metabolic activity increasing the germination potential, vigor, and stress tolerance. They contribute to systemic acquired resistance (SAR) against environmental stressors such as drought, salinity, high temperature, and against biotic stressors such as pathogen attack (El-Shetehy et al., 2015; Priatama et al., 2022). Overall, RONS brings up multiple biochemical changes and its role against stress is explained in the next section.

#### 3.1.3.2 Plant Hormones

Phytohormones are organic signaling molecules that regulate various physiological processes in plants, including growth, development, defense, and stress responses. The family of phytohormones consists of nine major groups, each playing distinct yet interconnected roles. Auxins (AUX), gibberellins (GA), and cytokinins (CK) primarily function as growth promoters, regulating cell division, elongation, and differentiation, while ethylene (ETH) regulates fruit ripening, senescence, and mechanical stress responses. Abscisic acid (ABA) controls seed dormancy and activates stress-responsive genes under drought and salinity stress. Salicylic acid (SA) and jasmonic acid (JA) drive plant defense, triggering resistance against pests and disease. Brassinosteroids enhance growth, photomorphogenesis, stress tolerance, and immune priming. Strigolactones maintain plant architecture by suppressing excessive shoot branching and promoting symbiosis with arbuscular mycorrhizal fungi, aiding root development and nutrient uptake. Peptide hormones regulate cell-to-cell communication, influencing meristem activity, stomatal development, and defense signaling. Despite being present in very low concentrations, these hormones work synergistically or antagonistically to shape plant responses, making them essential for optimizing plant performance under both optimal settings and stressed conditions.

In cold-plasma-treated seeds, scientists have observed the breaking of seed dormancy, stimulation of seed germination, increase in biomass of root, and stress mitigation coinciding with physiological roles of phytohormones. Further analysis using high-performance liquid chromatography (HPLC) combined with mass spectrometry (LC-MS) has revealed that plasma exposure leads to a spike in GA/ABA ratio, explaining the dormancy-breaking effect and accelerated germination kinetics. The shift in phytohormone balance varies with seed color and storage time after harvest, but its effectiveness in breaking dormancy has been observed across multiple crop species (Degutytė-Fomins et al., 2020; Attri et al., 2021). Although the seed coat is a source of hormones for the developing seed, plasma treatment generates RONS, which trigger changes in phytohormone synthesis, degradation, and signaling pathways, influencing enzymes and genes involved in hormone metabolism. Studies have reported an increase in indole-3-acetic acid (AUX) and CK, along with a decrease in ABA, which collectively promote lateral root growth, stimulate cell division, and enhance dormancy release. ETH levels may also rise after plasma exposure, further supporting seedling growth. Additionally, changes in JA, SA, and ABA signaling contribute to enhanced plant defense against biotic and abiotic stresses. These plasma-induced modifications in hormone signaling persist into later growth stages, leading to improved plant morphology, root architecture, and xylem differentiation compared to untreated seeds (Adhikari et al., 2020; Bilea et al., 2024).

## 3.2 Mechanistic Perspectives of Cold Plasma Seed Treatment in Mitigating Environmental Stressors

Global climate change has been a widely discussed topic in recent years, with its impact becoming increasingly evident. As climate change progresses, plants are continuously adapting to these shifting conditions to ensure their survival. Rising global temperatures, irregular rainfall patterns, and extreme weather events such as prolonged droughts, severe heat waves, and soil salinization are posing significant threats to plant ecosystems. Additionally, soil degradation, including loss of fertility and hardpan formation, further hampers plant growth and productivity. Extreme weather events such as hurricanes and cyclones can cause widespread destruction, disrupting entire plant ecosystems. Climate change also influences the life cycles of pests and pathogens, causing variations in the way it affects the plant system. These evolving challenges drive plants to develop adaptive mechanisms, while breeders are working on different techniques to create stress-resistant crops to ensure agricultural sustainability (Anderson and Song, 2020; Naika et al., 2013).

Plants colonized Earth around 500 million years ago during the Cambrian Period and have since evolved intricate mechanisms to adapt to changing climates. To mitigate environmental stress, plants employ a combination of genetic, molecular, physiological, and biochemical adaptations. Environmental stressors disrupt cellular homeostasis, leading to impaired photosynthesis, altered respiration rates, and disruptions in the synthesis of essential biomolecules such as carbohydrates, proteins, and lipids. Stress conditions often trigger the excessive accumulation of ROS, which damage cellular components and further disrupt metabolic pathways. To counteract these effects, plants undergo physiological changes, such as modifications in plant architecture (e.g., increased root biomass, fleshy water-storing stems, C4 metabolism, etc.), accumulation of osmoprotectants, and regulation of stomatal activity. Additionally, plants utilize phytohormones (ABA, GA, AUX, SA, JA) and melatonin, which act as signaling molecules to orchestrate systemic responses, including stomatal regulation, altered metabolism, and enhanced water-use efficiency (Georgieva and Vassileva, 2023; Barah et al., 2016; Ali et al., 2023).

In case of abiotic stresses such as drought, extreme temperature, heavy metals stress, etc., plants activate stress-responsive genes that synthesize protective proteins. Omics studies have identified key transcription factors (TFs), including WRKY, DREB, and NAC, which regulate gene expression to enhance stress tolerance (Erpen et al., 2018). Furthermore, epigenetic modifications such as DNA methylation and RNA methylation/demethylation fine-tune gene expression, activating stress-responsive pathways for long-term adaptation. In combating biotic stresses, such as pathogen attacks and herbivory, plants utilize both preexisting structural barriers and induced defense responses. The thickened cell wall, reinforced with lignin and suberin, acts as a physical barrier, preventing pathogen invasion. Upon herbivory or mechanical damage, plants activate JA and SA signaling pathways, leading to the production of defense-related proteins and toxic secondary metabolites to deter attackers. Additionally, plants detect pathogen-associated molecular patterns (PAMPs) through pattern recognition receptors (PRRs), triggering immune responses that enhance resistance (Kaur et al., 2022).

Scientists all over the world are trying to understand the plant defense mechanism, with a significant focus on the most explored biochemical responses that plants employ to mitigate diverse environmental stressors. ROS, generated during plant stress responses, can cause significant damage to cellular components, disrupting essential functions.

To counteract ROS and prevent oxidative damage, plants possess sophisticated antioxidant defense systems. These systems encompass both enzymatic (superoxide dismutase, catalase, peroxidases) and non-enzymatic components (glutathione, tocopherols (vitamin E), ascorbate (vitamin C)) that work in a coordinated manner to maintain ROS homeostasis and enable them to respond actively to various environmental challenges.

In addition to enzymatic defense, plants synthesize a diverse array of secondary metabolites, notably phenols and flavonoids, which act as potent antioxidants. These compounds directly scavenge ROS and protect plants against multiple abiotic stresses, such as UV radiation and heavy metal toxicity. Furthermore, they contribute to enhanced seed vigor and longevity by safeguarding cellular integrity during seed development and storage. Signal transduction pathways involving phytohormones and ROS regulate physiological and developmental responses, ensuring stress adaptation. This interconnected biochemical network enables plants to perceive, respond to, and ultimately tolerate a wide range of environmental stressors (Chen et al., 2022).

Cold plasma seed treatment is emerging as a promising technique in agriculture for enhancing seed quality and stress tolerance. This innovative approach triggers a cascade of changes within the seed, mimicking natural plant adaptation mechanisms through a combination of physical, physiological, biochemical, and molecular alterations. The altered seed coat, resulting from processes such as etching and surface oxidation, plays a crucial role in breaking seed dormancy that may be caused by a thick seed coat and enhancing water imbibition and gas exchange by reducing physical barriers. This modification effectively aids in mitigating drought and hypoxia stress. Additionally, the increased permeability of the seed coat promotes faster germination and early seedling vigor by regulating endogenous hormones and activating hydrolytic enzymes. The stimulation of hydrolytic enzymes and the efficient transport of water ensure an adequate nutrient supply to the actively growing embryo. Furthermore, cold plasma treatment influences gene expression related to growth and development (XTH17, XTH9, COBL9, OBP4, LAX3), contributing to improved plant performance. It has also been shown to mitigate salinity stress by counteracting the osmotic stress imposed by saline conditions, particularly when combined with seed priming treatments (Song et al., 2020; Sheteiwy et al., 2019).

Environmental stress cascades a series of environmental signals into plants. It leads to increased production of ROS leading to oxidative damage, lipid peroxidation, protein degradation, and extreme cases of disruption of cellular processes and DNA damage. Plants perceive these environmental signals through stress receptors in the plasma membrane, which immediately initiate signal transduction to relay stress signals. This process activates the antioxidant defense system (CAT, SOD, GPX) and antioxidant-related genes (POD, PPO, GST), regulates hormonal pathways through signaling genes (AOX, OPR, PAL, NCEDs, ICE1, CBF1), and modulates protein biosynthesis and energy metabolism to mount a comprehensive stress response and promote survival (Kosová et al., 2014).

In comparison to untreated seeds, cold-plasma-treated seeds exhibit enhanced recovery when exposed to environmental stressors. This improved stress tolerance is attributed to several key factors. First, the RONS and ozone generated during plasma treatment trigger the production of more ROS within the cell, notably via NADPH oxidase, initiating signaling cascades. These cascades involve crucial second messengers such as calcium ions and vital signaling pathways such as the mitogen-activated protein kinase (MAPK) cascade, ultimately leading to phosphorylation events. RONS further act as signaling molecules, triggering protein kinases and making the seed more responsive in the event of future stress exposure (Figure 5). Second, plasma treatment influences the activation of transcription factors, which play a role in enhancing antioxidant production. This enhancement of the antioxidant defense system boosts the seed's capacity to scavenge environmental-stress-induced accumulated ROS and further prevent oxidative damage. Third, modulation of phytohormone signaling pathways by plasma treatment results in multiple physiological changes, including enhanced root growth and stomatal opening regulation. Hormonal crosstalk is observed, leading to an intricate network where ABA upregulation enhances the seed's ability to respond to abiotic stresses. During germination, a rapid increase in the GA/ABA ratio results in promoting germination under stress conditions, while JA and SA function against biotic stress, with the overall hormonal interplay carefully balancing growth and stress responses (Priatama et al., 2022; Mildaziene et al., 2022).

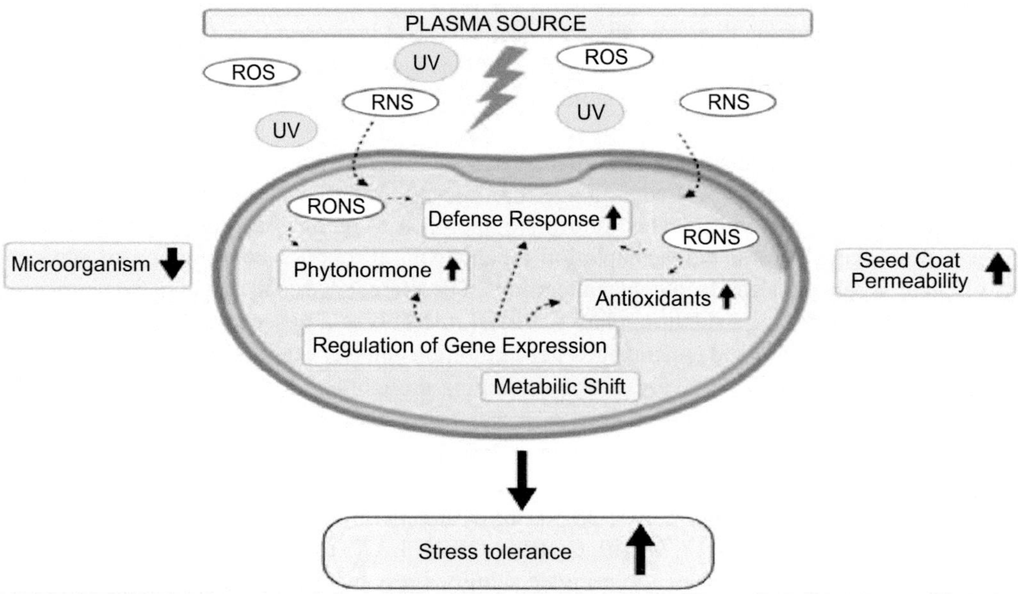


**Fig. 5** Mechanism of cold-plasma-induced stress tolerance in seeds or plants.

Cold-plasma-treated seeds also undergo significant metabolic shifts. Increased Adenosine Triphosphate (ATP) levels suggest increased anaerobic respiration, while higher concentrations of proteins and sugars result from increased reserve utilization enzymes. Plasma treatment can also change gene expression and induce epigenetic modifications. This includes activating primary metabolic pathways such as glycolysis, the pentose phosphate pathway (PPP), and the tricarboxylic acid (TCA) cycle to support growth, while simultaneously upregulating secondary metabolic pathways, such as the shikimate and phenylpropanoid pathways, for the production of protective compounds including pathogenesis-related proteins, phenolic compounds, and phytoalexins. These metabolic adjustments and defense responses are ultimately controlled by changes in gene expression, allowing the seed to adapt to the stress at a molecular level (Ling et al., 2014; Waskow et al., 2021).

The molecular changes brought by cold plasma treatment extend to gene expression, resulting in altered metabolic activity within the plant cell. Notably, an increase in histone methylation is observed, indicating epigenetic modifications that alter chromatin structure and gene accessibility, which is crucial for long-term stress adaptation. This treatment triggers the expression of new transcription factors, such as LEA chaperones and hydrolytic enzymes, such as pullulanase, as well as genes encoding ascorbate peroxidase and catalase. These changes play vital roles in coordinating stress responses and enhancing cellular protection (Priatama et al., 2022).

Cold plasma treatment plays a crucial role in mitigating biotic stress through direct pathogen inactivation and the activation of plant defense mechanisms. One of its primary benefits is the elimination of seed-borne pathogens, thereby preventing plant diseases. RONS along with ozone directly disinfect biological samples by breaking down microbial structures. Exposure time and plasma power are critical factors in determining the treatment's effectiveness in seed health studies. While shorter exposure times may be sufficient for activating stress tolerance and cellular antioxidant defense mechanism, longer exposure times allow RONS to penetrate deeper into the seed, treating pathogens residing within internal plant tissues. Cold plama treatment has demonstrated effectiveness against multiple pathogens, particularly bacterial and fungal diseases (Jiang et al., 2014; Rusu et al., 2018; Vatchev et al., 2025; Lee et al., 2021; Jo et al., 2014) and pests (Afsheen et al., 2019; Dilip et al., 2025; Kaur et al., 2020).

In addition to direct disinfection, cold plasma treatment primes the plant's innate immune system, enhancing its defense responses. RONS triggers a signaling cascade leading to upregulation of pathogen resistance genes. PAW has shown equal effectiveness in enhancing plant immunity. Both direct and indirect cold plasma treatment influence metabolic shift, gene expression regulation, and phytohormone signaling, particularly affecting JA and SA pathways, leading to systemic acquired resistance (SAR) and long-lasting, broad-spectrum protection against pests and pathogens. Cold plasma technology has a high potential in combating biotic stress in an eco-friendly and sustainable manner, but further research is needed to determine its ability to penetrate deep-seated pathogens within seeds. There is still limited understanding of its efficacy against viruses, nematodes, and deeply embedded pest eggs, as well as its potential to degrade toxins produced by pathogens. Standardizing plasma treatment parameters to address these challenges remains an important area for future research (Pańka et al., 2022)

## 4. Future Prospects and Conclusions

Cold plasma treatment is emerging as a transformative technology in agriculture, offering a sustainable and eco-friendly approach in priming seed quality and improving stress tolerance. Its versatility in being adapted to any of the present seed quality enhancement technologies such as seed priming, coating, pelleting, etc., makes it easier to adapt to the existing farming practices. It also has a synergistic effect when done in combination with fungicide, insecticide, or other plant protectant chemical treatments. It can be exclusively useful in modern agricultural practices such as precision agriculture, hydroponics, indoor agriculture, or others where plants need to acclimatize in limited resources.

Climate change phenomenon and policies focusing on "more yield per crop," "doubling farmers income," "organic farming," or "sustainable agriculture" create need for cold plasma treatment, presenting promising output for enhancing speed of germination, breaking dormancy, increase in nodule and/or root biomass, seedling vigor, and crop establishment under adverse environmental condition. It plays a dual role in both abiotic stress tolerance—by improving water uptake, regulating phytohormones, and activating antioxidant defense mechanisms—and biotic stress mitigation—by inactivating pathogens, inducing systemic resistance and complementing conventional disease management practices.

Laboratory findings on cold plasma treatment are widespread, showcasing its positive impact as a pre-sowing treatment, seed quality enhancement, in imparting stress tolerance to packaging of agricultural produce. The transfer of this technology to practical utility requires focused development in several key areas. First, standardization at the field level is paramount to ensure consistent results and facilitate reliable implementation of this technology across diverse agricultural settings. This necessitates a move beyond laboratory conditions and toward robust field trials. Second, scaling up the technology to meet industrial demands is crucial for widespread adoption. While small, portable plasma reactors may be suitable for high-value, less bulky crops such as vegetable seeds, field crops will likely require scaled-up technology within processing units. The development of portable plasma treatment vehicles, as exemplified by Interkorn Ltd. company, with their high throughput capacity of one tone per hour for pre-sowing seed treatment represents a significant step toward addressing this need. Similarly, small startup ventures such as Zayndu and Clean Crop Technologies are actively working on supplying cold-plasma-treated seeds, contributing to the growing commercial interest. However, continued attention to efficient and cost-effective upscaling strategies remains crucial for wider adoption. Furthermore, seed longevity studies are vital to fully understand the effective period of plasma treatment, informing optimal seed storage and marketing practices. Finally, in-depth research focusing on the underlying mechanisms related to seed vigor, seed health, and environmental stress tolerance is critical for a better understanding and subsequent standardization of cold plasma setups. Negative results must also be published to avoid repeating unsuccessful experiments, identify ineffective treatment parameters, and accelerate progress in the field by focusing research efforts on promising avenues.

In conclusion, cold plasma treatment stands out as a promising green technology in agriculture. Unlike traditional methods, it often does not require additional chemicals, yet it possesses the remarkable potential to trigger a cascade of beneficial physical, physiological, and molecular modifications within seeds and plants. Advancements in this technology are yielding crops with enhanced vigor and resilience to stress, paving the way for more cost-effective and environmentally sound agricultural practices.

## References

Adhikari, B., Adhikari, M. and Park, G. (2020). The effects of plasma on plant growth, development and sustainability. *Applied Sciences*, *10*(17), p.6045.

Afsheen, S., Fatima, U., Iqbal, T., Abrar, M., Muhammad, S., Saeed, A., Isa, M., Malik, M.F. and Shamas, S. (2019). Influence of cold plasma treatment on insecticidal properties of wheat seeds against red flour beetles. *Plasma Science and Technology*, *21*(8), p. 085506.

Ahmed, Z., Manzoor, M.F., Ahmad, N., Zeng, X.A., Din, Z.U., Roobab, U., Qayum, A., Siddique, R., Siddeeg, A. and Rahaman, A. (2020). Impact of pulsed electric field treatments on the growth parameters of wheat seeds and nutritional properties of their wheat plantlets juice. *Food Science & Nutrition*, *8*(5), pp. 2490–2500.

Ali, M., Pan, Y., Liu, H. and Cheng, Z. (2023). Melatonin interaction with abscisic acid in the regulation of abiotic stress in Solanaceae family plants. *Frontiers in Plant Science*, *14*, p. 1271137.

Alves Junior, C., de Oliveira Vitoriano, J., da Silva, D.L.S., de Lima Farias, M. and de Lima Dantas, N.B. (2016). Water uptake mechanism and germination of Erythrina velutina seeds treated with atmospheric plasma. *Scientific Reports*, *6*(1), p. 33722.

Anderson, J.T. and Song, B.H. (2020). Plant adaptation to climate change—Where are we?. *Journal of Systematics and Evolution*, *58*(5), pp. 533–545.

Anonymous. (2022). UNDESA. *World Population Prospects 2022: Summary of Results*; United Nations Department of Economic and Social Affairs: New York, NY, USA, 2022; Availableonline: https://www.un.org/development/desa/pd/sites/www.un.org.development.desa.pd/ files/wpp2022_summary_of_results.pdf (accessed on 1 October 2022).

Attri, P., Koga, K., Okumura, T. and Shiratani, M. (2021). Impact of atmospheric pressure plasma treated seeds on germination, morphology, gene expression and biochemical responses. *Japanese Journal of Applied Physics*, *60*(4), p. 040502.

Baek, E.J., Joh, H.M., Kim, S.J. and Chung, T.H. (2016). Effects of the electrical parameters and gas flow rate on the generation of reactive species in liquids exposed to atmospheric pressure plasma jets. *Physics of Plasmas*, *23*(7).

Bafoil, M., Jemmat, A., Martinez, Y., Merbahi, N., Eichwald, O., Dunand, C. and Yousfi, M. (2018). Effects of low temperature plasmas and plasma activated waters on Arabidopsis thaliana germination and growth. *PLoS One*, *13*(4), p. e0195512.

Barah, P., Jayavelu, N.D., Sowdhamini, R., Shameer, K. and Bones, A.M. (2016). Transcriptional regulatory networks in Arabidopsis thaliana during single and combined stresses. *Nucleic Acids Research*, *44*(7), pp. 3147–3164.

Barjasteh, A., Kaushik, N., Choi, E.H. and Kaushik, N.K., (2024). Cold atmospheric pressure plasma solutions for sustainable food packaging. International Journal of Molecular Sciences, *25*(12), p. 6638.

Bian, D. and Wu, Y. (2018). Enduring and stable surface dielectric barrier discharge (SDBD) plasma using fluorinated multi-layered polyimide. *Polymers*, *10*(6), p. 606.

Bilea, F., Garcia-Vaquero, M., Magureanu, M., Mihaila, I., Mildažienė, V., Mozetič, M., Pawłat, J., Primc, G., Puač, N., Robert, E. and Stancampiano, A. (2024). Non-thermal plasma as environmentally-friendly technology for agriculture: A review and roadmap. *Critical Reviews in Plant Sciences*, *43*(6), pp. 428–486.

Bogaerts, A. and Gijbels, R. (2000). Similarities and differences between direct current and radio-frequency glow discharges: a mathematical simulation. Invited Lecture Presented at the 2000 Winter Conference on Plasma Spectrochemistry, Fort Lauderdale, FL, USA, January 10–15, 2000. *Journal of Analytical Atomic Spectrometry*, *15*(9), pp. 1191–1201.

Bormashenko, E., Grynyov, R., Bormashenko, Y. and Drori, E. (2012). Cold radio frequency plasma treatment modifies wettability and germination speed of plant seeds. *Scientific Reports*, *2*(1), p. 741.

Butscher, D., Zimmermann, D., Schuppler, M. and von Rohr, P.R. (2016). Plasma inactivation of bacterial endospores on wheat grains and polymeric model substrates in a dielectric barrier discharge. *Food Control*, *60*, pp. 636–645.

Chauvin, J., Judée, F., Yousfi, M., Vicendo, P. and Merbahi, N. (2017). Analysis of reactive oxygen and nitrogen species generated in three liquid media by low temperature helium plasma jet. *Scientific Reports*, *7*(1), pp. 1–15.

Chen, D., Mubeen, B., Hasnain, A., Rizwan, M., Adrees, M., Naqvi, S.A.H., Iqbal, S., Kamran, M., El-Sabrout, A.M., Elansary, H.O. and Mahmoud, E.A. (2022). Role of promising secondary metabolites to confer resistance against environmental stresses in crop plants: Current scenario and future perspectives. *Frontiers in Plant Science*, *13*, p. 881032.

Chou, Y.J. and Ting, Y. (2023). Plasma-activated water regulated transcriptome gene expression leading to high germination and growth of mung beans. *Chemical and Biological Technologies in Agriculture*, *10*(1), p. 146.

Chuesaard, T., Peankid, P., Thaworn, S., Jaradrattanapaiboon, A., Veerana, M. and Panngom, K. (2023). Different effects of reactive species generated from chemical donors on seed germination, growth and chemical contents of Oryza sativa L. *Plants*, *12*(4), p. 765.

Corbella, C., Portal, S. and Keidar, M. (2023). Flexible cold atmospheric plasma jet sources. *Plasma*, *6*(1), pp. 72–88.

Čurila, V., Habuda–Stanić, M. and Marček, T. (2023). The effect of plasma-activated water on germination and plant growth under abiotic stresses. In *1. European GREEN Conference* (pp. 62–63).

Das, S., Mohapatra, S., Kar, S., Bhatt, S. and Pundir, S. (2022). Reactive species variation in cold atmospheric pressure plasma jet discharge under the influence of intrinsic parameters and its effect on E. coli inactivation. *Biointerphases*, *18*(6).

De Felice, G., Li, S., Gallucci, F., Pourali, N. and Rebrov, E. (2023). On design of plasma jet reactor for non-oxidative methane conversion. *Reaction Chemistry & Engineering*, *8*(5), pp. 1125–1133.

Degutytė-Fomins, L., Paužaitė, G., Žūkienė, R., Mildažienė, V., Koga, K. and Shiratani, M. (2020). Relationship between cold plasma treatment-induced changes in radish seed germination and phytohormone balance. *Japanese Journal of Applied Physics*, *59*(SH), p. SH1001.

Dilip, D., Modupalli, N., Rahman, M.M. and Kariyat, R. (2025). Atmospheric cold plasma alters plant traits and negatively affects the growth and development of fall armyworm in rice. *Scientific Reports*, *15*(1), p. 3680.

Dufour, T., Gutierrez, Q. and Bailly, C. (2021). Sustainable improvement of seeds vigour using dry atmospheric plasma priming: Evidence through coating wettability, water uptake *and* plasma reactive chemistry. *Journal of Applied Physics*, *129*(8).

El-Maarouf-Bouteau, H. (2022). The seed and the metabolism regulation. *Biology*, *11*(2), p. 168.

El-Shetehy, M., Wang, C., Shine, M.B., Yu, K., Kachroo, A. and Kachroo, P. (2015). Nitric oxide and reactive oxygen species are required for systemic acquired resistance in plants. *Plant Signaling & Behavior*, *10*(9), p. e998544.

Erpen, L., Devi, H.S., Grosser, J.W. and Dutt, M. (2018). Potential use of the DREB/ERF, MYB, NAC and WRKY transcription factors to improve abiotic and biotic stress in transgenic plants. *Plant Cell, Tissue and Organ Culture (PCTOC)*, *132*, pp. 1–25.

Fadhlalmawla, S.A., Mohamed, A.A.H., Almarashi, J.Q. and Boutraa, T. (2019). The impact of cold atmospheric pressure plasma jet on seed germination and seedlings growth of fenugreek (Trigonella foenum-graecum). *Plasma Science and Technology*, *21*(10), p. 105503.

Farouk, T., Farouk, B. and Fridman, A. (2010). Computational studies of atmospheric-pressure methane–hydrogen DC micro glow discharges. *IEEE Transactions on Plasma Science*, *38*(2), pp. 73–85.

Foroughbakhch Pournavab, R., Bacópulos Mejía, E., Benavides Mendoza, A., Salas Cruz, L.R. and Ngangyo Heya, M. (2019). Ultraviolet radiation effect on seed germination and seedling growth of common species from Northeastern Mexico. *Agronomy*, *9*(6), p. 269.

Georgieva, M. and Vassileva, V. (2023). Stress management in plants: Examining provisional and unique dose-dependent responses. *International Journal of Molecular Sciences*, *24*(6), p. 5105.

Gómez-Ramírez, A., López-Santos, C., Cantos, M., García, J.L., Molina, R., Cotrino, J., Espinós, J.P. and González-Elipe, A.R. (2017). Surface chemistry and germination improvement of quinoa seeds subjected to plasma activation. *Scientific Reports*, *7*(1), p. 5924.

Guo, D., Liu, H., Zhou, L., Xie, J. and He, C. (2021). Plasma-activated water production and its application in agriculture. *Journal of the Science of Food and Agriculture*, *101*(12), pp. 4891–4899.

Hasanuzzaman, M., Bhuyan, M.B., Parvin, K., Bhuiyan, T.F., Anee, T.I., Nahar, K., Hossen, M.S., Zulfiqar, F., Alam, M.M. and Fujita, M. (2020). Regulation of ROS metabolism in plants under environmental stress: A review of recent experimental evidence. *International Journal of Molecular Sciences*, *21*(22), p. 8695.

Hemathilake, D.M.K.S. and Gunathilake, D.M.C.C. (2022). Agricultural Productivity and Food Supply to Meet Increased Demands. In *Future Foods* (pp. 539–553). Academic Press.

Holubová, Ľ., Kyzek, S., Ďurovcová, I., Fabová, J., Horváthová, E., Ševčovičová, A. and Gálová, E. (2020). Non-thermal plasma—A new green priming agent for plants?. *International Journal of Molecular Sciences*, *21*(24), p. 9466–9485.

Ivankov, A., Zukiene, R., Nauciene, Z., Degutyte-Fomins, L., Filatova, I., Lyushkevich, V. and Mildaziene, V. (2021). The effects of red clover seed treatment with cold plasma and electromagnetic field on germination and seedling growth are dependent on seed color. *Applied Sciences*, *11*(10), p. 4676.

Jiang, J., Lu, Y., Li, J., Li, L., He, X., Shao, H. and Dong, Y. (2014). Effect of seed treatment by cold plasma on the resistance of tomato to Ralstonia solanacearum (bacterial wilt). *PloS One*, *9*(5), p. e97753.

Jo, Y.K., Cho, J., Tsai, T.C., Staack, D., Kang, M.H., Roh, J.H., Shin, D.B., Cromwell, W. and Gross, D. (2014). A non-thermal plasma seed treatment method for management of a seedborne fungal pathogen on rice seed. *Crop Science*, *54*(2), pp. 796–803.

Judée, F., Simon, S., Bailly, C. and Dufour, T. (2018). Plasma-activation of tap water using DBD for agronomy applications: Identification and quantification of long lifetime chemical species and production/consumption mechanisms. *Water Research*, *133*, pp. 47–59.

Kang, S.U., Kim, C.H., You, S., Lee, D.Y., Kim, Y.K., Kim, S.J., Kim, C.K. and Kim, H.K. (2023). Plasma surface modification of 3Y-TZP at low and atmospheric pressures with different treatment times. *International Journal of Molecular Sciences*, *24*(8), p. 7663.

Kaur, M., Hüberli, D. and Bayliss, K.L. (2020). Cold plasma: Exploring a new option for management of postharvest fungal pathogens, mycotoxins and insect pests in Australian stored cereal grain. *Crop and Pasture Science*, *71*(8), pp. 715–724.

Kaur, S., Samota, M.K., Choudhary, M., Choudhary, M., Pandey, A.K., Sharma, A. and Thakur, J. (2022). How do plants defend themselves against pathogens—Biochemical mechanisms and genetic interventions. *Physiology and Molecular Biology of Plants*, *28*(2), pp. 485–504.

Kitazaki, S., Sarinont, T., Koga, K., Hayashi, N. and Shiratani, M. (2014). Plasma induced long-term growth enhancement of Raphanus sativus L. using combinatorial atmospheric air dielectric barrier discharge plasmas. *Current Applied Physics*, *14*, pp. S149–S153.

Kletschkus, K., Haralambiev, L., Mustea, A., Bekeschus, S. and Stope, M.B. (2020). Review of innovative physical therapy methods: Introduction to the principles of cold physical plasma. *In Vivo*, *34*(6), pp. 3103–3107.

Kosová, K., Vítámvás, P. and Prášil, I.T. (2014). Proteomics of stress responses in wheat and barley—Search for potential protein markers of stress tolerance. *Frontiers in Plant Science*, *5*, p. 711.

Krapivina, S.A., Filippov, A.K., Levitskaya, T.N., and Bakhvalov, A. Gas Plasma Treatment of Plant Seeds. U.S. Patent No. US5281315A, 25 January 1994.

Kučerová, K., Henselová, M., Slováková, Ľ. and Hensel, K. (2019). Effects of plasma activated water on wheat: Germination, growth parameters, photosynthetic pigments, soluble protein content and antioxidant enzymes activity. *Plasma Processes and Polymers*, *16*(3), p. 1800131.

Kumar, S., Pipliya, S., Srivastav, P.P. and Srivastava, B. (2024). Exploring the role of various feed gases in cold plasma technology: A comprehensive review. *Food and Bioprocess Technology*, *17*(11), pp. 3367–3407.

Kumaş, Y.T., Kibar, H. and Soydemir, H.E. (2025). Investigating UV-C radiation as a non-chemical method to prevent quality loss in the pea seed during storage. *Journal of Stored Products Research*, *111*, p. 102532.

Lal, R. (2016). Feeding 11 billion on 0.5 billion hectares of area under cereal crops. *Food and Energy Security*, *5*(4), pp. 239–251.

Lee, Y., Lee, Y.Y., Kim, Y.S., Balaraju, K., Mok, Y.S., Yoo, S.J. and Jeon, Y. (2021). Enhancement of seed germination and microbial disinfection on ginseng by cold plasma treatment. *Journal of Ginseng Research*, *45*(4), pp. 519–526.

Ling, L., Jiafeng, J., Jiangang, L., Minchong, S., Xin, H., Hanliang, S. and Yuanhua, D. (2014). Effects of cold plasma treatment on seed germination and seedling growth of soybean. *Scientific Reports*, *4*(1), p. 5859.

Lukes, P., Dolezalova, E., Sisrova, I. and Clupek, M. (2014). Aqueous-phase chemistry and bactericidal effects from an air discharge plasma in contact with water: Evidence for the formation of peroxynitrite through a pseudo-second-order post-discharge reaction of H2O2 and HNO2. *Plasma Sources Science and Technology*, *23*(1), p. 015019.

Mareri, L., Parrotta, L. and Cai, G. (2022). Environmental stress and plants. *International Journal of Molecular Sciences*, *23*(10), p. 5416.

Mildaziene, V., Ivankov, A., Sera, B. and Baniulis, D. (2022). Biochemical and physiological plant processes affected by seed treatment with non-thermal plasma. *Plants*, *11*(7), p. 856.

Mildaziene, V., Pauzaite, G., Naucienė, Z., Malakauskiene, A., Zukiene, R., Januskaitiene, I., Jakstas, V., Ivanauskas, L., Filatova, I. and Lyushkevich, V. (2018). Pre-sowing seed treatment with cold plasma and electromagnetic field increases secondary metabolite content in purple coneflower (Echinacea purpurea) leaves. *Plasma Processes and Polymers*, *15*(2), p. 1700059.

Mizoi, K., Rodríguez-González, V., Sasaki, M., Suzuki, S., Honda, K., Ishida, N., Suzuki, N., Kuchitsu, K., Kondo, T., Yuasa, M. and Fujishima, A. (2022). Interactions between pH, reactive species and cells in plasma-activated water can remove algae. *RSC Advances*, *12*(13), pp. 7626–7634.

Murali, R., Evangelina, R., Samuel, J.P., Singh, P., Saha, S., Singhal, M. and Gandhirajan, R.K. (2024). Cold atmospheric plasma (CAP) in wound healing: Harnessing a dual-edged sword. *Redox Experimental Medicine*, *2024*(1).

Naika, M., Shameer, K. and Sowdhamini, R. (2013). Comparative analyses of stress-responsive genes in Arabidopsis thaliana: Insight from genomic data mining, functional enrichment, pathway analysis and phenomics. *Molecular BioSystems*, *9*(7), pp. 1888–1908.

Naika, M., Shameer, K., Mathew, O.K., Gowda, R. and Sowdhamini, R. (2013). STIFDB2: An updated version of plant stress-responsive transcription factor database with additional stress signals, stress-responsive transcription factor binding sites and stress-responsive genes in Arabidopsis and rice. *Plant and Cell Physiology*, *54*(2), pp e8(1–15).

Nishad, R., Ahmed, T., Rahman, V.J. and Kareem, A. (2020). Modulation of plant defense system in response to microbial interactions. *Frontiers in Microbiology*, *11*, p. 1298.

Nishanth, K., Madhukar, A. and Rajanikanth, B.S. (2020). "Estimation of Ozone Generation in Dielectric Barrier Discharge Plasma using Response Surface Methodology and Fuzzy Logic," *2020 IEEE International Conference on Power Systems Technology (POWERCON)*, Bangalore, India, pp. 1–5.

Nishime, T.M., Wannicke, N., Horn, S., Weltmann, K.D. and Brust, H. (2020). A coaxial dielectric barrier discharge reactor for treatment of winter wheat seeds. *Applied Sciences*, *10*(20), p. 7133.

Pańka, D., Jeske, M., Łukanowski, A., Baturo-Cieśniewska, A., Prus, P., Maitah, M., Maitah, K., Malec, K., Rymarz, D., Muhire, J.D.D. and Szwarc, K. (2022). Can cold plasma be used for boosting plant growth and plant protection in sustainable plant production? *Agronomy*, *12*(4), p. 841.

Parvathi, M.S., Nataraja, K.N., Reddy, Y.N., Naika, M.B. and Gowda, M.C. (2019). Transcriptome analysis of finger millet (Eleusine coracana (L.) Gaertn.) reveals unique drought responsive genes. *Journal of Genetics*, *98*(2), p. 46.

Pauzaite, G., Malakauskiene, A., Nauciene, Z., Zukiene, R., Filatova, I., Lyushkevich, V., Azarko, I. and Mildaziene, V. (2018). Changes in Norway spruce germination and growth induced by pre-sowing seed treatment with cold plasma and electromagnetic field: Short-term versus long-term effects. *Plasma Processes and Polymers*, *15*(2), p. 1700068.

Pawlat, J., Starek, A., Sujak, A., Kwiatkowski, M., Terebun, P. and Budzeń, M. (2018). Effects of atmospheric pressure plasma generated in GlidArc reactor on Lavatera thuringiaca L. seeds' germination. *Plasma Processes and Polymers*, *15*(2), p. 1700064.

Pérez-Pizá, M.C., Prevosto, L., Grijalba, P.E., Zilli, C.G., Cejas, E., Mancinelli, B. and Balestrasse, K.B. (2019). Improvement of growth and yield of soybean plants through the application of non-thermal plasmas to seeds with different health status. *Heliyon*, *5*(4).

Perrotti, V., Caponio, V.C.A., Muzio, L.L., Choi, E.H., Marcantonio, M.C.D., Mazzone, M., Kaushik, N.K. and Mincione, G. (2022). Open questions in cold atmospheric plasma treatment in head and neck cancer: A systematic review. *International Journal of Molecular Sciences*, *23*(18), p. 10238.

Piferi, C., Barni, R., Roman, H.E. and Riccardi, C. (2021). Current filaments in asymmetric surface dielectric barrier discharge. *Applied Sciences*, *11*(5), p. 2079.

Porto, C.L., Ziuzina, D., Los, A., Boehm, D., Palumbo, F., Favia, P., Tiwari, B., Bourke, P. and Cullen, P.J. (2018). Plasma activated water and airborne ultrasound treatments for enhanced germination and growth of soybean. *Innovative Food Science & Emerging Technologies*, *49*, pp. 13–19.

Priatama, R.A., Pervitasari, A.N., Park, S., Park, S.J. and Lee, Y.K. (2022). Current advancements in the molecular mechanism of plasma treatment for seed germination and plant growth. *International Journal of Molecular Sciences*, *23*(9), p. 4609.

Radchuk, V. and Borisjuk, L. (2014). Physical, metabolic and developmental functions of the seed coat. *Frontiers in Plant Science*, *5*, p. 510.

Randeniya, L.K. and de Groot, G.J. (2015). Non-thermal plasma treatment of agricultural seeds for stimulation of germination, removal of surface contamination and other benefits: a review. *Plasma Processes and Polymers*, *12*(7), pp. 608–623.

Rusu, B.G., Postolache, V., Cara, I.G., Pohoata, V., Mihaila, I., Topala, I. and Jitareanu, G. (2018). Method of fungal wheat seeds disease inhibition using direct exposure to air cold plasma. *Rom. J. Phys*, *63*, pp. 1–13.

Sajib, S.A., Billah, M., Mahmud, S., Miah, M., Hossain, F., Omar, F.B., Roy, N.C., Hoque, K.M.F., Talukder, M.R., Kabir, A.H. and Reza, M.A. (2020). Plasma activated water: The next generation eco-friendly stimulant for enhancing plant seed germination, vigour and increased enzyme activity, a study on black gram (Vigna mungo L.). *Plasma Chemistry and Plasma Processing*, *40*, pp. 119–143.

Shaban, M., Merkert, N., van Duin, A.C., van Duin, D. and Weber, A.P. (2024). Advancing DBD plasma chemistry: Insights into reactive nitrogen species such as NO2, N2O5 *and* N2O optimization and species reactivity through experiments and MD simulations. *Environmental Science & Technology*, *58*(36), pp. 16087–16099.

Shameer, K., Naika, M.B., Shafi, K.M. and Sowdhamini, R. (2019). Decoding systems biology of plant stress for sustainable agriculture development and optimized food production. *Progress in Biophysics and Molecular Biology*, *145*, pp. 19–39.

Shelar, A., Singh, A.V., Dietrich, P., Maharjan, R.S., Thissen, A., Didwal, P.N., and Shinde, M. (2022).

Sheteiwy, M.S., An, J., Yin, M., Jia, X., Guan, Y., He, F. and Hu, J. (2019). Cold plasma treatment and exogenous salicylic acid priming enhances salinity tolerance of Oryza sativa seedlings. *Protoplasma*, *256*, pp. 79–99.

Song, J.S., Kim, S.B., Ryu, S., Oh, J. and Kim, D.S. (2020). Emerging plasma technology that alleviates crop stress during the early growth stages of plants: a review. *Frontiers in Plant Science*, *11*, p. 988.

Soni, A., Choi, J. and Brightwell, G. (2021). Plasma-activated water (PAW) as a disinfection technology for bacterial inactivation with a focus on fruit and vegetables. *Foods*, *10*(1), p. 166.

Srii, V.S., Sanam, T. and Phartyal, S.S. (2023). Influence of hormonal seed priming on seedling growth, development and potential antioxidant performance under abiotic stress. In *Hormonal Cross-Talk, Plant Defense and Development* (pp. 273–294). Academic Press.

Stefanello, R., Barreto, R.A.M., Müller, G.L., Rodrigues, A.H.S., da Silva Garcia, W.J. and Dorneles, L.S. (2023). UV-B and UV-C radiation on the germination of soybean seeds. *Revista Brasileira de Ciências Agrárias*, *18*(2), pp. e2964–e2964.

Su, J., Liu, Y., Han, F., Gao, F., Gan, F., Huang, K. and Li, Z. (2024). ROS, an important plant growth regulator in root growth and development: Functional genes and mechanism. *Biology*, *13*(12), p. 1033.

Suschek, C.V. (2024). Plasma applications in biomedicine: A groundbreaking intersection between physics and life sciences. *Biomedicines*, *12*(5), p. 1029.

Thakur, N. and Vasudevan, S.N. (2021). Plasma treatment and seed quality advancement: A review. *Agricultural Reviews*, *42*(2), pp. 197–202.

Thirumdas, R., Kothakota, A., Kiran, K.C.S.S., Pandiselvam, R. and Prakash, V. (2017). Exploitation of cold plasma technology in agriculture. *Advances in Research*, *12*(4), pp. 1–7.

Thiyagarajan, M., Sarani, A. and Gonzales, X. (2013). Atmospheric pressure resistive barrier air plasma jet induced bacterial inactivation in aqueous environment. *Journal of Applied Physics*, *113*(9).

Tian, P., Kenney, J., Rauf, S., Korolov, I. and Schulze, J. (2024). Uniformity of low-pressure capacitively coupled plasmas: Experiments and two-dimensional particle-in-cell simulations. *Physics of Plasmas*, *31*(4).

Tuteja, N., Singh, M.B., Misra, M.K., Bhalla, P.L. and Tuteja, R. (2001). Molecular mechanisms of DNA damage and repair: Progress in plants. *Critical Reviews in Biochemistry and Molecular Biology*, *36*(4), pp. 337–397.

Umar, O.B., Ranti, L.A., Abdulbaki, A.S., Abdulhamid, A.K., Biola, M.R. and Victor, K.O. (2021). Stresses in Plants: Biotic and Abiotic. In *Current Trends in Wheat Research*. IntechOpen.

Vatchev, T., Yanashkov, I.T., Marinova, P. and Benova, E. (2025). Pre-sowing seed treatments with cold atmospheric plasma for the control of seedling blights of winter wheat. *Processes*, *13*(3), p. 632.

Wang, J., Cheng, J.H. and Sun, D.W. (2023). Enhancement of wheat seed germination, seedling growth and nutritional properties of wheat plantlet juice by plasma activated water. *Journal of Plant Growth Regulation*, *42*(3), pp. 2006–2022.

Wang, J., Song, H., Song, Z., Lu, Y., Yan, Y. and Li, F. (2020). Effect of positive and negative corona discharge field on vigour of millet seeds. *IEEE Access*, *8*, pp. 50268–50275.

Washington, H. and Kopnina, H. (2022). Discussing the silence and denial around population growth and its environmental impact. How do we find ways forward?. *World*, *3*(4), pp. 1009–1027.

Waskow, A., Butscher, D., Oberbossel, G., Klöti, D., Rudolf von Rohr, P., Büttner-Mainik, A., Drissner, D. and Schuppler, M. (2021). Low-energy electron beam has severe impact on seedling development compared to cold atmospheric pressure plasma. *Scientific Reports*, *11*(1), p. 16373.

Waskow, A., Howling, A. and Furno, I. (2021). Mechanisms of plasma-seed treatments as a potential seed processing technology. *Frontiers in Physics*, *9*, p. 617345.

Yan, D., Malyavko, A., Wang, Q., Ostrikov, K., Sherman, J.H. and Keidar, M. (2021). Multi-modal biological destruction by cold atmospheric plasma: Capability and mechanism. *Biomedicines*, *9*(9), p. 1259.

Zhang, Y., Xu, J., Li, R., Ge, Y., Li, Y. and Li, R. (2023). Plants' response to abiotic stress: Mechanisms and strategies. *International Journal of Molecular Sciences*, *24*(13), p.10915.

Zhao, Y.M., Ojha, S., Burgess, C.M., Sun, D.W. and Tiwari, B.K. (2021). Inactivation efficacy of plasma-activated water: Influence of plasma treatment time, exposure time and bacterial species. *International Journal of Food Science & Technology*, *56*(2), pp.721–732.

Zhou, R., Zhou, R., Wang, P., Xian, Y., Mai-Prochnow, A., Lu, X., Cullen, P.J., Ostrikov, K.K. and Bazaka, K. (2020). Plasma-activated water: Generation origin of reactive species and biological applications. *Journal of Physics D: Applied Physics*, *53*(30), p. 303001.

CHAPTER 11

# Bioengineering Starch and Amylose-lipid Complex: From Genetic to Sustainable Industrial Application

Mohammed Saba Rahim[1], Dinesh Arya[2,3], Ram Kumar Sharma[1] and Joy Roy[2,3*]

## 1. Introduction

Starch, a natural biopolymer is globally used across various industries, including both food and non-food sectors. Its versatile and renewable functional properties make it unique and distinguish it from other biopolymers. Moreover, starch is an easily available and valuable source of raw material. Notably, due to these features, starch has attracted considerable interest from researchers in the new arena of biotechnology, such as pathway engineering, nanobiosciences, biopolymer & bioplastic production, pharmaceutical industries, and food biotechnology. Based on established research finding, we assure that the biocompatibility and modifiability of starch can be harnessed to tailor industrial application and raise economic profit. Pathway engineering of starch genes has opened new avenues, revolutionizing industrial applications. For decades, efforts by biotechnology researchers have been focused on ameliorating the physicochemical properties of starch, particularly its thermal stability, solubility, pasting behavior, and mechanical strength. The objective of these research efforts is to ensure the quality of final products, whether they are foods with desirable textures and stability that meet consumer demands, or non-food goods such as adhesives, bioplastics, biodegradable packaging, pharmaceuticals, paper, 3D printing materials, coatings, and fillers. In consequence, starch continues plays a pivotal role as a sustainable and functional material in modern industries. In view of that, pathway engineering of starch, in combination with sustainable and green chemistry practices, needs a comprehensive understanding of its native versatile nature, chemical properties, and the underlying genetic pathways. This chapter addresses the structural properties of starch,

1 Molecular Genetic and Genomics Laboratory, Biotechnology Division, CSIR-Institute of Himalayan Bioresource Technology, Palampur, Himachal Pradesh, India.

2 BRIC-National Agri-Food & Biomanufacturing Institute (NABI), Mohali, Punjab, India.

3 Regional Centre for Biotechnology (RCB), NCR Biotech Science Cluster, 3rd Milestone, Faridabad–Gurgaon Expressway, Faridabad, Haryana, India.

* Corresponding author: joykroy@nabi.res.in

DOI: 10.1201/9781003584100-11

recent innovations in starch bioengineering through various biotechnological approaches, and the challenges associated with green chemistry and nanotechnology. To highlight the significance of this topic, the chapter also discusses the role of starch in food and non-food industries, the consumption of healthy starch and health benefits (resistant starch), and future scopes in this field.

## 1.1 Structural Properties of Starch

Starch is the primary renewable carbohydrate stored as an energy reservoir in plants and is considered the major fount of calories in human diet. Naturally, starch exists in granule form whose properties, such as structural and biochemical properties including the shape, size, and the proportion of amylose and amylopectin, depend on the source of plants. The differences in starch granule density, size, arrangement, and amylose (AM) to amylopectin (AP) ratio influence the physicochemical and functional characteristics of starch (Geigenberger, 2011; Van Hung et al., 2006). Amylose comprises linear, helical chains of 500–20,000 α-D-glucose units via α(1,4)-glycosidic bonds, with minimal branching. Amylopectin differs from amylose due to the presence of additional α(1,6)-glycosidic bonds, which form branching at approximately 20 to 30 anhydroglucose units (Figure 1). Amylopectin is highly stable, exhibits high viscosity, offers high clarity, and binds weakly with iodine, leading to the formation of a red color (Hii et al., 2012).

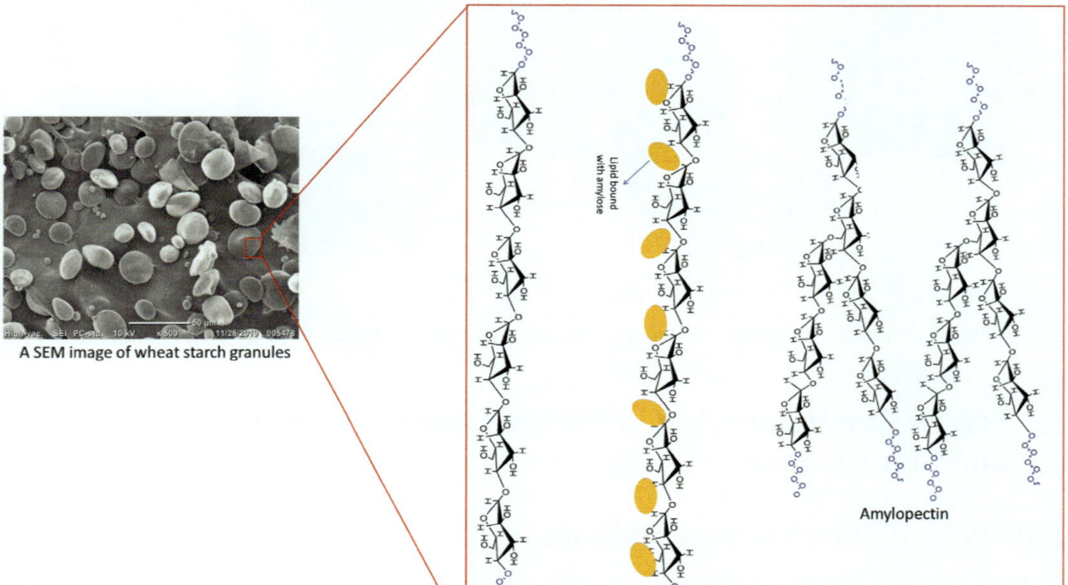


**Fig. 1** Starch granules and their composition: (A) Starch granule images visualized under scanning electron microscope (SEM) and (B) chemical composition of the three components—amylose, amylose–lipid complex, and amylopectin.

## 1.2 Starch Bioengineering: An Emerging Tool

Natural starch's characteristics can be enhanced by changing its composition and structure. Generally, the valuable attributes of starch are intent on the crystallinity, surface morphology of granules, the content and molecular weight of its composition (amylose, amylopectin, amylose–lipid complex), and the degree of branching in amylopectin (Vamadevan & Bertoft, 2018; Obadi & Xu, 2021; Zhu & Liu, 2020). Physical, chemical, and biological modification mainly relies on enzymes to change starch. Cereal starch showed promising applications in various food production and nonfood sectors, such as serving as a raw material in bakery, paper, and pharma sectors as a substrate for the synthesis of useful organic acids, disposable edible films, and highly efficient emulsifiers (Berton-Carabin & Schroën, 2015; Zhu, 2017) (Figure 2).

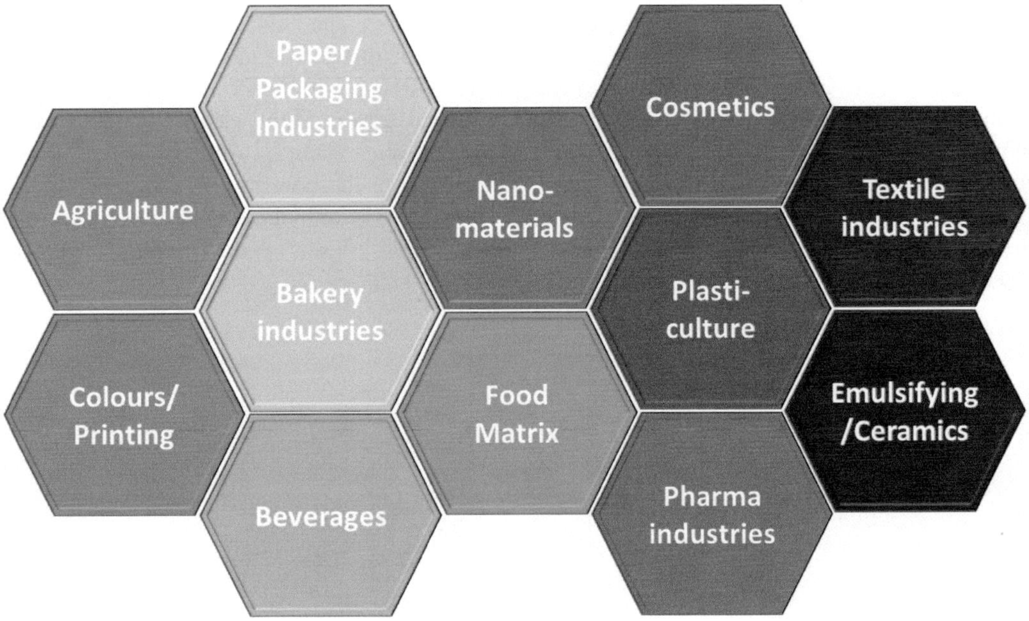


**Fig. 2** Versatile applications and industrial uses of starch in food and nonfood processing sectors.

# 2. Starch Bioengineering in Future Foods: Health Benefits and Significance

## *2.1 Role of Starch in Food Systems*

Carbohydrates such as starch comprise 65–75% of wheat grain, and their modification through various approaches enables their extensive application in the food sectors. Both low- and high-amylose content affect the texture of cooked foods, influenced by moisture content, swelling and solubility power, gelatinization, and pasting properties (Rahim et al., 2020a). These variations affect processing and cooking characteristics of grain-based food products (Singh et al., 2010; Kim et al., 2003). Commercially available starches derived from corn, potato, and wheat are gluten-free, making them suitable for individuals with gluten allergies, and are now essential raw materials in the food industry. Starch-rich crops such as barley, corn, wheat, potatoes, and rice have diverse applications in food production.

Barley (*Hordeum vulgare* L.) is a gluten-free flour source, yielding bread with a denser texture than wheat-based bread (Przetaczek-Rożnowska, 2017; Kallman et al., 2015). It is crucial in malt and beer production and serves as animal feed. Barley's functional properties support brewing, livestock energy, and human gut health. Studies suggest that substituting wheat flour with barley bran enhances the water absorption capacity of dough by approximately 71.5%, improving chapati quality (Shevkani et al., 2017). The production of quality wheat-based baked goods majorly depends on the physicochemical behavior of starch and its interaction with other molecules such as lipids and proteins. However, excessive wheat consumption is linked to health risks such as obesity and type 2 diabetes.

## 2.2 Health Benefits and Significance of Bioengineered Starch

Cereal crops also contain health-promoting constituents, including resistant starch (RS) and polysaccharide fibers. However, RS percentage is low in cereals, insisting on bioengineering of starch to fulfill the commercial demands and reach the target of nutritional security (Roy et al., 2024). Carbohydrate digestion causes the release of several sugars, but the release of glucose into the bloodstream impacts blood sugar levels. Recent research has shown that starchy foods raise blood glucose as well as affect insulin responses in both healthy and diabetic patients. On the other hand, it has been proven that the consumption of healthy starch has a positive impact on nutrition-associated metabolic dysregulation (Rahim et al., 2022a). However, the response is determined by the pace of the human system at which it digests starch.

A food's glycemic response is determined by several parameters, including the size of the particle, the extent of processing, starch structure, cooking method, and the amount of food components such as protein, fat, and dietary fibers (Lockyer & Nugent, 2017; Venn & Green, 2007). The glycemic index (GI) of different starchy crops is mentioned in Table 1. In cereal grain, lipid molecules such as fatty acids and monoglycerides bound with the helical groove of amylose starch to form a complex structure of starch called the amylose–lipid complex; due to its slow digestion in the gut, it is categorized under resistant starch type-5 (Rahim et al., 2020b; Rahim et al., 2022b; Madhawan et al., 2023). The fatty acid hydrocarbon chain can easily fit into the hydrophobic cavity created by the amylose helix to generate amylose–lipid complexes. Cereal crops differ in the lipid makeup of the lipids linked to starch. Lipids in oats and rye starches are primarily phospholipids, while free fatty acids are the predominant starch-bound lipids in maize (Farooq et al., 2021).

**Table 1** Resistant starch percentage along with carbohydrate content, glycemic index, and glycemic load of some major cereal and vegetable crops.

| *Crop type* | *Crop name* | *Carbohydrate content (%)* | *Glycemic index* | *Resistant starch (%)* | *References* |
|---|---|---|---|---|---|
| Cereal | Maize | 72–77 | 72–74 | <5 | Jocelyne et al., 2020; Bjorck & Elmståhl, 2003; Reddappa et al., 2023 |
| | Rice | 70–82 | 55–85 | 0.5–2.5 | Kumar et al., 2018 |
| | Wheat | 35–75 | 52–75 | 0.5–2.5 | Birt et al., 2013 |
| | Barley | 78–83 | 60–70 | 1–15 | Henry, 1988; Shu et al., 2012 |
| | Millet | 68–75 | 42–62 | 13.41–26.07 | Jocelyne et al., 2020; Anitha et al., 2021; Balli et al., 2023 |
| Vegetable | Sweet potato | 70–82.15 | 44–78 | 1.5–3.19 | Niu et al., 2019 |
| | Potato | 40–82 | 55–111 | 0.75–2.5 | Lal et al., 2021<br>Nayak et al., 2024 |
| | Taro | 26.8 | 54.43 | 3.47–3.57 | Kumar et al., 2024; Hedges & Lister, 2006 |
| | Cassava | 85–90 | 55.64 | 2.88 | Kumar et al., 2024 |

In recent years, amylose inclusion complexes have become increasingly important in food applications due to their ability to package bioactive molecules and ingredients. These complexes serve as potential delivery systems, offering health benefits such as reducing cancer, cardiovascular, and inflammatory risks, promoting eye health, and providing antioxidant protection.

As the world population growth rate continues to rise and environmental challenges are going to be bigger, the demand for sustainable food production has become increasingly urgent. The three most vital aspects of human life—food security, health, and environmental sustainability—have been addressed through traditional methods such as breeding, recombinant DNA technology, and genetic modification (Sharma et al., 2022). Bioengineered food or genetic engineering offers significant potential to meet these needs by improving crop resilience, boosting nutritional content, and increasing yields. Furthermore, the use of bioengineered crops can minimize pressure on natural ecosystems and encourage more effective land use, helping to achieve long-term sustainability objectives (Keskin Gundogdu et al., 2016). Various bioengineered crops, such as genetically modified organisms (GMOs) including soybeans, cotton, brinjal, corn, and canola, are widely grown and consumed across the globe. By creating crops capable of thriving in various environmental conditions and resisting pests and diseases, bioengineering plays a vital role in securing consistent food production against emerging difficulties (Yadav et al., 2024). Bioengineered crops minimize the reliance on chemical inputs such as pesticides and fertilizers, leading to less soil degradation and reduced water contamination. Furthermore, bioengineering plays a crucial role in cutting down greenhouse gas emissions. In conclusion, bioengineering can be considered a game-changer in sustainable agriculture.

## 3. Scope of Bioengineered Starch in Nonfood Industries: Green Chemistry

### *3.1 Eco-Friendly Plasticulture*

The emergence of plasticulture-based commodities affects our ecosystem and humankind's health. Mismanagement and nonbiodegradability are the problematic threats in agriculture sector and fisheries due to soil and water contamination. Hence, eco-friendly polymers from renewable sources would be a pragmatic gateway. Among these, starch is one of the biopolymers in agriculture. Starch has unique thermoplastic properties enabling its diverse application in nonfood industries as well as in agriculture. The various desired starchy products can be processed by several modifications in their physicochemical properties (Rahim et al., 2020b) and surface chemistry (Gamage et al., 2022). It can be applied as an additive in plasticulture for various uses such as filler, blending, plasticizer, and as a carrier in transporting and releasing of biocides (Gamage et al., 2022). Thus, the multidimensional properties of starch, including biodegradability, absorbency, nontoxicity, and biocompatibility, make it unique. Nevertheless, the large fraction of food and agriculture waste that is underutilized and can provide a substantial amount of starch would be a reliable solution for green chemistry in plasticulture. Moreover, it can be well utilized in other sectors such as pharma, textile, paper, and cosmetic industries.

Starch endures multi-physicochemical reactions during processing in the presence of substantial amounts of water such as diffusion, expansion and swelling of granules, melting, gelatinization, crystallization, and decomposition. Wherein, its philic nature toward water and brittleness behavior limited its application in membrane formation. Therefore, it is subjected to several chemical modifications to get the anticipated quality of starch in terms of hydrophobicity, solubility, thermal stability, pasting quality, amphiphilicity, strength, texture, adhesion, and retrogradation useful for desired manufacturing in industry. Thus, the amalgamation of different forms of starch may be one of the realistic techno-commercial tactics for farming and other industries favorable to natural abundance, low cost, and degradation readily in nature.

In this section, we documented various biotechnological approaches in the formation of starch composites and their application in different modes.

### 3.2 Physical Modification

Physical treatment includes thermal (pre-gelatinization, dry-heating, heat-moisture, and osmotic pressure treatment) and nonthermal (pulsed electric field, high-pressure, ultrasonic, and freeze-drying treatment) approaches to improve the solubility/swelling and to change in the crystallinity of the starch (Chibuogwu et al., 2019).

### 3.3 Chemical Modification

Chemical modification is done by the introduction of various functional groups in the native starch backbone by various approaches namely grafting, acetylation, acid hydrolysis, oxidation, and cross-linking, to achieve the unique properties of starch (Chibuogwu et al., 2019).

### 3.4 Enzymatic Modification

Enzymatic modification is done to modify the starch, making it effective for pharma industries by formulating fast-releasing nanoparticles to ensure rapid dissolution and absorption of drugs. This approach is known as immediate release (IR) tablet formulation (Chibuogwu et al., 2019). Previously, rice starch was modified by the method of oxidation and carboxylation and used for metronidazole drug formulation (Guan et al., 2008).

### 3.5 Hybrid Starch

Hybrid starch or hybrid polysaccharide (HP) is developed by several enzymatic polymerizations in vitro, and the approach is called "enzymatic polymerization to unnatural HP" (Ohmae et al., 2007). The enzymes used in this approach mainly include α-glucans such as dextran composed of $\alpha(1 \rightarrow 6)$ glycosidic linkage, mutan composed of $\alpha(1 \rightarrow 3)$ glycosidic linkage; and alternan composed of $\alpha(1 \rightarrow 3, 1 \rightarrow 6)$ glycosidic linkage. All these enzymes are experimentally used in potato-tuber-altered starch granule structure but further investigation is needed to verify the formation of hybrid starch having a manifest of covalently linked starch units (Kok-Jacon et al., 2007, 2005; Robyt et al., 2008).

## 4. Starch Bioengineering in Planta

### 4.1 Natural Chemical Modification of Starch

Natural phosphorylated starch mostly occurs in tubers such as potatoes. However, cereal grains have also phosphorylated starch (Hebelstrup et al., 2015). In plants, starch phosphorylation is catalyzed by the enzyme GWD1 forming G6P ester (Ritte et al., 2006), and GWD3/PWD forms G3P esters in the starch (Baunsgaard et al., 2005; Kotting et al., 2005). However, the regulatory role of phosphorylation in the biosynthesis of starch is still underneath (Shaik et al., 2014; Skeffington et al., 2014) and needs to be explored. Overexpression of GWD1 (glucan water dikinase 1) affects starch functional properties in several cereal crops, mainly in wheat (Schewe et al., 2004) and corn (Lanahan & Basu, 2006) improving the viscosity and pasting quality of starch. However, the increased phosphorylation by GWD1 initiates the degradation of starch (Blennow et al., 2013). Thus, both the enzymes would be an asset in modulating starch phosphate in crops.

### 4.2 Starch Hydrolysis

Monosaccharide or disaccharide sugars can be formed by hydrolysis of α-1,4 linkages of starch by α-amylases such as maltodextrins. Further, glucoamylases hydrolyze maltodextrins, maltose, and β-limit dextrins to form glucose. Such hydrolytic enzymes can be targeted for gene expression in cereal crop organs based on the concept of "self-processing plants" (Santa-Maria et al., 2011). Interestingly, self-processing corn induced by α-amylase enzyme resulted in a reduction in total greenhouse gas emission (11%), water usage (7.7%), natural gas usage (8.9%), and increment in fermenter content capacity (5%) (Urbanchuk et al., 2009).

### *4.3 Branching and Debranching Enzymes*

Branching and debranching of starch are regulated by several enzymes and play a crucial role in starch biosynthesis; particularly modifying α-1,6 linkages of starch and modulating amylopectin. Starch branching enzymes (SBEs) include α-glucanotransferases such as 4-α-glucanotransferases and cyclodextrin glucanotransferases (Kralj et al., 2011) used for the formation of cyclodextrin and cycloamylose in bioengineering processes. While, in natural conditions, SBEs act as catalysts during starch biosynthesis. Suppression of SBEs resulted in increased chain length of amylose as well as amylopectin units (Regina et al., 2012). In addition, debranching enzymes, including pullulanase and isoamylase, hydrolyze α-1,6 linkages, shortening the amylopectin chains in crop species (Utsumi et al., 2011; Streb & Zeeman, 2014). Thus, these enzymes have focused on starch bioengineering to modulate chain length in crop systems.

## 5. Starch Bioengineering in Polymer Preparation: Green Chemistry

### *5.1 Starch: A Biobased Biodegradable Polymer*

Biobased biodegradable polymers are prepared or extracted from materials derived from natural sources of biomass such as plants, animals, microbes, biogenic wastes/residues, etc. Herein, polysaccharides are the burgeoning biomacromolecules used as raw material in food as well as nonfood sectors. Likewise, starch and its derivatives (chitosan, chitin, cyclodextrin, etc.) have rising attention in industry-oriented research due to their dynamic physicochemical properties, bonding strength, eco-friendliness, nontoxicity, low-cost extraction, and abundance (Apopei et al., 2012; Abdul-Raheim et al., 2016). For green chemistry synthesis, modified starch is vastly used in plastic industries in the formation of eco-friendly fillers, blends, and composites. Thermoplastic starch (TPC) is one of the main components in polymer bends (Hamad et al., 2014).

### *5.2 Starch Blending*

Starch–polymer blends bring indispensable roles in enhancing the functional properties of biodegradable polymers, particularly polyvinyl alcohol (PVA), poly-ε-caprolactone (PCL), polylactic acid (PLA), polyhydroxybutyrate-hydroxyvalerate (PHB-HV), and polybutylene succinate (PBS). Loading starch in PVA blends proves biodegradable matrices; however, the biodegradability of PVA is influenced by hydrolysis degree and molecular weight (Lu et al., 2019; Azahari et al., 2011). Many plasticizers and compatibilizers have been used to recover the compatibility between starch and PVA (Priya et al., 2014), such as glycerol medium is used in PVA and starch to form thermoplastic material (Zhou et al., 2009). Starch in PLA or PCL blend improves the qualities and biodegradability, and these are the main objectives in industries (Soares et al., 2014; Martin & Avérous, 2001; Murariu & Dubois, 2016) as starch-extended hydrolysis reactions (Vikman et al., 1999). Thus, starch blended with PCL/PLA is more eco-friendly and could be the most definite biodegradable polymer for agriculture, biomedical, industrial, and packaging purposes. On a note, the hydrophilic nature or low water resistance of starch is challenging, which limits its application without modification (Singh et al., 2003).

In addition, the blend quality of PHB-HV has been examined with corn starch and showed poor interfacial adhesion between them (Reis et al., 2008). However, some modifications have been made in the formulation to reduce the aggregation between PHB-HV and starch (Wang et al., 2003). PBS having good mechanical properties is a commercially available thermoplastic polyester, while it has a low crystallinity which may result in a slow breakdown and affect the biodegradation rate.

Indeed, PBS has myriad importance in manufacturing and processing applications due to its better water resistance and excellence in decent chemical resistance and thermal stability (Suchao-In et al., 2014). Moreover, adding starch to PBS can accelerate biodegradation and flexibility, as well as used in packaging and other flushable/degradable products (Wang et al., 2003; Babu et al., 2013).

### *5.3 Impregnated Starch Blend in Essential Oils*

Captivatingly, the infusion of starch blends with essential oils enhances antimicrobial activity (Song et al., 2018; Souza et al., 2013). The thermal behavior and microstructure of starch blends can be improved by mixing essential oils. As noted, to serve organic horticulture, starch blends impregnated with essential oils are employed in active drug delivery systems in crop cultivation as a biocontrol (Chang et al., 2022; Varona et al., 2010). Previous reports have proven the successful utilization of rosemary (de Barros Fernandes et al., 2014), oregano (Caetano et al., 2017), cinnamon (de Barros Fernandes et al., 2014), and lavender (Varona et al., 2011) essential oils in starch blends. However, the infusion of essential oils is challenging, and therefore, precise microencapsulation and supercritical processes are critical (Varona et al., 2011; Glenn et al., 2010).

### *5.4 Starch in Green Nanotechnology*

Synthesis of the green nanoparticles is a way forward in the green biology frame. Green nanotechnology is an emerging approach and conformity to the sustainability challenges. Further, the use of nanofiller in the starch matrix makes it less hydrophilic, and improves its mechanical properties (Agustin et al., 2014). Several nanofillers such as chitin and cellulose (Salaberria et al., 2014; Orue et al., 2016; Ghanbari et al., 2018), nanocrystals of chitin (Salaberria et al., 2014), nanoclay (Harunsyah et al., 2018; Wahyuningtiyas & Suryanto, 2018), nano-$CaCO_3$ (Syafri et al., 2017), and nano-$TiO_2$ (Oleyaei et al., 2016) have been extensively used for nontoxic, eco-friendly plasticulture. Green synthesis of nanosystems using metallic nanoparticles, polymer nanoparticles, liposomes, and emulsions has a wide range of revolutionary applications that could be low cost, renewable, and eco-friendly. Previously, it was verified that the embodying of starch-based nanocrystals in blends and PVA improved its physical attributes vital for industrial application (Tang & Alavi, 2011).

## 6. Challenges and Research Gaps in Starch Bioengineering

The major challenges in the synthesis of biopolymer amalgamated with starch are hydrophilic nature, immiscibility, and phase-separated and poorly dispersed composites. Therefore, for commercial utilization of starch in agricultural and other industrial applications, several modifications must be intended to improve its hydrophobicity, thermal stability, retrogradation characteristics, paste behavior, and other physicochemical and physicomechanical attributes. Starch needs to be bioengineered in vivo in a plant system or in vitro through a chemical or structural modification that could expedite its applications in starch–polymer blends and plasticulture and make it feasible for other commercial usage such as in paper, pharmaceutical, cosmetics, and textile industries.

There are several chemical and physical methods used in starch modification. Wherein, the physical methods are osmotic pressure treatment, instantaneous controlled pressure, superheating, iterated syneresis, dry heating, pulsed electric fields treatment, micronization, and the mechanical drop (DIC) process (Zia-ud-Din et al., 2017; BeMiller, 2018; Pashkuleva et al., 2010). Further, the chemical modifications are oxidation, etherification, esterification, suspension co-polymerization, grafting from, and grafting onto (Masina et al., 2017; Singh et al., 2007). However, starch modification is still challenging due to the inclusion of proteins and fats in plant sources. It is reported that cassava and potato starch are substantial favorable sources of starch having comparatively low protein and fat (BeMiller & Whistler, 2009). Thus, the plant species sources define its application in agriscience and in different production processes.

On the other hand, in vivo starch bioengineering is crucial and uncompromising, and generally negatively impacts genetic gain. Previously, it has been shown that upon knockdown of SBE in barley, significant loss yields up to 20% (Carciofi et al., 2012). Likewise, starch bioengineering in cereal crops can jeopardize the seed sprouting rate (Shaik et al., 2014). However, in tuber starch bioengineering, expression of hyperthermophilic α-amylase gives better results without yield penalty (Santa-Maria et al., 2011). Likewise, starch bioengineering in sorghum with other debranching enzyme pullulanase improved its digestibility devoid of deleterious effects on crops (Gilding et al., 2013).

## 7. Conclusive Remarks

Undoubtedly, starch is a versatile biomolecule used as a raw material in a wide range of industrial sectors, including food and nonfood industries. However, the application and process are dependent on the plant species sources.

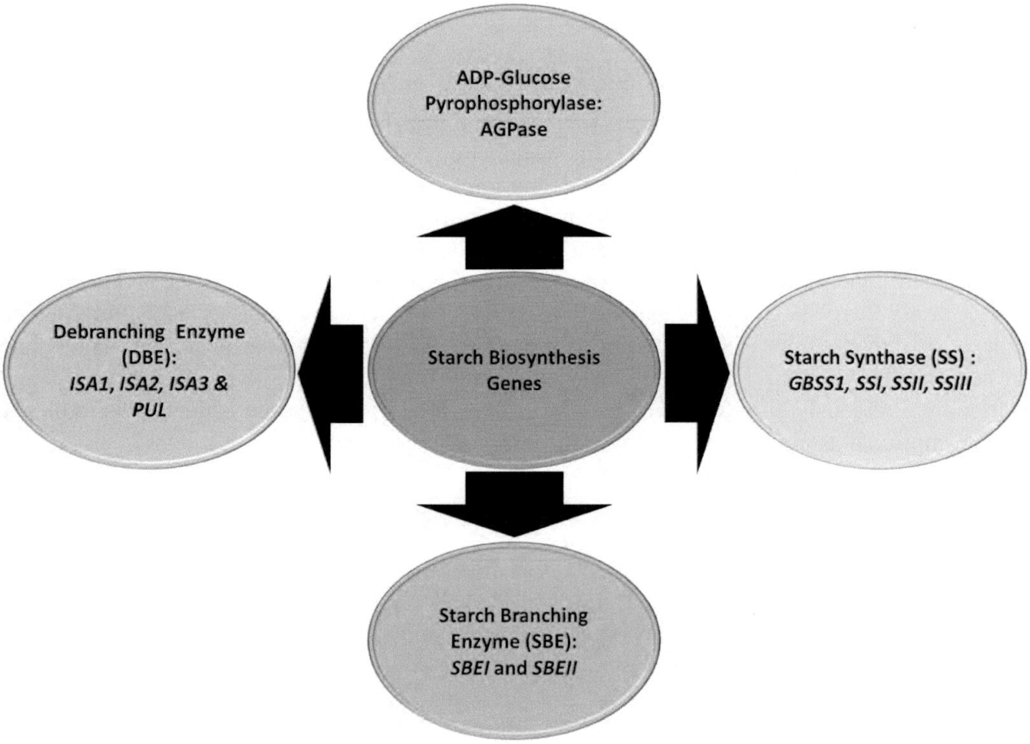


**Fig. 3** Major genes involved in starch biosynthesis in wheat can be targeted for bioengineering.

Remarkable advancements in sciences and engineering have made it possible to understand how starch is biosynthesized. Nevertheless, there are many challenges in planta starch bioengineering. Limited studies have been successfully utilized in agriculture sectors on a commercial level (Hebelstrup et al., 2015). On a note, a deeper understanding of the relationship between starch structure, function, and molecular-physiological behavior of amyloplasts is needed for the successful development of GM crops altering starch biosynthesis enzymes. Previous reports have shown that

SBEIIb mutant lines exhibited a notable increase in RS content (Sun et al., 2017). Employing multiplex CRISPR/Cas9 genome editing to simultaneously target multiple SBE genes effectively developed transgene-free rice crops with high resistant starch and high amylose content (Biswas et al., 2023). Likewise, high amylose (resistant starch type II) content cassava crop has been developed using CRISPR/Cas9-mediated mutagenesis of the SBE2 gene (Luo et al., 2022). Targeting mono-enzyme strategies is not very effective due to its complex genetic regulation (Figure 3), like in the maize bioengineering in enzyme isoamylase, *sugary-1* affects the activity of SBEII (Dinges et al., 2001).

Similarly, in barley, bioengineering in *sex6* resulted in the loss of SSIIa and also influenced the activity of other enzymes SBEIIa, SBEIIb, and SSI (Morell et al., 2003). Therefore, we can say still there is a gap in starch biosynthesis knowledge, and comprehending discoveries can open new strategies relating to starch bioengineering in planta or chemical engineering.

## References

Abdul-Raheim, A.R.M., El-Saeed Shimaa, M., Farag, R.K., Abdel-Raouf Manar, E., Reem, K.F. and Abdel-Raouf Manar, E. (2016). Low cost biosorbents based on modified starch iron oxide nanocomposites for selective removal of some heavy metals from aqueous solutions. Advanced Materials Letters, 7(5), 402–409.

Agustin, M.B., Ahmmad, B., Alonzo, S.M.M. and Patriana, F.M. (2014). Bioplastic based on starch and cellulose nanocrystals from rice straw. *Journal of Reinforced Plastics and Composites*, 33(24), 2205–2213.

Anitha, S., Kane-Potaka, J., Tsusaka, T.W., Botha, R., Rajendran, A., Givens, D.I., Parasannanavar, D.J., Subramaniam, K., Prasad, K.D.V., Vetriventhan, M. and Bhandari, R.K. (2021). A systematic review and meta-analysis of the potential of millets for managing and reducing the risk of developing diabetes mellitus. *Frontiers in Nutrition*, 8, 687428.

Apopei, D.F., Dinu, M.V., Trochimczuk, A.W. and Dragan, E.S. (2012). Sorption isotherms of heavy metal ions onto semi-interpenetrating polymer network cryogels based on polyacrylamide and anionically modified potato starch. *Industrial & Engineering Chemistry Research*, 51(31), 10462–10471.

Azahari, N.A., Othman, N. and Ismail, H. (2011). Biodegradation studies of polyvinyl alcohol/corn starch blend films in solid and solution media. *Journal of Physical Science*, 22(2), 15–31.

Babu, R.P., O'connor, K. and Seeram, R. (2013). Current progress on bio-based polymers and their future trends. *Progress in Biomaterials*, 2, 1–16.

Balli, D., Bellumori, M., Masoni, A., Moretta, M., Palchetti, E., Bertaccini, B., Mulinacci, N. and Innocenti, M. (2023). Proso millet (Panicum miliaceum L.) as alternative source of starch and phenolic compounds: a study on twenty-five worldwide accessions. *Molecules*, 28(17), 6339.

Baunsgaard, L., Lütken, H., Mikkelsen, R., Glaring, M.A., Pham, T.T. and Blennow, A. (2005). A novel isoform of glucan, water dikinase phosphorylates pre-phosphorylated α-glucans and is involved in starch degradation in Arabidopsis. *The Plant Journal*, 41(4), 595–605.

BeMiller, J.N. and Whistler, R.L. (2009). Starch: Chemistry and Technology. In: *Food Science and Technology International Series*. Academic Press, USA.

BeMiller, J.N. (2018). Physical Modification of Starch. In: *Starch in Food*. Woodhead Publishing, 223–253.

Berton-Carabin, C.C. and Schroën, K. (2015). Pickering emulsions for food applications: background, trends, and challenges. *Annual Review of Food Science and Technology*, 6(1), 263-297.

Birt, D.F., Boylston, T., Hendrich, S., Jane, J.L., Hollis, J., Li, L. and Whitley, E.M. (2013). Resistant starch: promise for improving human health. *Advances in Nutrition*, 4(6), 587–601.

Biswas, S., Ibarra, O., Shaphek, M., Molina-Risco, M., Faion-Molina, M., Bellinatti-Della Gracia, M. and Septiningsih, E.M. (2023). Increasing the level of resistant starch in 'Presidio' rice through multiplex CRISPR–Cas9 gene editing of starch branching enzyme genes. *The Plant Genome*, 16(2), e20225.

Bjorck, I. and Elmståhl, H.L. (2003). The glycaemic index: importance of dietary fibre and other food properties. *Proceedings of the Nutrition Society*, 62(1), 201–206.

Blennow, A., Jensen, S.L., Shaik, S.S., Skryhan, K., Carciofi, M., Holm, P.B. and Tanackovic, V. (2013). Future cereal starch bioengineering: cereal ancestors encounter gene technology and designer enzymes. *Cereal Chemistry*, 90(4), 274–287.

Caetano, K.D.S., Hessel, C.T., Tondo, E.C., Flôres, S.H. and Cladera-Olivera, F. (2017). Application of active cassava starch films incorporated with oregano essential oil and pumpkin residue extract on ground beef. *Journal of Food Safety*, 37(4), e12355.

Carciofi, M., Blennow, A., Jensen, S.L., Shaik, S.S., Henriksen, A., Buléon, A. and Hebelstrup, K.H. (2012). Concerted suppression of all starch branching enzyme genes in barley produces amylose-only starch granules. *BMC Plant Biology*, 12, 1–16.

Chang, Y., Harmon, P.F., Treadwell, D.D., Carrillo, D., Sarkhosh, A. and Brecht, J.K. (2022). Biocontrol potential of essential oils in organic horticulture systems: From farm to fork. *Frontiers in Nutrition*, 8, 805138.

Chibuogwu, C., Amadi, B., Anyaegbunam, Z., Emesiani, B. and Ofoefule, S. (2019). Application of starch and starch derivatives in pharmaceutical formulation. In Chemical Properties of Starch, Emeje, M. (Ed.), IntechOpen, 1–10.

de Barros Fernandes, R.V., Borges, S.V. and Botrel, D.A. (2014). Gum arabic/starch/maltodextrin/inulin as wall materials on the microencapsulation of rosemary essential oil. *Carbohydrate Polymers*, 101, 524–532.

Dinges, J.R., Colleoni, C., Myers, A.M. and James, M.G. (2001). Molecular structure of three mutations at the maize sugary1 locus and their allele-specific phenotypic effects. *Plant Physiology*, 125(3), 1406–1418.

Farooq, M.A., Murtaza, M.A., Aadil, R.M., Arshad, R., Rahaman, A., Siddique, R. and Haq, A.U. (2021). Investigating the structural properties and in vitro digestion of rice flours. *Food Science & Nutrition*, 9(5), 2668–2675.

Gamage, A., Liyanapathiranage, A., Manamperi, A., Gunathilake, C., Mani, S., Merah, O. and Madhujith, T. (2022). Applications of starch biopolymers for a sustainable modern agriculture. *Sustainability*, 14(10), 6085.

Geigenberger, P. (2011). Regulation of starch biosynthesis in response to a fluctuating environment. *Plant Physiology*, 155(4), 1566–1577.

Ghanbari, A., Tabarsa, T., Ashori, A., Shakeri, A. and Mashkour, M. (2018). Preparation and characterization of thermoplastic starch and cellulose nanofibers as green nanocomposites: Extrusion processing. *International Journal of Biological Macromolecules*, 112, 442–447.

Gilding, E.K., Frere, C.H., Cruickshank, A., Rada, A.K., Prentis, P.J., Mudge, A.M. and Godwin, I.D. (2013). Allelic variation at a single gene increases food value in a drought-tolerant staple cereal. *Nature Communications*, 4(1), 1483.

Glenn, G.M., Klamczynski, A.P., Woods, D.F., Chiou, B., Orts, W.J. and Imam, S.H. (2010). Encapsulation of plant oils in porous starch microspheres. *Journal of Agricultural and Food Chemistry*, 58(7), 4180–4184.

Guan, Y., Qian, L., Xiao, H. and Zheng, A. (2008). Preparation of novel antimicrobial-modified starch and its adsorption on cellulose fibers: Part I. Optimization of synthetic conditions and antimicrobial activities. *Cellulose*, 15, 609–618.

Hamad, K., Kaseem, M., Ko, Y.G. and Deri, F. (2014). Biodegradable polymer blends and composites: An overview. *Polymer Science Series A*, 56, 812–829.

Harunsyah, Sariadi and Raudah. (2018, January). The effect of clay nanoparticles as reinforcement on mechanical properties of bioplastic base on cassava starch. In *Journal of Physics: Conference Series* (Vol. 953, p. 012021). IOP Publishing.

Hebelstrup, K.H., Sagnelli, D. and Blennow, A. (2015). The future of starch bioengineering: GM microorganisms or GM plants? *Frontiers in Plant Science*, 6, 247.

Hedges, L.J. and Lister, C.E. (2006). Health attributes of roots and tubers. New Zealand Institute for Crop & Food Research Limited: Christchurch, New Zealand.

Henry, R.J. (1988). The carbohydrates of barley grains—A review. *Journal of the Institute of Brewing*, 94(2), 71–78.

Hii, S.L., Tan, J.S., Ling, T.C. and Ariff, A.B. (2012). Pullulanase: role in starch hydrolysis and potential industrial applications. *Enzyme Research*, 2012(1), 921362.

Jocelyne, R.E., Behiblo, K. and Ernest, A.K. (2020). Comparative study of nutritional value of wheat, maize, sorghum, millet, and fonio: some cereals commonly consumed in Côte d'Ivoire. *European Scientific Journal ESJ*, 16(21), 118–131.

Kallman, A., Vamadevan, V., Bertoft, E., Koch, K., Seetharaman, K., Aman, P. and Andersson, R. (2015). Thermal properties of barley starch and its relation to starch characteristics. *International Journal of Biological Macromolecules*, 81, 692–700.

Keskin Gundogdu, T., Deniz, I., Çalışkan, G., Şahin, E.S. and Azbar, N. (2016). Experimental design methods for bioengineering applications. *Critical Reviews in Biotechnology*, 36(2), 368–388.

Kim, W., Johnson, J.W., Graybosch, R.A. and Gaines, C.S. (2003). Physicochemical properties and end-use quality of wheat starch as a function of waxy protein alleles. *Journal of Cereal Science*, 37(2), 195–204.

Kok-Jacon, G.A., Vincken, J.P., Suurs, L.C., Wang, D., Liu, S. and Visser, R.G. (2005). Production of dextran in transgenic potato plants. *Transgenic Research*, 14, 385–395.

Kok-Jacon, G.A., Vincken, J.P., Suurs, L.C., Wang, D., Liu, S. and Visser, R.G. (2007). Expression of alternansucrase in potato plants. *Biotechnology Letters*, 29, 1135–1142.

Kotting, O., Pusch, K., Tiessen, A., Geigenberger, P., Steup, M. and Ritte, G. (2005). Identification of a novel enzyme required for starch metabolism in Arabidopsis leaves. The phosphoglucan, water dikinase. *Plant Physiology*, 137(1), 242–252.

Kralj, S., Grijpstra, P., van Leeuwen, S.S., Leemhuis, H., Dobruchowska, J.M., van der Kaaij, R.M. and Dijkhuizen, L. (2011). 4, 6-α-Glucanotransferase, a novel enzyme that structurally and functionally provides an evolutionary link between glycoside hydrolase enzyme families 13 and 70. *Applied and Environmental Microbiology*, 77(22), 8154–8163.

Kumar, A., Mahapatra, S., Nayak, L., Biswal, M., Sahoo, U., Lal, M.K. and Pati, K. (2024). Tuber crops could be a potential food component for lowering starch digestibility and estimated glycemic index in rice. *Journal of the Science of Food and Agriculture*, 104(14), 8519–8528.

Kumar, A., Sahoo, U., Baisakha, B., Okpani, O.A., Ngangkham, U., Parameswaran, C. and Sharma, S.G. (2018). Resistant starch could be decisive in determining the glycemic index of rice cultivars. *Journal of Cereal Science*, 79, 348–353.

Lal, M.K., Singh, B., Sharma, S., Singh, M.P. and Kumar, A. (2021). Glycemic index of starchy crops and factors affecting its digestibility: A review. *Trends in Food Science & Technology*, 111, 741–755.

Lanahan, M. and Basu, S. (2006). Modified starch, uses, methods for production thereof. Inventors: Syngenta Participations AG. U.S. Patent Application 10/556,276.

Lockyer, S. and Nugent, A.P. (2017). Health effects of resistant starch. Nutrition Bulletin, 42(1), 10–41.

Lu, D.R., Xiao, C.M. and Xu, S.J. (2009). Starch-based completely biodegradable polymer materials. *Express Polymer Letters*, 3(6), 366–375.

Luo, S., Ma, Q., Zhong, Y., Jing, J., Wei, Z., Zhou, W. and Zhang, P. (2022). Editing of the starch branching enzyme gene SBE2 generates high-amylose storage roots in cassava. *Plant Molecular Biology*, 1–14.

Madhawan, A., Bhunia, R. K., Kumar, P., Sharma, V., Sinha, K., Fandade, V. and Roy, J. (2023). Interaction between long noncoding RNA (lnc663) and microRNA (miR1128) regulates PDAT-like gene activity in bread wheat (Triticum aestivum L.). *Plant Physiology and Biochemistry*, 203, 108040.

Martin, O. and Avérous, L. (2001). Poly (lactic acid): plasticization and properties of biodegradable multiphase systems. Polymer, 42(14), 6209–6219.

Masina, N., Choonara, Y.E., Kumar, P., du Toit, L.C., Govender, M., Indermun, S. and Pillay, V. (2017). A review of the chemical modification techniques of starch. *Carbohydrate Polymers*, 157, 1226–1236.

Morell, M.K., Kosar-Hashemi, B., Cmiel, M., Samuel, M.S., Chandler, P., Rahman, S. and Li, Z. (2003). Barley sex6 mutants lack starch synthase IIa activity and contain a starch with novel properties. *The Plant Journal*, 34(2), 173–185.

Murariu, M. and Dubois, P. (2016). PLA composites: From production to properties. *Advanced Drug Delivery Reviews*, 107, 17–46.

Nayak, B., Berrios, J.D.J. and Tang, J. (2014). Impact of food processing on the glycemic index (GI) of potato products. *Food Research International*, 56, 35–46.

Niu, S., Li, X. Q., Tang, R., Zhang, G., Li, X., Cui, B. and Haroon, M. (2019). Starch granule sizes and degradation in sweet potatoes during storage. *Postharvest Biology and Technology*, 150, 137–147.

Obadi, M. and Xu, B. (2021). Review on the physicochemical properties, modifications, and applications of starches and its common modified forms used in noodle products. *Food Hydrocolloids*, 112, 106286.

Ohmae M., Makino A. and Kobayashi S. (2007). Enzymatic polymerization to unnatural hybrid polysaccharides. *Macromolecular Chemistry and Physics*, 208, 1447–1457.

Oleyaei, S.A., Zahedi, Y., Ghanbarzadeh, B. and Moayedi, A.A. (2016). Modification of physicochemical and thermal properties of starch films by incorporation of TiO2 nanoparticles. *International Journal of Biological Macromolecules*, 89, 256–264.

Orue, A., Corcuera, M.A., Peña, C., Eceiza, A. and Arbelaiz, A. (2016). Bionanocomposites based on thermoplastic starch and cellulose nanofibers. *Journal of Thermoplastic Composite Materials*, 29(6), 817–832.

Pashkuleva, I., Marques, A.P., Vaz, F. and Reis, R.L. (2010). Surface modification of starch based biomaterials by oxygen plasma or UV-irradiation. *Journal of Materials Science: Materials in Medicine*, 21, 21–32.

Priya, B., Gupta, V.K., Pathania, D. and Singha, A.S. (2014). Synthesis, characterization and antibacterial activity of biodegradable starch/PVA composite films reinforced with cellulosic fibre. *Carbohydrate Polymers*, 109, 171–179.

Przetaczek-Rożnowska, I. (2017). Physicochemical properties of starches isolated from pumpkin compared with potato and corn starches. *International Journal of Biological Macromolecules*, 101, 536–542.

Rahim, M.S., Chauhan, A., Madhawan, A., Mishra, A., Parveen, A., Garg, M. and Roy, J. (2020a). Development and evaluation of chapatti quality of high amylose wheat mutants on the basis of physicochemical, textural and sensory characteristics. *LWT-Food Science and Technology*, 133, 110051.

Rahim, M.S., Kumar, V. and Roy, J. (2022b). Genetic dissection of quantitative traits loci identifies new genes for gelatinization parameters of starch and amylose–lipid complex (Resistant starch 5) in bread wheat. *Plant Science*, 325, 111452.

Rahim, M.S., Kumar, V., Mishra, A., Fandade, V., Kumar, V., Bishnoi, M. and Roy, J. (2022a). High resistant starch mutant wheat 'TAC 35' reduced glycemia and ameliorated high fat diet induced metabolic dysregulation in mice. *Journal of Cereal Science*, 105, 103459.

Rahim, M.S., Mishra, A., Katyal, M., Thakur, S., Sharma, M., Kumar, P. and Roy, J. (2020b). Marker-trait association identified candidate starch biosynthesis pathway genes for starch and amylose–lipid complex gelatinization in wheat (*Triticum aestivum* L.). *Euphytica*, 216, 1–22.

Reddappa, S.B., Muthusamy, V., Zunjare, R.U., Chhabra, R., Talukder, Z.A., Maman, S., Chand, G., Pal, D., Kumar, R., Mehta, B.K. and Guleria, S.K. (2023). Composition of kernel-amylose and-resistant starch among subtropically adapted maize. *Journal of Food Composition and Analysis*, 119, 105236.

Regina, A., Blazek, J., Gilbert, E., Flanagan, B.M., Gidley, M.J., Cavanagh, C. and Morell, M.K. (2012). Differential effects of genetically distinct mechanisms of elevating amylose on barley starch characteristics. *Carbohydrate Polymers*, 89(3), 979–991.

Reis, K.C., Pereira, J., Smith, A.C., Carvalho, C.W.P., Wellner, N. and Yakimets, I. (2008). Characterization of polyhydroxybutyrate-hydroxyvalerate (PHB-HV)/maize starch blend films. *Journal of Food Engineering*, 89(4), 361–369.

Ritte, G., Heydenreich, M., Mahlow, S., Haebel, S., Kötting, O. and Steup, M. (2006). Phosphorylation of C6-and C3-positions of glucosyl residues in starch is catalysed by distinct dikinases. FEBS letters, *580*(20), 4872–4876.

Robyt J.F., Yoon S.H. and Mukerjea R. (2008). Dextransucrase and the mechanism for dextran biosynthesis. *Carbohydrate Research*. 343, 3039–3048.

Roy, J., Madhawan, A., Rahim, M.S. and Kumar, P. (2024). Starch Quality: High Energy or Low Calories–Both Matter. In: *Biofortification for Nutrient-Rich Crops*. CRC Press, 185–195.

Salaberria, A.M., Labidi, J. and Fernandes, S.C. (2014). Chitin nanocrystals and nanofibers as nano-sized fillers into thermoplastic starch-based biocomposites processed by melt-mixing. *Chemical Engineering Journal*, 256, 356–364.

Santa-Maria, M.C., Yencho, C.G., Haigler, C.H., Thompson, W.F., Kelly, R.M. and Sosinski, B. (2011). Starch self-processing in transgenic sweet potato roots expressing a hyperthermophilic α-amylase. *Biotechnology Progress*, 27(2), 351–359.

Schewe, G., Knies, P., Amati, S.F., Lorz, H., Becker, D., Landschutze, V. and Pilling, J. (2004). Monocotyledon plant cells and plants which synthesise modified starch. *Inventers: Bayer Crop Science AG*. U.S. Patent 6,734,340.

Shaik, S.S., Carciofi, M., Martens, H.J., Hebelstrup, K.H. and Blennow, A. (2014). Starch bioengineering affects cereal grain germination and seedling establishment. *Journal of Experimental Botany*, 65(9), 2257–2270.

Sharma, P., Bano, A., Singh, S.P., Atkinson, J.D., Lam, S.S., Iqbal, H.M. and Tong, Y.W. (2022). Nanomaterials as highly efficient photocatalysts used for bioenergy and biohydrogen production from waste toward a sustainable environment. *Fuel*, 329, 125408.

Shevkani, K., Singh, N., Bajaj, R. and Kaur, A. (2017). Wheat starch production, structure, functionality and applications—a review. *International Journal of Food Science & Technology*, 52(1), 38–58.

Shu, X., Backes, G. and Rasmussen, S.K. (2012). Genome-wide association study of resistant starch (RS) phenotypes in a barley variety collection. *Journal of Agricultural and Food Chemistry*, 60(41), 10302–10311.

Singh, J., Dartois, A. and Kaur, L. (2010). Starch digestibility in food matrix: a review. *Trends in Food Science & Technology*, 21(4), 168–180.

Singh, J., Kaur, L. and McCarthy, O.J. (2007). Factors influencing the physico-chemical, morphological, thermal and rheological properties of some chemically modified starches for food applications – A review. *Food Hydrocolloids*, 21(1), 1–22.

Singh, R.P., Pandey, J.K., Rutot, D., Degée, P. and Dubois, P. (2003). Biodegradation of poly (ε-caprolactone)/starch blends and composites in composting and culture environments: the effect of compatibilization on the inherent biodegradability of the host polymer. *Carbohydrate Research*, 338(17), 1759–1769.

Skeffington, A.W., Graf, A., Duxbury, Z., Gruissem, W. and Smith, A.M. (2014). Glucan, water dikinase exerts little control over starch degradation in Arabidopsis leaves at night. *Plant Physiology*, 165(2), 866–879.

Soares, F.C., Yamashita, F., Müller, C.M. and Pires, A.T. (2014). Effect of cooling and coating on thermoplastic starch/poly (lactic acid) blend sheets. *Polymer Testing*, 33, 34–39.

Song, X., Zuo, G. and Chen, F. (2018). Effect of essential oil and surfactant on the physical and antimicrobial properties of corn and wheat starch films. *International Journal of Biological Macromolecules*, 107, 1302–1309.

Souza, A.C.D., Goto, G.E.O., Mainardi, J.A., Coelho, A.C.V. and Tadini, C.C. (2013). Cassava starch composite films incorporated with cinnamon essential oil: Antimicrobial activity, microstructure, mechanical and barrier properties. *LWT-Food Science and Technology*, 54(2), 346–352.

Streb, S. and Zeeman, S.C. (2014). Replacement of the endogenous starch debranching enzymes ISA1 and ISA2 of Arabidopsis with the rice orthologs reveals a degree of functional conservation during starch synthesis. *PLoS One*, 9(3), e92174.

Suchao-In, K., Koombhongse, P. and Chirachanchai, S. (2014). Starch grafted poly (butylene succinate) via conjugating reaction and its role on enhancing the compatibility. *Carbohydrate Polymers*, 102, 95–102.

Sun, Y., Jiao, G., Liu, Z., Zhang, X., Li, J., Guo, X. and Xia, L. (2017). Generation of high-amylose rice through CRISPR/Cas9-mediated targeted mutagenesis of starch branching enzymes. *Frontiers in Plant Science*, 8, 298.

Syafri, E., Kasim, A., Abral, H. and Asben, A. (2017). Effect of precipitated calcium carbonate on physical, mechanical and thermal properties of cassava starch bioplastic composites. *International Journal on Advanced Science Engineering Information Technology*, 7(5), 1950–1956.

Tang, X. and Alavi, S. (2011). Recent advances in starch, polyvinyl alcohol based polymer blends, nanocomposites and their biodegradability. *Carbohydrate Polymers*, 85(1), 7–16.

Urbanchuk, J.M., Kowalski, D.J., Dale, B. and Kim, S. (2009). Corn amylase: improving the efficiency and environmental footprint of corn to ethanol through plant biotechnology. *AgBioForum*, 12(2), 149–154.

Utsumi, Y., Utsumi, C., Sawada, T., Fujita, N. and Nakamura, Y. (2011). Functional diversity of isoamylase oligomers: the ISA1 homo-oligomer is essential for amylopectin biosynthesis in rice endosperm. *Plant Physiology*, 156(1), 61–77.

Vamadevan, V. and Bertoft, E. (2018). Impact of different structural types of amylopectin on retrogradation. *Food Hydrocolloids*, 80, 88–96.

Van Hung, P., Maeda, T. and Morita, N. (2006). Waxy and high-amylose wheat starches and flours—characteristics, functionality and application. *Trends in Food Science & Technology*, 17(8), 448–456.

Varona, S., Kareth, S., Martín, Á. and Cocero, M.J. (2010). Formulation of lavandin essential oil with biopolymers by PGSS for application as biocide in ecological agriculture. *The Journal of Supercritical Fluids*, 54(3), 369–377.

Varona, S., Rodríguez-Rojo, S., Martín, Á., Cocero, M.J. and Duarte, C.M. (2011). Supercritical impregnation of lavandin (Lavandula hybrida) essential oil in modified starch. *The Journal of Supercritical Fluids*, 58(2), 313–319.

Venn, B.J. and Green, T.J. (2007). Glycemic index and glycemic load: measurement issues and their effect on diet–disease relationships. *European Journal of Clinical Nutrition*, 61(1), S122–S131.

Vikman, M., Hulleman, S.H.D., Van Der Zee, M., Myllärinen, P. and Feil, H. (1999). Morphology and enzymatic degradation of thermoplastic starch–polycaprolactone blends. *Journal of Applied Polymer Science*, 74(11), 2594–2604.

Wahyuningtiyas, N.E. and Suryanto, H. (2018). Properties of cassava starch based bioplastic reinforced by nanoclay. *Environment*, 2(1), 20–26

Wang, X.L., Yang, K.K. and Wang, Y.Z. (2003). Properties of starch blends with biodegradable polymers. *Journal of Macromolecular Science*, Part C: Polymer Reviews, 43(3), 385–409.

Yadav, V., Pal, D. and Poonia, A.K. (2024). A study on genetically engineered foods: need, benefits, risk, and current knowledge. *Cell Biochemistry and Biophysics*, 82(3), 1931–1946.

Zhou, X.Y., Cui, Y.F., Jia, D.M. and Xie, D. (2009). Effect of a complex plasticizer on the structure and properties of the thermoplastic PVA/starch blends. *Polymer-Plastics Technology and Engineering*, 48(5), 489–495.

Zhu, F. (2017). Encapsulation and delivery of food ingredients using starch based systems. *Food Chemistry*, 229, 542–552.

Zhu, F. and Liu, P. (2020). Starch gelatinization, retrogradation, and enzyme susceptibility of retrograded starch: Effect of amylopectin internal molecular structure. *Food Chemistry*, 316, 126036.

Zia-ud-Din, Xiong, H. and Fei, P. (2017). Physical and chemical modification of starches: A review. *Critical Reviews in Food Science and Nutrition*, 57(12), 2691–2705.

# CHAPTER 12

# High-Throughput Phenotyping Tools for Sustainable Agriculture

Pawan Saini[1*], Danishta Aziz[2], Chhattar Pal[3], Basangouda Gonal[1], Gulab Khan Rohela[1], Pooja Saini[4], Zakiya Saif[5], Plabani Roy[1], S Syam[1], Shivam Bhardwaj[1], Gurjeet Singh[6] and Belaghihalli N Gnanesh[7]

## 1. Introduction

Presently, world agriculture is facing the challenges of climate change, the growing global population at a very high pace, and limited available resources for feeding the 10 billion human population in coming few decades, that is, 2050 (Saini et al., 2020). Among these, climate vulnerability acts as the biggest challenge for agriculture community as meteorological parameters such as temperature, rainfall, and relative humidity are changing rapidly, thus creating issues of water stress and chilling and freezing effects. Besides these, the edaphic factors are also increasing such as salinity, acidity, alkalinity, toxicity, mineral deficiency, etc., which hamper crop productivity severely. The combined effect of edapho-climatic factors has drastically affected the crop productivity. Therefore, to meet out the food requirement for feeding 10 billion population by 2050, it becomes indispensable to develop climate-resilient crop cultivars to increase the crop productivity by 70% above the current level (Hunter et al., 2017).

For achieving the increment in crop productivity, the best way is genetic improvement of food crops by utilizing the advanced genomic breeding methodologies (Varshney et al., 2020). Nowadays, it has become very easy and cost effective to obtain the genome sequence information through advanced genome sequencing platforms of next-generation sequencing (NGS) [Roche/454 sequencing, Illumina/Solexa Genome Analyzer (GA), and Applied Biosystems (ABI) Sequencing using Oligonucleotide Ligation and Detection (SOLiD) system]; nanopore sequencing, PacBio RS

1 CSB-Central Sericultural Research & Training Institute (CSR&TI), Pampore 192 121, Jammu & Kashmir (UT), India.
2 Division of Entomology, Faculty of Agriculture, Sher-e-Kashmir University of Agricultural Sciences & Technology – Kashmir (SKUAST-K), Wadura, Sopore, Jammu & Kashmir (UT), India.
3 CSB-Regional Sericultural Research Station (RSRS), Sahaspur, Dehradun, Uttarakhand, India.
4 Department of Genetics, Plant Breeding and Biotechnology, Dr. Khem Singh Gill College of Agriculture, Eternal University, Baru Sahib, Himachal Pradesh, India.
5 College of Temperate Sericulture (CoTS), Sher-e-Kashmir University of Agricultural Sciences & Technology - Kashmir (SKUAST-K), Mirgund, Jammu & Kashmir (UT), India.
6 Texas A&M University, Agrilife Research Center at Beaumont, Texas 77713, USA.
7 Genetics & Plant Breeding, ICAR JSS – Krishi Vigyan Kendra, Suttur, Mysore, Karnataka, India.
Email: daniaziz9974@gmail.com[2]; chhattar@gmail.com[3]; sainipooja858@gmail.com[4]; saifzakiya786@gmail.com[5]; gurjeet.singh@ag.tamu.edu[6]; gnaneshbn@gmail.com[7]
* Corresponding author: pawansaini-coapbg@pau.edu

DOI: 10.1201/9781003584100-12

device; exome sequencing; de novo sequencing; and shotgun sequencing (Satam et al., 2023). By utilizing genome sequencing approaches, vast gene sequence pool has been identified and cloned for introgression in crop cultivars for genetic improvement. But sometimes, due to lack of proper phenotyping data, some of the important genomic loci might get missed while doing mapping studies. The integration of genome sequence with accurate quantification of plant traits is very useful in dissection of genetics of economically important traits of crop cultivars (Mir et al., 2019).

The in-hand occurrence of high-throughput phenotyping (HTP) facilities is very helpful in determining the contribution of genetic factors in phenotype. The plant phenome is total of all of its genes, as it results from the interaction between its genotype and the surrounding environment where it dwells; even plants possessing similarity in genotype can behave differently in expressing phenotype. Phenotyping involves examining a plant's physical and biological traits, while phenomics enhances this process by utilizing advanced imaging technology and computational power to accelerate data collection and analysis (Zavafer et al., 2023). The advanced phenotyping tools aim for recording the data on canopy architecture, biomass, growth & development, abiotic & biotic stresses, physiological traits, etc., in a single day for thousands of plants (Rahman et al., 2015). Plant phenomics platforms enable the efficient screening of extensive plant populations, germplasm collections, breeding materials, and mapping populations with enhanced precision and accuracy in phenotypic trait assessment. This is achieved through automation, remote control, and advanced data analysis pipelines, reducing labor requirements while supporting high-throughput processing with improved accuracy and reproducibility of the results. This enhanced throughput is particularly important for areas such as germplasm screening, breeding programs, and mapping studies, where phenotypic data can now be gathered at much higher resolution and with far less human intervention (Chawade et al., 2019).

HTP is becoming increasingly crucial for sustainable agriculture, as it facilitates the rapid and precise identification of valuable plant traits, which is essential for developing crops that can thrive under varying environmental conditions, use resources efficiently, and contribute to global food security. HTP allows breeders to evaluate large populations of crops for desirable traits (e.g., mitigating drought stress, efficient utilization of nutrients, resistance to pests and diseases) with much greater speed and precision than traditional methods. This speeds up the breeding program and increases the chances of developing crops that are well suited to particular environmental conditions (Jangra et al., 2021). HTP aids in identifying crops that thrive with minimal inputs such as water, fertilizers, and pesticides, without sacrificing yield. This contributes to sustainable agriculture by reducing resource overuse and mitigating issues such as soil degradation, water scarcity, and chemical pollution (Fu & Jiang, 2022). HTP systems, using remote sensing, unmanned aerial vehicles (UAVs or drones), and other technologies, can generate large amounts of high-resolution data (e.g., canopy temperature, leaf area, photosynthetic efficiency). This data allows farmers and agronomists to implement precision agriculture techniques, optimizing inputs and reducing waste (Gaju et al., 2021). HTP is reshaping sustainable agriculture by inculcating precision and enhancing effectiveness of crop breeding, improving resource-use efficiency, and enabling climate-smart agriculture. The amalgamation of advanced data analytics, remote sensing, and genomics with phenotyping technologies paves the way for more adaptable and ecologically balanced farming systems, having the ability to address the issues posed by rising global population and climate vulnerability.

## 2. Types of HTP Tools

HTP tools are critical in rapidly assessing and quantifying plant or organism traits in large populations, facilitating advancements in fields such as agriculture, genomics, and functional biology. There are many types of HTP tools used for crop plants (Table 1).

**Table 1** Details of high-throughput plant phenotyping platforms.

| *Platform* | *Traits recorded* | *Crop* | *References* |
|---|---|---|---|
| PHENOPSIS | Response of plant to water-related stress condition | Arabidopsis (*Arabidopsis thaliana*) | Granier et al. (2006) |
| PHENODYN | Levels of drought stress (Soil water content), rate of leaf elongation, and micro-meteorological variables | Rice (*Oryza sativa*) and maize (*Zea mays*) | Sadok et al. (2007) |
| GROWSCREEN FLUORO | Detection of stress tolerance level by assessing leaf growth and chlorophyll fluorescence | Arabidopsis (*Arabidopsis thaliana*) | Jansen et al. (2009) |
| Field monitoring support system | Presence of rice bug infestation | Rice (*Oryza sativa*) | Fukatsu et al. (2012) |
| BreedVision | Lodging, moisture content in plant, plant yield, and biomass or tiller density | Triticale (x*Triticosecale* Wittmack L.) | Busemeyer et al. (2013a) |
| LemnaTec 3D Scanalyzer system | Traits governing tolerance to salinity | Rice (*Oryza sativa*) | Hairmansis et al. (2014) |
| HyperArt | Leaf-related characters (disease intensity or leaf chlorophyll level) | Barley (*Hordeum vulgare*), maize (*Zea mays*), tomato (*Lycopersicon esculentum*), and field mustard (*Brassica rapa*) | Bergsträsser et al. (2015) |
| Automated video tracking platform | Resistance to insects with piercing and sucking mouth parts like aphids | Lettuce (*Lactuca sativa*) and Arabidopsis (*Arabidopsis thaliana*) | Kloth et al. (2015) |
| RhizoTubes | Root-related characters | Medicago (*Medicago truncatula*), pea (*Pisum sativum*), rapeseed (*Brassica napus*), grapes (*Vitis vinifera*), wheat (*Triticum aestivum*) | Jeudy et al. (2016) |
| RADIX | Traits governing root and shoot phenotype | Maize (*Zea mays*) | Le Marié et al. (2016) |
| PhenoBox | Showing presence of head smut fungus and corn smut on Brachypodium and maize, and response to salt stress in tobacco | Brachypodium (*Brachypodium distachyon*), maize (*Zea mays*), tobacco (*Nicotiana benthamiana*) | Czedik-Eysenberg et al. (2018) |
| PHENOVISION | Detection of stress due to drought and tolerance | Maize (*Zea mays*) | Asaari et al. (2019) |
| PhenoField | Abiotic stress characterization | Wheat (*Triticum aestivum*) | Beauchêne et al. (2019) |
| Liaphen | Expansion of leaves and rate of transpiration under water stress | Sunflower (*Helianthus annuus*) | Gosseau et al. (2019) |
| PlantScreen robotic XYZ system | Traits governing tolerance to drought | Rice (*Oryza sativa*) | Kim et al. (2020) |
| PHENOTIC | Evaluation of virulence of pathogens and plants ability to tolerate it | Horticultural crops | Boureau (2020) |

*Contd.*

**Table 1** *Contd.*

| *Platform* | *Traits recorded* | *Crop* | *References* |
|---|---|---|---|
| PhenoImage | Response of crop to water-related stress | Wheat (*Triticum aestivum*), sorghum (*Sorghum bicolor*) | Zhu et al. (2021) |
| Plant root monitoring platform (PlaROM) | Traits related to root | Arabidopsis (*Arabidopsis thaliana*) | Yazdanbakhsh & Fisahn (2009) |
| Phenoscope | Complex growth-related traits and involved in response toward environmental factors | Arabidopsis (*Arabidopsis thaliana*) | Tisné et al. (2013) |
| High-throughput rice phenotyping facility (HRPF) | Agronomic traits | Rice (*Oryza sativa*) | Yang et al. (2014) |
| Zeppelin NT aircraft | Leaf area index, leaf biomass, vigor, height in plants | Maize (*Zea mays*) | Liebisch et al. (2015) |
| Rhizoponics | Traits related to roots | Arabidopsis (*Arabidopsis thaliana*) | Mathieu et al. (2015) |
| Phenocart | Morphological traits | Wheat (*Triticum aestivum*) | Crain et al. (2016) |
| PhenoArch phenotyping platform | Light interception and radiation-use efficiency | Maize (*Zea mays*) | Cabrera-Bosquet et al. (2016) |
| Phenovator | Growth, photosynthesis, and spectral reflectance | Arabidopsis (*Arabidopsis thaliana*) | Flood et al. (2016) |
| PHENOARCH | For evaluating growth of maize plant and its silk | Maize (*Zea mays*) | Brichet et al. (2017) |
| Phenobot 1.0 | Traits governing biomass | Sorghum (*Sorghum bicolor*) | Salas Fernandez et al. (2017) |
| Field Scanalyzer | Morphological traits | Wheat (*Triticum aestivum*) | Virlet et al. (2016) |
| CropQuant | Traits related to overall performance | Wheat (*Triticum aestivum*) | Zhou et al. (2017) |
| PhenoRoots | Traits related to roots | Cotton (*Gossypium hirsutum* L.) | Martins et al. (2019) |
| Self-propelled electric HTPP Platform | Height of plant | Wheat (*Triticum aestivum*) | Pérez-Ruiz et al. (2020) |
| MVS-Pheno | Traits related to height of plant and leaf | Maize (*Zea mays*) | Wu et al. (2020) |
| Rhizoslides | Root-related development traits | Maize (*Zea mays*) | Le Marié et al. (2014) |
| GrainScan | Grain size and color measurement | Wheat (*Triticum aestivum*), brachypodium (*Brachypodium distachyon*) | Whan et al. (2014) |
| Digital still color camera | Chlorophyll levels and nitrogen content in leaf | Rice (*Oryza sativa*) | Wang et al. (2014) |
| LEAF-E | Grass leaf growth measurement | Grasses such as maize (*Zea mays*), *Miscanthus* spp., and brachypodium (*Brachypodium distachyon*) | Voorend et al. (2014) |
| Time series RGB images | Flowering characters | Rice (*Oryza sativa*) | Guo et al. (2015) |

*Contd.*

**Table 1** *Contd.*

| *Platform* | *Traits recorded* | *Crop* | *References* |
|---|---|---|---|
| Stereo imaging | Leaf and canopy characters | Sugar beet (*Beta vulgaris*) | Müller-Linow et al. (2015) |
| X-ray, CT, and MRI | Root system architecture (RSA) | Common bean (*Phaseolus vulgaris*) | Metzner et al. (2015) |
| RGB digital imaging | Growth of shoot, biomass generation, adaptation to cold | Pea (*Pisum sativum*) | Humplík et al. (2015) |
| Phenoplant | Chlorophyll fluorescence | Arabidopsis (*Arabidopsis thaliana*) | Rousseau et al. (2015) |
| TriP | Motion estimation in leaf | Arabidopsis (*Arabidopsis thaliana*) | Greenham et al. (2015) |
| Remote sensing, UAP | Nitrogen levels | Maize (*Zea mays*) | Zaman-Allah et al. (2015) |
| CropSight | Crop performance and microclimate databases | Wheat (*Triticum aestivum*) | Reynolds et al. (2019) |
| HTFP-3D CSM | Height of plant | Wheat (*Triticum aestivum*) | Volpato et al. (2021) |
| 3D LiDAR platform | Seedling phenotyping | Maize (*Zea mays*) | Forero et al. (2022) |
| MVS-Pheno V2 | Height of plant, area covered by leaves, projected area, volume of shoot, compactness | Wheat (*Triticum aestivum*) | Wu et al. (2022) |
| FT-MIR | Total fatty acids | Chick pea (*Cicer arietinum* L.) | Madurapperumage et al. (2023) |
| MVS-Pheno | Genetic makeup and plant architecture | Maize (*Zea mays*) | Wu et al. (2024) |
| PREPs | Growth of plant, varietal difference on crop height, coverage, and volume index | Sugar beet (*Beta vulgaris*), potato (*Solanum tuberosum*) | Itoh et al. (2024) |
| FT-MIR | Nutritional and agronomic traits | Pulses | Madurapperumage et al. (2025) |

## 2.1 Optical Sensors

Optical sensors, including hyperspectral, multispectral, and RGB cameras, are central to HTP in plant research, offering noninvasive methods for capturing detailed plant traits and environmental conditions. Hyperspectral sensors collect data across a broad range of wavelengths, often hundreds of bands, providing fine-grained spectral information that can reveal subtle variations in plant physiology, such as the amount of chlorophyll, nutrient levels, stress due to increased or decreased water level, and symptoms due to diseases. These sensors enable early detection of stress before visible symptoms emerge, allowing for precise, timely interventions (Banerjee et al., 2020). Hyperspectral imagery is primarily useful in phenotyping complex characters such as canopy structure and photosynthetic efficiency by analyzing specific light reflectance patterns that are linked to metabolic and biochemical processes (Thenkabail et al., 2014).

Multispectral sensors, on the other hand, capture fewer wavelengths—typically under specific regions of spectrum, that is, visible and near-infrared (NIR), which makes them more affordable and suitable for large-scale field monitoring. These sensors are widely used in computing different vegetation indices (VIs) such as Normalized Difference Vegetation Index (NDVI), which provides a reliable measure of plant health and biomass (Huang et al., 2021). Multispectral data can quickly reveal information on crop vigor, canopy cover, and water stress, aiding in precision irrigation and fertilizer management. While they offer lower spectral resolution than hyperspectral sensors, multispectral cameras are valuable carrying out regular field inspections pertaining to their cost-effectiveness along with ease of integration into drone or UAV systems.

RGB cameras, which capture images in the standard RGB (red, green, and blue) color codes, are mainly brought to use in HTP to capture high quality images of external plant characters like height of plant, area under leaves, and number of leaves. These cameras are often integrated into drones or mobile platforms and are essential for monitoring growth dynamics in real time (Kim & Chung, 2021). While RGB imaging lacks the detailed spectral information of hyperspectral or multispectral sensors, it is invaluable for morphological assessments and is frequently combined with other sensor types to give a clear view of plant's overall well-being and development.

Together, these optical sensors enable efficient, high-throughput monitoring of plant performance across various environments, driving precision agriculture and sustainable farming methods by producing on-time, actionable information for effective field management.

## 2.2 Nonoptical Sensors

HTP plays a vital role in advancing farming practices plus improving crop yields. Nonoptical sensors, such as temperature, moisture, and mechanical sensors, are crucial in this context. They provide valuable data that can help researchers and farmers understand plant responses to environmental conditions, assess growth, and optimize resource usage.

### 2.2.1 Temperature Sensors

Temperature sensors are vital in monitoring plant microenvironments. They can be employed in various forms, including thermocouples and infrared sensors. These devices provide real-time data on ambient temperatures, leaf temperatures, and soil temperatures, which are critical for understanding plant stress responses and growth patterns. For example, thermal imaging, while optical, can complement temperature sensors by revealing heat variations that indicate water stress or disease. It has been found that integrating temperature data with phenotypic analyses can lead to more accurate predictions of crop performance under varying climate scenarios (Lee et al., 2021).

### 2.2.2 Moisture Sensors

Soil moisture sensors are essential in precision farming, allowing for on-time examination of water availability in the root zone. Technologies such as capacitance sensors and tensiometers provide crucial data on soil moisture dynamics. This information is vital for optimizing irrigation strategies, especially in water-scarce regions. Yin et al. (2021) demonstrated that soil moisture data could be effectively integrated into crop management systems, enhancing decision-making processes and improving yields by up to 30%. Moreover, these sensors can also be linked with weather data and other environmental factors to predict irrigation needs more accurately.

### 2.2.3 Mechanical Sensors

Mechanical sensors, including tensile and compressive sensors, measure plant traits related to structural integrity, growth patterns, and response to mechanical stimuli. These sensors can quantify traits such as stem stiffness and leaf area index, providing insights into plant resilience and adaptability. Current progress in flexible and lightweight sensor technologies enables continuous monitoring of plant's overall growth and mechanical stress response (Moulia et al., 2015). The integration of mechanical sensors with robotic platforms enhances the efficiency of phenotyping processes, allowing for rapid and precise measurements across large areas.

### 2.2.4 Integration and Data Processing

The integration of nonoptical sensors with cutting-edge data processing methods, such as machine learning (ML) and data analytics, is revolutionizing HTP, which helps in processing of vast datasets collected from various sensors, allowing researchers to unleash patterns and relationships which were previously difficult to detect. For instance, by combining data from temperature, moisture, and mechanical sensors, it becomes easy to give an overall scenario of plant's well-being and developmental condition, facilitating the inclusion of predictive models for crop performance (Lee et al., 2021).

## 2.3 UAVs and Satellite Imaging

UAVs and satellite imaging have emerged as powerful tools for this purpose, providing unique perspectives and capabilities that enhance our understanding of crop performance in real-time and over large areas (Araus & Cairns, 2014).

### 2.3.1 UAVs in Crop Phenotyping

UAVs, or drones, have become increasingly popular in agricultural research because of their flexibility, high spatial resolution, and ability to capture data at various wavelengths (Olson & Anderson, 2021; Tsouros et al., 2019). These can be outfitted with a number of sensors, such as RGB cameras, multi- and hyper-spectral sensors, and thermal cameras, allowing for comprehensive monitoring of crop health and development (Mulla, 2013; Deng et al., 2018).

- **Data Collection and Analysis:** UAVs can efficiently collect data on various phenotypic characters in plants such as height, leaf area index, and biomass. Meiyan et al. (2023) demonstrated effectiveness of UAV-based multispectral imaging in assessing nitrogen status in maize crops. By analyzing specific spectral indices, researchers were able to correlate the data with agronomic parameters, thus improving nitrogen management practices.
- **Mapping and Monitoring:** UAVs facilitate high-resolution mapping of crop fields, allowing for the identification of spatial variability within fields. This capability is crucial for precision agriculture, enabling targeted interventions. Zhou et al. (2021) highlighted how UAVs could identify stress zones in soybean fields, providing insights into irrigation and fertilization needs.
- **Temporal Analysis:** UAVs can conduct regular flights over the same area, allowing for the temporal analysis of crop growth and development. This ongoing monitoring helps researchers to track phenological stages, assess the impact of surrounding environment, and also to evaluate the effectiveness of overall management over time (Li et al., 2022).

## 2.4 Satellite Imaging in Crop Phenotyping

Satellite imaging complements UAV technology by offering extensive coverage and repeatability over larger geographic areas. With advancements in satellite technology, researchers can access high-resolution imagery at different spectral bands, which is instrumental for large-scale agricultural monitoring.

- **Large-Scale Monitoring:** Satellites can capture data over vast areas, making them ideal for assessing crop performance at regional or national scales. For instance, the Sentinel-2 satellite, a component of the European Space Agency's Copernicus program, provides high-resolution multispectral imagery that has been successfully utilized in monitoring growth habit in wheat and predicting yield (Goh et al., 2022).
- **Vegetation Indices:** Satellite imagery allows for computation of vegetation indices, including NDVI, that helps in assessing biomass, chlorophyll levels, and overall plant health. Nazir et al. (2021) showcased the utility of satellite-derived NDVI in their research for assessing the health

of rice crops, correlating the data with yield estimates and provided critical insights into crop management.

- **Integration with Ground Data:** The pooling of satellite visuals with ground-based measurements enhances accuracy of phenotypic assessments. Combining UAV and satellite data allows for more robust models that can predict crop performance based on environmental variables. Shen et al. (2022) illustrated how integrating satellite imagery with UAV data improved the prediction accuracy of wheat yield models, demonstrating the synergistic potential of these technologies.

While UAVs and satellite imaging offer significant advantages for high-throughput crop phenotyping, several challenges remain. These include data processing complexities, the need for advanced analytical tools, and the high costs associated with certain satellite platforms (Yang et al., 2020). Moreover, weather conditions can limit the effectiveness of both UAV and satellite imaging (Maes & Steppe, 2019). Future advancements are likely to focus on improving technologies related to sensors and fusion of data, and ML algorithms for enhanced data analysis. With the continuous evolution of remote sensing field, the amalgamation of UAVs with satellite imaging in crop phenotyping will become increasingly important for promoting sustainable agriculture and ensuring global security of food (Tripodi et al., 2018).

UAVs and satellite imaging are transforming high-throughput crop phenotyping by providing innovative tools for monitoring plant health and growth. Their combined capabilities allow for detailed analysis at multiple scales, improving insights into crop performance and supporting data-driven agricultural decision-making. With further progress in technology, these tools will play a major role in shaping the future of agricultural research and practices.

## *2.5 Ground-Based Phenotyping Systems*

Ground-based phenotyping systems tend to be essential in assessing various plant traits in controlled environments and field conditions. These systems utilize a range of technologies to gather detailed data on crop characteristics, aiding in breeding and research efforts.

### *2.5.1 Automated Phenotyping Platforms*

Automated phenotyping platforms consist of robotic systems that can measure multiple traits simultaneously. Automated phenotyping platforms are integral to high-throughput plant phenotyping, enabling efficient data collection on various traits while minimizing human labor and errors. These systems use advanced robotics, imaging technologies, and sensors to gather detailed phenotypic information. The Phenoscope is an advanced phenotyping platform designed for high-throughput plant phenotyping. It ensures consistent environmental conditions across samples, which is crucial for reliable data collection (Tisné et al., 2013). LemnaTec offers a range of automated phenotyping systems, including the Scanalyzer, which utilizes imaging and environmental control to assess plant growth and responses. It can measure traits such as biomass, root structure, and leaf morphology (Marko et al., 2018). A cutting-edge autonomous field robot, BoniRob, has been developed for precise phenotyping of individual plants. It allows for detailed data collection on various plant traits in real-world conditions, enhancing the understanding of plant growth and development. Key features include mobility across different terrains, advanced sensors for measuring attributes such as height and leaf area, and its versatility to operate under widely varying agricultural settings (Ruckelshausen et al., 2009). There has been development of many systems pertaining to field-based phenotyping, combining UAVs and ground sensors to collect comprehensive data on crop performance (Table 1). Automated phenotyping platforms represent a significant advancement in agricultural research, allowing for high-throughput collection of data and its valid analysis. These systems enhance overall efficiency and precision of phenotypic assessments, driving progress in crop improvement and sustainable agricultural practices.

### 2.5.2 3D Imaging Systems

The 3D imaging technologies capture detailed morphological information about plants, allowing researchers to analyze growth patterns and structural traits. The use of structured light scanning to generate 3D models of plants can provide insights into leaf architecture and overall plant structure (Vázquez-Arellano et al., 2016). There are several commercially available image sensing technologies that facilitate 3D mapping of plants. By the help of LiDAR (Light Detection and Ranging) technology, laser pulses are being utilized to calculate distances, creating much precise 3D models of plants and their environments. It's effective for capturing complex structures, including canopy layers. Stereo Vision, with the aid of equal to or more than two cameras, simulates human binocular vision, allowing the system to calculate depth and create 3D models based on the disparity between images. Ground penetrating radar (GPR) typically used for subsurface imaging can provide insights into root structures and soil properties, which are important for understanding plant health. Photonic mixer devices work on the principle of calculating the time light rays take in to and fro travel from target, enabling depth perception and 3D mapping. Time-of-Flight (ToF) cameras work by emitting light signals and calculate time period for the reflection back, producing depth information and facilitating 3D imaging of plant structures. Consumer–gaming interfaces originally designed for gaming, these devices can also capture 3D data and are used in some plant phenotyping applications due to their low cost and handiness (Li et al., 2014).

Laser scanning systems calculate the time period a laser beam takes to get reflected back from target object for capturing detailed 3D representations of plant structures, and are specifically important in assessing complex shapes, such as root systems (Paulus et al., 2014). Photogrammetry involves taking multiple 2D images of an object from all possible dimensions in order to build a three-dimensional model by the aid of different softwares. Under field-based phenotyping, this technique has gained popularity because of its flexibility and cost-effectiveness (An et al., 2017). CT scanning allows for nondestructive imaging of plant structures, and high-resolution 3D imaging of internal and external traits, providing detailed insights into internal morphology, particularly for root systems. These systems utilize X-ray technology to generate detailed cross-sectional images, enabling researchers to visualize complex root architectures and other morphological features without damaging the plants. CT scanning clarifies root development under varying environmental conditions and its potential for correlating internal structures with phenotypic traits. Zhang et al. (2023) highlighted the applications of CT in plant phenotyping, emphasizing its effectiveness in studying root morphology and understanding plant–soil interactions. Furthermore, the amalgamation of CT with ML algorithms is poised to enhance trait analysis, offering more accurate assessments of crop performance. Some advanced systems integrate multiple imaging technologies (e.g., RGB, hyperspectral, and 3D imaging) to give a comprehensive information of plant well-being and structure. The 3D imaging systems are pivotal in high-throughput phenotyping, enabling researchers to capture detailed morphological data and better understand plant responses to environmental factors.

### 2.5.3 Phenotyping Greenhouses

Phenotyping greenhouses are specialized environments designed to facilitate the high-throughput assessment of different characters in plants under protected environment. These facilities utilize progressive technologies such as automated imaging systems, sensors, climate control mechanisms for monitoring overall plant well-being, growth, developmental phases, and its response to various environmental factors. This controlled condition makes it possible to manipulate light, temperature, relative humidity, and fertility levels, making it possible to study plant behavior under specific conditions. Recent advancements include the introduction of robotics and ML, which improve collection of data and its analysis. These systems are crucial for accelerating plant breeding programs and improving crop resilience in changing climates.

## 2.6 Field-Based Platforms

Field-based phenotyping platforms are designed to assess crop traits in real-world agricultural settings, enabling researchers for monitoring overall well-being of plant and its growth under diverse environments and vast fields. These platforms often integrate various technologies, including UAVs, ground-operated sensors, mobile imaging systems, to collect high-resolution data on plant morphology, physiology, and stress responses. For example, systems such as the field phenotyping platform combine aerial imaging with ground sensors for giving an image of overall performance of crop, facilitating trait analysis such as canopy cover, biomass, and water stress. These platforms are useful in enhancing breeding programs and optimizing agronomic practices (Liu et al., 2016).

### 2.6.1 Handheld Phenotyping Devices

Handheld phenotyping devices have emerged as valuable tools in agricultural research and practice, enabling the quick evaluation of plant characteristics in field or controlled environments. These portable instruments facilitate immediate data collection, enabling researchers, agronomists, and farming community to make informal decisions using real-time plant health metrics.

#### 2.6.1.1 Chlorophyll Meters

Chlorophyll meters are popularly brought in use for measuring the chlorophyll levels in leaves, providing insights into plant health and nutrient availability. These devices operate by measuring the transmittance or reflectance of light in specific wavelengths that correlate with chlorophyll concentration. Instruments such as the Soil Plant Analysis Development (SPAD) meter provide quick, nondestructive assessments, allowing for timely interventions in nutrient management (Jhanji & Sekhon, 2018).

#### 2.6.1.2 Spectrometers

Handheld spectrometers measure the spectral reflectance of plant tissues, providing insights into biochemical compositions such as nitrogen, carotenoids, and water content. These devices capture data across multiple wavelengths, enabling the evaluation of plant health and stress responses based on spectral signatures (Ecarnot et al., 2013; Dinish et al., 2023).

#### 2.6.1.3 Leaf Area Meters

Leaf area meters measure the area of leaves quickly and nondestructively, providing essential data for calculating photosynthetic capacity and growth rates. Devices such as the Li-3100C can assess leaf area in various species, aiding in studies of leaf morphology and canopy structure (Yau et al., 2021).

#### 2.6.1.4 Thermal Imaging Cameras

Handheld thermal imaging devices measure leaf and canopy temperatures, providing insights into plant moisture level and temperature stress. By sensing temperature changes, these devices can identify areas of drought stress or disease, guiding irrigation and overall management (Pineda et al., 2020).

Handheld phenotyping devices are increasingly integrated into precision agriculture practices. Generating real-time data on plant health enables farming community to include location-specific management practices that optimize inputs for improving crop yield. For example, chlorophyll meters can guide nitrogen applications, ensuring that plants receive adequate nutrients without excess fertilizer use, thereby reducing environmental impacts. Furthermore, the integration of handheld devices with mobile apps and data management systems enhances data collection and its analysis, enabling improved implementation of decisions (Botta et al., 2022). By evaluating the data collected from these devices, farmers can monitor trends in crop health, assess the effectiveness of management practices, and make timely adjustments to their strategies. Handheld phenotyping devices are revolutionizing agricultural research and practices. Their ability to provide immediate, accurate measurements of plant traits is invaluable for optimizing crop management and improving yields. As technology advances, these devices will likely become even more sophisticated, integrating artificial intelligence (AI) and data analytics to provide core view of plant's well-being and performance. The ongoing development and adoption of handheld phenotyping tools will act as a valuable component in advancing sustainability of agriculture and food security.

Ground-based phenotyping systems are integral to modern agricultural research, enabling comprehensive assessments of plant traits under various conditions. These systems are continuously evolving, incorporating advanced technologies that enhance their accuracy and efficiency, thereby contributing significantly to crop improvement efforts.

## 2.7 Root Phenotyping Platforms

Root phenotyping techniques are essential for understanding plant's root system and its effect on nutrient utilization, water-use efficiency, and general plant well-being (Sharma et al., 2024).

### 2.7.1 Shovelomics

Shovelomics involves the physical excavation of soil to analyze root systems *in situ*. This method allows researchers to directly measure root depth, density, and morphology. The study carried out by Salungyu et al. (2020) showcased the vital relationship of root traits in maize plantlets with their performance under drought conditions, with notable relationships between seminal and crown root numbers and negative correlations in branching patterns. These findings suggest that seedling root traits can predict mature root characteristics and plant performance. Utilizing high-throughput root phenotyping tools could enhance breeding efficiency, particularly in developing countries. Arifuzzaman et al. (2019) phenotyped 216 different genotypes over two years, evaluating 5 root architectural traits using the shovelomics approach. A genome-wide association study (GWAS) found 31 significant marker loci related to characteristics such as diameter of taproot, 1° root branches, root angle, and root score. The results indicate that taproot diameter is a key determinant of canola root architecture, with several markers linked to multiple traits. Burridge et al. (2016) designed adaptable high-throughput platforms for root architectural phenotyping in common bean and cowpea, incorporating both lab and field evaluations. The research demonstrated strong correlations between lab and field measurements for key traits, with heritability for 51 traits varying significantly, particularly for whorl numbers in basal root and large hypocotyl roots. The integrated approach, which combined visual scoring, manual measurements, and image analysis, proved effective in differentiating genotypes and can be applied to other annual legumes and dicotyledonous crops.

### 2.7.2 X-Ray Computed Tomography (CT)

X-ray computed tomography is a noninvasive imaging technique that generates detailed 3D representations of root systems. This technique enables researchers to visualize root architecture without disturbing the soil. Plant phenomics is essential for linking the genotype to phenotype in breeding, with progressive imaging inventions such as X-ray CT offering noninvasive 3D root scans. To enhance data collection on complex root traits, Herrero-Huerta et al. (2022) introduced 4DRoot, an open-source software, designed to analyze 4D data from temporal X-ray CT scans, automatically extracting key root architectural characteristics. In this study, we investigated 3D root scans of black walnut tree to assess spatial and temporal root architecture. 4DRoot aims to assist breeders and root biologists in quantifying carbon sequestration and improving crop yields while reducing farming inputs.

Teramoto et al. (2020) designed an HTP workflow for 3D visualization of rice root system architecture using X-ray CT, focusing on radicle and crown roots. This process employs uniform calcined clay particles to minimize non-root visualization, along with optimized X-ray settings and advanced image processing techniques for efficient root isolation. The analysis takes only 10 minutes for scanning and reconstruction, enabling genetic analysis linked to 3D root system architecture (RSA) phenotyping, with 85%–100% detection of radicle and crown roots in two-week-old seedlings, respectively. Additionally, we successfully tracked daily RSA development over three weeks, demonstrating the feasibility of monitoring 4D root architecture without hindering rice growth due to low X-ray exposure.

### *2.7.3 Rhizotron Systems*

Rhizotron systems are specialized installations that enable researchers to get a complete overview of root growth process over time *in situ*. These systems consist of transparent tubes or pits where roots can be monitored under controlled conditions. The outcome of the study conducted by Michels et al. (2024) indicated that optimized spatially resolved electrical impedance tomography (sEIT) can effectively provide detailed insights about root biomass and surface area across various root systems. It was observed that there is strong relationship among total polarization strength, root biomass, and root surface area, suggesting that the polarization signature is influenced by cell-scale processes. In addition, the resolution characteristics of the measurement scheme significantly affect the accuracy of tomographic reconstructions of root traits. The findings highlight EIT to be a promising tool for high-throughput root phenotyping trials, offering a potential alternative to traditional, time-consuming root characterization methods. By doing simple root phenotyping with the help of rhizotrons, Gandullo et al. (2021) investigated the RSA of various tomato cultivars under salinity stress. The findings revealed high root plasticity in tomato seedlings, with significant modifications in root architecture, particularly in the positioning of lateral roots in response to salt accumulation at the soil surface. Among the cultivars studied, H1015 demonstrated the highest tolerance to salinity at both developmental stages. They also found root traits significantly correlated with aboveground growth parameters, suggesting that the ionic regulation and root architecture are crucial for imparting tolerance in plants to salt stress.

### *2.7.4 Hydroponic Systems*

Hydroponic systems make it easy to study the growth of roots under controlled nutrient solution, providing insights into root response to specific nutrients without soil interference. For example, researchers have used hydroponics to analyze the root architecture of legumes, studying how different nutrient concentrations affect root development. Qiao et al. (2019) revealed significant difference in root architectural characteristics among diverse maize genotypes, 28 days after transplanting, with 16 traits showing coefficients of variation exceeding 0.25, particularly for local root characteristics. Key characteristics such as total length of root, root depth, total shoot mass, and nodal root angle are recommended for inclusion in maize breeding programs. The findings provide a foundation for selecting genotypes with contrasting root architectures to enhance adaptation to specific environments in future breeding efforts. Salim et al. (2021) investigated RSA and shoot traits in 30 genotypes of soybean, using a semi-hydroponic system at 35 days of sowing, identifying significant variation in 11 of 13 traits measured. Validation in 1.5 m-deep rhizoboxes revealed positive significant correlation between root and shoot traits, with larger-rooted genotypes exhibiting higher harvest indices and increased leaf area and shoot dry weight. Notably, genotypes with larger root systems had longer flowering times and demonstrated phenotypic plasticity in traits such as depth of taproot and average root diameter. The findings suggest that genotypes with distinct root architectures could enhance soybean adaptation to drought and specific environmental conditions in future studies.

### *2.7.5 Root Scanning and Imaging*

Automated root scanning systems, such as RhizoVision, enable to rapidly assess root morphology by capturing images of root and analyzing characteristics such as total root length, surface area, and volume. This technique has found its place in various studies for evaluating root traits across different plant species and conditions. The RhizoVision Crown platform enables high-throughput phenotyping of crop root crowns, essential for marker-assisted breeding and genetic mapping. It combines an imaging unit with free, open-source software for efficient image capture and analysis, optimized for extracting measurements from numerous root crowns. Validation using extensive simulated images confirmed the platform's reliability, with 2,799 soybean and 1,753 wheat root crowns successfully phenotyped. Results showed heritability values of 0.74 for soybean and 0.22 for wheat, emphasizing the need to consider species and population differences in root trait studies (Seethepalli et al., 2020). The RhizoChamber-Monitor is a prototype robotic platform designed for automatic analysis of plant root growth patterns. It features an automatic imaging system that captures sequential images of roots growing on a substrate made of cloth in selected rhizoboxes, along with an irrigation system, which works without manual intervention and flexible shading arrangement. Customized image processing software analyzes the root growth dynamics spatially as well as temporally, quantifying overall growth and extracting detailed traits such as length and diameter of primary and lateral roots. Testing with two cotton genotypes validated both the hardware and software capabilities. This integrated platform aims to enhance root geometry quantification and support high-throughput root phenotyping for crop breeding focused on optimizing root architecture (Wu et al., 2018).

### *2.7.6 Field-Based Root Phenotyping*

This approach combines traditional field methods with modern technologies such as UAVs and ground-penetrating radar to assess root traits in real agricultural settings. Researchers have employed these technologies to correlate root traits with yield performance and environmental resilience. The breeding of novel crop varieties with deeper root system can significantly enhance resilience to climate change by improving water acquisition, nitrogen fixation, and carbon sequestration. To address this challenge of quantification of root phenotypes in field-grown crops, Digital Imaging of Root Traits (DIRT)/3D platform was introduced, capable of measuring 18 architectural traits in mature maize root crowns extracted using the shovelomics technique. DIRT/3D demonstrated high reliability, with a coefficient of determination ($r^2 > 0.84$) and broad-sense heritability ($H^2$ mean $> 0.6$) for nearly all traits, validated against manual measurements. The platform effectively distinguished genotypes based on the mean value of the traits and a comprehensive root architecture descriptor. Ultimately, DIRT/3D supports breeders and root biologists in enhancing crop resilience and food security amid the challenges of climate vulnerability (Liu et al., 2021). Fang et al. (2022) developed an image-based semi-automatic root phenotyping method for field-grown crops, enabling the extraction of five global and 40 local root traits. The method includes image acquisition, denoising, segmentation, trait extraction, and data analysis, demonstrating a high consistency in first-order lateral root branching with an $R^2$ of 0.97 compared to visual counts. Results indicated that interspecific advantages in maize were primarily observed within 5 cm from the root base, while soybean inhibition was notable within 20 cm. The approach offers high-throughput and accurate analysis of root morphology and branching features in field research. Ultimately, it supports 3D reconstruction of root systems to enhance data inputs for models simulating root growth and nutrient acquisition.

These root phenotyping techniques provide clear view of root development and functionality, ultimately leading to breeding of novel varieties in various crops with enhanced performance under varying environmental conditions.

## *2.8 Machine Learning and Artificial Intelligence Applications*

High-throughput crop phenotyping is revolutionizing agricultural research and practice, enabling scientists to gather pool of information related to different plant traits and growth conditions efficiently. The amalgamation of ML and AI into this field enhances data analysis, facilitates pattern recognition, and ultimately drives more informed decision-making in overall management of crops.

### *2.8.1 Improving Data Analysis with Machine Learning*

ML algorithms speed up at processing vast datasets, in order to make them ideal for analyzing the complex data generated in phenotyping studies. Techniques such as supervised and unsupervised learning are employed to identify patterns in phenotypic data, enabling researchers to correlate traits with genetic and environmental factors.

- **Trait Prediction:** A study by Duc et al. (2023) focused on using digital image analysis to predict 100 seed weight in soybean by measuring various seed architectural traits (i-traits) such as area size, perimeter length, and circularity. Seven machine learning algorithms were employed, with random forest and multiple linear regression emerging as the most accurate models for predicting seed weight, achieving coefficients of determination ($R^2$) of 0.98 and 0.94, respectively. The findings depict that the identified traits and genotypes could be valuable for soybean improvement programs.
- **Phenotypic Trait Extraction:** AI techniques, particularly convolutional neural networks (CNNs), are increasingly utilized for evaluating phenotypic traits from images captured via UAVs and other imaging technologies. The Median Normalized-Vectors Growth (MNVG) method was brought into use for testing thirty maize samples at different plant heights and developmental stages. The segmentation achieved high point-level accuracy, with recall, precision, F-score, and overall accuracy values of 0.92, 0.93, 0.92, and 0.93, respectively. Accuracy for phenotypic trait extraction varied from 0.81 to 0.95 for leaves, 0.64 to 0.97 for stems, and reached 1.0 for individual levels.
- **Disease Detection:** AI models can identify plant diseases from images, allowing for early intervention. Mohanty et al. (2016) highlighted the importance of deep learning in disease detection across various crops, achieving accuracies over 90%. Such capabilities are essential for managing crop health and minimizing losses.
- **Integration of Multi-Source Data:** Machine learning can integrate diverse datasets, including genomic, phenotypic, and environmental information, to enhance crop modeling. Pinto et al. (2023) used ML techniques to integrate soil moisture, temperature, and satellite imagery data to predict drought stress in wheat. This multifaceted approach allowed for more accurate assessments of crop resilience under varying environmental conditions.

### *2.8.2 AI for Decision Support in Crop Management*

Artificial intelligence enhances decision-making in agriculture by providing actionable insights based on data analysis. AI-powered decision support systems (DSS) are increasingly brought into use for optimizing crop management practices, improving efficiency and sustainability. AI applications in precision agriculture involve analyzing data from various sources, such as IoT sensors and UAV imagery, to provide real-time recommendations for irrigation, fertilization, and pest management. Kamilaris & Prenafeta-Boldú (2018) highlighted how AI can analyze environmental data to optimize water usage in vineyards, leading to improved yield and resource efficiency. AI technologies are also employed in automated phenotyping platforms that combine robotics with advanced imaging and data analytics. A study by Tripathi et al. (2023) devised an automatic system using AI for monitoring plant growth in controlled environments. The system utilized computer vision and ML algorithms for tracking growth metrics, allowing for timely interventions in plant care. AI is revolutionizing crop breeding by identifying genetic markers associated with breeding associated traits. Current progress in deep learning has enabled scientists to analyze genomic data alongside phenotypic traits, accelerating the breeding process. Chen et al. (2023) showcased how deep learning models can predict the phenotype from genotypic data, streamlining the breeding pipeline for improved crop varieties.

Despite the promising applications of ML and AI in high-throughput phenotyping, several challenges remain. Data quality and availability can hinder model performance, particularly in diverse agricultural environments. In addition, the complexity of complex ML models puts up challenges for workers aiming to uncover the biological mechanisms behind observed phenotypic traits. Further studies must prioritize creating reliable models that can generalize across different environments and crops. Enhancements in sensor technology, data collection methods, and computational power will further empower ML and AI applications in crop phenotyping. ML and AI are transforming high-throughput crop phenotyping, enabling researchers to analyze complex datasets, enhance decision-making, and accelerate crop improvement efforts. With the continuous evolution of these technologies, their integration into agricultural practices can significantly increase crop productivity and sustainability, addressing the rising need of global food security in the era of climate change.

## 3. Applications in Sustainable Agriculture

### *3.1 Crop Monitoring and Stress Detection*

Crop monitoring is indeed a powerful approach that integrates agricultural practices with progressive technologies to improve overall crop production. By utilizing sensing technologies, farmers can gather timely information related to soil, moisture levels, and crop well-being, allowing for proper management. Geographic information systems (GIS) help in mapping and analyzing spatial data, enabling better decision-making regarding land use and resource allocation (Abobatta, 2021). The Internet of Things (IoT) facilitates the connection of various devices and sensors, enabling continuous monitoring and data collection (Suma & Madhumathy, 2019). Robotics and drones can do tasks of crop sowing, spraying, and harvesting without manual intervention, significantly bringing down cost of cultivation and increasing efficiency. Together, these technologies, in addition of improving crop yield can also promote sustainability in farming activities by precise usage of resources and minimization of wastes. This holistic approach is revolutionizing agricultural field into a more data-driven and efficient industry (Rayhana et al., 2021, 2022).

HTP is crucial for monitoring crop health and detecting stress across various crops. HTP technologies allow for rapid and extensive data collection on various crops, providing insights into growth patterns and health status. It has been particularly effective in detecting abiotic stress in cereals (wheat, rice, barley), such as drought and nutrient deficiencies, through physiological metrics (Araus & Cairns, 2014). In legumes such as soybean, pea, etc., HTP technologies have been used to assess traits such as nodulation and drought tolerance, providing insights for breeding programs (Varshney et al., 2018). HTP is employed in fruit and vegetable crops to monitor growth stages and stress responses, optimizing conditions for yield (Abebe et al., 2023). HTP techniques are being used in root crops for analyzing development of roots and their response toward stress, crucial for breeding programs related to yield improvement and resilience (Bucksch et al., 2014). HTP methods are also applicable in ornamental crops, helping to monitor plant health and optimize growth conditions for esthetic and commercial value (Yang et al., 2020). There are several HTPs for detecting the crop health, such as PHENOPSIS that is used to assess stress related to soil water availability (Granier et al., 2006) and GROWSCREEN FLUORO for estimating growth of leaves and chlorophyll fluorescence (Jansen et al., 2009) in *Arabidopsis*. Recently, in wheat crop, PhenoField was used for monitoring various types of abiotic stresses (Beauchêne et al., 2019). Likewise in rice crop, PlantScreen$^{TM}$ Robotic XYZ system was used for evaluating traits governing tolerance to drought (Kim, 2020). Recent studies highlight the importance of integrating HTP data across various crop types to enhance understanding of stress responses and breeding strategies. Overall, high-throughput phenotyping is a transformative approach that enhances our ability to monitor crops effectively, detect stress early, and ultimately enhance agricultural productivity and sustainability.

### *3.2 Yield Prediction and Forecasting*

HTP performs a key role in predicting yield by providing detailed information related to phenotype that shows correlation with yield outcomes. This data helps in timely decision-making by associated community and researchers to optimize management process, and improve breeding programs. HTP technologies allow to collect large amount of information related to phenotype, such as crop height, leaf area, and plant canopy, which helps in identifying relationship between different traits and yield potential. Advanced statistical models and ML algorithms can be used to predict yield based on phenotypic data collected through HTP. These models can incorporate environmental data to enhance accuracy (Washburn et al., 2020; Zhang et al., 2024; Khaki, 2022). In wheat, along with drought resistance, Zhang et al. (2024) developed an advanced grain yield prediction model using 24 i-traits through machine learning approach. HTP allows to assess how crops survive under different types of stress conditions, which is essential for predicting how these stresses might impact overall yield. Sabag et al. (2024) investigated the plant height, leaf area index, and vegetation indices by combining HTP and random regression model and correlated the genomic prediction models with plant traits for multi-trait genomic prediction. The UAV-generated multispectral images were investigated through PROSAIL model for linking the VIs of Miscanthus and yield predictions made using the VIs time series for enhancing the sustainable biomass production (Impollonia et al., 2022). By integrating HTP into breeding programs, researchers can select the traits associated with higher yield potential, resulting in the formation of improved cultivars, more resilient to environmental challenges (Bhat et al., 2020). HTP supports precision agriculture practices by providing real-time data that helps in optimizing inputs such as water, fertilizers, and pesticides based on predicted yield outcomes (Sishodia et al., 2020). HTP is revolutionizing yield prediction and forecasting in agriculture. By providing detailed data on phenotypic traits, enabling advanced modeling techniques, and integrating with precision agriculture, HTP enhances the ability to predict crop yields accurately.

## *3.3 Plant Breeding and Genetics*

The main goal of plant breeding is to create high-yielding, superior crop varieties, which could fit under diverse climatic conditions without yield penalty. HTP acts as a transformative agent in plant breeding and genetics, by allowing quick and detailed assessment of phenotypic characters, which can be correlated with genetic data. This integration enhances selection efficiency and accelerates the development of improved crop varieties. It is a nondestructive approach and generates huge temporal and spatial data of plant traits through machines and gadgets, that is, cameras, sensors, rays, etc. (Sheikh et al., 2024). Nowadays, the use of AI and ML learning approaches provides great advantages in combination with HTP for crop improvement. These approaches are helpful in precise plant phenotyping for complex traits such as abiotic stress and yield, which are highly influenced by environmental factors, agronomic management practices, genotype, and human interference. HTP allows for the rapid measurement of various phenotypic traits, facilitating the evaluation of large populations and reducing the time required for breeding cycles. The data generated through high HTPs could be utilized for many purposes based upon the approaches such as linkage mapping, association mapping, genomic selection, QTLseq, etc., for identifying the specific quantitative trait loci (QTLs)/alleles/genes or biochemical/physiological pathways linked with agronomic traits, stress resistance/tolerance, etc. (Fig. 1). In barley, HTP facility "The Plant Accelerator, Adelaide, Australia" was used for phenotyping of drought stress, and further QTLs governing tolerance toward drought stress were highlighted from 47 wild juvenile barley introgression lines (Honsdorf et al., 2014). The stable QTLs linked with grain protein in rice were identified through high-throughput phenotyping and genotyping (Chattopadhyay et al., 2019). Similarly, Yang et al. (2014), using rice automatic phenotyping (RAP) platform, phenotyped the Recombinant Inbred Lines (RIL) population and further identified 141 loci with GWAS techniques. Zhang et al. (2017) phenotyped RIL population of maize (n=167) using RAP platform and 988 QTLs for 106 traits were investigated. It enables the establishment of genotype–phenotype relationships, helping to link specific genetic variations with observable traits. By providing extensive phenotypic data, HTP supports marker-assisted selection and genomic selection, enhancing the efficiency of breeding programs. HTP allows for the assessment of disease-resistance traits in breeding populations, helping to identify resilient genotypes. HTP provides data that can refine breeding strategies by identifying traits that contribute most significantly to yield and quality. High-throughput phenotyping is reshaping plant breeding and genetics by facilitating rapid and precise assessments of phenotypic traits. This technology enhances genotype–phenotype mapping, improves genetic selection, and aids in disease resistance assessment, ultimately leading to more efficient and effective breeding programs. The integration of HTP with genomic approaches is vital for making climate-adaptable and high-yielding crop cultivars.

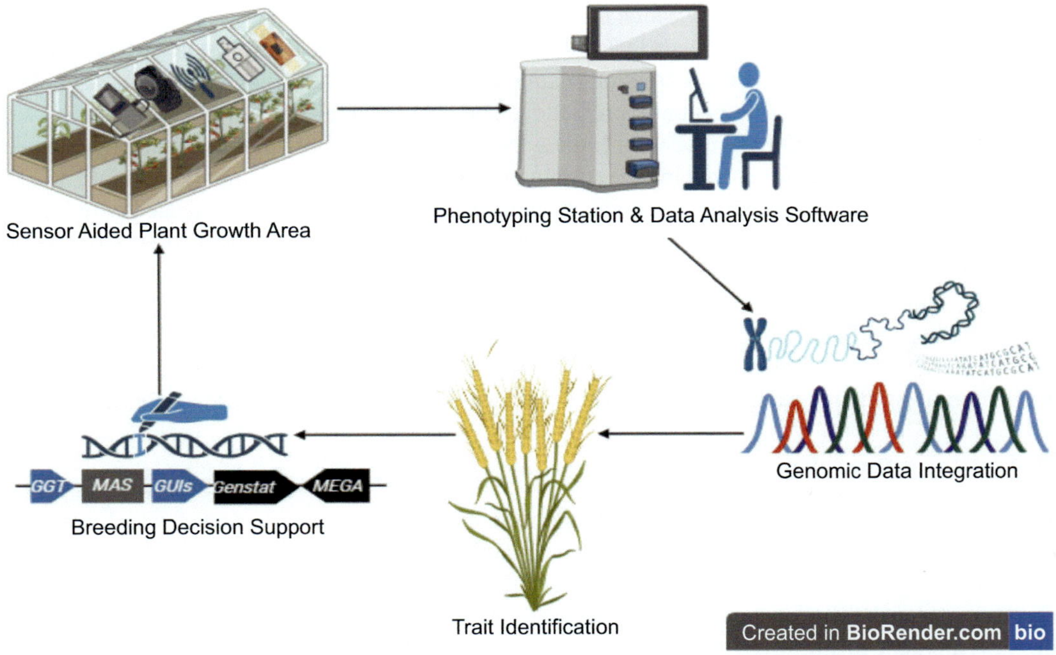


**Fig. 1** Accelerated crop improvement through the integration of advanced phenotyping platforms and molecular approaches.

## *3.4 Precision Agriculture*

HTP acts as a key player in precision agriculture and decision support system by providing detailed and timely phenotypic data that enhance management practices and inform decision-making. This integration helps optimize resource use, improve crop monitoring, and ultimately increase agricultural productivity and sustainability. In comparison to plant breeding, the need for phenotyping for precision agriculture is different. HTP technologies facilitate the collection of timely information related to plant well-being, growth, and development, enabling timely interventions. HTP aids in the precise application of inputs such as irrigation water, fertilizers and manures, and pesticides as per detailed crop needs, optimizing resource use, while in plant breeding HTPs assist the plant breeders in making a selection of suitable genotypes. This reduces the risk of overapplication, minimizing environmental impact, and lowering costs for farmers. For instance, through the use of remote sensing, farmers can identify zones of the field where nutrient deficiencies or water stress are present, enabling targeted irrigation and fertilization strategies. HTP systems can detect variations in soil moisture, plant transpiration, and canopy temperature (Mulla, 2013). This allows farmers to implement precision irrigation, where irrigation is done only if required and that too at specific locations in field, conserving water resources and ensuring crops get optimal conditions (Sishodia et al., 2020). Prior detection of abiotic stress (drought, salinity, or nutrient stress) and biotic stress (insect pest attack and diseases) is made possible using HTP (Kashyap & Kumar, 2021). By monitoring plant health in real-time with sensors and imaging technologies, farmers can detect stress at an early stage and take timely action to prevent yield loss, reducing the need for broad pesticide or herbicide application (Sishodia et al., 2020). HTP is integral in advancing precision farming. By providing timely information and enabling precise resource management, HTP enhances crop health monitoring, informs decision-making, and improves yield predictions. This approach enhances sustainability and efficiency in agriculture, tackling the issues of rising world population and shift in global climate.

### 3.5 Decision Support System

HTP performs a pivotal role in DSS for sustainable agriculture by enabling precise, real-time monitoring of plant traits at large scales. Through advanced imaging technologies such as drones, hyperspectral sensors, and remote sensing platforms, HTP provides detailed insights into crop health, growth patterns, and response of plants to different kinds of stress-causing agents such as drought, salinity, and pests. This phenotypic data can be integrated into DSS to use resources in a judicious way, promoting precision agriculture. For instance, HTP allows for early detection of nutrient deficiencies, enabling targeted fertilization, thus reducing the overapplication of fertilizers and minimizing environmental pollution (Fiorani & Schurr, 2013). Additionally, by assessing plant stress and disease symptoms early, DSS can recommend precise interventions, such as reduced pesticide use, preserving beneficial organisms, and minimizing chemical exposure (Kuska & Mahlein, 2018).

HTP also aids in breeding programs by identifying varieties more tolerant to abiotic stresses than others, thus supporting climate-resilient agriculture (Naik et al., 2017). This becomes more important in the changing global climate era, as crops must adapt to shifting environmental conditions. Furthermore, HTP-based DSS contribute to sustainable farm management by providing actionable insights that improve yield predictions and optimize input applications, leading to cost savings and reduced environmental impacts (Carroll, 2023). Ultimately, HTP within DSS in addition of enhancing productivity also ensures sustainability in farming activities by minimizing wastage of resources, reducing environmental degradation, and improving farm profitability (Zhai et al., 2020).

## 4. Challenges and Limitations for HTP Tools

HTP tools have significantly advanced plant research and breeding, but they also face several challenges and limitations that can impact their effectiveness.

### 4.1 Data Management and Analysis

The pool of information gathered by HTP tools can be overwhelming. Effective management, processing, and analysis of large datasets require advanced computational resources and expertise in data analytics. Inadequate data management can lead to difficulties in extracting meaningful insights from the data collected (Fiorani & Schurr, 2013).

### 4.2 Cost and Accessibility

Many advanced phenotyping systems, such as UAVs and imaging technologies, can be prohibitively expensive. This high cost can limit access for smaller research institutions and farmers, creating disparities in technology availability. This issue restricts the adoption of HTP tools, especially in developing regions (Hassler & Baysal-Gurel, 2019).

### 4.3 Environmental Variability

Phenotyping outcomes are under significant influence of surrounding environment such as soil type, moisture, and weather conditions, which could result in inappropriate outcomes and complicate the identification of reliable trait associations (Yang et al., 2020).

### 4.4 Calibration and Standardization

Different phenotyping tools often require specific calibration protocols to ensure accurate measurements. The lack of standardized calibration methods can result in variations in data quality, complicating comparisons across studies and limiting reproducibility (Araus & Cairns, 2014).

### 4.5 Complexity of Plant Traits

The complexity of plant traits, influenced by multiple genetic and environmental factors, poses a challenge for HTP tools. These tools may struggle to capture intricate interactions among traits, particularly those related to root architecture and physiological responses, making it difficult to link specific traits to desired outcomes (Pauli et al., 2016).

### 4.6 User Expertise

The result-oriented use of HTP technology often requires specialized training in data analysis and imaging techniques. Unskilled persons can deter the successful introduction of these tools in research and agricultural practices (Yang et al., 2020).

### 4.7 Integration with Breeding Programs

While HTP tools provide valuable data, integrating this information into traditional breeding programs can be challenging. Breeders may need to adjust their methodologies and workflows to effectively incorporate phenotyping data into their selection processes (Varshney et al., 2021).

Despite the significant advancements offered by high-throughput phenotyping tools, coping up with these issues is vital for utilizing their potential in agriculture and plant research. Continued development in data management, standardization, and user training will be needed to fully reap the benefits of HTP technologies in breeding resilient and high-yielding crop varieties.

## 5. Future Directions

HTP tools are pivotal in advancing sustainable agriculture by enhancing our knowledge about different plant characteristics and their response to environmental challenges. As these technologies evolve, several future directions can significantly contribute to their effectiveness in promoting sustainable practices.

### 5.1 Integration of AI and ML

The integration of artificial intelligence and machine learning algorithms can enhance data analysis capabilities, allowing for the extraction of actionable insights from large datasets generated by HTP tools. These technologies can improve the prediction of plant performance and help in identifying optimal traits for breeding programs. For instance, using AI-driven models can facilitate more accurate trait associations and selection criteria (Gupta et al., 2021).

### 5.2 Multi-Omics Approaches

Integrating high-throughput phenotyping with genomics, transcriptomics, and metabolomics can give a comprehensive image of plant responses to stressors. This multi-omics approach can help identify key genetic loci associated with desirable traits, enabling breeders to develop more resilient crop varieties. Research shows that combining HTP with genomic data can enhance selection efficiency (Varshney et al., 2021).

### 5.3 Enhanced Sensor Technologies

Advancements in sensor technologies, including hyperspectral and thermal imaging, will improve the precision and accuracy of phenotyping. These tools can gather timely information related to overall well-being of plant, water status, and nutrient levels, allowing timely interventions. Continuous improvements in sensor miniaturization and affordability will broaden their accessibility to farmers and researchers (Fiorani & Schurr, 2013).

### 5.4 Field-Based and Remote Sensing Technologies

The integration of UAVs and satellite imaging in high-throughput phenotyping allows for large-scale assessments of crop performance across diverse environments. Future research should focus on enhancing the spatial and temporal resolution of remote sensing technologies to give deeper insights into crop dynamics. This approach can facilitate precision agriculture practices, helping farmers make data-driven decisions (Feng et al., 2021).

### 5.5 User-Friendly Interfaces and Decision Support Systems

Developing user-friendly interfaces and DSSs that integrate phenotyping data can empower farmers to take necessary actions for managing crops. These tools offer insights into optimal planting times, nutrient applications, and irrigation strategies, promoting more sustainable practices (Sanchez, 2024).

### 5.6 Emphasis on Soil–Root Interactions

Future phenotyping efforts should place greater emphasis on understanding soil–root interactions and their role in plant health and productivity. By integrating root phenotyping technologies, researchers can better assess how root architecture influences nutrient uptake and tolerance to drought, leading to the introduction of crop varieties, better suited for varying soil types and conditions (Sharma et al., 2024).

The progress of HTP tools lies in their ability to integrate with other technologies and disciplines, providing comprehensive insights into plant performance and resilience. By focusing on AI integration, multi-omics approaches, advanced sensor technologies, and user-friendly systems, HTP can significantly contribute to sustainable agricultural practices, enhancing food security while minimizing environmental impacts.

## 6. Conclusion

HTP is getting more popular as a vital component in achieving sustainability goal in agriculture. By enabling quick and precise understanding of plant traits, HTP tools provide critical insights that support timely decision-making in crop management and breeding programs. HTP facilitates the uncovering of genes associated with stress tolerance, drought resistance, and nutrient efficiency. By breeding for these traits, crop varieties performing well in changing climatic conditions can be developed, which could alleviate the challenges of global food security. The ability to monitor plant health in real time allows for site-specific management strategies that minimize inputs such as water and fertilizers. This optimization boosts agricultural productivity and in addition reduces the environmental footprint of farming practices. HTP, when combined with genomic and molecular tools, enhances the validity of breeding programs. This integration enables the rapid unleashing of genetic loci associated with desirable traits, accelerating the production of crops, adaptable in diverse environments. HTP tools empower farmers with data-driven insights, allowing for precise interventions that improve crop performance. This approach fosters sustainable practices, ensuring that agricultural systems are both productive and environmentally responsible. By understanding how crops respond to various stressors, stakeholders can develop strategies that align with sustainable development goals and enhance global food security. High-throughput phenotyping is essential for advancing sustainable agriculture. By leveraging the capabilities of HTP, we can address critical issues such as changing climate, resource depletion, and food scarcity. As these technologies continue to evolve, their integration into agricultural practices will be helpful in shaping a more sustainable and balanced food system for the future.

## References


Abebe, A.M., Kim, Y., Kim, J., Kim, S.L. and Baek, J. (2023). Image-based high-throughput phenotyping in horticultural crops. *Plants*, 12(10), 2061.

Abobatta, W.F. (2021). Precision agriculture: A new tool for development. In: *Precision Agriculture Technologies for Food Security and Sustainability* (pp. 23–45). IGI Global.

An, N., Welch, S.M., Markelz, R.C., Baker, R.L., Palmer, C.M., Ta, J. and Weinig, C. (2017). Quantifying time-series of leaf morphology using 2D and 3D photogrammetry methods for high-throughput plant phenotyping. *Computers and Electronics in Agriculture*, 135, 222–232.

Araus, J.L. and Cairns, J.E. (2014). Field high-throughput phenotyping: the new crop breeding frontier. *Trends in plant science*, 19(1), 52–61.

Arifuzzaman, M., Oladzadabbasabadi, A., McClean, P. and Rahman, M. (2019). Shovelomics for phenotyping root architectural traits of rapeseed/canola (Brassica napus L.) and genome-wide association mapping. *Molecular Genetics and Genomics*, *294*, 985–1000.

Asaari, M.S.M., Mertens, S., Dhondt, S., Inzé, D., Wuyts, N. and Scheunders, P. (2019). Analysis of hyperspectral images for detection of drought stress and recovery in maize plants in a high-throughput phenotyping platform. *Computers and Electronics in Agriculture*, 162, 749–758.

Banerjee, B.P., Joshi, S., Thoday-Kennedy, E., Pasam, R.K., Tibbits, J., Hayden, M. and Kant, S. (2020). High-throughput phenotyping using digital and hyperspectral imaging-derived biomarkers for genotypic nitrogen response. *Journal of Experimental Botany*, 71(15), 4604–4615.

Beauchêne, K., Leroy, F., Fournier, A., Huet, C., Bonnefoy, M., Lorgeou, J., DeSolan, B., Piquemal, B., Thomas, S. and Cohan, J.P. (2019). Management and characterization of abiotic stress via PhénoField®, a high-throughput field phenotyping platform. *Frontiers in Plant Science*, 10, 904.

Bergsträsser, S., Fanourakis, D., Schmittgen, S., Cendrero-Mateo, M.P., Jansen, M., Scharr, H. and Rascher, U. (2015). HyperART: non-invasive quantification of leaf traits using hyperspectral absorption-reflectance-transmittance imaging. *Plant Methods*, 11, 1–17.

Bhat, J.A., Deshmukh, R., Zhao, T., Patil, G., Deokar, A., Shinde, S. and Chaudhary, J. (2020). Harnessing high-throughput phenotyping and genotyping for enhanced drought tolerance in crop plants. *Journal of Biotechnology*, 324, 248–260.

Botta, A., Cavallone, P., Baglieri, L., Colucci, G., Tagliavini, L. and Quaglia, G. (2022). A review of robots, perception, and tasks in precision agriculture. *Applied Mechanics*, 3(3), 830–854.

Boureau, T. (2020). PHENOTIC platform. Portail Data INRAE. https://doi.org/10.15454/u2bwf

Brichet, N., Fournier, C., Turc, O., Strauss, O., Artzet, S., Pradal, C. and Cabrera-Bosquet, L. (2017). A robot-assisted imaging pipeline for tracking the growths of maize ear and silks in a high-throughput phenotyping platform. *Plant Methods*, 13, 1–12.

Bucksch, A., Burridge, J., York, L.M., Das, A., Nord, E., Weitz, J.S. and Lynch, J.P. (2014). Image-based high-throughput field phenotyping of crop roots. *Plant Physiology*, 166(2), 470–486.

Burridge, J., Jochua, C.N., Bucksch, A. and Lynch, J.P. (2016). Legume shovelomics: high—throughput phenotyping of common bean (Phaseolus vulgaris L.) and cowpea (Vigna unguiculata subsp, unguiculata) root architecture in the field. *Field Crops Research*, 192, 21–32.

Busemeyer, L., Mentrup, D., Möller, K., Wunder, E., Alheit, K., Hahn, V. and Ruckelshausen, A. (2013). BreedVision—A multi-sensor platform for non-destructive field-based phenotyping in plant breeding. *Sensors*, 13(3), 2830–2847.

Cabrera-Bosquet, L., Fournier, C., Brichet, N., Welcker, C., Suard, B. and Tardieu, F. (2016). High-throughput estimation of incident light, light interception and radiation-use efficiency of thousands of plants in a phenotyping platform. *New Phytologist*, 212(1), 269–281.

Carroll, M.E. (2023). Development of digital phenotyping methods for breeding applications in stress phenotyping, spatial adjustments, and end season yield prediction (Doctoral dissertation, Iowa State University).

Chattopadhyay, K., Behera, L., Bagchi, T.B., Sardar, S.S., Moharana, N., Patra, N.R., Chakraborti, M., Das, A., Marndi, B.C., Sarkar, A. and Ngangkham, U. (2019). Detection of stable QTLs for grain protein content in rice (*Oryza sativa* L.) employing high throughput phenotyping and genotyping platforms. *Scientific Reports*, 9(1), 3196.

Chawade, A., van Ham, J., Blomquist, H., Bagge, O., Alexandersson, E. and Ortiz, R. (2019). High-throughput field-phenotyping tools for plant breeding and precision agriculture. *Agronomy*, 9(5), 258.

Chen, L., Chen, Z., Zhang, Y., Liu, Y., Osman, A.I., Farghali, M. and Yap, P.S. (2023). Artificial intelligence-based solutions for climate change: a review. *Environmental Chemistry Letters*, 21(5), 2525–2557.

Crain, J.L., Wei, Y., Barker III, J., Thompson, S.M., Alderman, P.D., Reynolds, M. and Poland, J. (2016). Development and deployment of a portable field phenotyping platform. *Crop Science*, 56(3), 965–975.

Czedik-Eysenberg, A., Seitner, S., Güldener, U., Koemeda, S., Jez, J., Colombini, M. and Djamei, A. (2018). The 'PhenoBox', a flexible, automated, open-source plant phenotyping solution. *New Phytologist*, 219(2), 808–823.

Deng, L., Mao, Z., Li, X., Hu, Z., Duan, F. and Yan, Y. (2018). UAV-based multispectral remote sensing for precision agriculture: A comparison between different cameras. *ISPRS Journal of Photogrammetry and Remote Sensing*, 146, 124–136.

Dinish, U.S., Teng, M.T.J., Xinhui, V.T., Dev, K., Tan, J.J., Koh, S.S. and Olivo, M. (2023). Miniaturized Vis–NIR handheld spectrometer for non-invasive pigment quantification in agritech applications. *Scientific Reports*, 13(1), 9524.

Duc, N.T., Ramlal, A., Rajendran, A., Raju, D., Lal, S.K., Kumar, S. and Chinnusamy, V. (2023). Image-based phenotyping of seed architectural traits and prediction of seed weight using machine learning models in soybean. *Frontiers in Plant Science*, 14, 1206357.

Ecarnot, M., Compan, F. and Roumet, P. (2013). Assessing leaf nitrogen content and leaf mass per unit area of wheat in the field throughout plant cycle with a portable spectrometer. *Field Crops Research*, 140, 44–50.

Fang, H.U.I., Zi-Wen, X.I.E., Hai-Gang, L.I., Yan, G.U.O., Bao-Guo, L.I., Yun-Ling, L.I.U. and Yun-Tao, M.A. (2022). Image-based root phenotyping for field-grown crops: An example under maize/soybean intercropping. *Journal of Integrative Agriculture*, 21(6), 1606–1619.

Feng, L., Chen, S., Zhang, C., Zhang, Y. and He, Y. (2021). A comprehensive review on recent applications of unmanned aerial vehicle remote sensing with various sensors for high-throughput plant phenotyping. *Computers and Electronics in Agriculture*, 182, 106033.

Fiorani, F. and Schurr, U. (2013). Future scenarios for plant phenotyping. *Annual Review of Plant Biology*, 64(1), 267–291.

Flood, P.J., Kruijer, W., Schnabel, S.K., van der Schoor, R., Jalink, H., Snel, J.F. and Aarts, M.G. (2016). Phenomics for photosynthesis, growth and reflectance in Arabidopsis thaliana reveals circadian and long-term fluctuations in heritability. *Plant Methods*, 12, 1–14.

Forero, M.G., Murcia, H.F., Méndez, D. and Betancourt-Lozano, J. (2022). LiDAR platform for acquisition of 3D plant phenotyping database. *Plants*, 11(17), 2199.

Fu, X. & Jiang, D. (2022). High-throughput phenotyping: The latest research tool for sustainable crop production under global climate change scenarios. *In*: *Sustainable Crop Productivity and Quality Under Climate Change* (pp. 313–381). Academic Press.

Fukatsu, T., Watanabe, T., Hu, H., Yoichi, H. and Hirafuji, M. (2012). Field monitoring support system for the occurrence of Leptocorisa chinensis Dallas (Hemiptera: Alydidae) using synthetic attractants, field servers, and image analysis. *Computers and Electronics in Agriculture*, *80*, 8–16.

Gaju, O., Basso, B. and Rosales, M.A. (2021). Utilizing high-throughput phenotyping in precision agriculture: Challenges and perspectives. *Agriculture*, 11(5), 451.

Gandullo, J., Ahmad, S., Darwish, E., Karlova, R. and Testerink, C. (2021). Phenotyping tomato root developmental plasticity in response to salinity in soil rhizotrons. Plant Phenomics.

Goh, B.B., King, P., Whetton, R.L., Sattari, S.Z. and Holden, N.M. (2022). Monitoring winter wheat growth performance at sub-field scale using multitemporal Sentinel-2 imagery. *International Journal of Applied Earth Observation and Geoinformation*, 115, 103124.

Gosseau, F., Blanchet, N., Varès, D., Burger, P., Campergue, D., Colombet, C. and Langlade, N. (2019). Heliaphen, an outdoor high-throughput phenotyping platform for genetic studies and crop modeling. *Frontiers in Plant Science*, 9, 1908.

Granier, C., Aguirrezabal, L., Chenu, K., Cookson, S.J., Dauzat, M., Hamard, P., Thioux, J.J., Rolland, G., Bouchier-Combaud, S., Lebaudy, A. and Muller, B. (2006). PHENOPSIS, an automated platform for reproducible phenotyping of plant responses to soil water deficit in Arabidopsis thaliana permitted the identification of an accession with low sensitivity to soil water deficit. *New Phytologist*, 169(3), 623–635.

Greenham, K., Lou, P., Remsen, S.E., Farid, H. and McClung, C.R. (2015). TRiP: Tracking rhythms in plants, an automated leaf movement analysis program for circadian period estimation. *Plant Methods*, 11, 1–11.

Guo, W., Fukatsu, T. and Ninomiya, S. (2015). Automated characterization of flowering dynamics in rice using field-acquired time-series RGB images. *Plant Methods*, 11, 1–15.

Gupta, R., Srivastava, D., Sahu, M., Tiwari, S., Ambasta, R. K. and Kumar, P. (2021). Artificial intelligence to deep learning: machine intelligence approach for drug discovery. *Molecular Diversity*, 25, 1315–1360.

Hairmansis, A., Berger, B., Tester, M. and Roy, S.J. (2014). Image-based phenotyping for non-destructive screening of different salinity tolerance traits in rice. *Rice*, 7, 1–10.

Hassler, S.C. and Baysal-Gurel, F. (2019). Unmanned aircraft system (UAS) technology and applications in agriculture. *Agronomy*, 9(10), 618.

Herrero-Huerta, M., Raumonen, P. and Gonzalez-Aguilera, D. (2022). 4DRoot: Root phenotyping software for temporal 3D scans by X-ray computed tomography. *Frontiers in Plant Science*, 13, 986856.

Honsdorf, N., March, T.J., Berger, B., Tester, M. and Pillen, K. (2014). High-throughput phenotyping to detect drought tolerance QTL in wild barley introgression lines. *PLoS One*, 9(5), e97047.

Huang, S., Tang, L., Hupy, J.P., Wang, Y. and Shao, G. (2021). A commentary review on the use of normalized difference vegetation index (NDVI) in the era of popular remote sensing. *Journal of Forestry Research*, 32(1), 1–6.

Humplík, J.F., Lazár, D., Fürst, T., Husičková, A., Hýbl, M. and Spíchal, L. (2015). Automated integrative high-throughput phenotyping of plant shoots: a case study of the cold-tolerance of pea (Pisum sativum L.). *Plant Methods*, 11, 1–11.

Hunter, M.C., Smith, R.G., Schipanski, M.E., Atwood, L.W. and Mortensen, D.A. (2017). Agriculture in 2050: recalibrating targets for sustainable intensification. *Bioscience*, 67(4), 386–391.

Impollonia, G., Croci, M., Ferrarini, A., Brook, J., Martani, E., Blandinières, H., ... and Amaducci, S. (2022). UAV remote sensing for high-throughput phenotyping and for yield prediction of miscanthus by machine learning techniques. *Remote Sensing*, 14(12), 2927.

Itoh, A., Njane, S. N., Hirafuji, M. and Guo, W. (2024). PREPs: An open-source software for high-throughput field plant phenotyping. *Plant Phenomics*, 6, 0221.

Jangra, S., Chaudhary, V., Yadav, R.C. and Yadav, N.R. (2021). High-throughput phenotyping: a platform to accelerate crop improvement. *Phenomics*, 1(2), 31–53.

Jansen, M., Gilmer, F., Biskup, B., Nagel, K.A., Rascher, U., Fischbach, A., Briem, S., Dreissen, G., Tittmann, S., Braun, S. and De Jaeger, I. (2009). Simultaneous phenotyping of leaf growth and chlorophyll fluorescence via GROWSCREEN FLUORO allows detection of stress tolerance in Arabidopsis thaliana and other rosette plants. *Functional Plant Biology*, 36(11), 902–914.

Jeudy, C., Adrian, M., Baussard, C., Bernard, C., Bernaud, E., Bourion, V. and Salon, C. (2016). RhizoTubes as a new tool for high throughput imaging of plant root development and architecture: test, comparison with pot grown plants and validation. *Plant Methods*, 12, 1–18.

Jhanji, S. and Sekhon, N.K. (2018). Evaluation of potential of portable chlorophyll meter to quantify chlorophyll and nitrogen contents in leaves of wheat under different field conditions. *Indian Journal of Experimental Biology*, 56, 750–758.

Kamilaris, A. and Prenafeta-Boldú, F.X. (2018). Deep learning in agriculture: A survey. *Computers and Electronics in Agriculture*, 147, 70–90.

Kashyap, B. and Kumar, R. (2021). Sensing methodologies in agriculture for monitoring biotic stress in plants due to pathogens and pests. *Inventions*, 6(2), 29.

Khaki, S. (2022). *Novel deep learning-based methods for high-throughput image-based plant phenotyping and large scale crop yield prediction* (Doctoral dissertation, Iowa State University).

Kim, J. and Chung, Y.S. (2021). A short review of RGB sensor applications for accessible high-throughput phenotyping. *Journal of Crop Science and Biotechnology*, 24, 495–499.

Kim, J.Y. (2020). Roadmap to high throughput phenotyping for plant breeding. *Journal of Biosystems Engineering*, 45, 43–55.

Kim, S.L., Kim, N., Lee, H., Lee, E., Cheon, K.S., Kim, M. and Kim, K.H. (2020). High-throughput phenotyping platform for analyzing drought tolerance in rice. *Planta*, 252(3), 38.

Kloth, K.J., Ten Broeke, C.J., Thoen, M.P., Hanhart-van den Brink, M., Wiegers, G.L., Krips, O.E. and Jongsma, M.A. (2015). High-throughput phenotyping of plant resistance to aphids by automated video tracking. *Plant Methods*, 11, 1–14.

Kuska, M.T. and Mahlein, A.K. (2018). Aiming at decision making in plant disease protection and phenotyping by the use of optical sensors. *European Journal of Plant Pathology*, 152, 987–992.

Le Marié, C., Kirchgessner, N., Flütsch, P., Pfeifer, J., Walter, A. and Hund, A. (2016). RADIX: rhizoslide platform allowing high throughput digital image analysis of root system expansion. *Plant Methods*, 12, 1–15.

Le Marié, C., Kirchgessner, N., Marschall, D., Walter, A. and Hund, A. (2014). Rhizoslides: paper-based growth system for non-destructive, high throughput phenotyping of root development by means of image analysis. *Plant Methods*, 10, 1–16.

Lee, G., Wei, Q. and Zhu, Y. (2021). Emerging wearable sensors for plant health monitoring. *Advanced Functional Materials*, 31(52), 2106475.

Li, L., Zhang, Q. and Huang, D. (2014). A review of imaging techniques for plant phenotyping. *Sensors*, 14(11), 20078–20111.

Li, Y., Yan, W., An, S., Gao, W., Jia, J., Tao, S. and Wang, W. (2022). A spatio-temporal fusion framework of UAV and satellite imagery for winter wheat growth monitoring. *Drones*, 7(1), 23.

Liebisch, F., Kirchgessner, N., Schneider, D., Walter, A. and Hund, A. (2015). Remote, aerial phenotyping of maize traits with a mobile multi-sensor approach. *Plant Methods*, 11, 1–20.

Liu, J., Zhao, C., Yang, G., Yu, H., Zhao, X., Xu, B. and Niu, Q. (2016). Review of field-based phenotyping by unmanned aerial vehicle remote sensing platform. *Transactions of the Chinese Society of Agricultural Engineering*, 32(24), 98–106.

Liu, S., Barrow, C.S., Hanlon, M., Lynch, J.P. and Bucksch, A. (2021). DIRT/3D: 3D root phenotyping for field-grown maize (Zea mays). *Plant Physiology*, 187(2), 739–757.

Madurapperumage, A., Johnson, N., Thavarajah, P., Tang, L. and Thavarajah, D. (2023). Fourier-transform mid-infrared (FT-MIR) spectroscopy as a high-throughput phenotyping tool for measuring total fatty acids in chickpea (Cicer arietinum L.) flour. *ACS Food Science & Technology*, 3(9), 1568–1576.

Madurapperumage, A., Naser, M.Z., Boatwright, L., Bridges, W., Vandemark, G. and Thavarajah, D. (2025). High-throughput phenotyping platforms for pulse crop biofortification. *Plants, People, Planet*, 7(1), 49–61.

Maes, W.H. and Steppe, K. (2019). Perspectives for remote sensing with unmanned aerial vehicles in precision agriculture. *Trends in Plant Science*, 24(2), 152–164.

Marko, D., Briglia, N., Summerer, S., Petrozza, A., Cellini, F. and Iannacone, R. (2018). High-throughput phenotyping in plant stress response: Methods and potential applications to polyamine field. *Polyamines: Methods and Protocols*, 373–388.

Martins, S.M., Brito, G.G.D., Gonçalves, W.D.C., Tripode, B.M.D., Lartaud, M., Duarte, J.B. and Giband, M. (2019). PhenoRoots: an inexpensive non-invasive phenotyping system to assess the variability of the root system architecture. *Scientia Agricola*, 77, e20180420.

Mathieu, L., Lobet, G., Tocquin, P. and Périlleux, C. (2015). "Rhizoponics": a novel hydroponic rhizotron for root system analyses on mature Arabidopsis thaliana plants. *Plant Methods*, 11, 1–8.

Meiyan, S., Jinyu, Z., Xiaohong, Y., Xiaohe, G., Baoguo, L. and Yuntao, M. (2023). A spectral decomposition method for estimating the leaf nitrogen status of maize by UAV-based hyperspectral imaging. *Computers and Electronics in Agriculture*, 212, 108100.

Metzner, R., Eggert, A., van Dusschoten, D., Pflugfelder, D., Gerth, S., Schurr, U. and Jahnke, S. (2015). Direct comparison of MRI and X-ray CT technologies for 3D imaging of root systems in soil: potential and challenges for root trait quantification. *Plant Methods*, 11, 1–11.

Michels, V., Chou, C., Weigand, M., Wu, Y. and Kemna, A. (2024). Quantitative phenotyping of crop roots with spectral electrical impedance tomography: a rhizotron study with optimized measurement design. *Plant Methods*, 20(1), 118.

Mir, R.R., Reynolds, M., Pinto, F., Khan, M.A. and Bhat, M.A. (2019). High-throughput phenotyping for crop improvement in the genomics era. *Plant Science*, 282, 60–72.

Mohanty, S.P., Hughes, D.P. and Salathé, M. (2016). Using deep learning for image-based plant disease detection. *Pattern Recognition*, 54, 25–32.

Moulia, B., Coutand, C. and Julien, J.L. (2015). Mechanosensitive control of plant growth: bearing the load, sensing, transducing, and responding. *Frontiers in Plant Science*, 6, 52.

Mulla, D.J. (2013). Twenty-five years of remote sensing in precision agriculture: Key advances and remaining knowledge gaps. *Biosystems Engineering*, 114(4), 358–371.

Müller-Linow, M., Pinto-Espinosa, F., Scharr, H. and Rascher, U. (2015). The leaf angle distribution of natural plant populations: assessing the canopy with a novel software tool. *Plant Methods*, 11, 1–16.

Naik, H.S., Zhang, J., Lofquist, A., Assefa, T., Sarkar, S., Ackerman, D. and Ganapathysubramanian, B. (2017). A real-time phenotyping framework using machine learning for plant stress severity rating in soybean. *Plant Methods*, 13, 1–12.

Nazir, A., Ullah, S., Saqib, Z.A., Abbas, A., Ali, A., Iqbal, M.S. and Butt, M.U. (2021). Estimation and forecasting of rice yield using phenology-based algorithm and linear regression model on sentinel-ii satellite data. *Agriculture*, 11(10), 1026.

Olson, D. and Anderson, J. (2021). Review on unmanned aerial vehicles, remote sensors, imagery processing, and their applications in agriculture. *Agronomy Journal*, 113(2), 971–992.

Pauli, D., Chapman, S.C., Bart, R., Topp, C.N., Lawrence-Dill, C.J., Poland, J. and Gore, M.A. (2016). The quest for understanding phenotypic variation via integrated approaches in the field environment. *Plant Physiology*, 172(2), 622–634.

Paulus, S., Dupuis, J., Riedel, S. and Kuhlmann, H. (2014). Automated analysis of barley organs using 3D laser scanning: An approach for high throughput phenotyping. *Sensors*, 14(7), 12670–12686.

Pérez-Ruiz, M., Prior, A., Martínez-Guanter, J., Apolo-Apolo, O.E., Andrade-Sanchez, P. and Egea, G. (2020). Development and evaluation of a self-propelled electric platform for high-throughput field phenotyping in wheat breeding trials. *Computers and Electronics in Agriculture*, 169, 105237.

Pineda, M., Barón, M. and Pérez-Bueno, M.L. (2020). Thermal imaging for plant stress detection and phenotyping. *Remote Sensing*, 13(1), 68.

Pinto, F., Zaman-Allah, M., Reynolds, M. and Schulthess, U. (2023). Satellite imagery for high-throughput phenotyping in breeding plots. *Frontiers in Plant Science*, 14, 1114670.

Qiao, S., Fang, Y., Wu, A., Xu, B., Zhang, S., Deng, X. and Chen, Y. (2019). Dissecting root trait variability in maize genotypes using the semi-hydroponic phenotyping platform. *Plant and Soil*, 439, 75–90.

Rahman, H., Ramanathan, V., Jagadeeshselvam, N., Ramasamy, S., Rajendran, S., Ramachandran, M., ... & Muthurajan, R. (2015). Phenomics: technologies and applications in plant and agriculture. In PlantOmics: The omics of plant science (pp. 385–411). New Delhi: Springer India.

Rayhana, R., Xiao, G. and Liu, Z. (2021). RFID sensing technologies for smart agriculture. *IEEE Instrumentation & Measurement Magazine*, 24(3), 50–60.

Rayhana, R., Xiao, G.G. and Liu, Z. (2021). Printed sensor technologies for monitoring applications in smart farming: A review. *IEEE Transactions on Instrumentation and Measurement*, 70, 1–19.

Reynolds, D., Ball, J., Bauer, A., Davey, R., Griffiths, S. and Zhou, J. (2019). CropSight: a scalable and open-source information management system for distributed plant phenotyping and IoT-based crop management. *Gigascience*, 8(3), giz009.

Rousseau, C., Hunault, G., Gaillard, S., Bourbeillon, J., Montiel, G., Simier, P. and Boureau, T. (2015). Phenoplant: a web resource for the exploration of large chlorophyll fluorescence image datasets. *Plant Methods*, 11, 1–12.

Ruckelshausen, A., Biber, P., Dorna, M., Gremmes, H., Klose, R., Linz, A. and Weiss, U. (2009). BoniRob: an autonomous field robot platform for individual plant phenotyping. In *Precision Agriculture'09* (pp. 841–847). Wageningen Academic.

Sabag, I., Bi, Y., Sahoo, M.M., Herrmann, I., Morota, G. and Peleg, Z. (2024). Leveraging genomics and temporal high-throughput phenotyping to enhance association mapping and yield prediction in sesame. *The Plant Genome*, 17(3), e20481.

Sadok, W., Naudin, P., Boussuge, B., Muller, B., Welcker, C. and Tardieu, F. (2007). Leaf growth rate per unit thermal time follows QTL-dependent daily patterns in hundreds of maize lines under naturally fluctuating conditions. *Plant, Cell & Environment*, 30(2), 135–146.

Saini, P., Saini, P., Kaur, J.J., Francies, R.M., Gani, M., Rajendra, A.A. and Chauhan, S.S. (2020). Molecular approaches for harvesting natural diversity for crop improvement. *Rediscovery of Genetic and Genomic Resources for Future Food Security*, 67–169.

Salas Fernandez, M.G., Bao, Y., Tang, L. and Schnable, P.S. (2017). A high-throughput, field-based phenotyping technology for tall biomass crops. *Plant Physiology*, 174(4), 2008–2022.

Salim, M., Chen, Y., Ye, H., Nguyen, H.T., Solaiman, Z.M. and Siddique, K.H. (2021). Screening of soybean genotypes based on root morphology and shoot traits using the semi-hydroponic phenotyping platform and rhizobox technique. *Agronomy*, 12(1), 56.

Salungyu, J., Thaitad, S., Bucksch, A., Kengkanna, J. and Saengwilai, P.J. (2020). From lab to field: Open tools facilitating the translation of maize root traits. *Field Crops Research*, 255, 107872.

Sanchez, J.R. (2024). *Automated Unmanned Systems and Data Analytics for In-field Digital Crop Phenotyping* (Doctoral dissertation, University of Georgia).

Satam, H., Joshi, K., Mangrolia, U., Waghoo, S., Zaidi, G., Rawool, S., Thakare, R.P., Banday, S., Mishra, A.K., Das, G. and Malonia, S.K. (2023). Next-generation sequencing technology: current trends and advancements. *Biology*, 12(7), 997.

Seethepalli, A., Guo, H., Liu, X., Griffiths, M., Almtarfi, H., Li, Z. and York, L.M. (2020). RhizoVision crown: an integrated hardware and software platform for root crown phenotyping. *Plant Phenomics*, 2020, 1–15.

Sharma, A., Saini, P., Saini, P., Tyagi, V., Sharma, S., Ahmed, N., Dhaliwal, H.S. and Sheikh, I. (2024). Root system architecture in cereals: exploring different perspectives of the hidden half. *Brazilian Journal of Botany*, 1–19.

Sheikh, M., Iqra, F., Ambreen, H., Pravin, K.A., Ikra, M. and Chung, Y.S. (2024). Integrating artificial intelligence and high-throughput phenotyping for crop improvement. *Journal of Integrative Agriculture*, 23(6), 1787–1802.

Shen, Y., Mercatoris, B., Cao, Z., Kwan, P., Guo, L., Yao, H. and Cheng, Q. (2022). Improving wheat yield prediction accuracy using LSTM-RF framework based on UAV thermal infrared and multispectral imagery. *Agriculture*, 12(6), 892.

Sishodia, R.P., Ray, R.L. and Singh, S.K. (2020). Applications of remote sensing in precision agriculture: A review. *Remote Sensing*, 12(19), 3136.

Suma, M.R. and Madhumathy, P. (2019). Acquisition and mining of agricultural data using ubiquitous sensors with Internet of Things. In: *International Conference on Computer Networks and Communication Technologies: ICCNCT 2018* (pp. 249–261). Springer Singapore.

Teramoto, S., Takayasu, S., Kitomi, Y., Arai-Sanoh, Y., Tanabata, T. and Uga, Y. (2020). High-throughput three-dimensional visualization of root system architecture of rice using X-ray computed tomography. *Plant Methods*, 16, 1–14.

Thenkabail, P.S., Gumma, M.K., Teluguntla, P. and Irshad, A.M. (2014). Hyperspectral remote sensing of vegetation and agricultural crops. *Photogrammetric Engineering & Remote Sensing (TSI)*, 80(8), 695–723.

Tisné, S., Serrand, Y., Bach, L., Gilbault, E., Ben Ameur, R., Balasse, H. and Loudet, O. (2013). Phenoscope: an automated large-scale phenotyping platform offering high spatial homogeneity. *The Plant Journal*, 74(3), 534–544.

Tripathi, S., Sharma, R. and Kumar, A. (2023). Automated phenotyping platforms for crop growth monitoring using AI. *Sensors,* 23(2), 657.

Tripodi, P., Massa, D., Venezia, A. and Cardi, T. (2018). Sensing technologies for precision phenotyping in vegetable crops: current status and future challenges. *Agronomy*, 8(4), 57.

Tsouros, D.C., Bibi, S. and Sarigiannidis, P.G. (2019). A review on UAV-based applications for precision agriculture. *Information*, 10(11), 349.

Varshney, R.K., Bohra, A., Yu, J., Graner, A., Zhang, Q. and Sorrells, M.E. (2021). Designing future crops: genomics-assisted breeding comes of age. *Trends in Plant Science*, 26(6), 631–649.

Varshney, R.K., Sinha, P., Singh, V.K., Kumar, A., Zhang, Q. and Bennetzen, J.L. (2020). 5Gs for crop genetic improvement. *Current Opinion in Plant Biology*, 56, 190–196.

Varshney, R.K., Thudi, M., Pandey, M.K., Tardieu, F., Ojiewo, C., Vadez, V. and Bergvinson, D. (2018). Accelerating genetic gains in legumes for the development of prosperous smallholder agriculture: integrating genomics, phenotyping, systems modelling and agronomy. *Journal of Experimental Botany*, 69(13), 3293–3312.

Vázquez-Arellano, M., Griepentrog, H.W., Reiser, D. and Paraforos, D.S. (2016). 3-D imaging systems for agricultural applications—a review. *Sensors*, 16(5), 618.

Virlet, N., Sabermanesh, K., Sadeghi-Tehran, P. and Hawkesford, M.J. (2016). Field Scanalyzer: An automated robotic field phenotyping platform for detailed crop monitoring. *Functional Plant Biology*, 44(1), 143–153.

Volpato, L., Pinto, F., González-Pérez, L., Thompson, I.G., Borém, A., Reynolds, M. and Rodrigues Jr, F.A. (2021). High throughput field phenotyping for plant height using UAV-based RGB imagery in wheat breeding lines: Feasibility and validation. *Frontiers in Plant Science*, 12, 591587.

Voorend, W., Lootens, P., Nelissen, H., Roldán-Ruiz, I., Inzé, D. and Muylle, H. (2014). LEAF-E: a tool to analyze grass leaf growth using function fitting. *Plant Methods*, 10, 1–13.

Wang, Y., Wang, D., Shi, P. and Omasa, K. (2014). Estimating rice chlorophyll content and leaf nitrogen concentration with a digital still color camera under natural light. *Plant Methods*, 10, 1–11.

Washburn, J.D., Burch, M.B. and Franco, J.A.V. (2020). Predictive breeding for maize: Making use of molecular phenotypes, machine learning, and physiological crop models. *Crop Science*, 60(2), 622–638.

Whan, A.P., Smith, A.B., Cavanagh, C.R., Ral, J.P.F., Shaw, L.M., Howitt, C.A. and Bischof, L. (2014). GrainScan: a low cost, fast method for grain size and colour measurements. *Plant Methods*, 10, 1–10.

Wu, J., Wu, Q., Pagès, L., Yuan, Y., Zhang, X., Du, M. and Li, Z. (2018). RhizoChamber-Monitor: a robotic platform and software enabling characterization of root growth. *Plant Methods*, 14, 1–15.

Wu, S., Wen, W., Gou, W., Lu, X., Zhang, W., Zheng, C. and Guo, X. (2022). A miniaturized phenotyping platform for individual plants using multi-view stereo 3D reconstruction. *Frontiers in Plant Science*, 13, 897746.

Wu, S., Wen, W., Wang, Y., Fan, J., Wang, C., Gou, W. and Guo, X. (2020). MVS-Pheno: A portable and low-cost phenotyping platform for maize shoots using multiview stereo 3D reconstruction. *Plant Phenomics*, 2020, 1848437.

Wu, S., Zhang, Y., Zhao, Y., Wen, W., Wang, C., Lu, X. and Zhao, C. (2024). Using high-throughput phenotyping platform MVS-Pheno to decipher the genetic architecture of plant spatial geometric 3D phenotypes for maize. *Computers and Electronics in Agriculture*, 225, 109259.

Yang, W., Feng, H., Zhang, X., Zhang, J., Doonan, J.H., Batchelor, W.D. and Yan, J. (2020). Crop phenomics and high-throughput phenotyping: past decades, current challenges, and future perspectives. *Molecular Plant*, 13(2), 187–214.

Yang, W., Guo, Z., Huang, C., Duan, L., Chen, G., Jiang, N. and Xiong, L. (2014). Combining high-throughput phenotyping and genome-wide association studies to reveal natural genetic variation in rice. *Nature Communications*, 5(1), 5087.

Yau, W.K., Ng, O.E. and Lee, S.W. (2021). Portable device for contactless, non-destructive and in situ outdoor individual leaf area measurement. *Computers and Electronics in Agriculture*, 187, 106278.

Yazdanbakhsh, N. and Fisahn, J. (2009) High throughput phenotyping of root growth dynamics, lateral root formation, root architecture and root hair development enabled by PlaRoM. *Functional Plant Biology,* 36, 938–946.

Yin, H., Cao, Y., Marelli, B., Zeng, X., Mason, A.J. and Cao, C. (2021). Soil sensors and plant wearables for smart and precision agriculture. *Advanced Materials*, 33(20), 2007764.

Zaman-Allah, M., Vergara, O., Araus, J.L., Tarekegne, A., Magorokosho, C., Zarco-Tejada, P. J. and Cairns, J. (2015). Unmanned aerial platform-based multi-spectral imaging for field phenotyping of maize. *Plant Methods*, 11, 1–10.

Zavafer, A., Bates, H., Mancilla, C. and Ralph, P.J. (2023). Phenomics: conceptualization and importance for plant physiology. *Trends in Plant Science*, 28(9), 1004–1013.

Zhai, Z., Martínez, J.F., Beltran, V. and Martínez, N.L. (2020). Decision support systems for agriculture 4.0: Survey and challenges. *Computers and Electronics in Agriculture*, 170, 105256.

Zhang, H., He, H., Gao, Y., Mady, A., Filipović, V., Dyck, M. and Liu, Y. (2023). Applications of computed tomography (CT) in environmental soil and plant sciences. *Soil and Tillage Research, 226,* 105574.

Zhang, X., Huang, C., Wu, D., Qiao, F., Li, W., Duan, L., ... and Yan, J. (2017). High-throughput phenotyping and QTL mapping reveals the genetic architecture of maize plant growth. *Plant physiology*, 173(3), 1554–1564.

Zhang, Z., Qu, Y., Ma, F., Lv, Q., Zhu, X., Guo, G. and Song, C.P. (2024). Integrating high-throughput phenotyping and genome-wide association studies for enhanced drought resistance and yield prediction in wheat. *New Phytologist*, 243(5), 1758–1775.

Zhou, J., Mou, H., Zhou, J., Ali, M.L., Ye, H., Chen, P. and Nguyen, H.T. (2021). Qualification of soybean responses to flooding stress using UAV-based imagery and deep learning. *Plant Phenomics*, 2021, 9892570.

Zhou, J., Reynolds, D., Cornu, T.L., Websdale, D., Orford, S., Lister, C. and Griffiths, S. (2017). CropQuant: an automated and scalable field phenotyping platform for crop monitoring and trait measurements to facilitate breeding and digital agriculture. BioRxiv, 161547.

Zhu, F., Saluja, M., Dharni, J.S., Paul, P., Sattler, S.E., Staswick, P. and Yu, H. (2021). PhenoImage: An open-source graphical user interface for plant image analysis. *The Plant Phenome Journal*, 4(1), e20015.

CHAPTER 13

# Utilizing Bioinformatics Tools and Machine Learning Techniques to Advance Sustainable Crop Improvement

Mahantesha BN Naika[1*], Mamatha YS[2], Dwijesh Chandra Mishra[2] and Belaghihalli N Gnanesh[3]

## 1. Introduction

The continuously increasing human population and the threat of change in climatic factors are negatively affecting the agriculture. The need of the hour is to adopt sustainable practices and improve crop yields resilient to environmental stresses (Dormatey et al., 2020). Advances in big data and machine learning are pivotal in addressing these challenges, heralding a new era in precision agriculture (Woźniak & Ijaz, 2024). Bioinformatics and big data will transform agriculture, increasing productivity, sustainability, and food security around the world (Sahu et al., 2023). This chapter provides an in-depth exploration of how cutting-edge technologies and data-driven approaches are reshaping the landscape of crop improvement. The integration of next-generation sequencing (NGS) technologies has revolutionized the understanding of genetic and molecular factors underlying crop traits. These technologies enable detailed genomic analyses at unprecedented speed and accuracy, providing the foundation for molecular breeding and genetic modification. Simultaneously, machine learning techniques harness complex datasets to predict, model, and understand crop performance under various environmental conditions. By analyzing patterns in large-scale genomic, transcriptomic, and phenotypic data, machine learning aids in the rapid identification of traits, optimization of breeding strategies, and management of biotic and abiotic stresses, ultimately leading to healthier and more productive agricultural systems. Through a series of detailed sections, this chapter discusses the synergy between bioinformatics tools and machine learning applications in the context of modern agricultural demands. Each segment, including the critical examination of NGS technologies, unfolds the current capabilities, practical applications, and prospects of these technologies in sustainable crop improvement, ensuring food security in an eco-friendly manner (Figure 1). Amid rising global food demands and the pressing need for sustainable agricultural practices, this chapter serves as an essential guide for understanding the current trends in crop

1 Centre for Biotechnology Research, Department of Biotechnology and Crop Improvement, College of Horticulture, University of Horticultural Sciences, Bagalkot 587104, Karnataka, India.
2 ICAR – Indian Agricultural Statistics Research Institute, PUSA, New Delhi 110012, India.
3 Genetics & Plant Breeding, ICAR JSS – Krishi Vigyan Kendra, Suttur 571129, Mysore, Karnataka, India.
* Corresponding author: mahantesha.bnn@uhsbagalkot.edu.in

DOI: 10.1201/9781003584100-13

**Fig. 1** Machine learning and its applications.

breeding and genetics. The chapter also delves deep into how cutting-edge genomic technologies and sophisticated machine learning algorithms are being harnessed to enhance crop productivity and resilience. From the detailed workings of NGS to the predictive powers of artificial intelligence (AI), this comprehensive resource lays out both the scientific foundations and the practical applications of these transformative technologies. The chapter covers the bioinformatics and machine learning resources and tools, and also presents the utilization of such tools for the analysis of stress responsive genes in plants.

## 2. Current Status of Climate Change and Its Influence on Food Quality and Quantity

The rapidly advancing climate change is posing significant risks to global food systems. Its impact is twofold: affecting both the quantity and quality of food production. Raising global temperatures, shifting precipitation patterns, and the increasing frequency of extreme weather events are already disrupting agricultural systems, with worse consequences expected in the future. The impact of climate change on crop yields varies depending on the crop's tolerance to heat and its geographic location. Some regions might experience temporary benefits, such as extended growing seasons or the ability to cultivate new crops that typically thrive in warmer climates. Mirón et al. (2023) reviewed the influence of climate change on food production and safety.

Rezaei et al. (2023) examined yield responses in wheat, maize, millet, sorghum, and rice to higher temperatures, $CO_2$, and water scarcity and noticed yield losses ranging from 7% to 23%.

Furthermore, for higher yields and better capture and analysis of yield responses to climate change, multifactor experiments, the integration of process-based and data-driven models, the development of biotic stress modeling methods, and the use of integrated impact assessments are the prerequisites.

# 3. Crop Improvement Programs

Crop improvement programs are initiatives designed to enhance the genetic quality, yield, resilience, and nutritional value of crops. These programs utilize various techniques ranging from traditional breeding to modern biotechnological approaches. Here are some key types of crop improvement programs.

## *3.1 Conventional Breeding Programs*

- **Hybridization:** Crossing two different parent plants to combine desirable traits such as disease resistance, drought tolerance, or high yield (Rieseberg et al., 1998).
- **Selection Breeding:** Selecting the best-performing plants over generations based on desired traits, such as size or quality (Lorenz et al., 2011).
- **Mutation Breeding:** Exposing seeds or plant parts to chemicals or radiation to create genetic variations that can lead to beneficial traits (Toker et al., 2007).
- **Backcross Breeding:** Used to introduce a specific trait from one plant into an already high-performing variety by repeated crossing (Vogel, 2009).

## *3.2 Marker-Assisted Selection (MAS) Programs*

MAS programs use molecular markers to track the inheritance of specific genes related to desirable traits. Promising examples of MAS in crop breeding are improving disease-resistant varieties in rice, wheat, and maize by pyramiding resistance genes against pathogens, enhancing stress tolerance in legumes from drought or salinity, and improving nutritional quality in tomato and pepper by integrating gene specific markers for desired traits. This accelerates the breeding process by allowing early selection of plants with the desired genetic markers (Boopathi, 2020).

## *3.3 Genetic Engineering Programs*

- **Transgenic Crops:** Introducing genes from other species into a crop to confer specific traits, such as pest resistance (Bt cotton), herbicide tolerance (Roundup Ready), or nutritional enhancement (Golden Rice) (Raymond Park et al., 2011).
- **Gene Editing (CRISPR/Cas):** Clustered regularly interspaced short palindromic repeats (CRISPR)-associated proteins (CRISPR/Cas) enable precise editing of crop genomes to introduce or modify traits such as drought resistance, improved shelf life, or disease resistance (Wang et al., 2016).

## *3.4 Genomic Selection Programs*

These programs utilize genomic data to predict the breeding value of plants by associating specific traits and selecting those with the best genetic potential for yield, tolerance to environmental stress, or nutrient-use efficiency. This is a more data-driven approach to traditional breeding (Lorenz et al., 2011).

## *3.5 Biotechnology-Based Programs*

These programs involve plant tissue culture approach such as somatic hybridization and micropropagation to develop disease-free, high-quality planting material. Doubled haploids (DH) are used to rapidly develop pure breeding lines by doubling the chromosome number (Abinaya & Viswanathan, 2021).

## 3.6 Participatory Plant Breeding (PPB) Programs

These programs engage farmers in the breeding process, allowing them to select traits that are important in their specific environment, tolerant to pests and diseases, and the best type the community wants to eat. These programs promote economic viability and orient farmers toward conservation. This approach combines traditional knowledge with scientific breeding methods (Ceccarelli & Grando, 2009).

## 3.7 Sustainable Crop Improvement Programs

The programs focus on improving crops while ensuring sustainability through eco-friendly practices such as breeding for organic farming systems, developing drought-tolerant and water-efficient crops, and enhancing soil health through crops' rotation-friendly varieties (Dormatey et al., 2020).

## 3.8 Climate Resilience Breeding Programs

These programs are aimed at developing crops that can withstand the impacts of climate change, such as increased temperatures and erratic rainfall patterns along with evolving pest pressures. These programs include breeding for heat tolerance, flood resistance, and salt tolerance (Langridge et al., 2021).

# 4. Limitations of Crop Improvement Methods across Various Approaches

- **Time-Consuming:** Developing new crop varieties, especially using traditional breeding methods, can take many years or decades. Even modern techniques such as genomic selection or marker-assisted breeding, while faster, still require several growing seasons for validation.
- **High Costs:** Molecular breeding tools such as genetic engineering, gene editing (e.g., CRISPR), and genomic selection are often expensive. They require sophisticated laboratory infrastructure and skilled personnel.
- **Complex Traits:** Many important traits, such as yield, drought tolerance, or pest resistance, are controlled by multiple genes and are influenced by environmental factors. Improving these traits is challenging because they are not easily manipulated by single-gene approaches, and interactions between genes and the environment can complicate outcomes.
- **Environmental Dependency:** Crop performance can vary significantly across different environments. A variety that performs well in one region may not do as well in another due to differences in climate, soil conditions, and pest pressures.
- **Regulatory and Public Acceptance Issues:** Genetically modified organisms (GMOs) and, to some extent, gene-edited crops face strict regulatory scrutiny and can be met with public resistance due to concerns over safety, environmental impacts, and ethical considerations. This can delay the adoption and commercialization of new crop varieties.
- **Limited Genetic Diversity:** Many crop improvement programs are restricted by the available genetic diversity within a species. This can limit the ability to introduce novel traits and may lead to a narrowing of the genetic base, which can make crops more susceptible to diseases and changing environmental conditions.
- **Unintended Consequences:** Breeding for specific traits can sometimes result in unintended negative effects, such as reduced resistance to other stresses or loss of important agronomic characteristics. For example, increasing drought resistance may inadvertently reduce yield potential.

- **Slow Adaptation to Climate Change:** While crop improvement programs can develop varieties resilient to climate stresses (e.g., heat, drought, etc.), the pace of breeding may not keep up with the rapid changes in climate patterns, making it difficult to meet future agricultural demands.

These limitations highlight the need for careful consideration of each method's strengths and weaknesses, as well as a balanced approach to crop improvement that integrates multiple strategies to address challenges in global agriculture.

## 5. Challenges in Crop Improvement

Crop improvement faces numerous challenges due to environmental stress factors, as well as factors such as anti-nutritional compounds and heavy metal toxicity. An overview of these challenges is provided in the following text.

### *5.1 Biotic Stress*

Biotic stressors such as pests, diseases, and weeds can significantly impact crop yields and quality.

- **Pests:** Insects, nematodes, and other pests can cause substantial damage to crops, reducing both yield and quality. Developing pest-resistant varieties is a key priority, but pests can quickly adapt and evolve resistance to control measures, making ongoing research and breeding efforts necessary.
- **Diseases:** Crops are vulnerable to a wide range of fungal, bacterial, and viral pathogens, which can severely impact plant health. Breeding for disease resistance is complex due to the diversity of pathogens and their rapid adaptation, requiring constant development of new resistant varieties.
- **Weeds:** Compete with crops for essential resources such as light, water, and nutrients, leading to reduced yields. While herbicide-resistant crops have been developed to combat weeds, excessive reliance on these varieties can result in the emergence of herbicide-resistant weed populations, posing a long-term challenge.

### *5.2 Abiotic Stress*

Abiotic factors such as drought, heat, salinity, etc., can hinder crop growth and productivity.

- **Drought:** Water scarcity is a major challenge for crop growth, affecting yields across many regions. Developing drought-tolerant varieties is complex, as drought stress impacts multiple physiological and biochemical processes within plants.
- **Salinity:** High soil salinity impairs plant growth by disrupting nutrient uptake and causing physiological stress. Breeding for salt tolerance requires a deep understanding of complex traits and their genetic foundations, which can be difficult to achieve.
- **Temperature Extremes:** Extreme temperatures, whether high or low, can negatively affect seed germination, plant growth, and crop yields. Breeding crops that can withstand temperature fluctuations is especially critical in light of ongoing climate change.

### *5.3 Anti-Nutritional Compounds*

Anti-nutritional factors present in many crops can impact their nutritional value.

- **Presence of Anti-Nutrients:** Compounds such as phytic acid, tannins, etc., can inhibit nutrient absorption in humans and animals. While some of these compounds have health benefits, their presence in staple crops can reduce overall nutritional value.

- **Breeding for Reduced Anti-Nutritional Factors:** Developing varieties with lower levels of these compounds while preserving desirable traits such as taste and yield is challenging, as these anti-nutritional factors often coexist with beneficial nutrients.

### *5.4 Heavy Metal Toxicity*

Heavy metal contamination poses significant risks to agriculture and food safety.

- **Soil Contamination:** Heavy metals such as lead, cadmium, arsenic, and mercury can accumulate in soils due to industrial activities, mining, and certain agricultural practices. Crops can absorb these toxic elements, posing risks to human and animal health.
- **Breeding for Metal Exclusion:** Developing crops that can either tolerate or exclude the uptake of heavy metals is complex, requiring an understanding of plant mechanisms related to metal transport and detoxification.

### *5.5 Nutrient Deficiencies and Imbalances*

Addressing nutrient-related challenges is crucial for improving crop quality and agricultural sustainability.

- **Micronutrient Deficiencies:** Many crops lack adequate levels of essential nutrients such as iron and zinc, which are critical for human health. Biofortification enhancing the nutrient content of crops requires substantial breeding efforts to achieve desired nutritional levels.
- **Nutrient-Use Efficiency:** Improving the efficiency with which crops absorb and utilize nutrients is vital for sustainable agriculture.

### *5.6 Technological and Research Challenges*

The development and implementation of crop improvement strategies also face technical and research-related hurdles.

- **Data and Infrastructure Limitations:** In many regions, a lack of data on crop performance under diverse stress conditions hampers efforts to develop improved varieties tailored to specific environments.
- **Need for Multidisciplinary Approaches:** Multidisciplinary approaches are crucial for tackling complex, real-world problems by integrating diverse perspectives and expertise, fostering collaboration and innovation across fields such as omics, system biology, molecular breeding, and agronomy, and ultimately leading to more comprehensive and effective solutions. Addressing the complex challenges in crop improvement requires collaboration across fields. These challenges highlight the complexity of crop improvement efforts. A successful approach requires integrating traditional breeding methods with modern biotechnological tools, alongside a nuanced understanding of ecological, economic, and social factors that influence agricultural practices and outcomes.

## 6. Advanced Technologies Generating Big Data in Agriculture

Advanced high-throughput technologies generating big data in agriculture are instrumental in improving crop management, enhancing yields, and promoting sustainability. Key methods include omics platforms, imaging technologies, and precision farming approaches utilizing satellite-based data.

### *6.1 Omics Platforms*

Omics technologies enable a detailed analysis of prokaryotes to eukaryotes, offering insights into genetic, biochemical, and physiological processes in organisms. Genomics involves sequencing and analyzing crop genomes to identify genes associated with traits such as yield and disease resistance, thereby supporting MAS and genomic selection, which facilitate more targeted and efficient breeding programs (Boopathi, 2020). Transcriptomics focuses on studying RNA transcripts, revealing gene expression patterns under specific conditions, including stress responses. Proteomics examines the structure, functions, and interactions of proteins, providing insights into how plants respond to biotic and abiotic stresses. Mass spectrometry is commonly employed to study protein abundance and modifications. Metabolomics analyzes metabolites within a plant, helping to understand metabolic pathways and the physiological status of crops while identifying biomarkers for stress tolerance, nutritional content, and quality traits. Additionally, microbiomics studies the microbial communities in soil and around plants, highlighting the role of beneficial microbes in plant health and nutrient uptake. Understanding plant–microbe interactions can lead to enhanced crop resilience and sustainable farming practices (Kumar et al., 2017).

## 6.2 Imaging Technologies

Imaging technologies generate valuable data for assessing crop health, growth, and productivity. Remote sensing employs satellite or aerial imagery to monitor crop conditions, soil health, and environmental variables over large areas. Data such as the Normalized Difference Vegetation Index (NDVI) helps assess plant health by measuring chlorophyll content and photosynthetic activity (Mir et al., 2019). Drones and unmanned aerial vehicles (UAVs) are equipped with sensors (e.g., multispectral, thermal) that capture high-resolution images, providing real-time data on plant health, stress levels, and nutrient deficiencies. This technology enables precise monitoring of crop development and early detection of issues such as pests or diseases. Hyperspectral imaging captures a broad spectrum of wavelengths, offering detailed analysis of plant health and is particularly useful for detecting specific plant traits, including nutrient content, moisture levels, and early signs of disease (Harbinson et al., 2012).

## 6.3 Precision Farming

Precision farming utilizes data-driven techniques to optimize agricultural practices, improving efficiency and sustainability. Key components of this approach include soil and nutrient sensors, which measure soil moisture, nutrient levels, and pH, providing real-time data for more precise irrigation and fertilization. This data optimizes resource use, enhancing crop growth while reducing environmental impacts. Variable rate technology (VRT) allows for the targeted application of seeds, fertilizers, and pesticides based on field-specific data, increasing efficiency and reducing input waste; these decisions are often guided by data from soil sensors and satellite imagery (Saleem et al., 2023). Yield mapping gathers data from harvesting equipment to create detailed maps of crop productivity, enabling the identification of areas of low and high productivity and aiding in future management decisions (Belal et al., 2021).

## 6.4 Satellite-Based Data

Satellite technologies provide extensive data for precision agriculture. Crop monitoring via satellites tracks crop health, growth stages, and development across large areas, allowing for timely actions in pest control, irrigation, and fertilization. Frequent image updates help detect changes over time and enable trend analysis. Additionally, satellites offer essential data on weather conditions, precipitation, temperature, and humidity, all crucial for agricultural planning. Climate models integrated with satellite data can help predict future conditions, guiding adaptive agricultural practices. Furthermore, satellite imagery assists in analyzing land use changes and the environmental impact of agricultural activities, informing sustainable land management practices and supporting policy development.

These advanced techniques generate vast amounts of data that can transform agricultural practices, improve crop resilience, and enhance sustainability. By integrating omics platforms, imaging technologies, and precision farming approaches, researchers can make data-driven decisions in real-time, leading to more efficient and productive agricultural systems.

# 7. Bioinformatics and Machine Learning in Crop Improvement

Bioinformatics is central to crop improvement, merging genomic, transcriptomic, and metabolomic data to pinpoint desirable traits and genetic markers that are critical for breeding. It enables researchers to manage and interpret vast datasets, helping to identify genes linked to resistance against biotic and abiotic stresses, which in turn facilitates the creation of high-yielding crop varieties (Figure 1). Machine learning further enhances these efforts by using advanced algorithms to predict trait outcomes and refine breeding strategies (Figure 2). By analyzing complex biological data, machine learning models can detect hidden patterns, increasing the precision of trait selection and speeding up the breeding process. Together, bioinformatics and machine learning are transforming crop improvement, and fostering sustainable agriculture in the face of global challenges.

## 7.1 Bioinformatics Databases

Bioinformatics databases are essential tools in crop improvement, organizing genomic, transcriptomic, metabolomic, and proteomic data. They provide access to gene sequences, expression profiles, and functional annotations, wh ich are crucial for identifying candidate genes and expediting breeding programs (Table 1).

- **Genomics Databases:** Resources such as GenBank (https://www.ncbi.nlm.nih.gov/genbank/) and Phytozome (https://phytozome-next.jgi.doe.gov/) provide extensive genomic data, with over 4,161 plant genomes available as of April 2024. Databases such as the Rice Genome Annotation Project (https://rice.uga.edu/) support the development of more resilient crop varieties (Lai et al., 2012).
- **Transcriptomics Databases:** Databases such as the Gene Expression Omnibus (GEO) (https://www.ncbi.nlm.nih.gov/geo/) and the Sequence Read Archive (SRA) (https://www.ncbi.nlm.nih.gov/sra) offer insights into gene expression under various stress conditions, aiding in the study of stress-responsive genes (Chao et al., 2024).
- **Proteomics Databases:** Proteomics databases such as UniProt (https://www.uniprot.org/) and the Protein Data Bank (PDB) (https://www.rcsb.org/) help researchers identify stress-responsive proteins and potential genes associated with desirable traits, enhancing understanding of stress adaptation in crops (Ramalingam et al., 2015).
- **Metabolomics Databases:** Databases such as METLIN (https://metlin.scripps.edu/landing_page.php?pgcontent=mainPage) and Kyoto Encyclopedia of Genes and Genomes (KEGG) (https://www.genome.jp/kegg/) support the identification of metabolites and pathway analysis, contributing to the development of crops with improved stress tolerance and nutritional profiles (Tounta et al., 2021).

## 7.2 Types of Machine Learning in Crop Improvement

- **Supervised Learning:** Utilizes labeled datasets to train algorithms for tasks such as regression (predicting crop yields) and classification (identifying diseases) using methods such as linear regression and decision trees.
- **Unsupervised Learning:** Analyzes unlabeled data to discover hidden patterns, useful for clustering similar crop varieties and reducing data complexity with techniques such as k-means and principal component analysis (PCA).

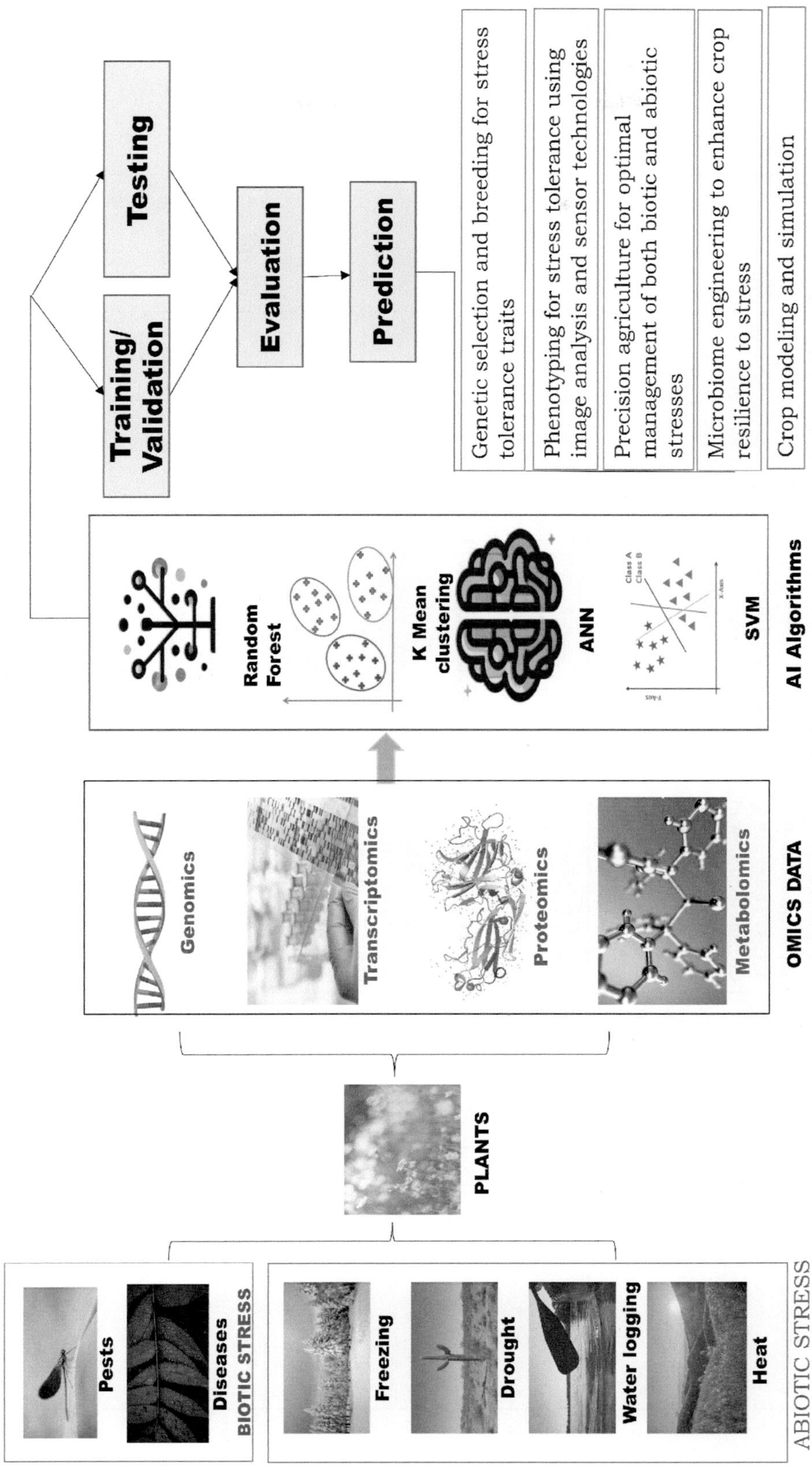


**Fig. 2** Integration of experimental, bioinformatics, and machine learning approaches for crop improvement programs (credits: Google images).

**Table 1** List of bioinformatics tools/databases utilized for various analysis.

| *S. No.* | *Omics data* | *Tools/method/ database used* | *Tool/method/database link* | *Functions of tools/methods/database* | *Reference* |
|---|---|---|---|---|---|
| 1. | Genomics<br>Transcriptomics | STEM software<br>CleaveLand 3.0 software<br>WGCNA<br>Cytoscape v.3.7.1<br>Soligo software | www.cs.cmu.edu/~jernst/stem/<br>https://sites.psu.edu/axtell/software/cleaveland4/<br>https://horvath.genetics.ucla.edu/html/CoexpressionNetwork/Rpackages/WGCNA/<br>https://cytoscape.org/release_notes_3_7_1.html<br>https://sfold.wadsworth.org/cgi-bin/soligo.pl | Clustering, comparing, and visualizing short time series gene expression data<br>Finding sliced miRNA and siRNA targets from degradome data<br>Finding clusters of highly correlated genes<br>Visualizing, integrating, and annotating molecular interaction networks and biological pathways<br>Targeting accessibility prediction | Wang et al., 2021 |
| 2. | Genomics<br>Transcriptomics<br>Proteomics | SAMtools<br>PlantCARE web tools<br>InterproScan5.21<br>LASTZ<br>DESeq2 software<br>KOBAS2.0<br>Primer premier software (version 5.0 | https://www.htslib.org/<br>https://bioinformatics.psb.ugent.be/webtools/plantcare/html/<br>https://www.ebi.ac.uk/Tools/services/web/toolform.ebi?tool=iprscan5&sequence=uniprot:KPYM_HUMAN<br>www.bx.psu.edu/~rsharris/lastz/<br>https://bioconductor.org/packages/release/bioc/html/DESeq2.html<br>http://kobas.cbi.pku.edu.cn/<br>http://www.premierbiosoft.com/primerdesign/oligo/oligo.html | Manipulating alignments<br>Plant cis-acting regulatory element database<br>Sequence search against the InterPro databases<br>DNA sequences alignment<br>Differential gene expression analysis<br>Gene annotation<br>Primer design | Shen et al., 2019 |
| 3. | Transcriptomics<br>Metabolomics | Subjunc aligner<br>R Limma package<br>Soybase GO term enrichment tool<br>BinBase 2.0<br>MetaboAnalyst 3.0<br>Primer3 software<br>EdgeR | http://subread.sourceforge.net/<br>https://bioconductor.org/packages/release/bioc/html/limma.html<br>https://www.soybase.org/goslimgraphic_v2/dashboard.php<br>https://sourceforge.net/projects/binbase/<br>https://www.metaboanalyst.ca/<br>https://primer3.ut.ee/<br>https://bioconductor.org/packages/release/bioc/html/edgeR.html | Identifying exon–exon junction and gene fusion<br>Differential expression for microarray data<br>SoyBase gene model data mining and analysis<br>MS data analysis<br>Metabolomics data analysis<br>Designing PCR primers<br>Differential expression analysis of RNA-seq expression profiles | Ranjan et al., 2019 |

*Contd.*

**Table 1** *Contd.*

| *S. No.* | *Omics data* | *Tools/method/ database used* | *Tool/method/database link* | *Functions of tools/methods/database* | *Reference* |
|---|---|---|---|---|---|
| 4. | Transcriptomics<br>Metabolomics<br>Proteomics | Spectrus software v.12.01<br>SIMCA-P+ v.13.0.3.0 software<br>Bowtie2 version 2.1.0.0<br>TopHat2 v2.0.8b<br>DESeq version 1.14.0<br>corrplot version 0.77<br>SoyBase databaseSoy Base database<br>t version 0.77 | https://www.acdlabs.com/products/spectrus-platform/spectrus-processor/<br>https://www.sartorius.com/en/products/process-analytical-technology/data-analytics-software/mvda-software/simca<br>http://bowtie-bio.sourceforge.net/bowtie2/index.shtml<br>https://ccb.jhu.edu/software/tophat/index.shtml<br>https://bioconductor.org/packages/release/bioc/html/DESeq2.html<br>https://cran.r-project.org/web/packages/corrplot/index.html<br>https://www.soybase.org/ | Spectroscopic data analysis<br>Multivariate data analysis<br>Aligning sequencing reads<br>Mapping RNA-seq<br>Differential gene expression analysis<br>Visualization of a correlation matrix<br>Database for soybean genetics and genomics | Copley et al., 2017 |
| 5. | Genomics<br>Transcriptomics<br>Metabolomics | Subio platform (Subio Inc.)<br>VitisNet<br>Blast2GO software<br>SIMCA-P 11.5 program (Umetrics)<br>KaPPA-View 4 | https://www.subioplatform.com/<br>https://openprairie.sdstate.edu/vitisnet-8x_files/<br>https://www.blast2go.com/<br>https://umetrics-simca-p.software.informer.com/11.5/<br>http://kpv.kazusa.or.jp/ | Data analysis for microarray, RNA-Seq, and other omics data<br>Collection of molecular networks for grapevine<br>Functional annotation and analysis of genomic datasets<br>Data mining, multivariate data analysis, and model interpretation<br>Plant pathway viewer | Suzuki et al., 2015 |
| 6. | Transcriptomics<br>Proteomics<br>Metabolomics | HISAT2<br>StringTie<br>edgeR<br>ChenomxNMR suite v7.6<br>SIMCA v13.0.3. | https://daehwankimlab.github.io/hisat2/<br>http://ccb.jhu.edu/software/stringtie/<br>https://bioconductor.org/packages/release/bioc/html/edgeR.html<br>https://www.chenomx.com/products/<br>https://www.sartorius.com/en/products/process-analytical-technology/data-analytics-software/mvda-software/simca | Spliced alignment of RNA-Seq reads<br>Assembler of RNA-Seq alignments<br>Empirical analysis of digital gene expression data<br>Identifying and quantifying metabolites in NMR spectra<br>Multivariate data analysis | Chin et al., 2021 |

*Contd.*

**Table 1** *Contd.*

| *S. No.* | *Omics data* | *Tools/method/ database used* | *Tool/method/database link* | *Functions of tools/methods/database* | *Reference* |
|---|---|---|---|---|---|
| 7. | Genomics<br>Transcriptomics<br>Metabolomics | DESeq2<br>Bowtie software<br>MultiExperiment Viewer software | https://bioconductor.org/packages/release/bioc/html/DESeq2.html<br>http://bowtie-bio.sourceforge.net/index.shtml<br>https://webmev.tm4.org/ | Differential gene expression analysis<br>Aligns short DNA sequences<br>Analysis, visualization, and stratification of RNA-seq and microarray data | Zhao et al., 2020 |
| 8. | Transcriptomics<br>Degradomics<br>Proteomics | Limma packages<br>CLC Genomics Workbench 8<br>miRPlant version 5<br>Agricolae R package<br>MaxQuant<br>Proteome Discoverer<br>SEQUEST<br>ta-si prediction<br>psRNATarget<br>CD-HIT | https://bioconductor.org/packages/release/bioc/html/limma.html<br>https://www.qiagenbioinformatics.com/products/clc-genomics-workbench<br>http://www.australianprostatecentre.org/research/software/mirplant<br>https://cran.r-project.org/web/packages/agricolae/index.html<br>https://www.maxquant.org/<br>https://www.thermofisher.com/order/catalog/product/OPTON-30945<br>https://proteomicsresource.washington.edu/protocols06/sequest.php<br>https://bio.tools/ta-si_prediction<br>https://www.zhaolab.org/psRNATarget/<br>http://weizhong-lab.ucsd.edu/cd-hit/ | Data analysis for microarray data<br>NGS analysis<br>Identification of plant MiRNA<br>Statistical analysis tool for agricultural research<br>Analysis of MS data<br>Identifying and quantifying proteins in complex biological samples<br>Correlates tandem mass spectra of peptides against peptide sequences from a sequence database<br>Prediction of phased ta-siRNAs in plant sRNA datasets<br>Analysis of plant small RNA target<br>Clustering and comparing protein/nucleic acid | Stare et al., 2019 |
| 9. | Transcriptomics<br>Genomics<br>Metabolomics | Glimmer (v.3.02)<br>RNAmmer (v.1.2)<br>tRNAscan-SE (v.2.0)<br>Anti-SMASH (v.5.1.1).<br>HISAT2 | http://ccb.jhu.edu/software/glimmer/index.shtml<br>https://services.healthtech.dtu.dk/service.php?RNAmmer-1.2<br>http://lowelab.ucsc.edu/tRNAscan-SE/<br>https://antismash.secondarymetabolites.org/#!/download<br>https://daehwankimlab.github.io/hisat2/ | Finding genes in microbial DNA<br>Prediction of ribosomal RNA sub-units<br>tRNA gene searching from genomic sequences<br>Antibiotics & secondary metabolite analysis<br>Spliced alignment of RNA-Seq reads | Zhang et al., 2022 |
| 10. | Trsnscriptomics<br>Genomics<br>Metabolomics | Inparanoid<br>QuickParanoid<br>MAFFT<br>RAxML | https://inparanoid.sbc.su.se/cgi-bin/index.cgi<br>http://pl.postech.ac.kr/QuickParanoid/<br>https://www.ebi.ac.uk/Tools/msa/mafft/<br>https://cme.h-its.org/exelixis/web/software/raxml/ | Genome comparison<br>Ortholog clustering and analysis<br>Multiple sequence alignment<br>Phylogenetic analysis | Brilli et al., 2018 |

*Contd.*

**Table 1** *Contd.*

| *S. No.* | *Omics data* | *Tools/method/ database used* | *Tool/method/database link* | *Functions of tools/methods/database* | *Reference* |
|---|---|---|---|---|---|
| 11. | Metabolomics<br>Proteomics<br>Genomics | MetaboAnalyst<br>MS-DIAL software<br>MONA<br>MassBank<br>UCHIME algorithm<br>UNITE (8.0) | https://www.metaboanalyst.ca/<br>http://prime.psc.riken.jp/compms/msdial/main.html<br>https://www.zbh.uni-hamburg.de/en/forschung/amd/software/mona.html<br>https://massbank.eu/MassBank/<br>https://www.drive5.com/usearch/manual/uchime_algo.html<br>https://unite.ut.ee/ | Metabolomics data analysis<br>MS data analysis<br>MONA visualizes large small-molecule datasets<br>High-quality mass spectral database<br>Detects chimeric sequences<br>rDNA ITS-based identification of eukaryotes | Liu et al., 2022 |
| 12. | Transcriptomics<br>Genomics<br>Proteomics | BowStrap<br>Multiple Array Viewer<br>WoLF PSORT | http://www.bio.anl.gov/molecular_and_systems_biology/bowstrap/bowstrap_download.html<br>https://webmev.tm4.org/#/about<br>https://wolfpsort.hgc.jp/ | Detects significant gene model expression<br>K-means clustering<br>Protein subcellular localization prediction | Larsen et al., 2016 |
| 13. | Genomics<br>Proteomics<br>Metabolomics | MetWare<br>Majorbio cloud platform<br>UPARSE | https://sourceforge.net/projects/metware/<br>http://www.majorbio.com/<br>https://drive5.com/uparse/ | Metabolomics data analysis<br>Multiomics analyses<br>Generating clusters for NGS reads | Pang et al., 2022 |
| 14. | Genomics<br>Transcriptomics<br>Metabolomics | MEGA-X<br>HMMER software<br>KEGG Mapper – Color<br>EBSeq R package<br>KOBAS 2.0<br>MetaboScape software | http://www.megasoftware.net/<br>http://hmmer.org/download.html<br>https://www.genome.jp/kegg/tool/map_pathwa2.html<br>https://bioconductor.org/packages/release/bioc/html/EBSeq.html<br>http://kobas.cbi.pku.edu.cn/<br>https://www.bruker.com/en/products-and-solutions/mass-spectrometry/ms-software/metaboscape.html | Phylogenetic analysis<br>Sequence alignment and homolog sequence search<br>Searches various KEGG objects<br>Gene and isoform differential expression analysis of RNA-seq data<br>Gene annotation<br>Compound identification from non-targeted workflows | Rizwan et al., 2021 |

*Contd.*

**Table 1** *Contd.*

| *S. No.* | *Omics data* | *Tools/method/ database used* | *Tool/method/database link* | *Functions of tools/methods/database* | *Reference* |
|---|---|---|---|---|---|
| 15. | Transcriptomics<br>Metabolomics | Salmon software<br>metaX<br>MetaboAnalyst | https://combine-lab.github.io/salmon/<br>https://bioconductor.statistik.tu-dortmund.de/packages/3.2/bioc/html/metaX.html<br>https://www.metaboanalyst.ca/ | Quantifying the expression of transcripts<br>Metabolomics data analysis<br>Metabolomics data analysis | Jiang et al., 2022a |
| 16. | Transcriptomics<br>Metabolomics | DAVID<br>Primer-BLAST | https://david.ncifcrf.gov/<br>https://www.ncbi.nlm.nih.gov/tools/primer-blast/ | Functional annotation of gene<br>primer identification | Men et al., 2021 |
| 17. | Transcriptomics<br>Metabolomics | DEGseq<br>ProteoWizard<br>CAMERA | https://www.bioconductor.org/packages/release/bioc/html/DEGseq.html<br>https://proteowizard.sourceforge.io/<br>http://camera.calit2.net/ | Differential gene expression analysis<br>Proteomics data analysis<br>Estimation of inter-gene correlation | Jiang et al., 2022b |
| 18. | Proteomics<br>Metabolomics | Proteome discoverer<br>STRING v10.0<br>Metabolon Inc<br>STITCHv5.0<br>Omicade4 package of R | https://thermo.flexnetoperations.com/control/thmo/login<br>https://string-db.org/<br>https://www.metabolon.com/get-started/<br>http://stitch.embl.de/<br>https://bioconductor.org/packages/release/bioc/html/omicade4.html | Protein identification<br>Pathway enrichment analysis of proteomics dataset<br>QC processing and peak identification<br>Metabolite—protein interaction on maize endogenous pathways<br>Integration of multiomics dataset | Mesnage et al., 2016 |

- **Semi-Supervised Learning:** Combines labeled and unlabeled data, enhancing predictive accuracy when labeled data is scarce, employing techniques such as self-training to leverage the structure of unlabeled data.
- **Reinforcement Learning:** Enables an agent to learn optimal decision-making through environmental interactions, applicable in resource management for irrigation or pesticide application, using algorithms such as Q-learning (Tables 2 and 3).

## 7.3 Bioinformatics Tools in Crop Improvement

Bioinformatics tools are critical for analyzing genomic data, identifying key traits and genetic markers, and speeding up the breeding of stress-resistant, high-yield crop varieties.

- **Alignment and Assembly Tools:** Genome assembly tools improve the accuracy of sequence reconstruction. Advances from first-generation to third-generation sequencing technologies, such as PacBio, have significantly enhanced assembly quality. Tools such as Nanopolish and HapCol are used for error correction and to better understand genetic variation (Ranjan et al., 2014).
- **Variant Calling Tools:** These tools are essential for identifying genetic variations through NGS. Platforms such as GATK (https://gatk.broadinstitute.org/hc/en-us) and Samtools (https://www.htslib.org/) efficiently detect single-nucleotide polymorphisms (SNPs) and insertions/deletions (INDELs), though challenges remain in analyzing polyploid genome (Yao et al., 2020).
- **Functional Annotation Tools:** Functional annotation tools help interpret genomic data more effectively. The gene ontology (GO) provides a standardized vocabulary for gene functions, while tools such as GOMAP (https://dill-picl.org/projects/gomap/) enhance the precision of gene function annotations, deepening our understanding of plant genomes (Wimalanathan & Lawrence-Dill, 2021).

## 7.4 Applications of Bioinformatics in Crop Improvement

### 7.4.1 Genomic Data Analysis

Bioinformatics plays a vital role in analyzing genomic and transcriptomic data through high-throughput sequencing technologies (HTST), particularly NGS. These technologies allow for the rapid generation of extensive genomic data from both model and non-model crop species. Bioinformatics tools and algorithms assist in genome assembly, variant calling, and marker identification. By understanding the genetic makeup of crops, researchers can identify genes linked to desirable traits (Bevan et al., 2017).

### 7.4.2 Trait Mapping and Marker-Assisted Selection

Trait mapping, a critical application of bioinformatics, involves identifying genomic regions associated with specific traits of interest. Common methodologies include genome-wide association studies (GWAS) and quantitative trait locus (QTL) mapping (Kumar et al., 2017). Once genomic regions are identified, bioinformatics tools facilitate the development of molecular markers for MAS. This enables breeders to select plants with desirable traits early in their development, thereby accelerating the breeding process and enhancing the efficiency of crop improvement programs (Xu & Crouch, 2008).

### 7.4.3 Genome Editing and CRISPR/Cas Technology

Recent advancements in genome editing technologies, particularly CRISPR/Cas systems, have revolutionized crop improvement efforts. Bioinformatics is essential in designing and optimizing

**Table 2** Types of machine learning in crop improvement.

| *Type* | *Algorithms* |
|---|---|
| Supervised | Linear Regression, Logistic Regression, Decision Trees, Random Forest, Support Vector Machines (SVM), Naive Bayes, K-Nearest Neighbors (KNN), Gradient Boosting Machines (GBM), Neural Networks, Ensemble Methods (e.g., AdaBoost, XGBoost, LightGBM, CatBoost) |
| Unsupervised | K-Means Clustering, Hierarchical Clustering, DBSCAN (Density-Based Spatial Clustering of Applications with Noise), Gaussian Mixture Models (GMM), Principal Component Analysis (PCA), t-Distributed Stochastic Neighbor Embedding (t-SNE), Self-Organizing Maps (SOM), Independent Component Analysis (ICA), A priori Algorithm (Association Rule Learning), Mean Shift Clustering |
| Semi-supervised | Self-training, Co-training, Tri-training, Generative models (e.g., Generative Adversarial Networks—GANs), Graph-based methods, Label propagation, Transductive Support Vector Machines (TSVM), Semi-supervised clustering, Entropy-based methods, Consistency-based methods |
| Reinforcement | Q-Learning, Deep Q-Networks (DQN), Policy Gradient Methods, Actor-Critic Methods, Deep Deterministic Policy Gradient (DDPG), Proximal Policy Optimization (PPO), Trust Region Policy Optimization (TRPO), SARSA (State-Action-Reward-State-Action), Temporal Difference Learning, Monte Carlo Methods |

CRISPR/Cas systems for precise genome editing in crops (Jaganathan et al., 2018). Researchers utilize bioinformatics tools to identify target sites within the crop genome, design guide RNAs (gRNAs), and predict potential off-target effects. These technologies allow for precise modifications, including gene knockouts, gene editing, and gene insertion, which facilitate the development of crops with improved traits, such as enhanced disease resistance, abiotic stress tolerance, and nutritional quality (Gao, 2021).

### *7.4.4 Functional Genomics and Systems Biology*

These support functional genomics studies that aim to elucidate the functions of genes and regulatory elements within crop genomes. By leveraging transcriptomics, proteomics, and metabolomics, researchers can investigate gene expression patterns, protein functions, and metabolic pathways under various conditions (Ganeshan et al., 2010). Integrating omics data with bioinformatics tools allows for the reconstruction of regulatory networks and metabolic pathways, offering insights into the molecular mechanisms that underlie complex traits. Systems biology approaches, which incorporate multi-omics data with computational models, enable a holistic analysis of crop systems and enhance predictions of crop performance under diverse environmental conditions (Pazhamala et al., 2021).

### *7.4.5 Data Integration and Knowledge Discovery*

These practices facilitate the integration of diverse biological data sources, including genomic, transcriptomic, proteomic, metabolomic, and phenotypic data. This integration provides researchers with a comprehensive understanding of crop biology. Data integration platforms and bioinformatics databases offer access to curated datasets and analytical tools for knowledge discovery. By employing machine learning and data mining techniques, bioinformatics enables the identification of novel gene-trait associations, biomarkers for crop improvement, and predictive models for crop performance. Additionally, bioinformatics contributes to the development of crop simulation

**Table 3** Machine learning applications.

| *S.No.* | *Utility* | *Reference* |
|---|---|---|
| 1 | Machine Learning for Gene Function Prediction | Identifies functional genomic sequences (e.g., protein-coding genes, microRNAs) using ML techniques like SVMs and Random Forests for enhanced accuracy (Mahood et al., 2020). |
| 2 | Protein-Coding Gene Prediction | Utilizes pipelines such as MAKER-P and ab initio gene finders (Augustus, SNAP) for accurate predictions in plant genomes (Chao et al., 2024). |
| 3 | Cis-Regulatory Element Prediction | Employs Convolutional Neural Network (CNN) to identify motifs in genomic sequences, enhancing prediction of transcription factor binding and gene expression (Li et al., 2018). |
| 4 | Gene Expression Prediction | Leverages features such as Cis-Regulatory Elements (CRE) and methylation to predict gene regulation, achieving high accuracy in plants (Washburn et al., 2019). |
| 5 | Subcellular Localization Prediction | Analyzes protein function using tools such as SignalP and TargetP, integrating features for improved localization predictions (Chou & Shen, 2010). |
| 6 | Prediction of Protein–Protein Interactions | Utilizes ML algorithms for binary classification of Protein Protein Interaction (PPI), integrating various protein features for accurate predictions (Sarkar & Saha, 2019). |
| 7 | Prediction of Biological Function | Predicts phenotypes from genotypes, offering insights into gene function and linking to metabolic pathways (Notin et al., 2024). |
| 8 | Prediction of Gene Ontology Category | Utilizes features such as gene expression and homology for automated gene annotation, improving accuracy in GO predictions (Schietgat et al., 2010) . |
| 9 | Prediction of Metabolic Pathways | Employs ML to analyze gene/protein structures and expression patterns, improving predictions of metabolic pathways (Dale et al., 2010) . |
| 10 | Prediction of Phenotypes and Biological Processes | Integrates diverse features for accurate phenotype predictions, showcasing adaptability across species (Costello & Martin, 2018). |
| 11 | Genomic Prediction | Uses genetic markers such as SNPs to predict complex traits, enhancing efficiency in crop and livestock breeding (Lee et al., 2020). |
| 12 | Disease Prediction | Applies ML algorithms on various data types for forecasting disease likelihood, aiding in early diagnosis and prevention (Shoaib et al., 2023). |
| 13 | Machine Learning in Biotic Stress Management | Early detection of diseases through imaging data, high-throughput phenotyping, and predictive modeling using ML models such as SVM and RF (Dormatey et al., 2020). |
| 14 | Machine Learning in Abiotic Stress Management | Identifies early stress indicators using high-throughput phenotyping, predictive modeling of crop responses, and optimizing water-use efficiency (Dormatey et al., 2020). |
| 15 | Deep Learning Enhancements in Stress Management | Integrates DL technologies (e.g., CNNs) with ML for stress detection and phenotypic screening, optimizing crop health management and sustainability (Shoaib et al., 2023). |

models and decision support systems, aiding breeders in optimizing crop management practices and improving agricultural productivity (Ngo & Kechadi, 2020).

### *7.4.6 Biotic and Abiotic Stresses Response in Plants*

To support and concentrate our experimental research on a variety of crops with biotic and abiotic pressures, we have integrated bioinformatics tools and resources for crop development activities and the numerous facets. A unique database on stress responsive genes in model plant *Arabidopsis thaliana* called STIFDB was developed (Khader et al., 2009). Further STIFDB was updated as STIFDB2 with additional stress signals (such as drought, cold, NaCl, ABA, aluminum, dehydration, high light, iron, osmotic stress, oxidative stress, UV-B, cold–drought–salt, and wounding, 31 transcription factors, and 5984 stress-responsive genes from *Arabidopsis thaliana*, *Oryza sativa* subsp. *japonica*, and *Oryza sativa* subsp. *indica*) with single and combined stress signals, stress-responsive genes, and transcription factor binding sites (TFBS) predicted using the inhouse Stress-responsive Transcription Factor (STIF) algorithm (Ambika et al., 2008); it also maintains functional annotation and orthologs information. As a unique plant stress regulatory genomics data platform, STIFDB2 (http://caps.ncbs.res.in/stifdb2) can be utilized for targeted as well as high-throughput experimental and computational studies to unravel principles of the stress regulome in dicots and Gramineae (Naika et al., 2013a). Available resources in the database are extensively utilized for genomics, bioinformatics, and experimental purposes. Enrichment analysis using 3091 genes in stress signals in *Arabidopsis thaliana* revealed several shared and distinct biological processes, cellular components and molecular functions, and stress-signal specific protein domains associated with stress-responsive genes. Also, phenomic characteristics were found to be shared by multiple stresses in both anatomy and temporal categories of plant ontology (Naika et al., 2013b).

Plant Ontology Enrichment Analysis Server (POEAS) (http://caps.ncbs.res.in/poeas.) performs phenomic enrichment analyses using plant genes as a complementary resource, permitting the adoption of PO-based phenomic analysis as part of analytical workflows (Khader et al., 2014). Transcriptional regulatory network using bioinformatic tools research was carried out to comprehend the general TFs' and condition-specific actions as well as their regulatory linkages with the target genes during single and combined stresses in *Arabidopsis thaliana* (Barah et al., 2015). By combining various bioinformatics tools, an automated pipeline known as the PUFs Annotation Server (PUFAS) (http://caps.ncbs.res.in/pufas/) was created for the quick and simple annotation of genes of proteins with unknown functions (Dhanyalakshmi et al., 2016). Using the molecular complex detection algorithm, the study discovered that the 26S proteasome subunits were a highly clustered module, and based on enrichment analysis, the network is linked to flower development process in *Arabidopsis thaliana* (Khader et al., 2016). Red fluorescence protein sequence analysis predicted its gene associated with the chbp gene, which is a member of the lipocalin gene family and is known to have anti-pathogenic properties (Manjunatha et al., 2018). Predicted stress-inducible transcription factors are ERF, NAC, bZIP, MYB, DREB, and WRKY, and antifungal related genes in coriander (Choudhary et al., 2019). Khader et al. (2019) reviewed on the developments in data science, genomics, bioinformatics, and CRISPR/Cas9-based genome engineering systems for comprehending plant mechanisms, creating climate-resilient crops and ensuring food security.

Analysis of transcriptome data in finger millet showed various levels of functional characteristics and pathways linked to serine/threonine protein phosphatase 2A (PP2A), calcineurin B-like interacting protein kinase 31 (CIPK31), signal recognition particle receptor α (SRPR α), farnesyl pyrophosphate synthase (FPS), etc. TATA-binding protein (TBP)-associated factors (TAFs) and other basal regulatory genes were discovered to be drought responsive (Parvathi et al., 2019). Understanding the structural and functional significance of the leucine-rich repeat, Gly-Asp-Ser-(Leu) motif (GDSL), universal stress protein, ankyrin repeat, and probable disease resistance in coriander was aided by bioinformatic tools (Choudhary et al., 2020). Total 10 SSR functional domain markers (FDM) linked with pathogen response were predicted and verified across Indian coriander cultivars, and their transferability was tested across fennel cultivars (Choudhary et al., 2021). The sequence of *Beauveria bassiana* UHSB-END1 was analyzed and found its phylogenetic relatedness with available sequence in the NCBI (Jamunarani et al., 2022). Das et al. (2022) developed database named "Knowledgebase for rice sheath blight information" (KRiShI) for rice sheath blight, which is very much useful for rice breeders. Gunnaiah and Naika (2023) reported on Targeting Induced Local Lesions IN Genomes (TILLING) importance in functional genomics and crop breeding. They also compiled several bioinformatic tools and databases available publically for functional annotation of genes and mining functional SNPs for efficient TILLING. Sidharthan et al. (2023) predicted and validated potential new viruses such as Coriandrum sativum deltapartitivirus 1, 2 (CsDPV1, 2) and Coriandrum sativum enamovirus (CsEV) using transcriptome data of Indian coriander genotypes; further predicted CsDPV2 presence was also confirmed in fruit samples using Real Time Polymerase Chain Reaction (RT-PCR) and sequencing. Naika et al. (2023) provide an overview of transposable elements (TEs), classification, and discuss the existing bioinformatic tools for predicting and analyzing the function of TEs in plants. Recently, Aishwarya et al. (2024) predicted and developed novel long non-coding RNAs (lncRNA)-derived SSR markers in carrot, and highlighted their potential application in crop improvement programs. Using bioinformatic methods, the influence of discovered "non synonyms of single-nucleotide polymorphism" (nsSNPs) on protein structure and stability of Fe-deficiency-tolerant genes in rice were investigated (Panda et al., 2024).

## 8. Future Prospects and Conclusion

Bioinformatics tools and machine learning techniques can significantly advance sustainable crop improvement by analyzing vast datasets, predicting traits, and optimizing breeding strategies, ultimately leading to more resilient and productive crops. ML algorithms can analyze vast genomic and phenotypic datasets to identify key genes and genomic regions associated with desirable traits, allowing for targeted breeding strategies. They can be used to predict crop performance under different environmental conditions, optimize resource use, and improve yield predictions. By leveraging ML for genomic selection and marker-assisted selection, breeders can shorten the time required to develop improved crop varieties. The integration of bioinformatics and machine learning is transforming plant breeding, offering powerful tools for developing sustainable and resilient crop varieties. By leveraging these technologies, researchers and breeders can accelerate crop improvement programs, improve food security, and promote sustainable agricultural practices. Continued research and development in this area will be crucial for maximizing the benefits of these technologies and meeting the challenges of a growing global population.

## References

Abinaya, R.V. and Viswanathan, P. (2021). Biotechnology-Based Therapeutics. In *Translational Biotechnology*. Elsevier. 27–52.

Aishwarya, H., Sarvamangala, S.C., Manjunath, H., Naika MBN, Bapurayagouda P. (2024). Characterization of Diverse Carrot (Daucus carota L.) Genotypes Using Long Noncoding RNA Based Microsatellite Markers. *Journal of Advances in Biology & Biotechnology*, 27(8), pp. 1316–1324.

Ambika S., Varghese S.M., Shameer K., Udayakumar M. and Sowdhamini R. (2008). STIF: Hidden Markov model-based search algorithm for the recognition of binding sites of stress-upregulated transcription factors and genes in Arabidopsis thaliana. *Bioinformation*, 2(10):431–437.

Barah,P., Naika, M.B.N., Jayavelu, N.D., Sowdhamini, R., Shameer, K., Bones, A.M. (2016). Transcriptional regulatory networks in Arabidopsis thaliana during single and combined stresses, *Nucleic Acids Research*, 44, (7), 20: 3147–3164.

Belal, A.A., El-Ramady, H., Jalhoum, M., Gad, A. and Mohamed, E.S. (2021). Precision Farming Technologies to Increase Soil and Crop Productivity. In M. Abu-hashim, F. Khebour Allouche, & A. Negm (Eds.), *Agro-Environmental Sustainability in MENA Regions*. 117–154.

Bevan, M.W., Uauy, C., Wulff, B.B.H., Zhou, J., Krasileva, K. and Clark, M.D. (2017). Genomic innovation for crop improvement. *Nature*, *543*(7645), 346–354.

Boopathi, N.M. (2020). Marker-Assisted Selection (MAS). In N. M. Boopathi, *Genetic Mapping and Marker Assisted Selection*. Springer, Singapore. 343–388.

Ceccarelli, S. and Grando, S. (2009). Participatory Plant Breeding. In M. J. Carena (Ed.), *Cereals*. Springer US. 395–414.

Chao, H., Zhang, S., Hu, Y., Ni, Q., Xin, S., Zhao, L., Ivanisenko, V.A., Orlov, Y.L. and Chen, M. (2024). Integrating omics databases for enhanced crop breeding. *Journal of Integrative Bioinformatics*, *20*(4), 20230012.

Chou, K.-C. and Shen, H.-B. (2010). Plant-mPLoc: A top-down strategy to augment the power for predicting plant protein subcellular localization. *PLoS One*, *5*(6), e11335.

Choudhary, S., Naika, M.B.N., Sharma, R., Meena, R.D., Singh, R. and Lal, G. (2019). Transcriptome profiling of coriander: A dual purpose crop unravels stem gall resistance genes. *Journal of Genetics*. 98:1, 1–19.

Choudhary, S., Naika, M.B.N. and Meena, R.D. (2020). Identification and expression analysis of candidate genes associated with stem gall disease in coriander (*Coriandrum sativum* L.) cultivars. *Molecular Biology Reports*. 47: 5403–5409.

Choudhary, S., Naika, M.B.N. and Meena, R.D. (2021). Development and characterization of genic SSR-FDM for stem gall disease resistance in coriander (*Coriandrum sativum* L.) and its cross species transferability. *Molecular Biology Reports*. 48 (5), 3963–3970.

Costello, Z. and Martin, H.G. (2018). A machine learning approach to predict metabolic pathway dynamics from time-series multiomics data. *NPJ Systems Biology and Applications*, *4*(1), 19.

Dale, J.M., Popescu, L. and Karp, P.D. (2010). Machine learning methods for metabolic pathway prediction. *BMC Bioinformatics*, 11(1), 15.

Das, A., Mishra, A., Kashyap, A., Naika, M.B.N. and Barah, P. (2022). KRiShI: a manually curated knowledgebase on rice sheath blight disease. *Functional & Integrative Genomics*. 22 :(6), 1403–1410.

Dhanyalakshmi, K.H., Naika, M.B.N., Sajeevan, R.S., Mathew, O.K., Safi, M., Sowdhamini, R. and Nataraja, N.K. (2016). Transcriptome sequencing of mulberry (*Morus alba* L.) in drought conditions and functional annotation of proteins of unknown function. *PLoS One.* 11(3):1–18.

Dormatey, R., Sun, C., Ali, K., Coulter, J.A., Bi, Z. and Bai, J. (2020). Gene pyramiding for sustainable crop improvement against biotic and abiotic stresses. *Agronomy*, *10*(9), 1255.

Ganeshan S., Sharma P. and Chibbar R.N. (2010). Functional Genomics For Crop Improvement. In S. M. Jain & D. S. Brar (Eds.), *Molecular Techniques in Crop Improvement*. Springer Netherlands (63–95).

Gao C. (2021). Genome engineering for crop improvement and future agriculture. *Cell*, *184*(6), 1621–1635.

Gunnaiah, R., Naika, M.B.N. (2023). Bioinformatics and Candidate Gene Mining for TILLING. In: Bhattacharya, A., Parkhi, V., Char, B. (eds) TILLING and Eco-TILLING for Crop Improvement. Springer, Singapore. 978–981.

Harbinson J., Prinzenberg A.E., Kruijer W. and Aarts M.G. (2012). High throughput screening with chlorophyll fluorescence imaging and its use in crop improvement. *Current Opinion in Biotechnology*, *23*(2), 221–226.

Jaganathan, D., Ramasamy, K., Sellamuthu, G., Jayabalan, S. and Venkataraman, G. (2018). CRISPR for crop improvement: An update review. *Frontiers in Plant Science*, 9, 985.

Jamunarani, G.S., Ramanagouda, S.H., Venkateshalu, B., Jayappa, J., Raghavendra, G., Rudresh, D.L., Kulkarni, M.S., Naika, M. B. N. and Gopali J. B. (2022). Isolation and evaluation of indigenous endophytic entomopathogenic fungus, Beauveria bassiana UHSB-END1 (Hypocreales: Cordycipitaceae), against Spodoptera litura Fabricius. *Egyptian Journal of Biological Pest Control*. 32 (120), 1–15.

Khader, S., Ambika, S., Varghese, S.M., Karaba, N., Udayakumar, M., and Sowdhamini, R. (2009). STIFDB-Arabidopsis stress responsive transcription factor data base. *International Journal of Plant Genomics.* 2009: 583429.

Khader, S., Naika, M.B.N., Mathew, O.K., Sowdhamini, R. (2014). POEAS: Automated plant phenomic analysis using plant ontology. *Bioinformatics and Biology Insights*, 8, 209–214.

Khader, S., Naika, M., Mathew, O.K. and Sowdhamini, R. (2016). Network modules driving plant stress response, tolerance and adaptation: A case study using abscisic acid induced protein–protein interactome of *Arabidopsis thaliana. BioRxiv. Cold Spring Harbor Labs Journals.* NewYork, USA. 1–21.

Khader, S., Naika, M.B.N., Shafi, K.M. and Sowdhamini, R. (2019). Decoding systems biology of plant stress for sustainable agriculture development and optimized food production. *Progress in Biophysics and Molecular Biology*. S0079-6107(17)30089-5.

Kumar, J., Gupta, D.S., Gupta, S., Dubey, S., Gupta, P. and Kumar, S. (2017). Quantitative trait loci from identification to exploitation for crop improvement. *Plant Cell Reports*, *36*(8), 1187–1213.

Lai, K., Lorenc, M.T. and Edwards, D. (2012). Genomic databases for crop improvement. *Agronomy*, *2*(1), 62–73.

Langridge, P., Braun, H., Hulke, B., Ober, E. and Prasanna, B.M. (2021). Breeding crops for climate resilience. *Theoretical and Applied Genetics*, *134*(6), 1607–1611.

Lee, J., Lee, S.M., Lim, B., Park, J., Song, K.L., Jeon, J.H., Na, C.S. and Kim, J.M. (2020). Estimation of variance components and genomic prediction for individual birth weight using three different genome-wide SNP platforms in yorkshire pigs. *Animals*, *10*(12), 2219.

Li, Y., Shi, W. and Wasserman, W.W. (2018). Genome-wide prediction of cis-regulatory regions using supervised deep learning methods. *BMC Bioinformatics*, 19(1), 202.

Lorenz, A.J., Chao, S., Asoro, F.G., Heffner, E.L., Hayashi, T., Iwata, H., Smith, K.P., Sorrells, M.E. and Jannink, J.L. (2011). Genomic selection in plant breeding. *Advances in Agronomy* (110), 77–123.

Mahood, E.H., Kruse, L.H. and Moghe, G.D. (2020). Machine learning: A powerful tool for gene function prediction in plants. *Applications in Plant Sciences*, 8(7), e11376.

Manjunatha, G.K.S., Peter, A., Naika, M.B.N., Niranjana, P. and Shamprasad, P. (2018). Identification of *in vitro* red fluorescent protein with antipathogenic activity from the midgut of the silkworm (*Bombyx Mori* L.). *Protein and Peptide Letter.* 25(3):302–313.

Mir, R.R., Reynolds, M., Pinto, F., Khan, M.A. and Bhat, M.A. (2019). High-throughput phenotyping for crop improvement in the genomics era. *Plant Science*, *282*, 60–72.

Mirón, I.J., Linares, C. and Díaz, J. (2023). The influence of climate change on food production and food safety, *Environmental Research*, 216, 114674.

Naika, M., Khader, S. and Sowdhamini, R., (2013). Comparative analyses of stress-responsive genes in *Arabidopsis thaliana*: Insights from genomic data mining, functional enrichment, pathway analysis and phenomics. *Molecular BioSystem.* 9(7):1888–1908.

Naika, M., Khader, S., Methew, O., Gowda, R. and Sowdhamini, R. (2013). STIFDB2: An updated version of plant stress responsive transcription factor data base with additional stress signals, stress responsive transcription factor binding sites and stress responsive genes in *Arabidopsis* and rice. *Plant and Cell Physiology*, 54(2):1–15.

Naika, M.B.N., Das, A., Paramesha, H.T., Barah, P. Chaitra, H.V. (2023). Plant Transposable Elements: Biology and Biotechnology, Deepu Pandita, Anu Pandita, (Eds.) Apple Academic Press, Inc. Co-published with CRC Press (Taylor & Francis). 41–57.

Ngo, V.M. and Kechadi, M.T. (2020). Crop Knowledge Discovery Based on Agricultural Big Data Integration. *Proceedings of the 4th International Conference on Machine Learning and Soft Computing*, 46–50.

Notin, P., Rollins, N., Gal, Y., Sander, C. and Marks, D. (2024). Machine learning for functional protein design. *Nature Biotechnology*, *42*(2), 216–228.

Pankaj, B., Naika, M.B.N., Naresh, J., Ramanathan, S., Khader, S. and Atle, B. (2015). Transcriptional regulatory networks in *Arabidopsis thaliana* during single and combined stresses. *Nucleic Acid Research*. 20;44(7):3147–3164.

Parvathi, M.S., Nataraja, K.N., Reddy, Y.A.N., Naika, M.B.N. and Gowda, M.V.C. (2019). Transcriptome analysis of finger millet (*Eleusine Corocana* (L.) Gaertn) reveals unique drought responsive genes. *Journal of Genetics*. 98:(2)1–12.

Pazhamala, L.T., Kudapa, H., Weckwerth, W., Millar, A.H. and Varshney, R.K. (2021). Systems biology for crop improvement. *The Plant Genome*, *14*(2), e20098.

Panda, S., Anandan, A., Shafi, M.K., Naika, M.B.N., Sowdhamini, R., Vanishree, G., Sarkar, S., Travis, A.J., Norton, G.J., Price, A. H. (2024). Genome-wide association study reveals effect of nsSNPs on candidate genes in rice during iron deficiency. *Functional and Integrative Genomics*. 24:198.

Rieseberg L.H. and Carney S.E. (1998). Plant hybridization. *New Phytologist*, 140(4), 599–624.

Ramalingam, A., Kudapa, H., Pazhamala, L.T., Weckwerth, W. and Varshney, R.K. (2015). Proteomics and metabolomics: Two emerging areas for legume improvement. *Frontiers in Plant Science*, 6.

Ranjan, A., Ichihashi, Y., Farhi, M., Zumstein, K., Townsley, B., David-Schwartz, R. and Sinha, N.R. (2014). De novo assembly and characterization of the transcriptome of the parasitic weed dodder identifies genes associated with plant parasitism. *Plant Physiology*, *166*(3), 1186–1199.

Raymond Park, J., McFarlane, I., Hartley Phipps, R. and Ceddia, G. (2011). The role of transgenic crops in sustainable development. *Plant Biotechnology Journal*, *9*(1), 2–21.

Rezaei, E.E., Webber, H., Asseng, S. et al. (2023). Climate change impacts on crop yields. *Nat. Rev. Earth Environ.* 4, 831–846.

Saleem, S.R., Zaman, Q. U., Schumann, A.W. and Abbas Naqvi, S.M.Z. (2023). Variable Rate Technologies. In *Precision Agriculture*, 103–122.

Sahu, S.K., Waseem, M. and Aslam, M.M. (2023). Editorial: Bioinformatics, big data and agriculture: a challenge for the future. *Frontier Plant Science*. 14:1271305.

Sarkar, D. and Saha, S. (2019). Machine-learning techniques for the prediction of protein–protein interactions. *Journal of Biosciences*, *44*(4), 104.

Schietgat, L., Vens, C., Struyf, J., Blockeel, H., Kocev, D. and Džeroski, S. (2010). Predicting gene function using hierarchical multi-label decision tree ensembles. *BMC Bioinformatics*, 11(1), 2.

Shoaib, M., Shah, B., EI-Sappagh, S., Ali, A., Ullah, A., Alenezi, F., Gechev, T., Hussain, T. and Ali, F. (2023). An advanced deep learning models-based plant disease detection: A review of recent research. *Frontiers in Plant Science*, 14, 1158933.

Sidharthan, V.K., Diksha, D., Singh, R., Choudhary, S., Naika, M.B.N., Baranwal, K.V. (2023). Identification of two putative novel deltapartivirus and an in coriander transcriptomes. *Archives of Microbiology*. 27; 205(10):342.

Toker, C., Yadav, S.S. and Solanki, I.S. (2007). Mutation Breeding. In S.S. Yadav, D.L. McNeil and P.C. Stevenson (Eds.), *Lentil*, 209–224.

Tounta, V., Liu, Y., Cheyne, A. and Larrouy-Maumus, G. (2021). Metabolomics in infectious diseases and drug discovery. *Molecular Omics*, *17*(3), 376–393.

Vogel, K.E. (2009). Backcross Breeding. In M. P. Scott (Ed.), *Transgenic Maize* (526), 161–169.

Wang, H., La Russa, M. and Qi, L.S. (2016). CRISPR/Cas9 in genome editing and beyond. *Annual Review of Biochemistry*, *85*(1), 227–264.

Wang, S., Liu, L., Mi, X., Zhao, S., An, Y., Xia, X., Guo, R., Wei, C. (2021). Multi-omics analysis to visualize the dynamic roles of defense genes in the response of tea plants to gray blight. *The Plant Journal* 106, 862–875.

Washburn, J.D., Mejia-Guerra, M.K., Ramstein, G., Kremling, K.A., Valluru, R., Buckler, E.S. and Wang, H. (2019). Evolutionarily informed deep learning methods for predicting relative transcript abundance from DNA sequence. *Proceedings of the National Academy of Sciences*, 116(12), 5542–5549.

Wimalanathan, K. and Lawrence-Dill, C.J. (2021). Gene ontology meta annotator for plants (GOMAP). *Plant Methods*, 17(1), 54.

Woźniak, M. and Ijaz, M.F. (2024). Editorial: Recent advances in big data, machine, and deep learning for precision agriculture. *Frontier Plant Science.* 15:1367538.

Xu, Y. and Crouch, J.H. (2008). Marker-assisted selection in plant breeding: From publications to practice. *Crop Science*, 48(2), 391–407.

Yao, Z., You, F.M., N'Diaye, A., Knox, R.E., McCartney, C., Hiebert, C.W., Pozniak, C. and Xu, W. (2020). Evaluation of variant calling tools for large plant genome re-sequencing. *BMC Bioinformatics*, 21(1), 360.

# Index


**A**

Abiotic 1, 3, 5, 6, 9, 11, 16, 20, 37, 48, 53, 110, 143-145, 149, 155, 180, 181, 186, 190, 198, 208, 210, 212, 234, 246, 263, 274, 275

Abiotic stress 4, 9, 14, 19, 37, 53, 103, 155, 181, 190, 199, 208, 210, 246, 247, 263, 275

Amylose–lipid complex 218, 219, 221

Antioxidants 9, 10, 123, 145, 147, 159, 165, 185, 190, 207, 209

Automation 81, 190, 233

**B**

Bioavailability 64, 112, 120, 121, 123, 124, 126, 127, 129, 131, 133, 147, 148

Biofortification 63, 65, 89, 91, 103, 112, 113, 120, 122, 123, 126, 132, 133, 264

Biofortified crops 131, 133, 150

Bioinformatics 40, 87, 111, 259, 260, 266, 267, 270, 273-275, 277

Biotechnology 38, 41, 148, 187, 190, 218, 219, 261

**C**

Cereals 1, 6, 9, 12, 14, 19, 21, 34, 37, 55, 58, 91, 101, 102, 106, 125, 141, 143, 144, 147, 221, 246

Climate change 1, 7, 20, 31, 41, 53, 55, 56, 62, 78, 79, 101, 102, 104, 106, 107, 110, 113, 125, 143, 144, 149, 165, 181, 208, 211, 232, 244, 246, 260, 262, 263

Climate resilience 41, 107, 262

Climate-resilient 15, 111, 121, 141-143, 232, 249, 276

Climate-resilient crop varieties 15

CRISPR/Cas9 18, 19, 32, 36, 37, 51, 57, 62, 106, 107, 113, 129, 130, 133, 227, 276

Crop enhancement 102, 190

Crop improvement 9, 18, 31-34, 37, 41, 42, 48, 53, 59, 102, 111, 113, 131, 166-168, 239, 242, 246-248, 259, 261-264, 266, 267, 273, 274, 277

**D**

Deep learning 20, 245, 275

Drone 20, 233, 237, 238, 246, 249, 265

**E**

Eco-friendly plasticulture 222, 225

Enzymatic 2-4, 10, 33, 35, 37, 132, 183, 209, 223

**F**

Food 1, 2, 14, 20, 31, 34, 37, 41, 46, 53, 54, 56, 58, 62, 63, 78-80, 93, 101-103, 106, 108, 111, 112, 120, 122, 131, 141, 150, 155, 165, 173, 180, 186, 198, 218, 219, 222, 233, 251, 259, 264

Food crop 41, 56, 59, 80, 120, 121, 125, 133, 141, 232

Food security 1, 2, 20, 31, 34, 41, 46, 78, 79, 101, 103, 106-108, 111, 113, 121, 122, 129, 143, 145, 148, 155, 159, 165, 180, 182, 190, 242, 244, 252, 259, 276

**G**

Genome editing 18, 21, 31, 35, 41, 51, 106, 108, 112, 113, 123, 129, 218, 227, 273, 274

Genome engineering 276

Genome estimated breeding values 79

Genomic selection 40, 48-50, 78-80, 83, 86, 90, 91, 94, 108, 109, 112, 113, 247, 161, 262, 265, 277

Genotyping 48, 79-83, 79, 87, 93, 94, 104, 105, 109, 111, 148, 247

Global food security 34, 41, 101, 111, 113, 121, 143, 155, 181, 184, 190, 198, 233, 246, 251, 252

Grafting innovations 180, 190

Green chemistry 218, 219, 222, 224

**I**

Iron 4, 63, 112, 120, 122, 124, 125, 127, 129, 130, 141, 147, 161, 164, 165, 264

**M**

Machine learning 84, 85, 89, 95, 238, 244, 245, 247, 250, 259, 260, 266, 274, 277

Malnutrition 65, 101, 106, 111, 112, 120, 122, 125, 130, 133, 154, 155, 165, 166, 177

Mangalore melon 154-158, 160-162, 165-166, 168, 177

MAS 15, 47, 48, 57, 62, 79, 81, 82, 102-106, 110-112, 130, 131, 188, 261, 265, 273

Millets 131, 141, 142, 144, 146, 148, 149

Molecular breeding 15, 34, 106, 112, 126, 145, 187, 188, 190, 259, 262, 264

**N**

Next-generation sequencing 38, 39, 81, 93, 111, 232, 259

Next-generation varieties 78, 86, 93, 95

Nutri-cereals 141

Nutrient-Dense Rice 47, 63

Nutritional quality 14, 63, 95, 111, 112, 121, 148, 180, 181, 183, 190, 261, 274
Nutritional security 78, 113, 154, 177, 221

**P**

Phenome 78, 145, 200, 211, 233
Phenotype 8, 19, 36, 47, 48, 50, 62, 79, 83, 85, 86, 88, 233, 234, 242, 244, 245, 246, 248, 275
Phenotyping 20, 47, 50, 59, 79, 80, 82, 87-89, 92, 93, 106, 111, 232, 233, 236, 239, 241, 244, 246, 247, 250, 251, 275
Plant defense mechanism 209, 211
Plant metabolism 3, 199
Precision agriculture 211, 233, 237, 238, 241, 245, 247, 248, 251, 259, 265

**R**

Reactive oxygen and nitrogen species (RONS) 202
Rootstock–scion interactions 187, 191

**S**

Seed quality enhancement 211, 212
Sensors 13, 20, 56, 205, 236-241, 245-249, 265
Spine gourd 154, 155, 157, 159, 160, 162, 163, 165, 166, 169, 170, 177,
Starch bioengineering 218-220, 223, 224, 226, 227
Stress management 10, 19, 275
Sustainability 101, 102, 113, 121, 143, 159, 208, 218, 222, 225, 242, 245, 246, 248, 249, 251, 259, 262, 264, 265, 275
Sustainable agriculture 58, 101, 103, 106, 166, 183, 190, 191, 211, 222, 233, 239, 246, 249, 250, 252, 264, 266

**U**

Underutilized 41, 59, 122, 148, 154, 155, 157, 165, 166, 170, 176, 218, 222

**V**

Vegetable crops 154, 155, 177, 180, 190, 221, 246

**W**

Wheat 5-13, 17-20, 36, 37, 48, 63, 79, 88, 101, 102, 104-107, 110-113, 127, 128, 220, 221, 234, 235, 244, 245, 247, 260, 261

**Y**

Yield 1, 3, 5, 8, 9, 11, 14, 16, 17, 20, 21, 31, 36, 37, 41, 46, 47, 49, 51, 54, 55, 57, 59, 78, 82, 91, 93, 104, 110, 130, 149, 180, 184, 199, 222, 242, 246, 247, 260, 261, 264, 266, 277

**Z**

Zinc 3, 4, 18, 63, 64, 101, 106, 112, 120, 124, 125, 128, 130, 141, 162, 185, 264

# About the Editors

**Dr. Belaghihalli N Gnanesh** is currently serving as Senior Scientist and Head at ICAR JSS Krishi Vigyan Kendra, Suttur, Mysuru, India. He previously held the position of Professor and Dean at the Sampoorna International Institute of Agricultural Science and Horticultural Technology under the University of Mysore. He earned his Ph.D. in Genetics and Plant Breeding from the International Crops Research Institute for the Semi-Arid Tropics (ICRISAT), Hyderabad, and the University of Agricultural Sciences (UAS), Bangalore. With several years of research experience, Dr. Gnanesh has made significant contributions in the areas of germplasm resource management, crop genomics, and disease-resistance breeding. He is a recipient of the prestigious SERB Ramanujan Fellowship and was awarded a Visiting Fellowship by the Natural Sciences and Engineering Research Council (NSERC) of Canada. During his tenure at Agriculture and Agri-Food Canada (AAFC), Winnipeg, he developed cost-effective, allele-specific diagnostic markers for oat crown rust resistance genes. Dr. Gnanesh's current interests span plant genomics, genome survey, SNP discovery, and linkage mapping. He is also deeply involved in agricultural training and extension, focusing on field demonstrations, technology trials and precision farming that promote sustainable and climate-resilient agriculture.

**Prof. Chittaranjan Kole** is an internationally reputed scientist with an illustrious professional career spanning over forty years and original contributions in the fields of plant genomics, biotechnology, and molecular breeding leading to the publication of more than 160 impactful research articles and reviews. He has edited over 200 books for the globally leading publishers. His scientific contributions and editing acumen have been appreciated by seven Nobel Laureates including Prof. Norman Borlaug, Arthur Kornberg, Werner Arber, Phillip Sharp, Günter Blobel, Lee Hartwell and Roger Kornberg. He has been honored with several Fellowships, Honorary Fellowships, and national and international awards including the Outstanding Crop Scientist award conferred by the International Crop Science Society. He was also awarded the Raja Ramanna Fellow by the Department of Energy, Government of India. He has served in all prestigious positions in academia including as Vice-Chancellor of Bidhan Chandra Krishi Viswavidyalaya, Project Coordinator of Indo-Russian Center of Biotechnology in India, and Director of Research of Institute of Nutraceutical Research of Clemson University, in USA. He also served as a Visiting Professor at the Pennsylvania State University and Clemson University. He is also heading the International Climate-Resilient Crop Genomics Consortium, International consortium for Phytomedomics and Nutriomics and Genome India International as their Founder President.